SCIENCEPOWER™ 7

SCIENCE • TECHNOLOGY • SOCIETY • ENVIRONMENT

McGraw-Hill Ryerson SCIENCEPOWER™ Program

SCIENCEPOWER™ 7

SCIENCEPOWER™ 8

SCIENCEPOWER™ 9

SCIENCEPOWER™ 10

Chenelière/McGraw-Hill OMNISCIENCES Program

OMNISCIENCES 7

OMNISCIENCES 8

OMNISCIENCES 9

OMNISCIENCES 10

This program is available directly from Chenelière/McGraw-Hill.

Teacher Support for Each Grade Level

Teacher's Resource Binder
Blackline Masters
Computerized Assessment Bank
Web site: *http://www.mcgrawhill.ca*
Videotape series

Our cover The praying mantis is a common insect in Canadian meadows, but have you ever seen one? If so, where did you see it and what was it doing? What are its habits? What living and non-living things are a part of its life? Answer these questions as best you can. As you study Unit 1 in this textbook, do some research about the life of the praying mantis, and expand your answers to the questions.

SCIENCEPOWER™ 7

SCIENCE • TECHNOLOGY • SOCIETY • ENVIRONMENT

Author Team

Don Galbraith
Ontario Institute for Studies in Education of the University of Toronto
Toronto, Ontario

David Gue
Crescent Heights High School
Medicine Hat, Alberta

Derek Bullard
Mount Klitsa Junior Secondary School
Parksville, British Columbia

Jean Bullard
Professional Writer
Parksville, British Columbia

Christina Clancy
Loyola Catholic Secondary School
Mississauga, Ontario

Betty Anne Kiddell
Acadia Junior High School
Winnipeg, Manitoba

Contributing Author

Sandy Wohl
Hugh Boyd Secondary School
Richmond, British Columbia

Senior Program Consultants

Malisa Mezenberg
Loyola Catholic Secondary School
Mississauga, Ontario

Douglas A. Roberts
University of Calgary
Calgary, Alberta

McGraw-Hill Ryerson

Toronto Montréal New York Burr Ridge Bangkok Bogotá Caracas
Lisbon London Madrid Mexico City Milan New Delhi
Seoul Singapore Sydney Taipei

McGraw-Hill Ryerson Limited
A Subsidiary of The **McGraw-Hill** Companies

COPIES OF THIS BOOK MAY BE OBTAINED BY CONTACTING:

McGraw-Hill Ryerson Ltd.

WEBSITE:
http://www.mcgrawhill.ca

E-MAIL:
Orders@mcgrawhill.ca

TOLL FREE FAX:
1-800-463-5885

TOLL FREE CALL:
1-800-565-5758

OR BY MAILING YOUR ORDER TO:
McGraw-Hill Ryerson
Order Department,
300 Water Street
Whitby, ON L1N 9B6

Please quote the ISBN and title when placing your order.

SCIENCEPOWER™ 7
Science • Technology • Society • Environment

0-07-560357-8

http//www.mcgrawhill.ca

8 9 0 TCP 08 07 06

Printed and bound in Canada

Canadian Cataloguing in Publication Data

Main entry under title:

Sciencepower 7: science, technology, society, environment

Includes index.

ISBN 0-07-560357-8

1. Science – Juvenile literature. I. Galbraith, Donald I. II Title: Sciencepower seven.

Q161.2.S384 1999 500 C99-931206-5

The SCIENCEPOWER™ Development Team

SCIENCE PUBLISHER: Trudy Rising
SENIOR DEVELOPMENTAL EDITOR: Sheila Fletcher
DEVELOPMENTAL EDITORS: Gerry De Iuliis, Jenna Dunlop, Lois Edwards, Tom Gamblin, Barbara Hehner, Dan Kozlovic, Jane McNulty, Elma Schemenauer, Mary Kay Winter
SENIOR SUPERVISING EDITOR: Nancy Christoffer
PROJECT CO-ORDINATORS: Nancy Landry, Kelli Legros, Crystal Shortt
ASSISTANT PROJECT CO-ORDINATOR: Janie Reeson
EDITORIAL ASSISTANT: Joanne Murray
FIELD-TEST CO-ORDINATOR: Jill Bryant
SPECIAL FEATURES: Jill Bryant, Jean Bullard, Christina Clancy, Trudee Romanek, Elma Schemenauer
COPY EDITOR: Paula Pettitt-Townsend
PERMISSIONS EDITORS: Ann Ludbrook, Jacqueline Donovan
SENIOR PRODUCTION CO-ORDINATORS: Yolanda Pigden, Brad Madill
COVER AND INTERIOR DESIGN: Pronk&Associates
ELECTRONIC PAGE MAKE-UP: Pronk&Associates
SET-UP PHOTOGRAPHY: Ian Crysler, Dave Starrett
SET-UP PHOTOGRAPHY CO-ORDINATOR: Jane Affleck
TECHNICAL ILLUSTRATIONS: Imagineering Scientific and Technical Artworks Inc./Pronk&Associates
ILLUSTRATIONS: Steve Attoe, Margo Davis Leclair, Tina Holdcroft, Jun Park, Theresa Sakno
COVER IMAGE: J.A. Wilkinson/Valan Photos

Acknowledgements

Our ability to offer you this high-quality resource is possible thanks to the honest and frank feedback from the individual consultants and reviewers listed below. In particular, Erminia Pedretti's special interest in STSE (science, technology, society, and the environment), Sylvia Constancio's ESL (English as a second language) specialty and her interest in providing a resource for non-specialist as well as specialist teachers, and Dan Forbes' understanding of students' comprehension at different levels were all critical to the development of ***SCIENCEPOWER***™ 7. Likewise, each and every pedagogical reviewer across the country, and our highly experienced safety reviewer, brought different perspectives based on their special interests, their regions, and their own teaching expertise. The authors, editors, senior program consultants, and publisher sincerely thank them all.

Consultants

Sylvia Constancio
Toronto District School Board
ESL, FSL Specialist
Toronto, Ontario

Dan Forbes
St. Anne Elementary School
St. Anne, Manitoba

Erminia Pedretti
Ontario Institute for Studies in Education of the University of Toronto
STSE Specialist
Toronto, Ontario

Pedagogical and Academic Reviewers

Dave Bekkers
Green Glade Senior Public School
Mississauga, Ontario

Pat Bright
Faculty of Education
University of Victoria
Victoria, British Columbia

Philip Capstick
Coldbrook & District School
Coldbrook, Nova Scotia

Audrey Cook
George Street Middle School
Fredericton, New Brunswick

Mike Elson
Dresden Area Central School
Dresden, Ontario

Professor Gregory C. Finn
Chair, Earth Sciences Department
Brock University
St. Catharines, Ontario

Jenni Foss
Hilltop Middle School
Etobicoke, Ontario

Annelies Groen
Deer Park Public School
Toronto, Ontario

Stephen Haberer
Trinity College School
Port Hope, Ontario

Bruce Hickey
Brother Rice High School
St. John's, Newfoundland

Roy Hughes
Breton Educational Centre
New Waterford, Nova Scotia

Marleen Kacevychius
Malcolm Munroe Memorial Junior High School
Sydney, Nova Scotia

David Knox
Canterbury High School
Ottawa, Ontario

Brenda Kusmenko
Humber Summit Middle School
Toronto, Ontario

Mia MacIntyre
Malcolm Munroe Memorial Junior High School
Sydney, Nova Scotia

Dale Makar
Montgomery Junior High School
Calgary, Alberta

Greg Mazanik
St. Cornelius School
Caledon, Ontario

Sean Marks
River Oaks Public School
Oakville, Ontario

Cedric McGrath
École Le Tremplin
Tracadie-Sheila, New Brunswick

Bob Mealey
Riverview Middle School
Riverview, New Brunswick

Bob Moulder
Montclair Public School
Oakville, Ontario

Professor David Pearson
Laurentian University
Sudbury, Ontario

Terry Quinlan
Charles P. Allen High School
Bedford, Nova Scotia

Bill Reynolds
Morning Glory Public School
Pefferlaw, Ontario

Hazen Savoie
École Dr Marquerite-Michaud
Buctouche, New Brunswick

Taunya Sheffield
Central Kings Rural High School
Cambridge Station, Nova Scotia

Alison Smith
Bayside Middle School
Saint John, New Brunswick

John Smith
Green Glade Senior Public School
Mississauga, Ontario

Jay Sugunan
Beaumonde Heights Junior Middle School
Etobicoke, Ontario

Lindsay Thierry
St. Michael's University Middle School
Victoria, British Columbia

Gary Turner
Cunard Junior High School
Halifax, Nova Scotia

Elgin Wolfe
Ontario Institute for Studies in Education of the University of Toronto
Toronto, Ontario

Safety Reviewer

Margaret Redway
Fraser Scientific & Business Services
Delta, British Columbia

Field Testing Acknowledgements

Imagine asking teachers, just at the end of a busy teaching year, if they would like to be involved in field testing a new science resource you are creating, not in beautifully polished final form, but as photocopied first draft, with stick figures sketched in by authors; now, more challengingly, try to imagine the generosity of the response we received. The following teachers were recommended to us by their boards. We asked them for their help and they provided it willingly and patiently, sharing with us their enthusiasm, their frustrations, and above all their invaluable recommendations for improvements prior to publication. We sincerely thank them and their students for helping us, through months of testing, to develop the most useful possible resource for you and your students in teaching and learning the grade 7 curriculum.

We wish to extend special thanks to Christina Clancy, who wrote the *Projects* for Units 1 and 2; to Jean Bullard, who wrote *An Issue to Analyze* for Units 2 and 3; and to Bob Moulder, for his invaluable assistance in providing props for set-up photographs.

Field Test Teachers

Dave Bekkers
Green Glade Senior Public School
Mississauga, Ontario

Tracy Berry
Terry Fox Public School
Cobourg, Ontario

Rosemary Caruso
Maplehurst School
Burlington, Ontario

Andrea Craig
Johnsview Village Public School
Thornhill, Ontario

Maggie Flynn
St. Elizabeth Seton Catholic School
Newmarket, Ontario

Kevin Freckelton
Chief Dan George Public School
Scarborough, Ontario

Annelies Groen
Deer Park Public School
Toronto, Ontario

Suzy Hall
Trinity College School
Port Hope, Ontario

Catherine Vanderburgh Kerr
Terry Fox Public School
Cobourg, Ontario

Jay Kilburn
Terry Fox Public School
Cobourg, Ontario

Brenda Kusmenko
Humber Summit Middle School
Toronto, Ontario

Kathy LeBlanc
St. Paul's Catholic School
Newmarket, Ontario

Sean Marks
River Oaks Public School
Oakville, Ontario

Bob Moulder
Montclair Public School
Oakville, Ontario

Judy Paterson
Centennial Middle School
Georgetown, Ontario

Dale Ripley
Queen Elizabeth II Public School
Chatham, Ontario

Zélia Tavares
Lawrence Heights Middle School
North York, Ontario

Contents

Unit 4 Earth's Crust . . . 276

1
Height = ?m
Length = ?m
2
Area of wall
= Height x width
of wall
3
Number of
10 cm x 10cm
squares needed
to fill 1m²
= 100
1 m
1 m
10cm
x 10cm
4
Number of
10 cm x 10cm squares
needed to fill wall
= area of wall
in m² x 100

To the Teacher

We are very pleased to have been part of the team of experienced science educators and editors working together to bring you and your students this new program — the ***SCIENCEPOWER™ 7-10*** series of textbooks, and its French equivalent, ***OMNISCIENCES** 7-10*. The ***SCIENCEPOWER™*** and ***OMNISCIENCES*** student and teacher resources were specifically developed to provide 100 percent congruence with the new curriculum. As the titles ***SCIENCEPOWER™*** and ***OMNISCIENCES*** suggest, these resources are designed to foster an appreciation of the power of scientific explanation as a way of understanding our world, and to empower students to critically examine issues and questions from a societal and environmental perspective.

*SCIENCEPOWER™ 7/**OMNISCIENCES*** 7 provide:

- A science inquiry emphasis, in which students address questions about the nature of science involving broad explorations as well as focussed investigations. Skill areas emphasized include: careful observing; questioning; proposing ideas; predicting; hypothesizing; making inferences; designing experiments; gathering, processing, and interpreting data; and explaining and communicating.
- A technological problem-solving emphasis, in which students seek answers to practical problems. Problem solving may either precede knowledge acquisition or provide students with opportunities to apply their newly acquired science knowledge in novel ways. Skill areas emphasized include: understanding the problem; setting and/or understanding criteria; developing a design plan, carrying out the plan; evaluating; and communicating.
- A societal decision-making emphasis, in which students draw upon those science and technology concepts and skills that will inform the question or issue under consideration. Students are encouraged to focus attention on sustainability and stewardship. Skill areas that are emphasized include: identifying the issue; identifying alternatives; researching, reflecting, and deciding; taking action; evaluating; and communicating.

The particular emphases within a unit are, in part, suggested by the topic itself. The primary and secondary emphases for ***SCIENCEPOWER™*** 7 and ***OMNISCIENCES*** 7 are listed in the table opposite.

Scientific literacy has become the goal in science education throughout the world, and this goal has been given expression in Canada in the *Common Framework of Science Learning Outcomes, K-12: Pan-Canadian Protocol for Collaboration on School Curriculum* (Council of Ministers of Education, Canada, 1997).

> "Scientific literacy is an evolving combination of the science-related attitudes, skills, and knowledge students need to develop inquiry, problem-solving, and decision-making abilities, to become lifelong learners, and to maintain a sense of wonder about the world around them. To develop scientific literacy,

students require diverse learning experiences which provide opportunity to explore, analyze, evaluate, synthesize, appreciate, and understand the interrelationships among science, technology, society, and the environment that will affect their personal lives, their careers, and their future."

	***SCIENCEPOWER™ 7/ OMNISCIENCES 7* Unit**	**Primary Emphasis**	**Secondary Emphasis**
Life Systems Interactions Within Ecosystems	Unit 1 Interactions Within Ecosystems	Science and Science Inquiry	Societal Decision Making
Matter and Materials Pure Substances and Mixtures	Unit 2 Pure Substances and Mixtures	Science and Science Inquiry	Technology and Technological Problem Solving
Energy and Control Heat	Unit 3 Thermal Energy and Heat Technology	Technology and Technological Problem Solving	Science and Science Inquiry
Earth and Space Systems The Earth's Crust	Unit 4 Earth's Crust	Science and Science Inquiry	Technology and Technological Problem Solving
Structures and Mechanisms Structural Strength and Stability	Unit 5 Structural Strength and Stability	Technology and Technological Problem Solving	Science and Science Inquiry Societal Decision Making

Through varied text features, ***SCIENCEPOWER*™ 7** enables students to understand and develop skills in the processes of scientific inquiry, and in relating science to technology, society, and the environment.

Like the other textbooks in our series, ***SCIENCEPOWER*™ 7** builds on the three basic goals of the curriculum and reflects the essential triad of knowledge, skills, and the ability to relate science to technology, society, and the environment (STSE). Science is approached both as an intellectual pursuit, and also as an activity-based enterprise operating within a social context.

Our extensive *Teacher's Resource Binder* provides essential planning and implementation strategies that you will find helpful and practical. Our *Blackline Masters* include materials that you can use for vocabulary building, skill building, and concept clarification, as well as alternative activities for multiple learning styles, forms for performance task assessment of student achievement that are specific to the unit of study, and forms for assessment that focus on larger encompassing skills of science, technology, and societal decision making. Our *Computerized Assessment Bank* will assist you in your full implementation of the ***SCIENCEPOWER*™ 7** program.

We feel confident that we have provided you with the best possible program to help ensure that your students achieve excellence and a high degree of scientific literacy through their course of study.

The Authors and Senior Program Consultants

A Tour of Your Textbook

Welcome to *SCIENCEPOWER*™ 7. This textbook introduces you to some of the wonders of science and technology in the world around you. To understand the book's structure, begin by taking the brief tour on the following pages. Then do the *Feature Hunt* on page xxiii to check your understanding of how to use this book.

Unit Opener

- *SCIENCEPOWER*™ 7 has five major units.
- Each unit opener provides a clear overview of the unit's contents.
- The unit opener sparks interest in the topic. It might suggest a problem to think about, present science ideas to consider, or highlight a societal issue to explore.
- The unit opener identifies the three chapters in the unit.

UNIT 4

Earth's Crust

What forces unleash the monstrous power of volcanoes and earthquakes? What processes form Earth's mountains, its rocks and boulders, and the minerals and fuels deep within it, which we mine and use? What processes form the fertile soils that produce our food?

The answers to these questions lie in Earth's crust — the thin, ever-changing, outermost layer of our planet. Throughout our history, scientists have been trying to understand the forces that shape and change Earth's crust. In this unit, you will see how theories about Earth's crust were developed and then discarded, as scientists made new observations and saw new connections between their past observations. You will see how recent technology, which allows scientists to see the sea floor and to "X-ray" Earth's interior, has led to the current theory of a shifting crust. You will also have a chance to think about how our demands for food, minerals, and fuel harm Earth and to consider how we might take better care of Earth's resources.

Unit Contents

Chapter 10 Minerals, Rocks, and Soils 277

Chapter 11 Earthquakes, Volcanoes, and Mountains 312

Chapter 12 The Story of Earth's Crust 342

276 277

Chapter Opener

- Each chapter opener gives you a clear idea of what the chapter is about.
- **Getting Ready** questions give you a chance to reflect on what you already know (or perhaps do not know) about the topics in the chapter.
- **Science Log** suggests ways to answer the *Getting Ready* questions. As well, it provides an opportunity to keep a record of what you learn. Scientists too keep a careful log of their observations and the results of their findings. (Your teacher may call the *Science Log* a *Science Journal* instead.)

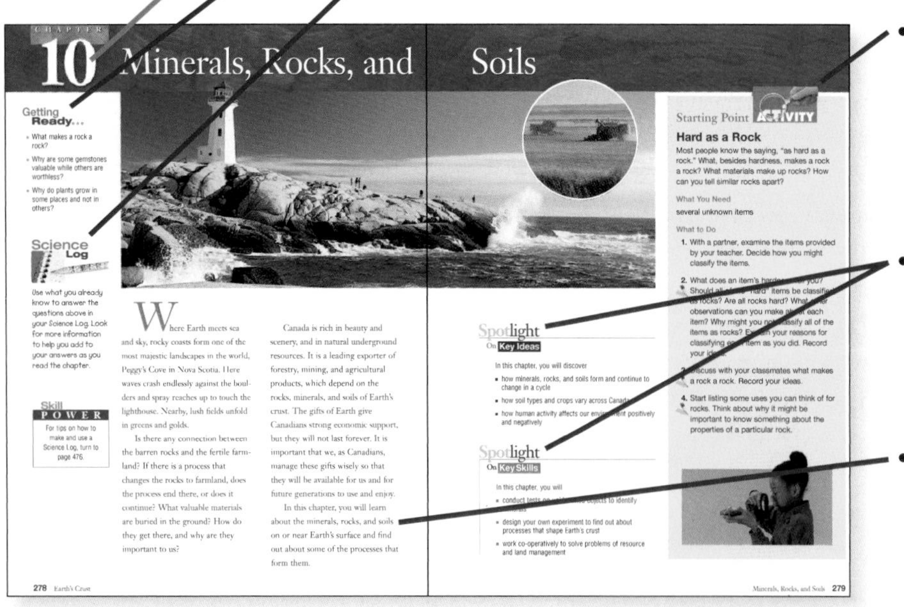
CHAPTER 10 Minerals, Rocks, and Soils

Getting Ready...

- What makes a rock a rock?
- Why are some gemstones valuable while others are worthless?
- Why do plants grow in some places and not in others?

Science Log

Use what you already know to answer the questions above in your Science Log. Look for more information to help you add to your answers as you read the chapter.

Skill POWER

For tips on how to make and use a Science Log, turn to page 476.

Where Earth meets sea and sky, rocky coasts form one of the most majestic landscapes in the world, Peggy's Cove in Nova Scotia. Here waves crash endlessly against the boulders and spray reaches up to touch the lighthouse. Nearby, lush fields unfold in greens and golds.

Is there any connection between the barren rocks and the fertile farmland? If there is a process that changes the rocks to farmland, does the process end there, or does it continue? What valuable materials are buried in the ground? How do they get there, and why are they important to us?

Canada is rich in beauty and scenery, and in natural underground resources. It is a leading exporter of forestry, mining, and agricultural products, which depend on the rocks, minerals, and soils of Earth's crust. The gifts of Earth give Canadians strong economic support, but they will not last forever. It is important that we, as Canadians, manage these gifts wisely so that they will be available for us and for future generations to use and enjoy.

In this chapter, you will learn about the minerals, rocks, and soils on or near Earth's surface and find out about some of the processes that form them.

Spotlight On Key Ideas

In this chapter, you will discover

- how minerals, rocks, and soils form and continue to change in a cycle
- how soil types and crops vary across Canada
- how human activity affects our environment positively and negatively

Spotlight On Key Skills

In this chapter, you will

- conduct tests [illegible] objects to identify [illegible]
- design your own experiment to find out about processes that shape Earth's crust
- work co-operatively to solve problems of resource and land management

Starting Point ACTIVITY

Hard as a Rock

Most people know the saying, "as hard as a rock." What, besides hardness, makes a rock a rock? What materials make up rocks? How can you tell similar rocks apart?

What You Need

several unknown items

What to Do

1. With a partner, examine the items provided by your teacher. Decide how you might classify the items.
2. What does an item's [illegible] you? Should [illegible] "hard" items be classif[illegible] as rocks? Are all rocks hard? What [illegible] observations can you make [illegible] each item? Why might you [illegible]ssify all of the items as rocks? [illegible] your reasons for classifying [illegible] item as you did. Record your [illegible]
3. [illegible]scuss with your classmates what makes a rock a rock. Record your ideas.
4. Start listing some uses you can think of for rocks. Think about why it might be important to know something about the properties of a particular rock.

278 Earth's Crust

Minerals, Rocks, and Soils 279

- The **Starting Point Activity** begins each chapter in a variety of ways. Like the *Getting Ready* questions, the *Starting Point Activity* helps you think about what you already know (or do not know) about the chapter's main topics.
- **Spotlight on Key Ideas** and **Spotlight on Key Skills** focus your attention on the major ideas and skills that you will be expected to know by the time you have completed the chapter.
- The **introductory paragraphs** of each chapter invite you to learn more about the topics. They tell you what you will be studying in the chapter.

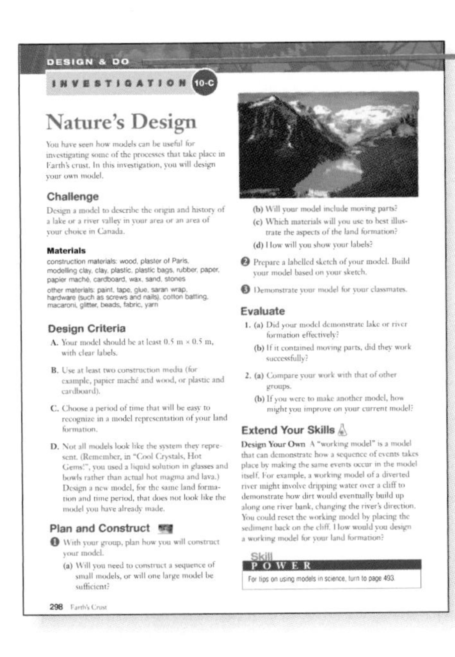

DESIGN & DO

INVESTIGATION 10-C

Nature's Design

You have seen how models can be useful for investigating some of the processes that take place in Earth's crust. In this investigation, you will design your own model.

Challenge

Design a model to describe the origin and history of a lake or a river valley in your area or an area of your choice in Canada.

Materials

construction materials: wood, plaster of Paris, modelling clay, clay, plastic, plastic bags, rubber, paper, papier maché, cardboard, wax, sand, stones

other materials: paint, tape, glue, saran wrap, hardware (such as screws and nails), cotton batting, macaroni, glitter, beads, fabric, yarn

Design Criteria

A. Your model should be at least 0.5 m × 0.5 m, with clear labels.

B. Use at least two construction media (for example, papier maché and wood, or plastic and cardboard).

C. Choose a period of time that will be easy to recognize in a model representation of your land formation.

D. Not all models look like the system they represent. (Remember, in "Cool Crystals, Hot Gems!", you used a liquid solution in glasses and bowls rather than actual hot magma and lava.) Design a new model, for the same land formation and time period, that does not look like the model you have already made.

Plan and Construct

1. With your group, plan how you will construct your model.
   (a) Will you need to construct a sequence of small models, or will one large model be sufficient?
   (b) Will your model include moving parts?
   (c) Which materials will you use to best illustrate the aspects of the land formation?
   (d) How will you show your labels?
2. Prepare a labelled sketch of your model. Build your model based on your sketch.
3. Demonstrate your model for your classmates.

Evaluate

1. (a) Did your model demonstrate lake or river formation effectively?
   (b) If it contained moving parts, did they work successfully?
2. (a) Compare your work with that of other groups.
   (b) If you were to make another model, how might you improve on your current model?

Extend Your Skills

Design Your Own A "working model" is a model that can demonstrate how a sequence of events takes place by making the same events occur in the model itself. For example, a working model of a diverted river might involve dripping water over a cliff to demonstrate how dirt would eventually build up along one river bank, changing the river's direction. You could reset the working model by placing the sediment back on the cliff. How would you design a working model for your land formation?

Skill POWER

For tips on using models in science, turn to page 493.

298 Earth's Crust

Design & Do Investigation

- These hands-on investigations set challenges to design and construct your own models, systems, or products. They teach design skills and blend science and technology in new and different ways.
- The co-operative group work icon signals that you will be doing these investigations in a team.
- The Design Criteria provide a way to evaluate your results.
- You and your team members are then on your own to design and construct!

Did You Know?

- These features present interesting facts that are related to science, technology, nature, and the universe.

Mathconnect

- These features review mathematics skills that you need in order to do activities or investigations.
- They make connections between your science studies and your mathematics studies.

Find Out Activity

- These are short, informal inquiries that usually involve hands-on exploration.
- They require simple materials and equipment.
- In these activities, as well as in the investigations, you will use important science inquiry skills, such as predicting, estimating, and hypothesizing.
- The pencil icon signals that you should make a written note of your predictions or observations.

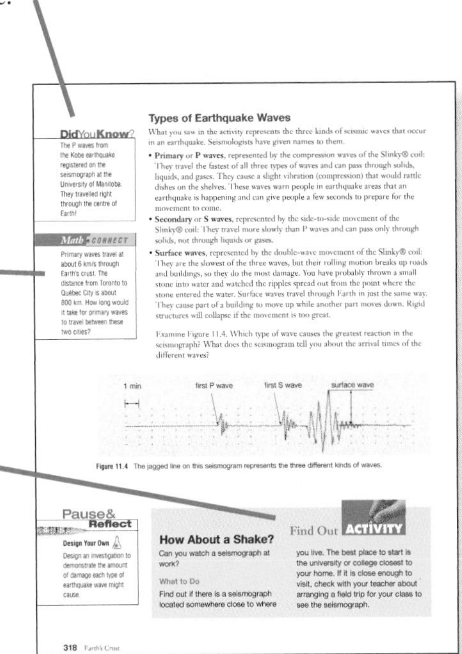

Did You Know?

The P waves from the Kobe earthquake registered on the seismograph at the University of Manitoba. They travelled right through the centre of Earth!

Math CONNECT

Primary waves travel at about 6 km/s through Earth's crust. The distance from Toronto to Québec City is about 800 km. How long would it take for primary waves to travel between these two cities?

Types of Earthquake Waves

What you saw in the activity represents the three kinds of seismic waves that occur in an earthquake. Seismologists have given names to them.

- **Primary** or **P waves**, represented by the compression waves of the Slinky® coil: They travel the fastest of all three types of waves and can pass through solids, liquids, and gases. They cause a slight vibration (compression) that would rattle dishes on the shelves. These waves warn people in earthquake areas that an earthquake is happening and can give people a few seconds to prepare for the movement to come.
- **Secondary** or **S waves**, represented by the side-to-side movement of the Slinky® coil: They travel more slowly than P waves and can pass only through solids, not through liquids or gases.
- **Surface waves**, represented by the double-wave movement of the Slinky® coil: They are the slowest of the three waves, but their rolling motion breaks up roads and buildings, so they do the most damage. You have probably thrown a small stone into water and watched the ripples spread out from the point where the stone entered the water. Surface waves travel through Earth in just the same way. They cause part of a building to move up while another part moves down. Rigid structures will collapse if the movement is too great.

Examine Figure 11.4. Which type of wave causes the greatest reaction in the seismograph? What does the seismogram tell you about the arrival times of the different waves?

Figure 11.4 The jagged line on this seismogram represents the three different kinds of waves.

Pause & Reflect

Design Your Own

Design an investigation to demonstrate the amount of damage each type of earthquake wave might cause.

Find Out ACTIVITY

How About a Shake?

Can you watch a seismograph at work?

What to Do

Find out if there is a seismograph located somewhere close to where you live. The best place to start is the university or college closest to your home. If it is close enough to visit, check with your teacher about arranging a field trip for your class to see the seismograph.

318 Earth's Crust

Conduct an Investigation

- One- to four-page "formal" labs provide an opportunity to develop science inquiry skills using various equipment and materials.
- These investigations provide a chance to ask questions about science, to make observations, and to obtain results.
- You then analyze your results to determine what they tell you about the topic you are investigating.
- Photographs showing each major step in the Procedure help you to carry out the investigation.
- Safety icons and Safety Precautions alert you to any special precautions you should take to help maintain a safe classroom environment.
- The pencil icon signals that you should make a written note of your predictions or observations.

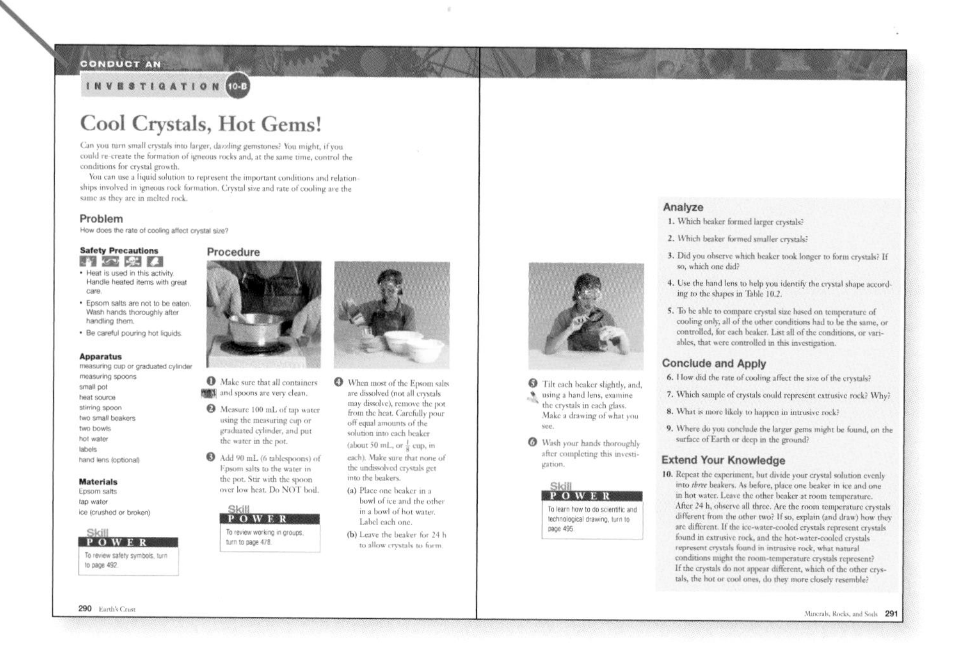

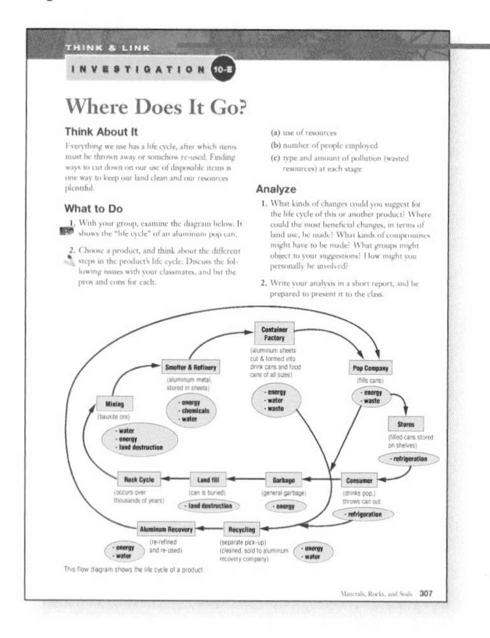

Think & Link Investigation

- One- to two-page "thought" or "paper-based" investigations let you explore ideas or connections that might be impractical or dangerous in the science classroom.
- These investigations emphasize a variety of skills. These skills include analyzing data, interpreting diagrams or photographs, and forming ideas, opinions, or recommendations based on analysis of a societal issue.
- They provide you with opportunities to "think and link" — to think about scientific results and issues that involve science and technology. You will increase your skills of analysis by doing these investigations.

At Home Activity

- These short, informal inquiry activities can be done at home using simple, everyday materials.

Cool Tools

- These features provide information about some of the equipment and instruments that have been invented to help humans explore the unknown.
- This information is often related to a variety of occupations and situations.

Check Your Understanding

- A set of review questions appears at the end of each numbered section in a chapter.
- These questions provide opportunities for ongoing self-assessment.
- **Apply**, **Design Your Own**, and **Thinking Critically** questions give you additional challenges.

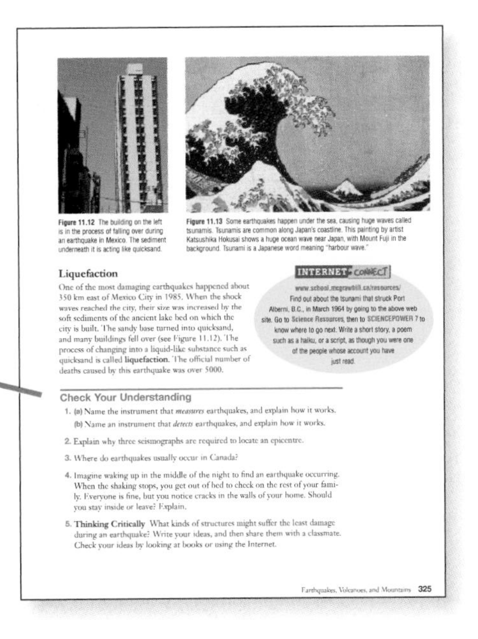

SkillPower

- Skill development tips refer you to the **Science and Technology Skills Guide** at the back of this textbook.
- These tips provide specific skill development methods and activities as they are needed, for example in estimating and measuring and in scientific drawing. (The Contents page of the *Science and Technology Skills Guide* is shown on page xxiii, "Wrapping Up the Tour.")

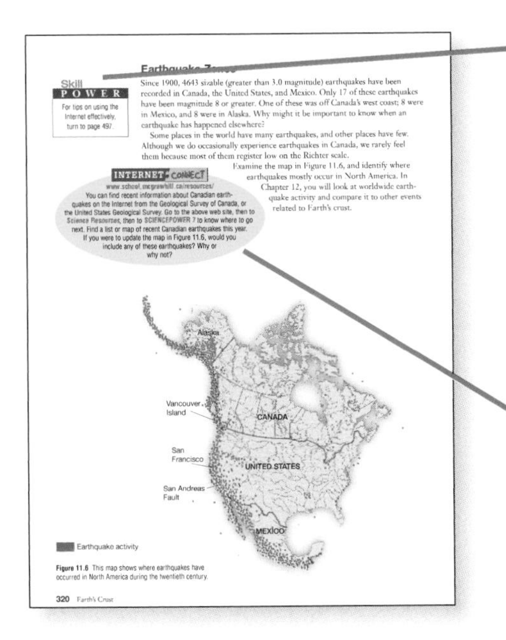

Internetconnect

- These features encourage productive use of the Internet by offering content-related sites.
- Web site suggestions will save you time as you do research.

Across Canada

- These "mini-essays" feature information on Canadian scientists involved in important research and discoveries.
- The essays increase awareness and appreciation of the work of Canadian scientists. They also provide role models for those of you who are interested in careers or further study in science.

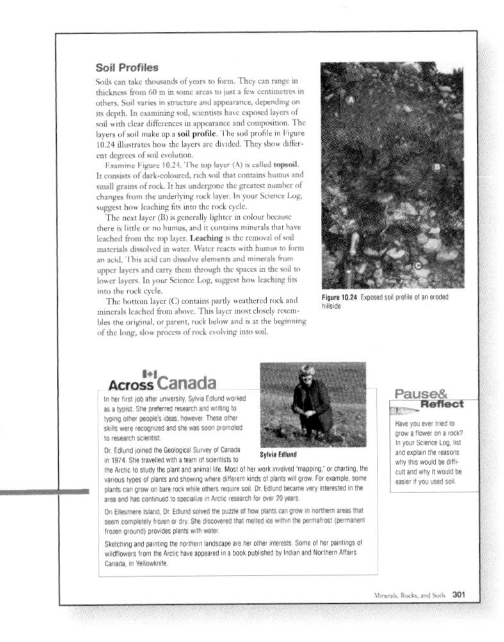

Chapter at a Glance

- Located at the end of each chapter, this page provides self-assessment opportunities as you look back at the chapter as a whole.
- It gives parents or guardians an overview of what you have accomplished.
- **Prepare Your Own Summary** encourages you to summarize your understanding in a variety of ways — using diagrams, flowcharts, concept maps, artwork, writing, or any approach you prefer.

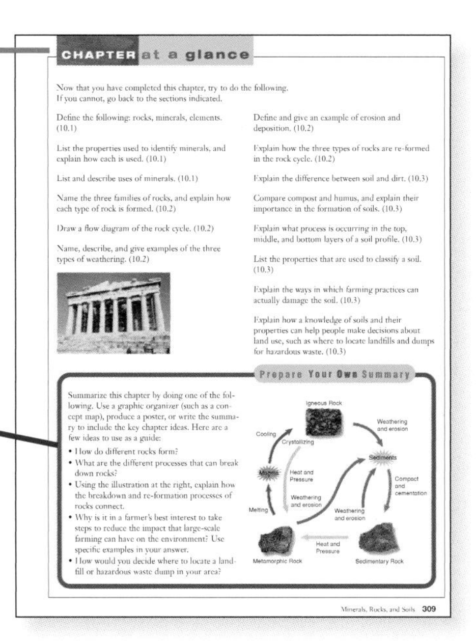

CHAPTER at a glance

Now that you have completed this chapter, try to do the following. If you cannot, go back to the sections indicated.

Define the following: rocks, minerals, elements. (10.1)

List the properties used to identify minerals, and explain how each is used. (10.1)

List and describe uses of minerals. (10.1)

Name the three families of rocks, and explain how each type of rock is formed. (10.2)

Draw a flow diagram of the rock cycle. (10.2)

Name, describe, and give examples of the three types of weathering. (10.2)

Define and give an example of erosion and deposition. (10.2)

Explain how the three types of rocks are re-formed in the rock cycle. (10.2)

Explain the difference between soil and dirt. (10.3)

Compare compost and humus, and explain their importance in the formation of soils. (10.3)

Explain what process is occurring in the top, middle, and bottom layers of a soil profile. (10.3)

List the properties that are used to classify a soil. (10.3)

Explain the ways in which farming practices can actually damage the soil. (10.3)

Explain how a knowledge of soils and their properties can help people make decisions about land use, such as where to locate landfills and dumps for hazardous waste. (10.3)

Prepare Your Own Summary

Summarize this chapter by doing one of the following. Use a graphic organizer (such as a concept map), produce a poster, or write the summary to include the key chapter ideas. Here are a few ideas to use as a guide:

- How do different rocks form?
- What are the different processes that can break down rocks?
- Using the illustration at the right, explain how the breakdown and re-formation processes of rocks connect.
- Why is it in a farmer's best interest to take steps to reduce the impact that large-scale farming can have on the environment? Use specific examples in your answer.
- How would you decide where to locate a landfill or hazardous waste dump in your area?

Minerals, Rocks, and Soils 309

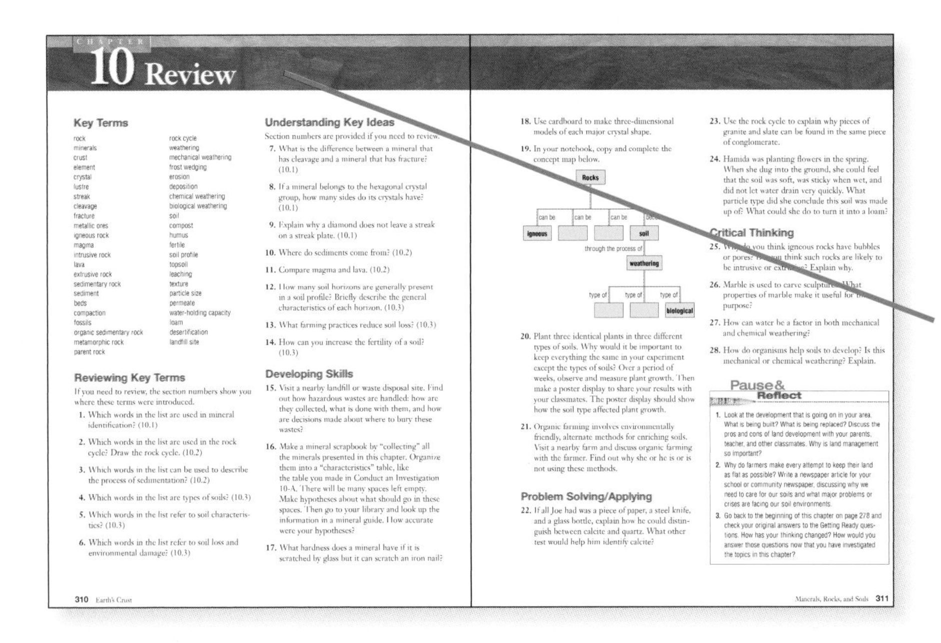

CHAPTER 10 Review

Chapter Review

- This final wrap-up of each chapter reviews basic concepts, skills of inquiry and communication, and skills relating science to technology, society, and the environment.
- These questions help you recall, think about, and apply what you have learned.

End-of-Unit Features

Ask an Expert

- Experts in every area of science and technology are working to understand better how the world "works" and to try to find solutions to difficult problems. The *Ask an Expert* feature at the end of each unit is an interview with one of these experts.
- After you read each interview, you will have a chance to do an activity that is related to the kind of work the expert does.

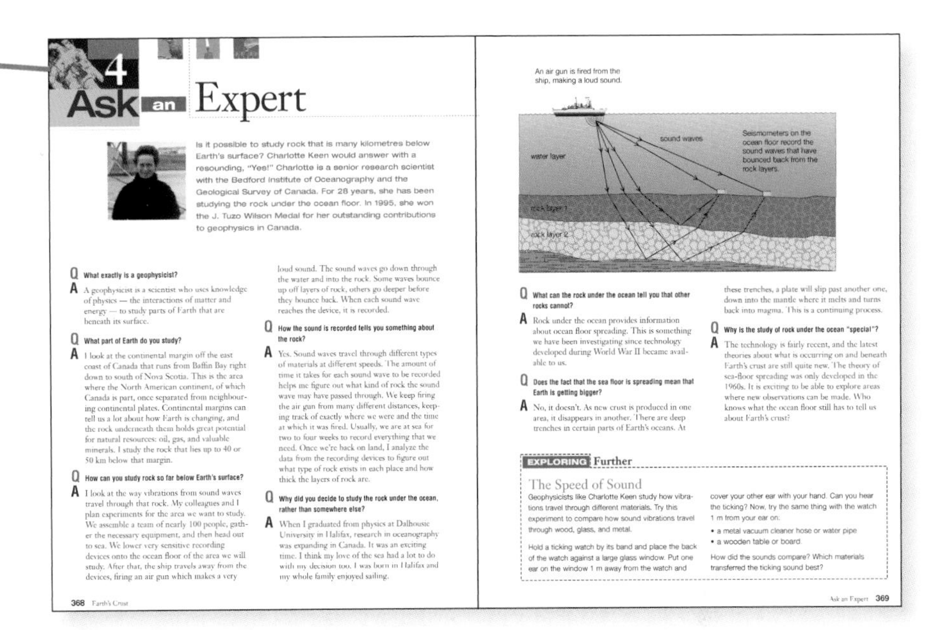

4 Ask an Expert

Is it possible to study rock that is many kilometres below Earth's surface? Charlotte Keen would answer with a resounding, "Yes!" Charlotte is a senior research scientist with the Bedford Institute of Oceanography and the Geological Survey of Canada. For 28 years, she has been studying the rock under the ocean floor. In 1995, she won the J. Tuzo Wilson Medal for her outstanding contributions to geophysics in Canada.

Q What exactly is a geophysicist?

A A geophysicist is a scientist who uses knowledge of physics — the interactions of matter and energy — to study parts of Earth that are beneath its surface.

Q What part of Earth do you study?

A I look at the continental margin off the east coast of Canada that runs from Baffin Bay right down to south of Nova Scotia. This is the area where the North American continent, of which Canada is part, once separated from neighbouring continental plates. Continental margins can tell us a lot about how Earth is changing, and the rock underneath them holds great potential for natural resources: oil, gas, and valuable minerals. I study the rock that lies up to 40 or 50 km below that margin.

Q How can you study rock so far below Earth's surface?

A I look at the way vibrations from sound waves travel through that rock. My colleagues and I plan experiments for the area we want to study. We assemble a team of nearly 100 people, gather the necessary equipment, and then head out to sea. We lower very sensitive recording devices onto the ocean floor of the area we will study. After that, the ship travels away from the devices, firing an air gun which makes a very loud sound. The sound waves go down through the water and into the rock. Some waves bounce up off layers of rock, others go deeper before they bounce back. When each sound wave reaches the device, it is recorded.

Q How the sound is recorded tells you something about the rock?

A Yes. Sound waves travel through different types of materials at different speeds. The amount of time it takes for each sound wave to be recorded helps me figure out what kind of rock the sound wave may have passed through. We keep firing the air gun from many different distances, keeping track of exactly where we were and the time at which it was fired. Usually, we are at sea for two to four weeks to record everything that we need. Once we're back on land, I analyze the data from the recording devices to figure out what type of rock exists in each place and how thick the layers of rock are.

Q Why did you decide to study the rock under the ocean, rather than somewhere else?

A When I graduated from physics at Dalhousie University in Halifax, research in oceanography was expanding in Canada. It was an exciting time. I think my love of the sea had a lot to do with my decision too. I was born in Halifax and my whole family enjoyed sailing.

368 Earth's Crust

An air gun is fired from the ship, making a loud sound.

sound waves

water layer

Seismometers on the ocean floor record the sound waves that have bounced back from the rock layers.

rock layer 1

rock layer 2

Q What can the rock under the ocean tell you that other rocks cannot?

A Rock under the ocean provides information about ocean floor spreading. This is something we have been investigating since technology developed during World War II became available to us.

Q Does the fact that the sea floor is spreading mean that Earth is getting bigger?

A No, it doesn't. As new crust is produced in one area, it disappears in another. There are deep trenches in certain parts of Earth's oceans. At these trenches, a plate will slip past another one, down into the mantle where it melts and turns back into magma. This is a continuing process.

Q Why is the study of rock under the ocean "special"?

A The technology is fairly recent, and the latest theories about what is occurring on and beneath Earth's crust are still quite new. The theory of sea-floor spreading was only developed in the 1960s. It is exciting to be able to explore areas where new observations can be made. Who knows what the ocean floor still has to tell us about Earth's crust?

EXPLORING Further

The Speed of Sound

Geophysicists like Charlotte Keen study how vibrations travel through different materials. Try this experiment to compare how sound vibrations travel through wood, glass, and metal.

Hold a ticking watch by its band and place the back of the watch against a large glass window. Put one ear on the window 1 m away from the watch and cover your other ear with your hand. Can you hear the ticking? Now, try the same thing with the watch 1 m from your ear on:

- a metal vacuum cleaner hose or water pipe
- a wooden table or board

How did the sounds compare? Which materials transferred the ticking sound best?

Ask an Expert 369

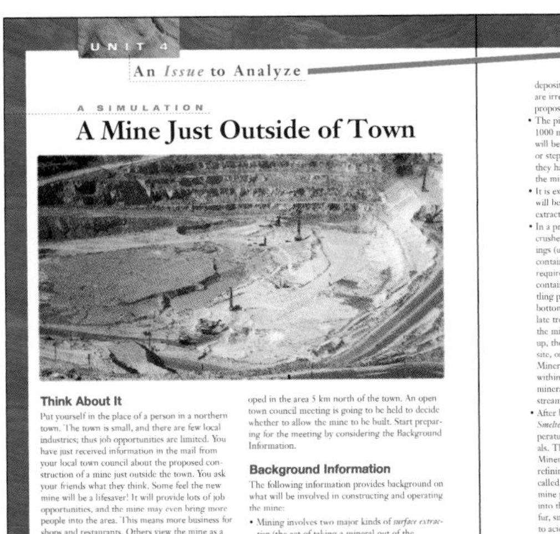

UNIT 4

An *Issue* to Analyze

A SIMULATION

A Mine Just Outside of Town

Think About It

Put yourself in the place of a person in a northern town. The town is small, and there are few local industries; thus job opportunities are limited. You have just received information in the mail from your local town council about the proposed construction of a mine just outside the town. You ask your friends what they think. Some feel the new mine will be a lifesaver! It will provide lots of job opportunities, and the mine may even bring more people into the area. This means more business for shops and restaurants. Others view the mine as a disaster. It will disturb, perhaps even destroy, part of the natural environment. Jobs are needed, but mines have created eyesores and major environmental problems in other places.

Mineralite Mining has presented a proposal to the town council for an open-pit mine to be developed in the area 5 km north of the town. An open town council meeting is going to be held to decide whether to allow the mine to be built. Start preparing for the meeting by considering the Background Information.

Background Information

The following information provides background on what will be involved in constructing and operating the mine:

- Mining involves two major kinds of *surface extraction* (the act of taking a mineral out of the ground). One is strip mining, and the other is open-pit mining. Strip mining does the least amount of damage, but it can be used only when a mineral lies in horizontal layers close to Earth's surface. Open pit mining is necessary when the deposits of minerals are close to the surface but are irregularly shaped. This is the case for the proposed mine outside your town.
- The pit that will be required will be more than 1000 m in diameter and 100 m to 200 m deep. It will be cut into solid rock in a series of terraces or steps. Trucks can use these steps as roads when they haul minerals and equipment in and out of the mine.
- It is expected that up to 3 to 4 tonnes of waste will be removed for every tonne of mineral that is extracted. What will happen to the waste?
- In a process called *beneficiation*, extractions will be crushed and ground in a mill to separate the tailings (unwanted rock materials) from the ore that contains the valuable minerals. This process will require huge amounts of water. The waste water, containing the tailings, will be transported to settling ponds, where the tailings will settle to the bottom. It is expected that the mine will recirculate treated water from the settling ponds back to the mill for re-use. When the settling ponds fill up, the waste will have to be moved to another site, or new settling ponds will be required. Mineralite Mining has promised to do everything within its power to ensure that chemicals and minerals from the tailings are not washed into streams and lakes by rainwater.
- After beneficiation comes the smelting process. *Smelters* are furnaces that reach very high temperatures and turn the ore concentrates into metals. The metals are then refined in refineries. Mineralite Mining recognizes that smelting and refining can cause serious pollution. Waste rock, called *slag*, is sometimes dumped in heaps on mine property, and fumes from smokestacks go into the air. Because many minerals contain sulfur, smelters and refineries are major contributors to acid rain. Mineralite Mining believes that these problems can be prevented or lessened. They propose to install state-of-the-art, tall smokestacks. Fumes will be blown out of the area before falling to the ground. As well, Mineralite Mining promises that when the area has been mined and the mine closes, they will reclaim the land. They will bury the slag, replace the topsoil, and turn the area into parkland.

Plan and Act

1. Plan to attend the town council meeting. These are some of the people who have asked to make a presentation at the meeting:

   Mineralite Mining executive

   town councillor

   Aboriginal representative

   unemployed town citizen
2. Consider these questions: What point of view do you think each of these people might have before the meeting? What are some concerns and considerations they might want to discuss? Do you think anything could change their points of view?
3. Your teacher will give your group the role of one of these people with additional information to help you plan your presentation. As a group, research your role, and be prepared to argue your case at the town council meeting.
4. At the town council meeting, your task, as a class, will be to decide whether to accept the mining company's proposal. Your teacher will provide you with a blackline master to show you the correct *Procedure* for a Public Hearing.

Analyze

1. What was your decision?
2. Were all points of view well researched and presented? If not, how do you think they could have been improved?
3. Were all participants fairly treated? If not, how could this be improved?
4. How did your understanding of science and technology help you make your decision?

370 Earth's Crust

An Issue to Analyze 371

An Issue to Analyze

- You, your community, and society in general face complex issues in today's world. Understanding science and technology cannot provide a "correct" answer to the problems these issues present, but understanding will lead to more informed decisions. *An Issue to Analyze* gives you a chance to start thinking now about how you can help make the best decisions for yourself and your community, today and in the future.
- Each issue analysis is in the form of a simulation, a debate, or a case study.

Unit Project

- A *Unit Project* gives you a chance to use key concepts and skills from the unit to design and create a device, system, or model of your own.
- Early in the unit, your teacher might ask you to begin to consider how you might design, plan, and complete your wrap-up project.
- You will complete the project as part of a team.
- *More Project Ideas* offers additional suggestions for enjoyable ways to demonstrate your learning.

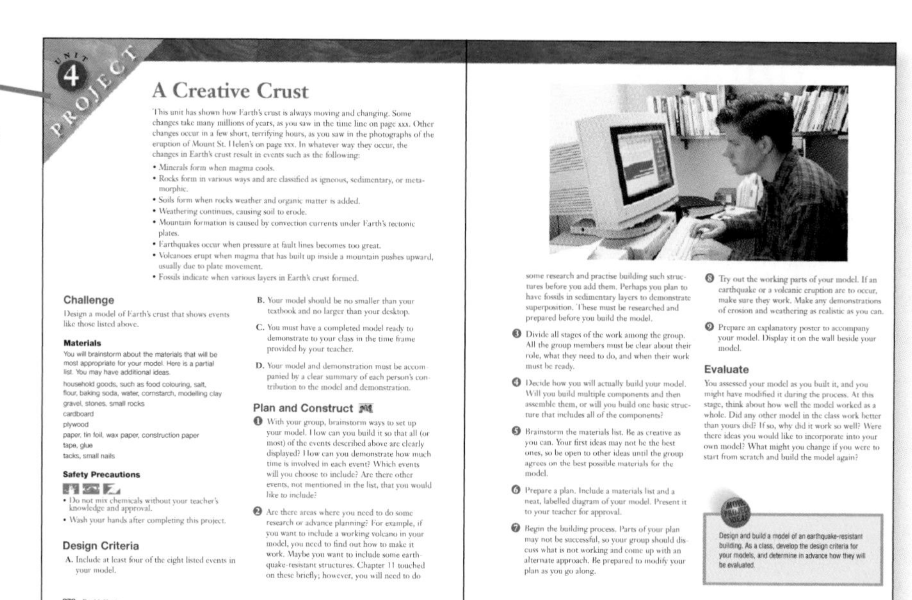
UNIT 4 PROJECT

A Creative Crust

This unit has shown how Earth's crust is always moving and changing. Some changes take many millions of years, as you saw in the time line on page xxx. Other changes occur in a few short, terrifying hours, as you saw in the photographs of the eruption of Mount St. Helen's on page xxx. In whatever way they occur, the changes in Earth's crust result in events such as the following:

- Minerals form when magma cools.
- Rocks form in various ways and are classified as igneous, sedimentary, or metamorphic.
- Soils form when rocks weather and organic matter is added.
- Weathering continues, causing soil to erode.
- Mountain formation is caused by convection currents under Earth's tectonic plates.
- Earthquakes occur when pressure at fault lines becomes too great.
- Volcanoes erupt when magma that has built up inside a mountain pushes upward, usually due to plate movement.
- Fossils indicate when various layers in Earth's crust formed.

Challenge

Design a model of Earth's crust that shows events like those listed above.

Materials

You will brainstorm about the materials that will be most appropriate for your model. Here is a partial list. You may have additional ideas.

household goods, such as food colouring, salt, flour, baking soda, water, cornstarch, modelling clay
gravel, stones, small rocks
cardboard
plywood
paper, tin foil, wax paper, construction paper
tape, glue
tacks, small nails

Safety Precautions

- Do not mix chemicals without your teacher's knowledge and approval.
- Wash your hands after completing this project.

Design Criteria

A. Include at least four of the eight listed events in your model.

B. Your model should be no smaller than your textbook and no larger than your desktop.

C. You must have a completed model ready to demonstrate to your class in the time frame provided by your teacher.

D. Your model and demonstration must be accompanied by a clear summary of each person's contribution to the model and demonstration.

Plan and Construct

1. With your group, brainstorm ways to set up your model. How can you build it so that all (or most) of the events described above are clearly displayed? How can you demonstrate how much time is involved in each event? Which events will you choose to include? Are there other events, not mentioned in the list, that you would like to include?
2. Are there areas where you need to do some research or advance planning? For example, if you want to include a working volcano in your model, you need to find out how to make it work. Maybe you want to include some earthquake-resistant structures. Chapter 11 touched on these briefly; however, you will need to do some research and practise building such structures before you add them. Perhaps you plan to have fossils in sedimentary layers to demonstrate superposition. These must be researched and prepared before you build the model.
3. Divide all stages of the work among the group. All the group members must be clear about their role, what they need to do, and when their work must be ready.
4. Decide how you will actually build your model. Will you build multiple components and then assemble them, or will you build one basic structure that includes all of the components?
5. Brainstorm the materials list. Be as creative as you can. Your first ideas may not be the best ones, so be open to other ideas until the group agrees on the best possible materials for the model.
6. Prepare a plan. Include a materials list and a neat, labelled diagram of your model. Present it to your teacher for approval.
7. Begin the building process. Parts of your plan may not be successful, so your group should discuss what is not working and come up with an alternate approach. Be prepared to modify your plan as you go along.
8. Try out the working parts of your model. If an earthquake or a volcanic eruption are to occur, make sure they work. Make any demonstrations of erosion and weathering as realistic as you can.
9. Prepare an explanatory poster to accompany your model. Display it on the wall beside your model.

Evaluate

You assessed your model as you built it, and you might have modified it during the process. At this stage, think about how well the model worked as a whole. Did any other model in the class work better than yours did? If so, why did it work so well? Were there ideas you would like to incorporate into your own model? What might you change if you were to start from scratch and build the model again?

Design and build a model of an earthquake-resistant building. As a class, develop the design criteria for your models, and determine in advance how they will be evaluated.

372 Earth's Crust

Project 373

Other Important Features

- These items feature intriguing situations, odd events, or weird facts.
- Ideas for connecting science with other curriculum areas are often included.

- These brain teasers are often related to mathematics.
- They draw upon your problem-solving skills and your imagination.

Pause& Reflect

- These features give you opportunities to reflect on what you know (or do not know) and to make connections among ideas throughout the textbook.
- They encourage you to construct your own learning on an ongoing basis and to keep track of how your knowledge is building.

Word CONNECT

- Word origins and a variety of language activities provide links to language arts.

Career CONNECT

- These features portray people with various levels of education, making practical use of science and technology in their jobs.

Computer CONNECT

- These features highlight opportunities where using spreadsheets or data base applications would be helpful.

Design Your Own

- Opportunities are given in investigations, activities, *Pause & Reflect* feaures, *Check Your Understanding* sections, and *Chapter Reviews* for you to plan, design, and conduct your own experimental investigations.

Wrapping Up the Tour

At the back of ***SCIENCEPOWER™ 7***, you will find some additional features to help you review and develop skills and knowledge that you will need to be successful in this course. Are you having trouble with graphing? Would you like help setting up a data table? Have you forgotten how to make a concept map? Do you need a reminder about the metric system? The *Science and Technology Skills Guide* will help you review or improve your skills. A *Glossary* provides all the key vocabulary for the whole course, and an *Index* will help you find your way to a topic.

Science and Technology Skills Guide

Special Icons

The co-operative group work icon alerts you to opportunities to work within a group, and the pencil icon signals you to record your predictions and observations on paper. The safety icons are extremely important because they alert you to any safety precautions you must take, such as wearing safety goggles or a lab apron. Other safety icons that are used in this book are shown on page 492. Make sure that you become familiar with what they mean, and remember to follow the precautions.

Instant Practice — Feature Hunt

To acquaint yourself further with your textbook before you start using it, see if you can find some of the features it contains. Work with a partner.

1. Find the following features. Briefly tell what each feature is about, and record the page number where you found it.
   (a) a *Stretch Your Mind* in Chapter 14
   (b) an *Off the Wall* in Chapter 2
   (c) a *Pause & Reflect* in Chapter 10
   (d) a *Did You Know?* in Unit 1
   (e) an *Across Canada* in Chapter 15
   (f) a *Find Out Activity* called "No Fishing Allowed" in Chapter 2
   (g) a *Design & Do Investigation* entitled "The Windproof Wonder" in Chapter 13
2. Find these words and their meanings.
   (a) force diagram (Unit 5)
   (b) symbiosis (Unit 1)
   (c) solution (Unit 2)
   (d) calibrate (Unit 3)
   (e) erosion (Unit 4)
   (f) concept map (*SkillPower 1*)
   (g) histogram (*SkillPower 5*)

Extensions

3. How did you find the words in question 2? Give one other way to find each word and its meaning.
4. Prepare your own feature hunt for a classmate to do.

Safety in Your Science Classroom

Become familiar with the following safety rules and procedures. It is up to you to use them, and your teacher's instructions, to make your activities and investigations in ***SCIENCEPOWER***™ 7 safe and enjoyable. Your teacher will give you specific information about any other special safety rules and procedures that need to be used in your school.

1. **Working with your teacher . . .**
   - Listen carefully to any instructions your teacher gives you.
   - Inform your teacher if you have any allergies, medical conditions, or other physical problems that could affect your work in the science classroom. Tell your teacher if you wear contact lenses or a hearing aid.
   - Obtain your teacher's approval before beginning any activity you have designed yourself.
   - Know the location and proper use of the nearest fire extinguisher, fire blanket, first-aid kit, and fire alarm.

2. **Starting an activity or investigation . . .**
   - Before starting an activity or investigation, read all of it. If you do not understand how to do any step, ask your teacher for help.
   - Be sure you have checked the safety icons and have read and understood the safety precautions.
   - Begin an activity or investigation only after your teacher tells you to begin.

3. **Dressing for success in science . . .**
   - When you are directed to do so, wear protective clothing, such as a lab apron and safety goggles. Always wear protective clothing when you are using materials that could pose a safety problem, such as unidentified substances, or when you are heating something.
   - Tie back long hair, and avoid wearing scarves, ties, or long necklaces.

4. **Acting responsibly . . .**
   - Work carefully with a partner, and make sure that your work area is clear.
   - Handle equipment and materials carefully.
   - Make sure that stools and chairs are resting securely on the floor.
   - If other students are doing something that you consider dangerous, report it to your teacher.

5. **Handling edible substances . . .**
   - Do not chew gum, eat, or drink in your science classroom.
   - Do not taste any substances or use your mouth to draw any materials into a tube.

6. Working in a science classroom . . .

- Make sure that you understand all the safety labels on school materials and materials you bring from home. Familiarize yourself with the WHMIS symbols and the special safety symbols used in this book (see page 492).
- When carrying equipment for an activity or investigation, hold it carefully. Carry only one object or container at a time.
- Be aware of others during activities and investigations. Make room for students who are carrying equipment to their work stations.

7. Working with sharp objects . . .

- Always cut away from yourself and others when using a knife or razor blade.
- Always keep the pointed end of scissors or any other sharp object facing away from yourself and others if you have to walk with it.
- If you notice sharp or jagged edges on any equipment, take special care with it and report it to your teacher.
- Dispose of broken glass as your teacher directs.

8. Working with electrical equipment . . .

- Make sure that your hands are dry when touching electrical cords, plugs, or sockets.
- Pull the plug, not the cord, when unplugging electrical equipment. Report damaged equipment or frayed cords to your teacher.
- Place electrical cords in places where people will not trip over them.

9. Working with heat . . .

- When heating something, wear safety goggles and any other safety equipment that the textbook or your teacher advises.
- Always use heatproof containers.
- Do not use broken or cracked containers.
- Point the open end of a container that is being heated away from yourself and others.
- Do not allow a container to boil dry.
- Handle hot objects carefully. Be especially careful with a hot plate that might look as though it has cooled down.
- If you use a Bunsen burner, make sure that you understand how to light it and use it safely.
- If you do receive a burn, inform your teacher and apply cold water to the burned area immediately.

10. Working with various chemicals . . .

- If any part of your body comes in contact with a substance, wash the area immediately and thoroughly with water. If you get anything in your eyes, do not touch them. Wash them immediately and continuously for 15 min, and inform your teacher.

- Always handle substances carefully. If you are asked to smell a substance, never smell it directly. Hold the container slightly in front of and beneath your nose, and waft the fumes toward your nostrils, as shown here.
- Hold containers away from your face when pouring a liquid, as shown below.

Use this method to smell a substance in the laboratory.

11. **Working with living things . . .**

On a field trip:

- Try to disturb the area as little as possible.
- If you move something, do it carefully and always replace it carefully.
- If you are asked to remove plant material, remove it gently and take as little as possible.

In the classroom:

- Treat living creatures with respect.
- Make sure that living creatures receive humane treatment while they are in your care.
- If possible, return living creatures to their natural environment when your work is complete.

12. **Cleaning up in the science classroom . . .**

- Clean up any spills, according to your teacher's instructions.
- Clean equipment before you put it away.
- Wash your hands thoroughly after doing an activity or an investigation.
- Dispose of materials as directed by your teacher. Never dispose of materials in a sink unless your teacher directs you to do this.

Hold containers away from your face when pouring liquids.

13. Designing, constructing, and experimenting with structures and mechanisms . . .

- Use tools safely to cut, join, and shape objects.
- Handle modelling clay correctly. Wash your hands after using it.
- Follow proper procedures when studying mechanical systems and the way they operate.
- Use special care when observing and working with objects in motion (for example, gears and pulleys, elevated objects, and objects that spin, swing, bounce, or vibrate).
- Do not use power equipment, such as drills, sanders, saws, and lathes, unless you have specialized training in handling such tools.

Instant Practice

Get to know the safety logos in your textbook. Understanding what they mean and following what they tell you to do will help protect everyone in your science classroom.

What You Need

notebook
pens/pencils
Bristol board
markers
this textbook

What to Do

1. Divide the textbook into sections, so that each person in your group is responsible for one section. Have each person find and list all the safety logos that appear in a particular section.

2. Make a common group list, which includes one of each logo. Discuss each logo in your group, and write what it means beside it. Check with your teacher to make sure that you understand what all the logos mean.

3. Choose at least one logo, and brainstorm what your group might illustrate about it. You could come up with your own ideas, or you could use the safety rules in the textbook that relate to the logo(s) you have chosen.

4. Make a poster to illustrate your logo(s). Your poster should show the right way and the wrong way to do an activity or handle equipment. Place a large red X across the wrong way, so that everyone who sees your poster will understand that this is not safe. You might want to make a rough copy of your poster and get your teacher's approval before making the final copy.

5. Give your poster to your teacher for display.

Introducing the *SCIENCEPOWER*™ 7

The world is full of questions. Everywhere you look there are questions to be asked and (sometimes) answers to be found. Why do some mountains explode in volcanic eruptions? How does salt melt ice on our streets and sidewalks? How can buildings be constructed to resist earthquakes? How and for what purpose do we construct dams? For answers to questions such as these, we can turn to science and technology.

The main goal of **science** is to investigate and better understand the nature of the universe, which includes all the living and non-living things that we can know about. Science is a way of thinking, a way of looking at things and asking questions that helps us find answers about the natural world. In other words, it is a special method of inquiry. The goal of **technology** is to design and construct devices, processes, and materials to solve practical problems and to satisfy human needs and wants. Look at the photographs and read the captions to see how science and technology are related.

A Science can help us to explore and explain the forces under Earth's crust that cause volcanoes. We observe natural events and ask, "What causes this event?"

B Technology helps us to construct buildings that can resist earthquake destruction. We could not engineer such buildings without an understanding of the science behind earthquake activity.

C Science helps us to explain how salt lowers the freezing point of water. We observe the ability of salt to melt ice and ask, "How and why does this happen?"

D Technology makes huge dams possible. The technology is based on scientific understanding of the energy of moving water.

Program

In your grade 7 course, you will take part in both scientific and technological investigations. The new skills and new knowledge you will develop will help you to understand the relationships among science, technology, society, and the environment (abbreviated **STSE**). Why is it important to understand such relationships? Are you really affected all that much by technology? Just think for a moment. Write down some of the things you have done today that involve science and technology. For example, did you take milk or juice out of your refrigerator this morning? You probably turned on a water tap, and you might have used a pencil or pen to jot an event on a calendar. Did you turn on a light? Did you come to school by bicycle, bus, or car? If you walked to school, were you in bare feet, or did you wear shoes? All of the things just mentioned are such a routine part of everyday life that we often take them for granted. They all, however, result from science and technology.

Have you ever flown a kite or built and flown a paper airplane? Then, you know lots about aerodynamics, and you have used the science to build a structure that "worked." Well, perhaps not. Like all humans since prehistoric times, you have probably used technology long before you knew about the science behind it. For example, people have been making music for thousands of years, and in the last few hundred years, people have made musical instruments that produce breathtaking sounds. All this occurred well before people knew the scientific explanation behind sound. Indeed, Antonio Stradivari (1644-1727) from Cremona, Italy, and his family made extraordinary violins that are regarded as the best ever made. Scientists still do not agree on what it is about a Stradivarius, such as the one shown here, that gives it such a beautiful sound.

Have you ever thought of working in the area of science or technology? You might! Even if you choose to pursue a career in another area, you will need to know the basic ideas and skills that you develop by studying science and technology. Start with the *Instant Practice* below, to see what you already know about these subjects.

Instant Practice

1. With a partner, examine the pictures shown here.

2. Identify which of the pictures relate specifically to science and which relate specifically to technology. Tell why you think so.

3. For the pictures you identified as relating specifically to science, write a question you might ask about each one.

4. For the pictures you identified as relating specifically to technology, write a note about
   (a) the problem the technology solved, and
   (b) the area of science you would need to know about in order to work with this type of technology.

5. Share your ideas with the rest of the class, and discuss how science and technology relate to each other.

Science, Technology, and Society

Usually, it is not possible to think about science and technology without thinking about issues, as well. **Issues** are problems, often arising from the use of technology, that affect society in various ways. There are no easy answers in response to these issues, but an alternative can usually be found that most of society can accept. For example, you probably identified the pop can, with its easy-open top, as an example of technology. Pop cans are convenient and easy to use, but they do litter our landscape and pose a threat to the environment. Should they be banned? Should people pay fines for littering? Should the cans be constructed from biodegradable material? What alternative would you suggest?

Examine the diagram below. It shows one way of thinking about the meaning of science, technology, and issues, and how these ideas are related to each other.

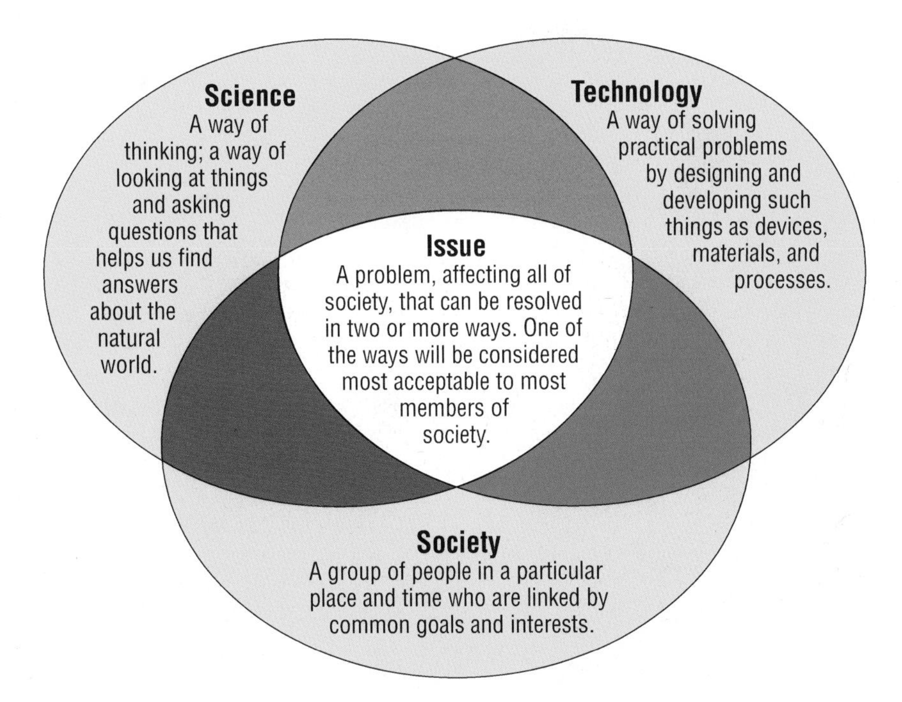

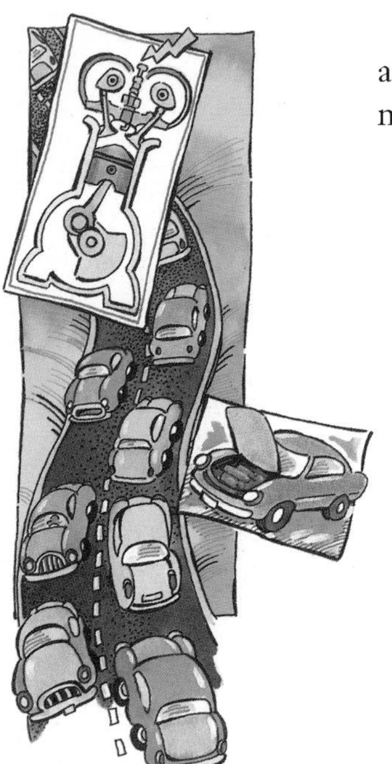

The following diagram illustrates one specific example of the relationships among science, technology, society, and the environment. Examine it with a partner, and then go on to the *Instant Practice*.

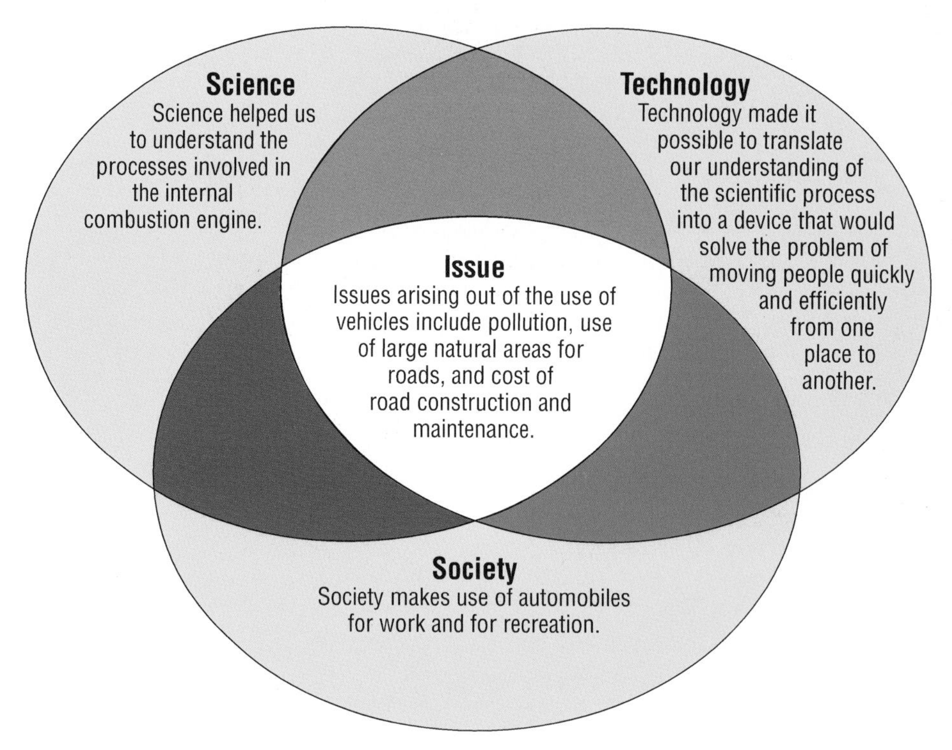

Instant Practice

1. In your notebook, draw three overlapping shapes like the ones shown in the diagram above. Colour your shapes as the ones in the text have been coloured.
2. With a partner, think about people who live in areas where earthquakes occur. In the shapes you have drawn, write notes about how science helps us to deal with earthquakes, ways in which technology can be used in such areas, and the issues that are raised.

Science and Science Inquiry

How is science related to a farmer's crop of canola?

How does science help this athlete ski faster and more safely?

What is science? If your assignment for today was to draw a picture that represented science, what would you draw? Would your picture look like either of the photographs above? It probably would not. Then why does a science textbook show photographs of a canola crop and a skier?

Science and the results of scientific research are evident in common items such as canola and ski equipment. Several varieties of canola, such as the variety shown in the photograph above, were developed in Canada. You have probably eaten salads with canola oil in the dressing, or food cooked in canola oil. Presently, scientists are doing genetic engineering experiments on canola. Someday, along with its oil, a crop of canola may produce human antibodies to help fight disease. Another canola crop may produce a blood-thinning substance to help treat heart attack victims. In the future, farms may be very important in producing medicines.

The photograph above right shows Canadian champion skier Mark Ludbrook competing at the Winter Paralympics in Nagano, Japan. Research scientists developed the materials in the boot, helmet, and racing suit that Mark is wearing. The plastic in the boot and helmet is hard — it resists impacts — and it offers protection from the extremely cold air. The suit reduces air friction and protects the skier from the cold. Mark worked with Maurice Lavoie, a metallurgist — a scientist who studies metals — to design his prosthetic leg, made of aluminum and titanium. Examine the photographs on the following pages to see more examples of what scientists are studying.

Did You Know?

The tongue of a humpback whale weighs more than a whole elephant.

If a whale swims too close to shore, the tide can wash it up onto the beach. A "beached" whale is unable to get back to the ocean and will die unless someone helps it return. Scientists are trying to learn how and why whales beach themselves in order to find a way to prevent this from happening.

DidYouKnow?

The heart of a blue whale is larger than a Volkswagen.

DidYouKnow?

A neutrino is so small that scientists do not know whether it has any mass at all.

The enormous tank in the photograph holds 1000 t (tonnes) of heavy water. It is located 2 km underground in a mine near Sudbury, Ontario. Scientists use this tank to study the smallest known particles, called neutrinos. The knowledge they gain about neutrinos will help them learn more about the origin of the universe and the possibility of the universe collapsing.

This athlete's clothing has reflective tags attached to it. The athlete is videotaped in action. Then the scientist analyzes and compares many videotapes of athletes. In this way, the scientist can learn about human motion and the types of motion that are likely to cause athletic injuries.

Karl von Frisch observed bees in hives. He discovered that bees do a dance, called a "waggle dance," to tell other bees in the hive where to find good nectar. If the flowers with the nectar are nearby, the bee "dances" in a circle. If the flowers are farther away, the bee "dances" in a figure eight to show the other bees the direction of the nectar. While making the figure eight, the bee waggles its body more quickly or more slowly to show how far away the nectar is located.

Science involves understanding and explaining nature. You do not have to work in a laboratory with test tubes and chemicals to be a scientist. As you can see from the photographs, scientists do research on a wide range of topics, in a variety of places. If there are so many different kinds of research, are there any particular characteristics that a person should have in order to be a scientist? Two of the most important qualities of a scientist are a curious mind and a strong desire to find answers to questions about nature.

Instant Practice

What aspect of nature interests you? Would you travel into space or to the bottom of the ocean to look for answers to questions? Maybe you would rather read about things that other people have studied.

Take a few minutes to skim through the pages of this book. Think of questions about the natural world that you would like to investigate. Write down four or five questions that spark your curiosity. Exchange your questions with a partner and discuss how you might find answers to them.

Categories of Knowledge in Science

Suppose that you became curious about one of the photographs above and you decide to learn more about the subject. For example, suppose that you are simply amazed that bees can communicate such detailed information to each other. You want to find out more about animal communication. Where would you look for this information? There are thousands of books about science and just as many web sites on the Internet. How do you know where to begin?

Categories of Knowledge in Science

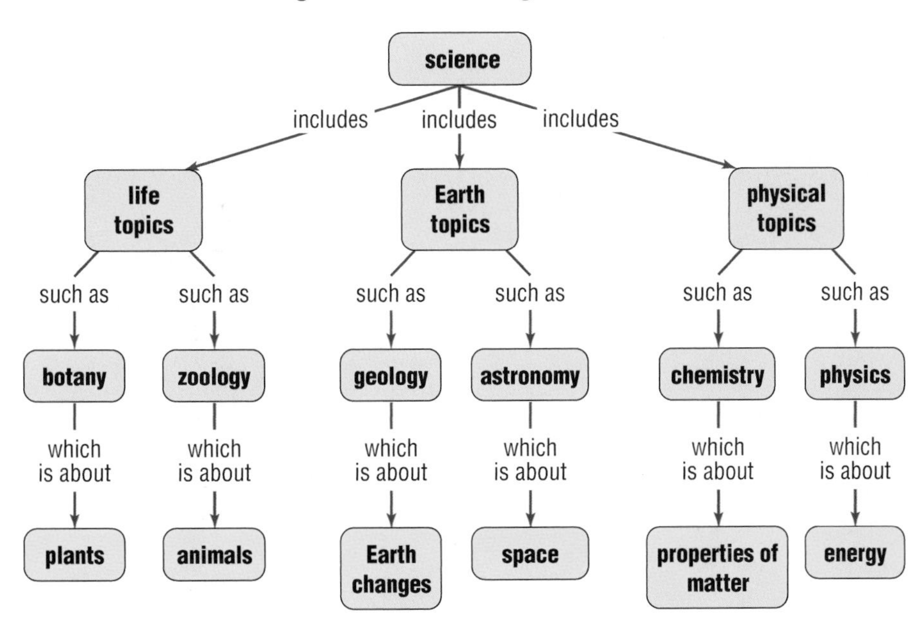

Topics in science are grouped into categories. Each category is subdivided into smaller, more specific categories. The concept map here shows one way to start grouping science topics. Each of the categories in the last row is subdivided many more times. You could use this concept map to start looking for information on animal communication. You would go to *life* topics and then to *zoology* (the study of animals). Soon you would discover a division of zoology called *animal behaviour*. (Some scientists call it *ethology*.) Nearly every book on animal behaviour has a chapter on animal communication.

By now, you have probably realized that there are a tremendous number of books and articles about nature and the universe. No one could ever learn all there is to know about science. Everyone can learn a reasonable amount about the general areas of science, however. In this course, you will learn about five different areas of science. The photographs on this page and the next give you a glimpse into the topics you will study in the five units of this textbook.

1 First, you will learn how animals and plants live together in communities. You will see how all living things depend on each other. You will also see how living things depend on air, water, and a proper balance of food and living space.

2 Next, you will learn about non-living substances. What makes a substance pure, and what is a mixture? You will see how living things depend on the solutions and mixtures they find in nature.

3 Then you will learn about temperature and heat, and how they differ. You will see how changes in temperature affect solids and liquids.

4 Next, you will learn about the substances that make up Earth. You will learn how Earth's crust is continually changing. You will see how Earth appeared millions of years ago and how it changed to reach its current form.

5 Finally, you will apply scientific principles to structures such as buildings and bridges. What makes some buildings collapse in a strong wind, while others remain stable? You will find answers in the last unit of your textbook.

In this science course and the courses that follow, you will build a strong foundation of knowledge and skills in science. You will learn to read and understand newspaper and magazine articles about science. Then you can participate in making the important decisions that are needed in society today. Maybe you will even find a subject that interests you so much that you will want to choose an area of science for your career.

Science Inquiry

You have read that curiosity is a very important quality of scientists. Is curiosity enough? For example, suppose that, while watching a basketball game, you started wondering exactly why balls bounce. Basketballs are large and filled with air, and they bounce quite well. Golf balls are very small and solid, and they bounce extremely well. Baseballs are intermediate in size and solid, but they do not bounce very well at all. What, precisely, determines how high a ball bounces? How would you begin to look for specific answers? How would you know if one answer was correct and another was false?

In order to answer questions such as these, scientists have developed a method for planning and carrying out scientific experiments. This method of science inquiry is outlined in the concept map shown here. By using this method, you can find reliable answers in an orderly way.

Soon you will have an opportunity to test this model, but first examine it in detail. To make the steps easier to understand, consider the example of a mass bouncing on the end of a spring. If you were doing the experiment with the students in the

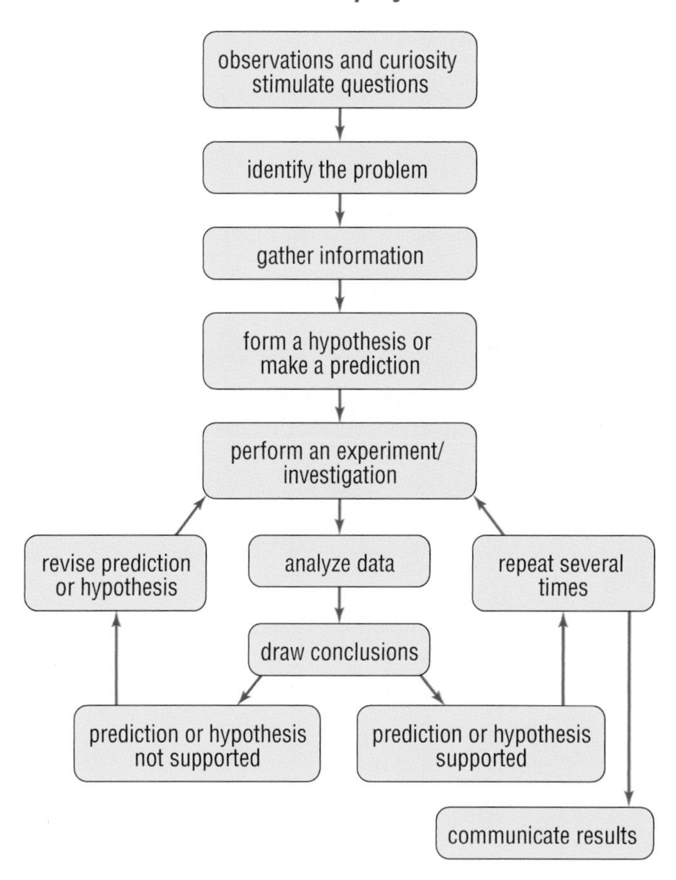

law: an action or condition that has been observed so consistently that scientists are convinced it will always happen. A law has no theoretical basis. For example, scientists do not know exactly why gravity exists, but they understand how it works. No observations have ever contradicted the law of gravity.

photograph, you would pull down on the mass and then let it go. You would notice that the bouncing motion of the mass is very rhythmic. Now look back at the first step in the science inquiry model. It indicates that observations should lead to questions. What questions would you ask?

The second step in scientific inquiry is to identify the problem. To get a specific answer, you must ask a specific question. A good question might be "What characteristics of the system determine the length of time it takes for the mass to make one complete up-and-down cycle?" This length of time is called the period of motion.

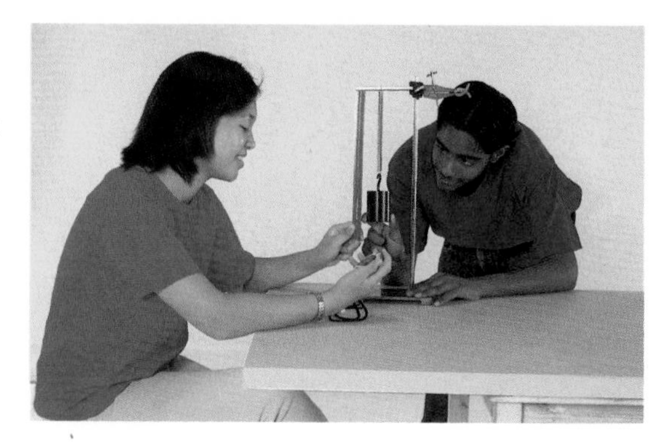

A mass bouncing on the end of a spring is a good model of a vibrating molecule. Some of the same concepts that are used to analyze the motion of the mass also apply to the vibrations of molecules. These concepts have practical applications. For example, the waves in a microwave oven cause water molecules in food to vibrate, and this motion heats the food. To learn more about models in science, turn to page 493.

model: a mental image, diagram, structure, or mathematical expression that attempts to explain a concept or hypothesis. A road map is an example of a model. So is a mathematical formula such as *Area = length x width*. Chemists use ball-and-stick models to represent molecules.

hypothesis: a possible explanation for a question or observation, stated in a way that makes the explanation testable.

Before doing an experiment, you should summarize any information that you already know about the motion of a mass on the end of a spring. You already know, for example, that when you pull the mass down and let go, the spring will pull it up but then gravity will pull it back down. The **law** of gravity is a fundamental law of science, and gravity usually affects the motion of objects.

Now that you have observed a **model**, it is time to form a **hypothesis** or prediction. For example, "I hypothesize (or predict) that the period of motion will be longer when the spring is stretched farther." To test this hypothesis, you would pull the mass down a certain distance, let it go, and measure the period of motion with a stopwatch. Then you would pull the mass down a greater distance, let it go, and, once again, measure the period of motion. Finally, you would compare the measured times.

When you are doing an experiment, it is important that you perform a **fair test**. To ensure that a test is fair, you must be certain that nothing is influencing your results except the specific part of the system you are testing. In the experiment with the mass and spring, what features of the system might affect the period of motion? Besides the distance that you stretched the spring, the spring itself and the size of the mass you attached to the spring might affect the period of motion. These features are called the **variables** of the experiment. Since you can independently choose variables such as the spring you wish to test, the mass you attach, and the distance that you stretch the spring, these are **independent** (or **manipulated**) **variables**. Since the period of motion might depend on any or all of these independent variables, the period is a **dependent** (or **responding**) **variable**.

fair test: an investigation carried out under strictly controlled conditions to ensure accuracy.

variable: any factor that might influence the outcome of an experiment.

Finally it is time to set up the apparatus and carry out the **experiment**. While performing an experiment, you should make several **observations** for each part. In the experiment with the mass and spring, for example, you would take three or four measurements for each distance that you stretched the spring, then average the results. This is important because you could make a slight error on one measurement, such as being slow to start or stop the stopwatch. After taking several measurements, you usually plot your data on a graph. (To learn more about graphing and its importance in experiments, turn to page 486.)

After you have collected and plotted your data, you analyze your data and draw a **conclusion**. A conclusion is a statement that indicates whether your results support or falsify your hypothesis. For example, your conclusion regarding the period of motion of the mass might be, "The distance that I stretched the spring had no effect on the period of the motion. Therefore, my results do not support my hypothesis."

If your results do not support your hypothesis, does this mean that your hypothesis was wrong? No, it does not. A hypothesis simply gives you a place to start and helps you design an experiment. The conclusion regarding the period of motion of the mass is a very important piece of information about the motion.

What do you do next if your results do not support your hypothesis? The science inquiry process indicates that you should formulate a new hypothesis. For example, you might now say, "I hypothesize that as the mass gets larger, the period of motion becomes longer." Then you would design and carry out another experiment. You would discover that your new results support this hypothesis. You are not finished yet, however. Even when your hypothesis appears to be supported, you should always repeat the experiment. You would probably want to test a different spring to be sure that all springs respond in the same way. In fact, you should test several different springs. These tests (or trials) would ensure that your results demonstrated the qualities of springs in general, and not just one particular spring.

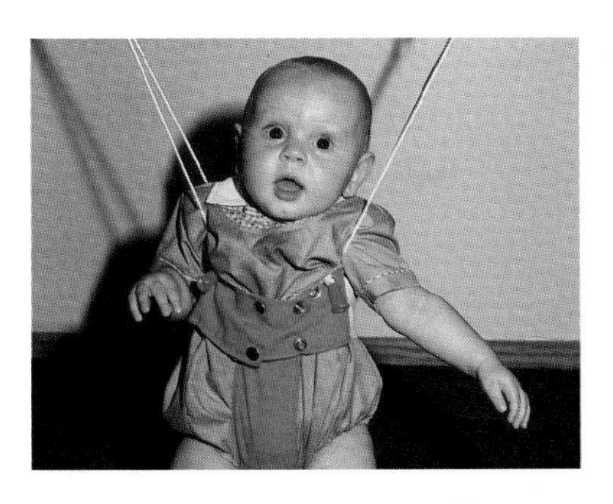

Some hypotheses are extremely important to many areas of science, and a large number of scientists test the same hypothesis in a variety of ways. Eventually, when a hypothesis has been thoroughly tested and nearly all scientists agree that the results support the hypothesis, it becomes a **theory**.

Throughout this course, you will be using the process of science inquiry when you do activities and investigations. To get some practice, try Conduct an Investigation IS-A: Tin-Can Telephones, as directed on the following pages.

experiment: an activity or procedure that is designed to falsify a hypothesis. It may seem strange to attempt to prove that something is wrong. It is not possible to prove that something is absolutely true, however, because there could always be one more experiment that will falsify the hypothesis. If your results do not falsify a hypothesis, then they support it.

observation: something seen and noted; the result of an experiment that you see and record

qualitative observation: an observation described using words only; for example, "No matter how far I stretched the spring, the period of motion was always the same."

quantitative observation: an observation that includes numbers: for example, "When I stretched the spring 5 cm, the period of motion was 0.8 s. When I stretched the spring 10 cm, the period of motion was still 0.8 s."

conclusion: an interpretation of the results of an experiment as it applies to the hypothesis being tested.

theory: an explanation of an observation or event that has been supported by consistent, repeated experimental results and has therefore been accepted by a majority of scientists. One example is the partich theory of matter.

Tin-Can Telephones

In this investigation, you will apply your knowledge of science inquiry to test the properties of "tin-can" telephones. You will think about variables that might affect how well the telephones carry sound. Then you will choose one variable and observe the effects of changing this variable on the ability of the telephones to transmit sound.

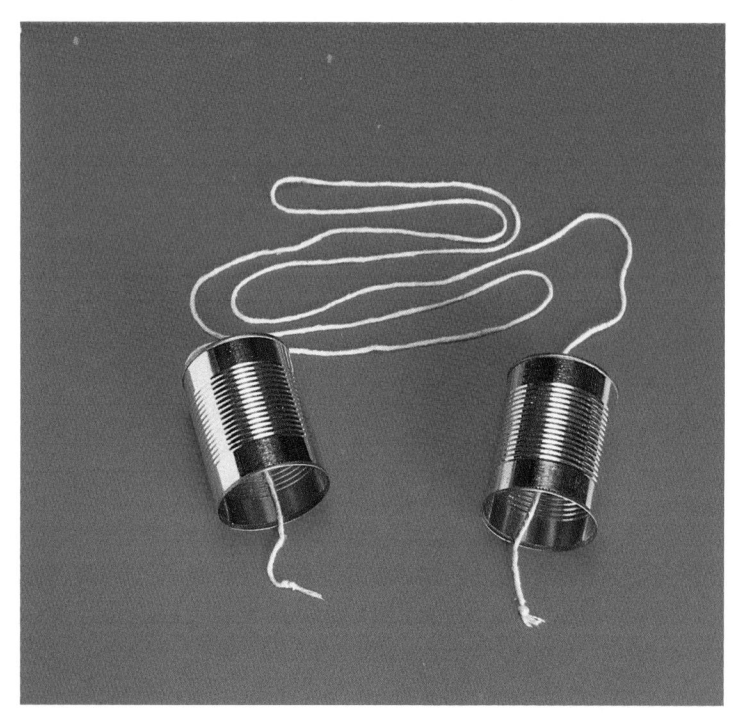

The design of the telephones is shown in the photograph above. You can make your telephones out of aluminum cans, paper cups, plastic cups, or Styrofoam™ cups. You may want to test different sizes of cups or cans, as well as different thicknesses and lengths of string.

Problem

Determine how changes in one variable affect the ability of your telephones to transmit sound.

Safety Precautions

- Be very careful when using sharp objects, such as scissors or nails.
- When using aluminum cans, make sure that there are no sharp edges.
- When using a hammer, be careful not to hit your finger.

Apparatus

scissors
metre stick or measuring tape
small nails
hammer

Materials

2 aluminum cans, paper cups, plastic cups, or Styrofoam™ cups
string (a variety, from fishing line to heavy twine)

Procedure

1. Choose your materials, and build two telephones.
   (a) You can make a hole in the closed end of each can or cup by pushing a small nail through it. If you use an aluminum can, you will need to hammer the nail through.
   (b) Cut a piece of string about 2 m long. Put each end of the string through the hole in the end of each can or cup. Tie a knot in the end of the string so the string will not slip back through the hole. If your string is very thin and the knot is not big enough, tie the end of your string to a larger piece of string and make a knot.
   (c) Test your telephones to ensure that they carry sound.
2. Make a list of all the possible variables that your group could test. As a group, select one of the variables to test.
3. Decide on four different versions of your variable to test. For example, if you want to test the effects of the thickness of the string, use strings that are four different thicknesses.
4. Formulate and write down a hypothesis. In your hypothesis, state how you think the variable you chose will affect the ability of your telephones to carry sound.

5. Test the different versions of your variable. When talking into the telephones, be sure that you are using the same tone and volume every time. Have several different people listen and evaluate (on a scale of 1 to 5, with 5 being "very clear") how well they can hear the person who is speaking.
6. Describe any changes in the level of the sound that you detected when you changed your variable.

Analyze

Did the changes in your variable make a difference in the ability of your telephones to carry sound?

Conclude and Apply

1. Did your results support or falsify your hypothesis? Explain how you interpreted your results.
2. What changes would you make in your telephones if you performed another set of tests?

Technology and Technological Problem Solving

This telephone is over 100 years old. To use it, the caller had to pick up the earpiece and turn the crank on the right. An operator would answer and ask whom the caller wanted to contact. Then the operator would plug in a wire and make the connection.

The photographs above represent some amazing changes in communication — a very important part of our lives. If you were asked to describe these changes in one word, you would probably say "technology." What exactly is technology, and how does it differ from science?

All technology has science behind it to explain how and why it works. If science is understanding and explaining nature, then the science behind modern communication involves understanding both electricity and sound. Technology is the use of scientific knowledge, as well as everyday experiences, to solve practical problems.

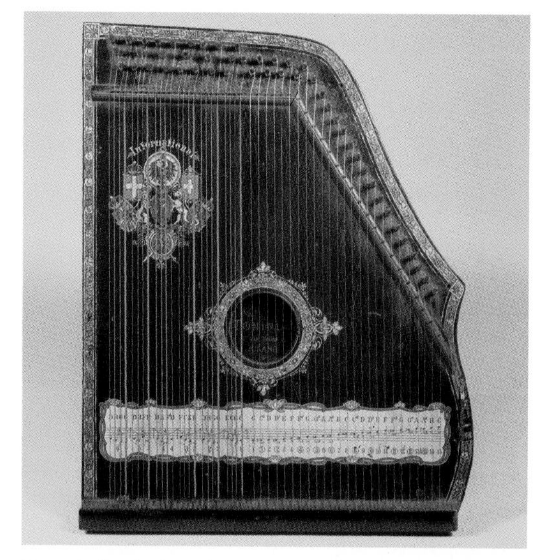

Some technologies rely only on experience. For example, people played musical instruments long before scientists figured out the details of how instruments create their unique sounds. The psaltery, shown here, was a popular instrument as early as 1275.

Cellular telephones have been available for just a few years. They enable callers to place calls at any time from any place, in or out of doors, and on buses, boats, and other vehicles.

To develop our modern telephones, for example, some scientists studied the characteristics of electric current and the way it carries signals. They also studied the nature of electromagnetic waves that carry energy through space. Other scientists learned about sound and the way it causes air particles, and even solid objects, to vibrate. When a scientific principle becomes clearly understood, it is often applied to develop new technology. The modern telephone converts sound vibrations into electrical signals that are carried over a telephone line. Cellular telephones convert sound vibrations into electromagnetic waves that travel through space.

Technological Problem Solving

When engineers and inventors design and build a new device, such as a telephone or a musical instrument, they go through a lot of trial and error. They build a device based on a design that, according to all the information they have gathered, should work. Then they test the device. The first attempt is usually not completely successful. It gives information for the next attempt, however.

As you probably know from your own experience, an orderly problem-solving method can save time and effort. Examine the steps in the problem-solving method shown here.

Solving a Technological Problem

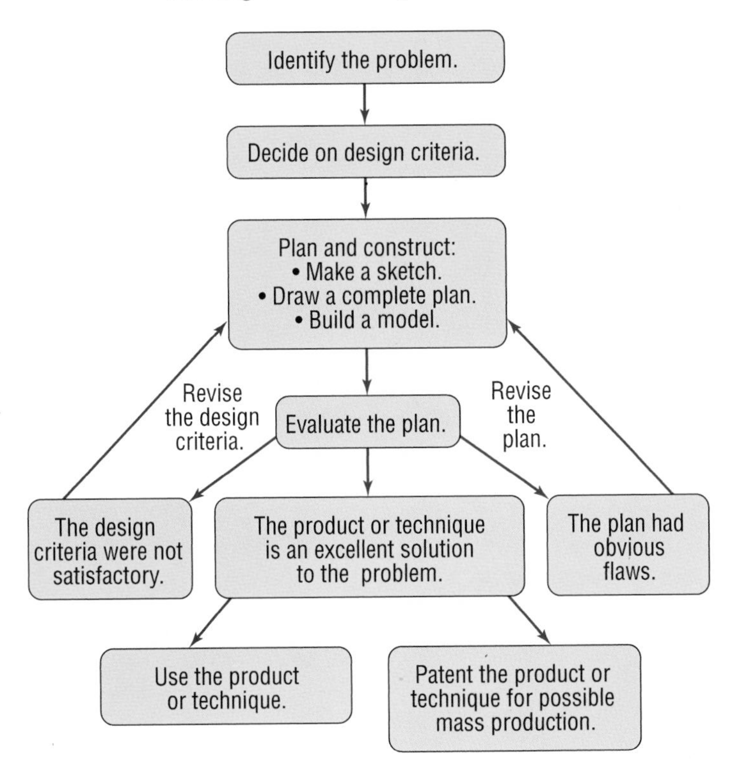

You and your classmates gathered useful information about tin-can telephones while performing Conduct an Investigation IS-A. Now you will put this information to use by carrying out the following Design & Do Investigation. You will also practise using the technological problem-solving method.

Talk It Over

Suppose that you and several friends are planning a camping trip. At your planning meeting, someone mentions that it rained heavily during the last camping trip. Everyone waited in the tents with little to do, wishing the rain would stop. The tents were too small for everyone to gather in one tent. To ensure that you will have a good time, even if it rains, you decide to go prepared. You will design, build, and test the very best tin-can telephones possible. Then, if it rains while you are camping, you can use the telephones to talk to friends in other tents. You might even invent a game to play over the telephones.

Challenge

Design, construct, and test tin-can telephones that can be used to communicate from tent to tent on a camping trip.

Safety Precautions

- Be very careful when using sharp objects, such as scissors or nails.
- When using aluminum cans, make sure that there are no sharp edges.
- When using a hammer, be careful not to hit your finger.

Apparatus

scissors
metre stick or measuring tape
small nails
hammer

Materials

aluminum cans, paper cups, plastic cups, or Styrofoam™ cups
string (a variety, from fishing line to heavy twine)
water

Design Criteria

A. The string between each telephone must be at least 4 m long.

B. Your telephones must allow you to hear another person's normal speaking voice, even if two people near you are talking in a normal voice.

C. Your telephones must work, even if the string is wet.

D. You must draw a sketch and label the parts. Include the materials used to make the parts.

Plan and Construct

1. Your teacher will summarize the class data for the tin-can telephones you made in Conduct an Investigation IS-A and give you this summary.

2. As a group, examine the class data. Discuss any data that seem to show that different people in your class did not get similar results. Discuss your own experiences working with the telephones. Choose materials that your group agrees will work best for this investigation.

3. Sketch a design that you believe will fit the criteria. Label your sketch, indicating exactly what materials you plan to use. Submit your design to your teacher for approval.

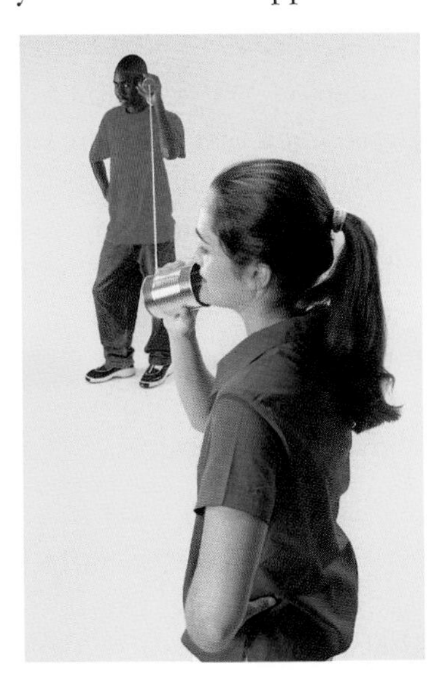

4. Construct and test your telephones.

Evaluate

1. Did your design meet the criteria? If not, which criterion did it fail to meet? Explain what your telephones were supposed to do that they were not able to do.

2. What did you learn while building and testing your telephones? How would you revise your design to make your telephones work better?

3. If you have time, build and test your revised design.

Societal Decision Making

What is your favourite television program? Which video games do you like to play? if you have access to the Internet, do you visit specific sites to find information about people that you admire in literature, history, politics, or the media?

All of these are examples of technology that we use and enjoy. What could be wrong with that? Some people would argue that there is much that is wrong. For example, while you might think that video games are challenging and exciting, someone else feels that they are violent and that they put players out of touch with the real world. Could that opinion be true? How could you find out? If it is true, what are the alternatives to violent video games? How could this issue be resolved? By a ban on video games? By regulation of time spent playing video games? By laws that force game designers to develop games that more closely resemble real-life situations with less violence? Would any of the above alternatives be acceptable to you? Why or why not?

Instant Practice

1. With a partner, think about television and the Internet and identify at least one positive and one negative thing about each technology.

2. Suggest some alternative ways of dealing with the negatives. Would some people be happy with your alternatives? Would others be unhappy? Explain your answers.

Every STSE issue that confronts us has positive and negative points. Almost every way of resolving an issue displeases someone. It is not possible to make people happy who hold opposing views and opinions. So what do you do? You have to find a compromise that everyone can live with. With some issues, it is not very diffcult, but with complex issues, it can be a major challenge.

Here is an example. Many years ago, roads started out as cart tracks between popular places. Everyone needed to travel those routes, and as the carts travelled over them, they wore a permanent track through the vegetation. They detoured around rocks and water, but at least there was a track for others to follow. It was more convenient than bumping over vegetation.

Modern technology allows us to build roads wherever we want them along routes that people seem to need. We can bulldoze trees, blast through huge piles of earth, build bridges over water, and tunnel through rock. The advantages seem clear, but as always, there are disadvantages to the use of the technology. Imagine if a highway were to be planned to go through a recreational greenbelt in a large city. What might be some of the advantages and disadvantages in such a project?

Advantages

- fast, efficient, convenient way to get from point A to point B
- direct route for trucks — freight can be moved more economically
- less pollution because traffic keeps moving. On smaller roads, there is more stopping, starting, and idling.
- costs will be recovered because the highway will not just be a bypass but will be an efficient access to the city.
- part of an entire system of highways. The entire system is inefficient without this important link.

Disadvantages

- cost to taxpayers is enormous
- natural areas are destroyed; those living nearby suffer noise pollution
- more pollution — a modern highway will attract more traffic, but there are fewer trees to handle exhaust gases.
- highway is not necessary — it will be a route for non-tax-paying trucks, not a means of improving the city's economy.
- there are other ways of linking up the highways if such a link really is necessary. Greenbelt needs preserving.

Think back to your favourite video game or television program. How did you feel about the idea of having it altered to suit someone else's ideas? People often feel very strongly about issues that need resolving, so such situations are rarely just about facts.

Consider the highway example. It takes many years to research, plan, and raise money to build a major highway. Perhaps planning for this one began during the years when much city and suburban planning was done with cars as the focus. Cars were the basis of our economy, and they gave their owners a feeling of independence and freedom. Now, many people feel very strongly that we have been wrong in building so many highways and allowing so many air-polluting vehicles to use them. Even though Canada makes sincere efforts to curb the pollution — in some other countries, people see no need to control the pollution that their vehicles produce — building, repairing, and maintaining highways cost taxpayers huge amounts of money.

It is important to be able to present the facts for whichever side of the issue you support, particularly when an issue is likely to be an emotional one. Science and technology are more effective in helping resolve issues if the citizens of a society understand how science works. Why? People who are able to recognize valid scientific information and decide whether "scientific" claims are really scientific will be better prepared to contribute positively to the decision-making process. This book will help you develop your knowledge and thinking skills so that you can appreciate and understand how science is related to current societal issues and those that might come up in the future. These same skills will be useful throughout your life in making important decisions about personal and community issues.

In order to make such decisions, you need a logical, thoughtful approach that will help you to see the issue clearly. The flowchart on the opposite page is designed to help you do just that. Examine it carefully with a partner. Your teacher may give you a blackline master with a flowchart that has been completed with information about the highway. Examine it carefully to see how the categories have been used to organize the information.

Throughout your *SCIENCEPOWER*™ 7 textbook you will have opportunities to develop an understanding of issue analysis, but two special features are specifically designed to help you in this respect. Many of the *Think & Link Investigations* deal with issues. All of the activities in *An Issue to Analyze* provide an issue to debate, to role-play using a simulation, or to analyze in a case study. You can use the "Developing Decision-Making Skills" flowchart as a guide to analyzing issues. There is an opportunity to use it in the following *Instant Practice*.

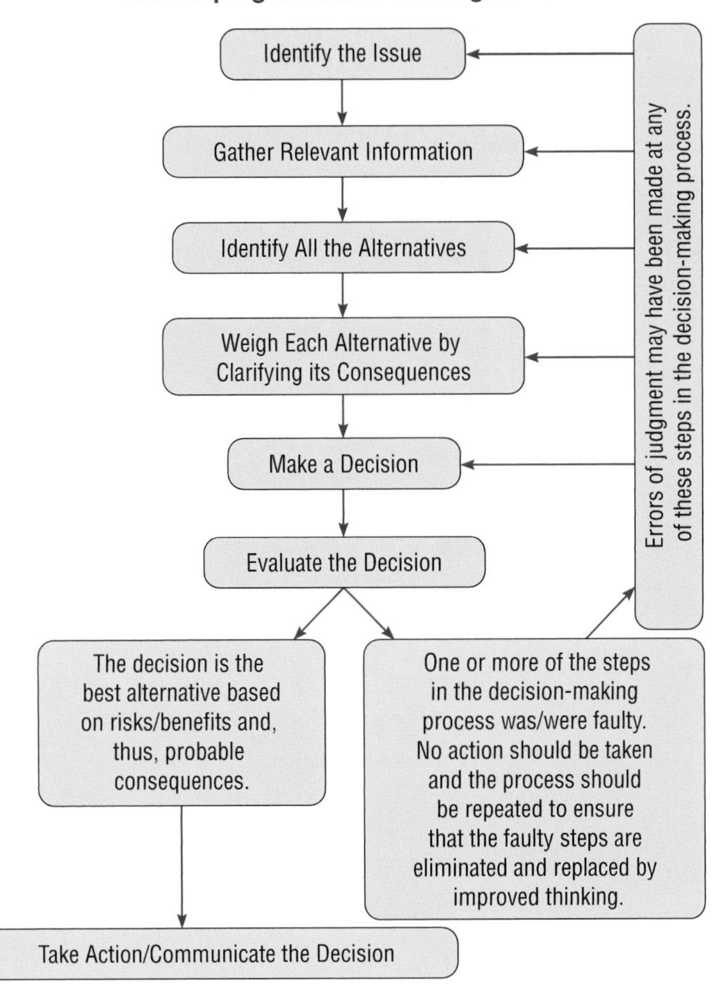

Instant Practice

1. Read the following brief news clipping:

British Columbia could soon become the first province to regulate the use of cellphones by drivers in order to cut automobile accidents. In a policy shift yesterday, B.C.'s attorney general said he will reconsider his opposition to such regulation, which has been strongly endorsed by the B.C. Medical Association. Ujjal Dosanjh said he was reacting to a tide of public support for such regulation, building since B.C.'s doctors called for a law at their annual meeting in June.

2. With a partner, work through the flowchart above, using the news clipping as a basis. First, write what you think the issue is in the news clipping.

3. Instead of gathering relevant information, write notes about the kind of information you would need to gather in order to come to a decision.

4. Follow through the rest of the chart. Be ready to share your decision with the rest of the class and to back up your decision and explain the alternatives you considered.

UNIT 1

Interactions Within Ecosystems

Even before these chicks learn how to fly, they dive off 15-storey-high cliffsides into the cold, dark water of Hudson Bay in Canada's North. This begins a two-month journey for the young, thick-billed murres, first swimming, and then flying, to Labrador where they spend their winters.

The scientist shown here is interested in learning about living things in their environment. In order to learn more about the lives of thick-billed murres, she has to band the chicks now, when they are easily caught because they have not yet learned to fly.

In the small photograph, peregrine falcon chicks are being placed in the nest of a pair of peregrine falcons that have not been able to produce their own young. Peregrine falcons were close to extinction in eastern Canada in the mid-1900s following the common use of the pesticide, DDT. Why? The use of this pesticide had some unfortunate side effects. One negative effect was that it caused the egg shells of many kinds of birds to become so thin and fragile that their chicks did not survive. DDT is no longer used in Canada. The ban on the use of DDT, and programs such as the one shown here, to help peregrine falcons achieve nesting success, are increasing the numbers of this majestic bird. Peregrine falcons are making a comeback and can be seen today nesting on tall buildings in Canadian cities. (Originally, peregrine falcons nested on cliffsides, but now they also use buildings for their nests — a human-made substitute.)

How do living things interact with one another? How do they interact with the non-living parts of their environment? How do we, as humans, fit in, and what should we do to ensure that Earth remains a planet full of life? These are topics you will explore in this unit.

Unit Contents

CHAPTER

1 The Big Picture:

Getting Ready...

- How can plants grow when no one feeds them?
- Why is the Sun important to plants, animals, *and* you?
- What impact can the tiniest insect have on the lives of other creatures?

Use words, a diagram, or a flowchart to answer the Getting Ready questions in your Science Log. Think about what you already know, and look for more information as you read this chapter.

For tips on how to make and use a Science Log, turn to page 476.

Imagine that you are the insect shown in the small photograph on the opposite page, living in the area shown in the photograph above. You will never, in your entire life, be able to explore the whole area where you live. Your "world" consists of the huge, thick grass stalks and the green leaves where you crawl and find your food. Whether you know about it or not, the larger area means life or death to you. What if there is a drought? What if the stream dries up, and the grasses and your food supply die? You and others like you will die off, and the birds that feed on you will no longer have a food source.

You are not a tiny insect, but your immediate world — made up of your school, your home, and your local community — is very small compared with the entire Earth. You may be unaware of events that are occurring in the larger world, but they may well affect you in some ways. This chapter will help you develop the tools you need to assess those larger world events as well as the effects they have on you.

Levels of Life

Spotlight On Key Ideas

In this chapter, you will discover

- how living things are adapted to survive in their environments
- how living things interact with each other and with non-living things
- why plants are important to Canada's economy
- why some species of living things are endangered, and how their disappearance can be prevented

Spotlight On Key Skills

In this chapter, you will

- do a comparative study to discover adaptations that organisms possess
- identify and evaluate human changes to a natural environment
- sample populations in an ecosystem and examine the interactions among those populations
- design an experiment to assess the impact of changing a variable in an ecosystem

Starting Point

Chain of Events

People have often thought about the impact of one tiny event on an entire chain of events. Sometimes the results can be surprising.

What to Do

1. With a partner, read the following poem and discuss what it means. If you have any difficulty, invite other pairs of students to share their ideas.

   For want of a nail, the shoe was lost;
   For want of a shoe, the horse was lost;
   For want of a horse, the rider was lost;
   For want of the rider, the battle was lost;
   For want of the battle, the kingdom was lost.
   And all from the want of a horseshoe nail.

2. Relate the ideas in the poem to the world around you and to the ways that living and non-living things depend on each other.

3. Think about your "wants" — the basic things that you, as a living creature, must have in order to live. List the "wants" that are essential for your survival.

Extensions

4. Make up your own chain of events starting with one small event. Include at least eight events in your chain. For example, you could start with the following event: there was no milk left when you went to have breakfast this morning, so . . . .

5. Write a poem about how different kinds of living and non-living things depend upon one another.

1.1 Individuals, Populations, and Communities

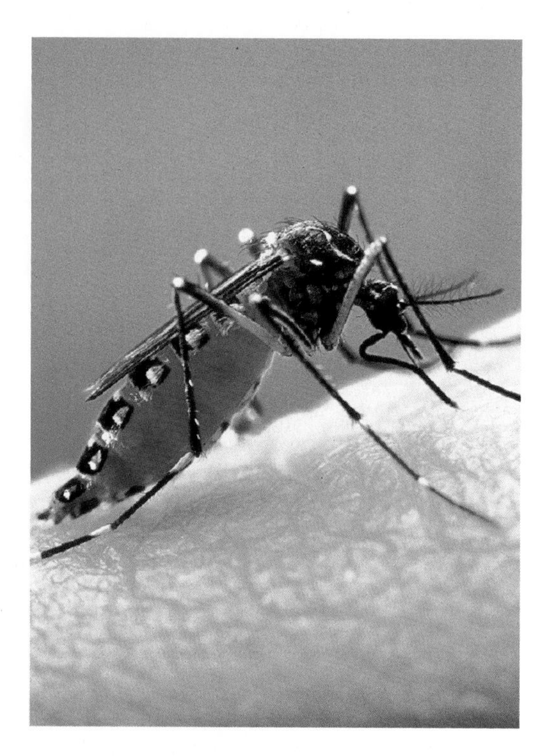

"Wow! There seem to be a lot more mosquitoes than usual this year." Have you ever made a comment like this? If so, you were not just making a comment; you were making an observation. For many hundreds of years, scientists have been making similar observations. Observations lead to questions, such as these: Why are there more mosquitoes this year? What conditions help them to breed? These questions lead to new knowledge and still more questions: What do male mosquitoes eat (since only female mosquitoes feast on human blood)? What **organisms** (living things) eat mosquitoes? What is the life span of mosquitoes? How many young do they produce at a time? And so on . . .

When scientists began to organize and classify observations, the first major division they made was between living and non-living things. The living things — humans, plants, birds, animals, insects — were called the **biotic** parts of the environment. The non-living things — air, water, soil — were called the **abiotic** parts of the environment. An **ecologist** is a scientist who studies the biotic parts of the environment and their interactions with the abiotic parts.

An **individual** is one organism. A group of individuals of the same species, living together in one place at one time, is called a **biological population**. Biological populations vary in size depending on the species. For example, a population of caribou may contain thousands of individuals, while a population of red bats may contain only two or three individuals. A **species** is a group of organisms who can successfully reproduce among themselves. A male and female of each of the species shown in Figure 1.1 on the next page can together produce young that will later have young of their own. The world is filled with a great variety of living things. If you were to stop right now and list the organisms you know, it would probably take you quite a long time. Appendix A on page 468 and 469 shows an overview of the organisms living on Earth.

DidYouKnow?

There are more than 73 different species of mosquitoes living in Canada. In order to hide from a mosquito, you would have to stop breathing, because the main thing that attracts mosquitoes to people is the carbon dioxide that humans exhale.

DidYouKnow?

A scientist named Rachel Carson (1907–1964) popularized the study of ecology by writing and speaking clearly about important ecological issues of her time. Her observations led to studies that showed that the pesticide, DDT, caused the death of thousands of young eagles and hawks. The pesticide was eventually banned from use in many countries, including Canada.

Figure 1.1 Each of these individuals is part of its own biological population.

Populations do not live in isolation. They share their environment and interact with populations of other species. A **biological community** is made up of all the populations that live in one area. For instance, your local park has a biological community made up of the population of squirrels and the population of trees on which they feed. Other populations in the park are also part of the community. They include the population of robins that nest in the trees — nests that the squirrels sometimes destroy — and the population of earthworms that the robins eat. In every community, each species has a particular place where it lives. This place is called its **habitat**. For example, the habitat of an earthworm is the soil.

There are different types of interactions between community members: for example, robins eat worms and nest in the trees. Sometimes individuals compete with each other for limited resources. For example, other birds must compete with the robins for nest sites and for worms.

When you look around your local park, you may be curious and have questions about individuals, populations, and communities. These are different levels of biological organization. For example, you are asking a question about an individual if you ask "What kind of tree is that?" You are asking a question about a population if you ask "How old are the trees in this park?" You are asking a question about the community where the trees are found if you ask "What organisms live in these trees?"

Pause& Reflect

Think of a kind of animal that lives in Canada. Write a brief story about this animal in your Science Log using the terms that are highlighted on pages 6 and 7.

Word CONNECT

"Don't bug me!"

"Have you had the bug that's going around?"

"There's a bug in my computer."

"Look at the bug on that leaf."

These are all different uses of the word "bug." Look up the words "insect" and "bug" in a dictionary. What differences do you see in their definitions? Make a point of using the word "insect" accurately. In your notebook, write why a scientist might choose the word "insect" over the word "bug."

Floria is a horse who spends her days in an enclosed pasture. Floria is an individual, and she is part of the pasture's population of horses. All of the horses, and the populations of grasses on which they feed, are part of the pasture's biological community. Other interacting populations in this community include flies, which lay their eggs in horse droppings, and cottontail rabbits, which compete with the horses to eat the grasses. Figure 1.2 shows these levels of biological organization.

Level 1: individual

Level 2: population

Level 3: community

Figure 1.2 The first three levels of biological organization

Organisms are usually very well suited to live in a particular environment. For instance, animals that are active at night, or live in very dark places, usually have very large eyes or hear very well. This good "fit" between the characteristics of the organism and the characteristics of the environment is the result of adaptation. An **adaptation** is an inherited characteristic that helps an organism survive and reproduce in its environment (see Figure 1.3). Sometimes characteristics that help animals survive in their environment are learned during the animal's lifetime. For example, humans learn to look both ways before crossing a street. This helps humans survive, but it is not an adaptation because it is not inherited; humans are not born knowing to look before crossing a street.

Figure 1.3 Robins' feet are an example of an adaptation. Like other perching birds, robins have feet with three front toes, one long hind toe, and a specialized tendon that automatically locks their hind toes around a branch when they land.

INVESTIGATION 1-A

Tools for the Task

Think About It

You have seen that organisms are adapted, or well-suited, to their environment. Adaptations take many forms and are often easy to identify once you begin to think about them.

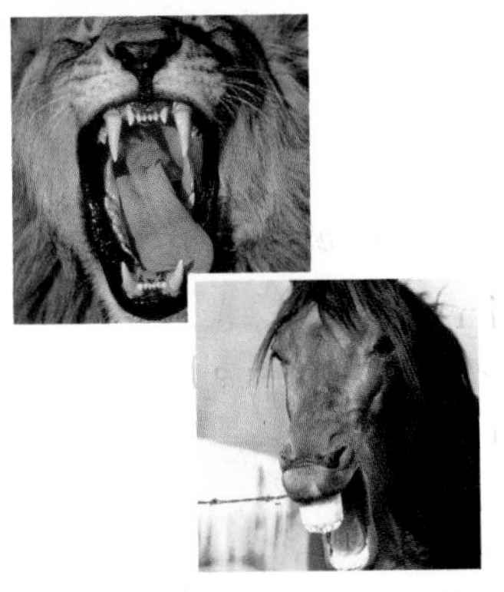

What to Do

1. Examine the photographs of a horse's and a lion's teeth. How are the teeth of each organism suited to its specific needs? Record your answers.

2. Examine the pelican's beak, and compare it to the eagle's beak. How do they differ? What could you hypothesize about the type of food each bird eats? What do you think each beak does best?

3. How do snakes move? How do alligators move? How is each organism adapted to the environment in which it moves around?

4. In what way are the duck's feet an adaptation for living in water? Imagine the foot in water. Compare the falcon's feet to the duck's feet. In what way are the falcon's feet adapted for catching food? Draw a sketch to illustrate how either the duck or the falcon uses its feet.

Skill POWER

To find out how to use resources and the Internet effectively, turn to page 497.

Analyze

1. What behaviours make a dog or a cat a good house pet? Are these adaptations? Why or why not?

2. Think about the ancestors of the pet from question 1. What behaviours did they have that helped them to survive in the wild? Are these adaptations? Why or why not?

3. Compare the behaviours described in questions 1 and 2. List the ways in which these behaviours are similar or the ways in which they are different.

Extend Your Skills

4. In a group, choose three different species not mentioned in this chapter. Use the Internet or library resources to research and briefly describe the adaptations shown by the species.

Skill POWER

For tips on working in groups, turn to page 478.

Check Your Understanding

1. Write the letters (a) through (h) in your notebook. Use the clues in parentheses ( ) to solve the x puzzle. That is, write the complete scientific terms, replacing each x with the letter that fits.
   - **(a)** bioxxx (living)
   - **(b)** xbioxxx (non-living)
   - **(c)** bioxxxxxx (scientist who studies living things)
   - **(d)** xxdivxxxxx (one organism)
   - **(e)** xxxxlatxxx (group of organisms of the same kind)
   - **(f)** sxxxxxs (group of closely related organisms that can produce young)
   - **(g)** xxmmxxxxx (interacting groups of different organisms, which share a common environment)
   - **(h)** xxxptaxxxx (inherited characteristic that helps an organism live successfully in its environment)
2. Look up "bio" in one or more dictionaries. Then explain how the puzzle's first three words are connected.
3. Give two examples of biological adaptation.
4. What might happen to an organism if a permanent change occurs in the environment to which the organism is adapted? Give an example.
5. Look back over the chapter so far. Choose five or more important words, and use them to make your own x puzzle like the one in question 1. Exchange puzzles with a classmate, and solve each other's puzzles.
6. **Thinking Critically** Think about the organisms you have classified and the relationships you have noted so far. Consider the kinds of interactions that might be important to explore next. Predict what the next level of organization is, and what it might include.

DidYouKnow?

Many of the bones in a bird's body are hollow. Why might this be considered an adaptation?

Skill POWER

For tips on using your textbook effectively, turn to page 473.

Word CONNECT

You may be familiar with some common names for populations of animals, such as a herd of elephants or a school of fish, but you may not be so familiar with less common population names. Did you know that gorillas live in bands, turtles live in bales, and goldfish live in charms? Do some research, and match the animals in column A with the population name in column B. Write your answers in your notebook.

A		B	
rhinoceroses	toads	trip	leap
goats	monkeys	skulk	knot
kangaroos	leopards	mob	troop
		crash	

1.2 Ecosystems

Figure 1.4 This beaver dam is part of an ecosystem that includes biotic and abiotic factors.

Pause& Reflect

Select an area that you think makes up an ecosystem. For example, is there a pond surrounded by fields, a school lawn surrounded by streets, or a park surrounded by houses? Identify the area in your Science Log, and explain why you consider it to be an ecosystem.

A community is filled with groups of organisms that interact with each other, but it does not exist on its own. It has boundaries, and it is affected by external conditions, such as the average amount of sun and rain, and the average temperature. Look at the photograph of the beaver dam in Figure 1.4. Imagine the community that lives there. What non-living, or abiotic, features can you think of that interact with the organisms in the community? Your list may include air, sunlight, rocks, rain, and soil. An **ecosystem** is the interactions between abiotic features of an area and the biotic community that lives in the area. An ecosystem includes individuals, populations, and communities (see Figure 1.5). and in this case, the ecosystem is the beaver dam.

An ecosystem can be large or small, but it must contain all the abiotic and biotic features of the area. For example, a rotting log is an ecosystem, as long as all the organisms living in or on the log and all the abiotic factors affecting the log are included. In the same way, a forest is an ecosystem, as long as it includes all the biotic organisms living in it, as well as the abiotic features affecting it. Usually the environment is the same across an ecosystem, so that it gets the same amount of rain or sunshine. An ecosystem also has fairly clear physical or environmental boundaries. For example, the edge of a ploughed field would physically mark the edge of a field ecosystem. The edge of a pond would physically mark the edge of a pond ecosystem. An environmental boundary could be the edge of an area that gets more sunshine or rain, or a change in altitude along the side of a mountain.

Figure 1.5 The ecosystem is the fourth level of biological organization.

Level 1: individual

Level 2: population

Level 3: community

Level 4: ecosystem

Figure 1.6 Plants interact with sunlight and air to make their own food, and they use water and nutrients in the soil to grow.

Abiotic-Biotic Interactions in an Ecosystem

In an ecosystem, the non-living and living parts interact and affect each other. Think about the following four abiotic factors as you consider an ecosystem.

- *Sun* The Sun provides energy that gives warmth to spiders, and other organisms, enabling them to survive. It provides light, which green plants like the one in Figure 1.6 use to make their own food. (You will learn more about this process in Chapter 3.) The number of hours of daylight triggers seasonal events, such as plants flowering and the migration of birds.
- *Air* The air contains oxygen, which animals breathe. It provides carbon dioxide, which plants use to make their own food.
- *Water* Plants need water to grow. Water is important to all organisms for life processes. These include distributing food particles through their bodies, breathing, and digesting food. The bodies of most organisms are 50 to 95 percent water. Some organisms, such as trout, whales, and algae, live in water not air. (Can you think how they obtain their oxygen?)
- *Soil* Soil contains both abiotic parts, such as minerals, and biotic parts, such as decaying bodies of dead organisms. Soil provides a home for many animals that live underground. For example, earthworms burrow in and overturn soil, allowing air and water to go down into the soil. They also eat organic material in the soil and pass the remains out behind them, often on the surface. In this way, they bring valuable nutrients to the topsoil (see Figure 1.7). Soil also provides minerals and other nutrients for plants. Plants, in turn, hold soil in place, helping to keep it from being blown or washed away. (In Chapter 10, you will learn more about soil.)

Figure 1.7 Earthworms are an important part of an ecosystem.

DidYouKnow?

Earthworms stay underground during the day. They breathe directly through their thin skin. They cannot survive the drying heat of the Sun because their skin must stay moist to breathe properly. When rain falls, it floods their burrows, and earthworms must come to the surface to breathe. If they stay underground, they will drown. If they are away from soil when the rain stops, however, they cannot dig back into their burrows, and they dry out and die. This is why there are so many dead earthworms on the pavement after a rain shower.

INVESTIGATION 1-B

Creating an Ecosystem

You have been learning about ecosystems and about their biotic and abiotic parts. Although you can observe ecosystems all around you, you can learn a great deal more if you focus on one ecosystem over time and examine it in detail.

Challenge

Design and make a model ecosystem for observation in your classroom.

Safety Precautions

Materials

masking tape
gravel or small rocks
seeds and/or small plants
one glass bowl or other transparent container
water
twigs
potting soil

Design Criteria

A. Your model ecosystem should contain at least four different plants.

B. Add at least one abiotic element (besides the essential ones) with which the biotic elements can interact.

Plan and Construct 

1 With your group, decide what type of ecosystem you will be creating. Research different parts of the ecosystem, to find out which biotic and abiotic features are usually found in this type of ecosystem. Plan how you will construct your model.

(a) What kinds of seeds and plants will you use? Why? Do they normally grow near each other in nature?

(b) What do the plants in your ecosystem need to stay alive? Must you add anything to your ecosystem over time, such as water?

(c) What location in your classroom will you choose for your ecosystem? Why? What might be the effect of choosing this location over another one?

2 Prepare a labelled sketch of your model. Write a hypothesis about what will happen in your ecosystem. After obtaining your teacher's approval, build your model. Label your model with your names using the masking tape.

3 Observe your model over an agreed-on period of time. Take careful notes on a regular basis. If you discover problems, try to solve them.

4 As you study your model, write any questions that occur to you about the interactions that are taking place. You might be able to find the answers when you go on your field trip later in this chapter.

Evaluate

1. Did your model succeed as you had planned in step 1? Why or why not? Did you experience any problems? If so, were you able to resolve them? How?

2. Compare your model with the models of other groups. Whose model worked best? Whose model seemed to experience the greatest number of problems? Why? How might you improve your model if you designed it again?

Skill POWER

To review the safety symbols used in this textbook, turn to page 492.

Skill POWER

To find out more about models in science, turn to page 493.

Finding Ways to Travel

Gardening can be fun. There are many different types of ornamental plants and flowers available to make all sorts of different gardens. However, most plants in our environment are not planted on purpose. Forests are full of wild trees and bushes, and flowers grow wild in sunny meadows. Some plants, usually called weeds, grow in our gardens even when we do not want them to.

How did these plants manage to start growing if they were not planted by humans? Many plants have special adaptations that help their seeds spread to new areas. Examine Figure 1.8 and see if you can tell how each of the seeds is spread. Plants also have other adaptations to succeed in their environments. The water lily, shown in Figure 1.9, and the bristlecone pine shown in Figure 1.10, are both well-suited to their environments.

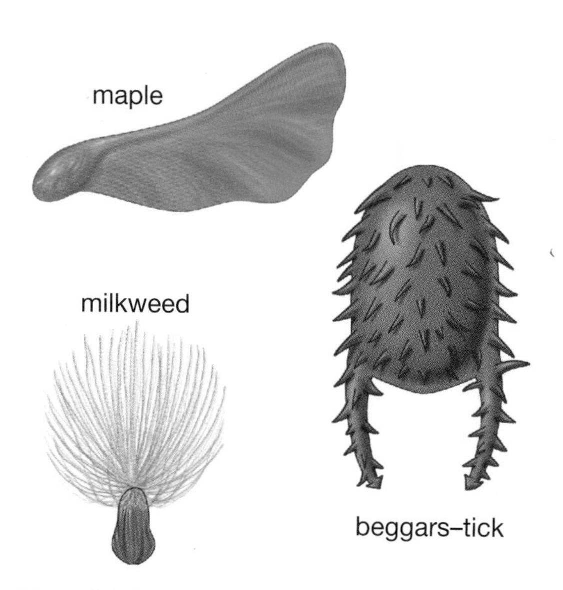

Figure 1.8 Each of these seeds is adapted to spread in a different way.

Figure 1.9 The water lily has a long flexible stem, which anchors it to the bottom of the pond. It will not break easily when the wind whips the water into waves.

Skill POWER

For tips on how to use the Internet effectively, turn to page 497.

INTERNET CONNECT

www.school.mcgrawhill.ca/resources/

You have looked at animal adaptations, and the photographs here show some plant adaptations. Find out about the special adaptations of cacti for their dry environments by going to the web site above. Go to **Science Resources**, then to **SCIENCEPOWER 7** to know where to go next. Write and sketch your findings.

Figure 1.10 The bristlecone pine leans against the wind. If it stood straight, it would be more likely to break. The bristlecone pine also has strong roots that cling to the soil even when strong winds blow.

Check Your Understanding

1. What are the main parts of an ecosystem? Give examples of each.
2. Think of five different ecosystems that might be found in an area such as a national park. List the biotic and abiotic factors each ecosystem might contain. (Hint: Remember that ecosystems can be large, like a forest, or small, like a rotting log, as long as they include all of the biotic and abiotic factors present.)
3. Make a chart, poster, or other representation to show abiotic-biotic interactions in an ecosystem.
4. **(a)** Give two or more examples of how plants are adapted for interactions with abiotic factors in their environments.

   (b) Think about common plants in your area. Choose one, and explain how it is adapted to its environment. (You might have to do some research.)

1.3 Sampling Ecosystem Populations

In the previous section, you created a model ecosystem and observed it. Did you have any unanswered questions as a result of your observations? In this section, you will examine an outdoor ecosystem and perhaps find some answers to your questions.

To study ecosystems, scientists must find out how many different types of biotic and abiotic features are present in an area. For very small areas, this is not too difficult. For large areas, however, such as forests and fields, counting the numbers of all the plants and animals can be impossible. It is possible, though, to estimate the numbers of organisms and plants in an area. Biologists estimate population sizes in ecosystems by first choosing a small area, such as a square metre, that is representative of that ecosystem. Then they count the features in that smaller area. Multiplying those results by the square metres in the whole area gives an estimate of population size. This method of estimation is called **sampling**. Biologists use many ways to sample ecosystems, but one of the most popular is the quadrat, a tool shown in Figure 1.11. Features inside the **quadrat** are counted and used to estimate the number of features in the whole ecosystem.

It can be very challenging to sample organisms in some ecosystems, such as a tropical rain forest, a pond, or a treetop. Scientists use many creative methods to see what is living in a particular ecosystem. What types of animals might get sampled by each of the following methods?

- pit traps (a cup is placed into the ground so that its rim is even with the ground)
- mist netting (nets with very fine, almost invisible string are set up like a volleyball net)
- aerial surveying (planes fly over an area and count the number of organisms).

Figure 1.11 A quadrat can be made from metre sticks that are taped together or from wooden pegs that are joined with string.

Skill POWER

For more practice estimating and measuring, turn to page 481.

Sampling Populations in a Lawn Ecosystem

In this investigation, you will sample the populations in a lawn ecosystem. You will use a quadrat to count the features in a small area of your ecosystem. This will give you a good estimate of the total populations in the larger system. The lawn ecosystem could be your school lawn, a park, or another similar ecosystem. When you know what is in an ecosystem, you can start looking at interactions among different parts of the ecosystem.

Problem

How could you sample the populations of a lawn ecosystem and examine the interactions among those populations?

Safety Precaution

Apparatus

4 metre sticks or 4 wooden pegs per group

several jars with caps that have holes in them

magnifying glass

clipboard

Materials

duct tape

string

paper for sketching

notebook

Procedure

1. In this study, you will count the populations of each kind of organism in a lawn ecosystem. Look at the illustrations here, as well as books of plants and insects to help you identify the organisms you may meet.

2. Make an observation sheet by drawing a chart on a piece of paper. One column should list the names of plant and animal species you are likely to encounter. Leave several rows blank for species you cannot immediately identify. Add an additional column with a separate heading for every quadrat you sample (quadrat 1, quadrat 2, etc.).

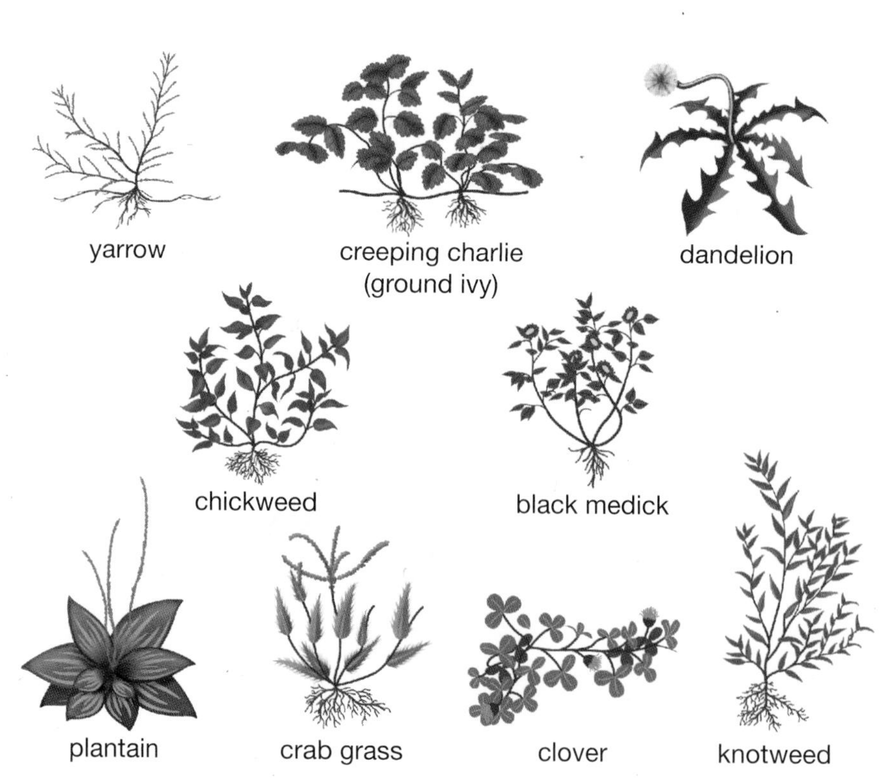

These typical lawn plants are not to scale.

3 Make a 1 m^2 quadrat using Figure 1.11 on page 15 as a guide. Choose a suitable location for your study. Toss your pencil at random onto the grass. Place a corner of your quadrat where the pencil lands.

(a) Count all of each kind of organism within your quadrat. If a plant is on the boundary line, only count it if more than half of it lies within the quadrat.

(b) Do not count grass, since the grass plants will probably outnumber everything else. Similarly, do not count every leaf of a plant such as plantain (see the illustration on the previous page).

(c) On your observation sheet, record the total number of each kind of organism.

4 Repeat step 3 for several more quadrat samples.

5 Average the totals for your quadrat samples to get the average number of each organism per square metre.

6 Multiply the average number of each organism per square metre by the number of square metres in the total ecosystem area. This will give you an estimate of the total population of each organism in the lawn ecosystem. An example is given in the sample calculation.

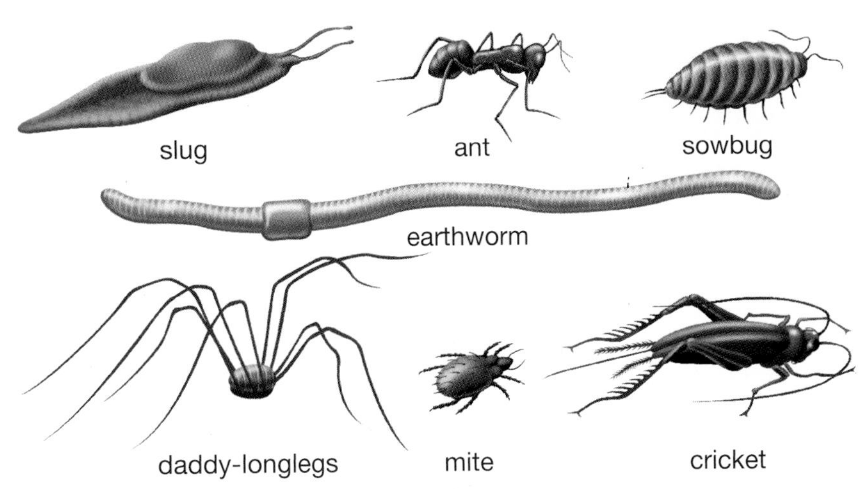

These common lawn organisms are not to scale.

Sample calculation:
The number of blue beetles counted in 5 quadrat (1 m^2) samples is 13, 15, 16, 14, 13.
Total number of blue beetles in 5 quadrats = 71.

$$\text{Average} = \frac{\text{total number of blue beetles}}{\text{total area of quadrats}}$$

$$= \frac{\text{71 blue beetles}}{\text{5 quadrants} \times 1\ \text{m}^2}$$

$$= \text{14.2 blue beetles/m}^2$$

$$\text{Total ecosystem area} = \text{length} \times \text{width of ecosystem}$$
$$= 50\ \text{m} \times 20\ \text{m}$$
$$= 1000\ \text{m}^2$$

Estimated total population of blue beetles in the ecosystem

$$= \frac{\text{14.2 blue beetles}}{\text{m}^2} \times 1000\ \text{m}^2$$

$$= \text{14 200 blue beetles}$$

Math CONNECT

To find an average, total the numbers obtained from the samples. Then divide those totals by the number of samples. For example, suppose you have sampled eight areas. The total number of ladybird beetles in each sample is 12, 16, 4, 7, 12, 8, 10, 18. Find the average number of ladybirds beetles per sample.

CONTINUED

7. If you find a weed you cannot name, make a sketch of it. Include an outline of the leaf. (This will help you to identify it later.) Use a letter-code, such as B, to identify the unknown specimen instead of its name.

8. If you find an insect or other animal you do not know, put one specimen in a jar, close the lid, and make a simple sketch. (Be sure to check the number of legs and wings, to help you identify it later.) Once you have completed your sketch, release the organism. Give your unknown specimen a number-code, such as 2, instead of its name.

Skill POWER

For tips on scientific drawing, turn to page 495.

Pause& Reflect

Think about the investigation of this lawn ecosystem. You sampled several different quadrats, in areas that were chosen by randomly throwing your pencil. Explain why you sampled more than one area. What reasons can you think of for choosing the areas by chance? Write your answers in your Science Log.

Analyze

1. Try to identify your unknown organisms by matching them with the illustrations shown on pages 16-17. You could also try matching them with pictures and descriptions in plant, insect, and animal books.
2. Replace the number-codes or letter-codes on your observation sheets with the names of the specimens you have identified.
3. Complete your calculations to determine the population numbers in the ecosystem for each kind of organism.

Conclude and Apply

4. What is the most common plant in the lawn ecosystem you studied?
5. What different species of animals did you find in this community?
6. Compare the number of individuals in the animal populations with the number of individuals in the plant populations.
7. What abiotic factors (non-living things) were present in your lawn ecosystem?

Extend Your Knowledge

8. What other factors might affect your lawn ecosystem?
9. What action could you take to protect or preserve this ecosystem?

Extend Your Skills

10. Wait for conditions to change. For example, if you did your first study in dry weather, do a second study after a night of rain. If you did your first study before most plants had produced seeds, do a second study after they have seeds.
11. **Design Your Own** Design your own experiment to examine the effect of changing the conditions in your ecosystem. Make sure your experiment is a fair test, and have your teacher approve your procedure before you carry out your experiment. (Hint: Your hypothesis should ask whether or not there will be a change in one variable, such as number of crawling insects, if another variable, such as a certain type of plant, is changed. You can then modify your study area to reflect your hypothesis.) Compare the results of your modified quadrat with the results from your original ecosystem experiment. What variable did you change? What was the effect in your quadrat?

Career CONNECT

What's the Count?

Linda Söber is an environmental biologist who helps governments and developers use their land in a way that preserves the existing wildlife. "I count each type of plant and animal I see. Once I know what's there, I can suggest ways to protect the natural environment." When doing a survey of an area, Linda does not look for just the animals themselves. She looks for tracks, droppings, nests or bedding sites, and fish eggs on plants along the water's edge or in a swamp. She also listens for identifying sounds of certain bird calls.

Developers often want to fill in wetland areas on their land, to make the areas solid ground on which they can build. Unfortunately this destroys the wetlands and almost everything that lives there. According to Linda, wetlands have more wildlife than either fields or forests.

Do some research at the library or on the Internet, talk to somebody at a wetland reserve if there is one in your area, or contact a wildlife organization, such as Ducks Unlimited. Identify ten animals and plants that live in wetlands (also called swamps, bogs, or marshes). For each animal or plant, write a sentence about how filling in the wetland will affect it. Will filling in the wetland remove its food supply or breeding ground? What are some other ways that wetland animals could be affected?

Word CONNECT

The name "dandelion" comes from the French *dents de lion*, meaning "a lion's teeth". From the shape of the leaves, can you suggest how this name could have come about?

Check Your Understanding

1. Why are quadrats used to carry out ecosystem studies?
2. How would you use a quadrat to estimate population sizes? Why is the result only an estimate, not an exact figure?
3. Why could it be useful to do two population studies of the same ecosystem, with a period of time between the studies?
4. Do you have a fish, hamster, or other living creature in your classroom or school? Observe it, and write some notes about its original (wild) habitat based on visible adaptations. (For example, what kind of teeth does it have? What about its feet? Can it run quickly?)
5. **Thinking Critically** Imagine you are a biologist, and the company you work for assesses the impact of development projects. There is a plan to build a new luxury resort on the shore of a large bay. Builders need to know what environmental impact the project will have on particular ecosystems. Your job is to estimate the number of organisms in these different ecosystems. How could you sample:
   (a) the numbers of insects in a large tree
   (b) the number of whales in a large bay off the coast
   (c) the number of fish in a small lake
   (d) the number of groundhogs living in a local golf course

1.4 Climates and Biomes

Figure 1.12A The grassland biome in Saskatchewan

People from other countries often associate Canada with a snowy winter climate. This association is so strong that some tourists visiting Canada in the summer are surprised that there is no snow. Although we live in a country that has a cold winter climate, we still check to see what the weather is like before going outside in the winter. The weather may be unusually warm, so we do not need a hat; or the weather may be very cold, so we need to wear extra clothes. This is the difference between weather and climate. **Weather** refers to local conditions that change from day to day, and even from hour to hour. For example, it may be raining when you wake up in the morning, but, by the afternoon, the rain may have cleared and the sun may be shining. **Climate** refers to the average weather pattern of a region. For example, the climate of northern Canada is very cold, with long harsh winters and short cool summers.

Figure 1.12B The grassland biome in Africa

Earth can be divided into distinct regions based on climate. For example, the high northern latitudes have very cold climates, while the regions around the equator have very hot climates. Similar climates have similar types of soil and get similar amounts of yearly precipitation. This encourages similar types of plants to grow in them. A **biome** is a large area with characteristic climate, soil, plants, and animals. Each type of biome can be found in different parts of the world. For example, imagine that you are in a small airplane flying over a large open area that is covered in grasses and small shrubs. Trees are growing, but only around sources of water. You can tell that you are in a grassland biome, such as the ones shown in Figures 1.12A and 1.12B, because of the characteristic grasses and small shrubs, but can you tell what country you are in? One way to find out where you are is to look for characteristic animals. If you see elephants, lions, and giraffes, you are in Africa. If you see prairie dogs, foxes, and burrowing owls, you are in Canada. The grassland biome also occurs in Australia, Brazil, and Russia.

Level 1: individual

Level 2: population

Level 3: community

Level 4: ecosystem

Level 5: biome

Figure 1.13 The biome is the fifth level of biological organization.

Biomes are the fifth level of biological organization (see Figure 1.13 on the previous page) they contain many different ecosystems. For example, within the grassland biome there could be a wetland ecosystem, a marsh ecosystem, and a grass ecosystem. Unlike ecosystems, there are not many different types of biomes in the world. The land on Earth can be divided into six major biomes. Figure 1.14 gives the names and locations of these biomes (the ice is not a biome). Look at Figure 1.15, and find which biome you are living in. Examine Figures 1.16A-D on the next page. They show examples of the biomes in Canada.

Tundra is cold, dry, and treeless, with less than 25 cm of precipitation per year. Tundra becomes wet and marshy in the summer because the ground stays permanently frozen (permafrost), and this does not allow water to drain into the soil. The boreal forest is the largest biome in Canada. The dominant organisms are evergreen trees, and this biome receives between 35-40 cm of precipitation per year. Most of the precipitation is in the form of snow. The temperate forest receives 75-150 cm of precipitation per year, and unlike the boreal forest, the dominant organism are deciduous trees. There are four very obvious seasons every year. Grasslands are dominated by grasses, and they receive 25-75 cm of rain per year.

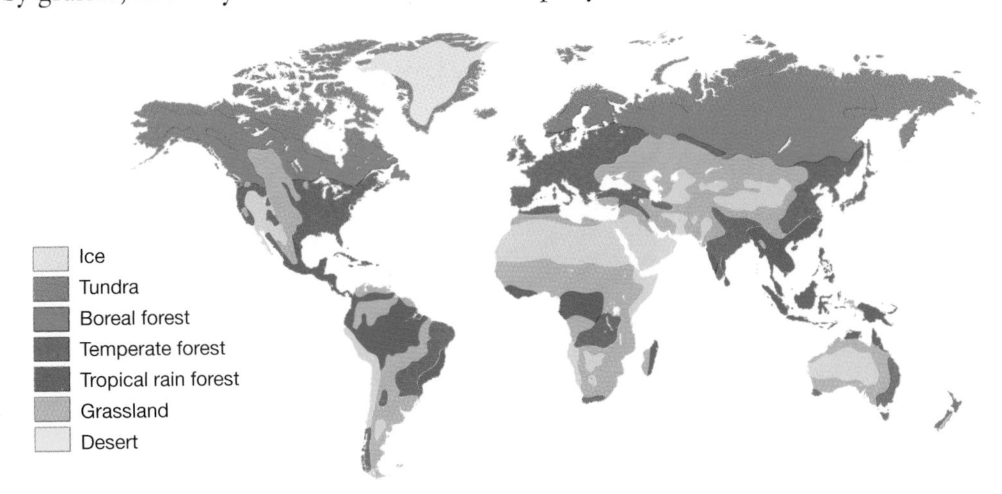

Figure 1.14 Names and locations of the world's biomes

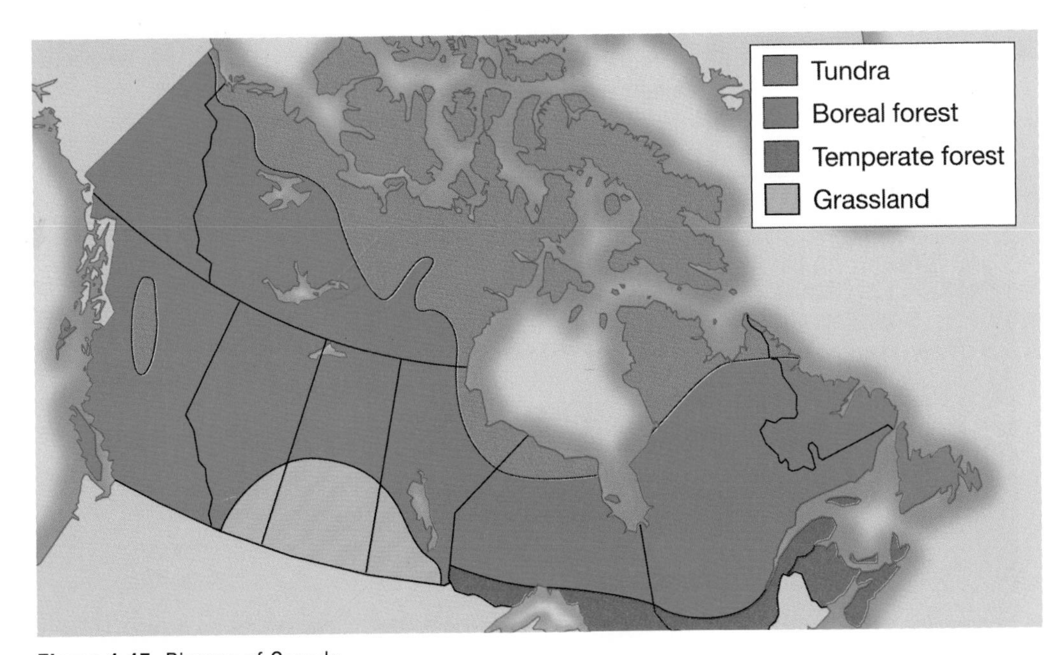

Figure 1.15 Biomes of Canada

All six types of biomes occur in North America, and four occur in Canada. Canada does not have a tropical rain forest, but it does have a temperate rain forest which receives more than 200 cm of precipitation per year. Huge western cedar, Douglas fir, and Sitka spruce trees grow in the mild, wet climate along the British Columbia coastline (see Figure 1.16E). These are Canada's biggest trees.

Figure 1.16A Tundra

Figure 1.16B Boreal forest

Figure 1.16C Temperate forest

Figure 1.16D Rolling grassland

Figure 1.16E Temperate rain forest

Canada does not have a true desert biome. It does, however, have semi-desert areas. One is British Columbia's sagebrush-dotted southern Okanagan Valley, which receives only about 25 cm of rain a year (see Figure 1.17). Warm moist air off the Pacific Ocean drops its moisture on the west-facing mountain slopes. As the air descends the east-facing slopes, it warms up and becomes even drier. This is what creates the Okanagan Valley's dry climate (see Figure 1.18).

Figure 1.17 The Okanagan Valley has semi-desert conditions.

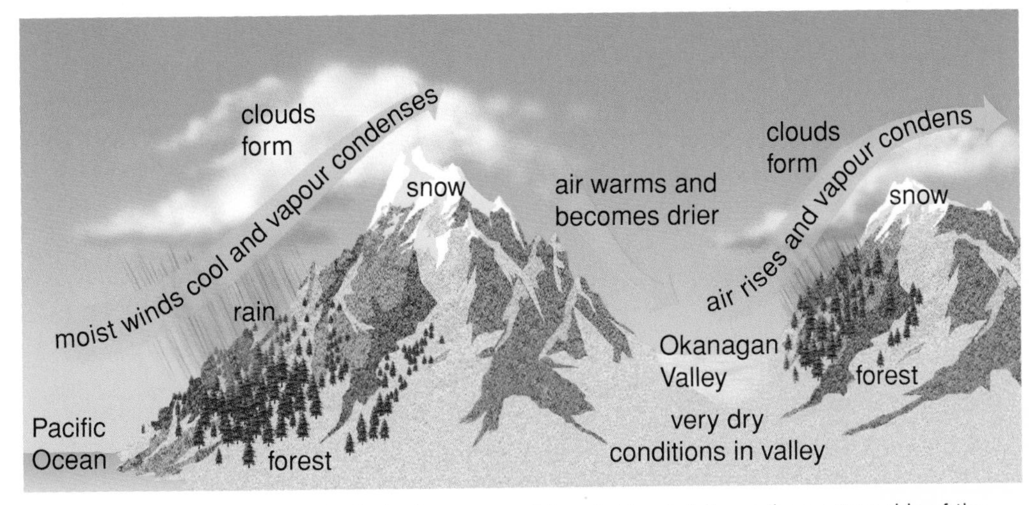

Figure 1.18 The Okanagan Valley is dry because moisture in the air falls on the western side of the mountains.

Across Canada

"Fat, black leeches, bloated with our blood, dropped out of our socks and off our necks," wrote Birutė Galdikas in her notebook. Tracking and observing orangutans in the leech-infested tropical rain forests of Indonesia has been her life for over 23 years. Fascinated by the human-like characteristics of orangutans since her childhood, this world-famous scientist has devoted her life to studying orangutan behaviour.

The calm nature of orangutans appeals to Dr. Galdikas, but it also makes them hard to find. Dr. Galdikas has learned that they like to be alone, high up in the treetops. One of the best ways to find an orangutan is to listen for the sound of falling fruit pits coming from a tree branch where the orangutan sits, munching on a snack. Dr. Galdikas spent two months sitting quietly in the rain forest before she spotted her first orangutans — a mother and a baby.

Conservation of the rain forest is another focus of Dr. Galdikas' work. As parts of the rain forest are destroyed, the orangutans' homes are also destroyed. The number of orangutans in the world is decreasing every year. In 1986 Dr. Galdikas set up the Orangutan Foundation International to raise money for educational programs to teach people about orangutans and to try to save their natural habitat.

Dr. Birutė Galdikas

Employed at Simon Fraser University in Vancouver, the Lithuanian-born professor lives and works for most of the year among the orangutans on the island of Borneo in Indonesia. Dr. Galdikas is a professor of anthropology, which is the study of early humans. She believes that orangutans are very similar to early humans; therefore, much can be learned about early humans by studying orangutans.

Biome Boundaries Overlap

Boundaries between biomes are not as clear as those shown in Figures 1.14 and 1.15. If you have ever travelled long distances in Canada, you may have crossed different biomes. The change from one biome to another is not sudden. Instead, the biomes gradually blend together over broad areas. For example, the transition from temperate forest into boreal forest is not sudden. Slowly more and more of the oaks and maples of the deciduous forest give way to the spruces, firs, and evergreens of the boreal forest. Finally there are no more deciduous forest species left. Most Canadian provinces consist of more than one biome, and in all cases there are broad regions of overlap.

With your teacher's help, choose to do either of the two following activities that feature industries in two of Canada's biomes. You can then communicate your results to the rest of the class.

Find Out **ACTIVITY**

From the Forest

Lumber is an important industry in Canada. Perhaps you live in a part of Canada that contributes to the economy through the lumber industry. What processes does the lumber undergo in order to turn into a finished product?

What to Do

1. Examine the illustrations below. In which three provinces is forestry especially important to the economy?
2. How are the harvested trees used?
3. Choose one aspect of the lumber industry, and research how raw lumber becomes a finished product.
4. There are many conservation issues associated with the lumber industry, particularly as large tracts of forests are scheduled to be harvested. Research the point of view of either a logger or a conservationist on the West Coast of Canada. What are the issues related to conservation versus logging? What are the short-term and long-term effects of logging on the environment? What if no new lumber were harvested, and there was no wood to use for construction? Prepare an illustrated report to present to your class, or debate both sides of the issue in class.

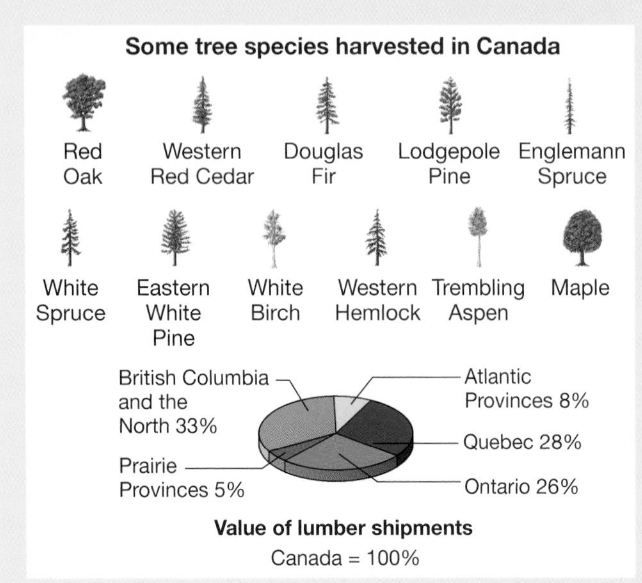

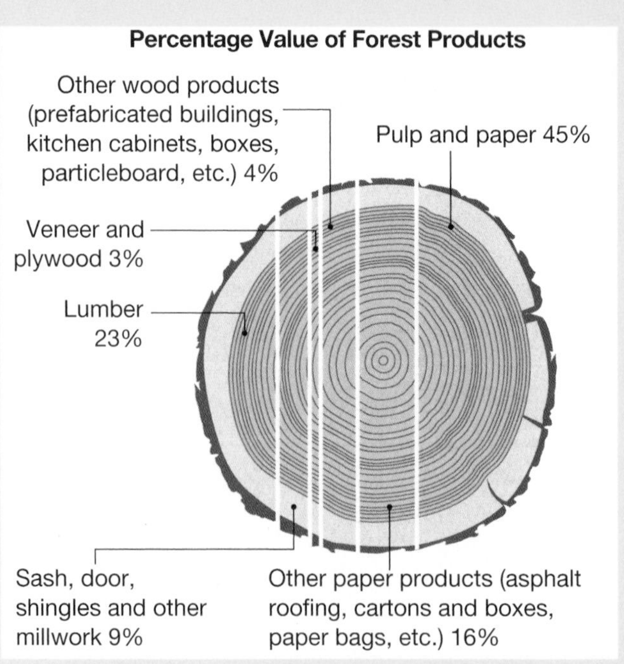

From the Fields

Canada is famous worldwide for its wheat exports, but wheat is only one of a wide variety of agricultural products grown in Canada. What other products come from Canada's fields?

What to Do

1. Examine the graph. Which three plants are most important to the Canadian economy, in dollar values?
2. Choose a product that is grown in your biome. Find out all you can about it, and prepare an illustrated report for the rest of the class. How does the product affect your life? How does it affect your local economy? What happens when this plant is out of season or when there are problems that cause crop shortages?

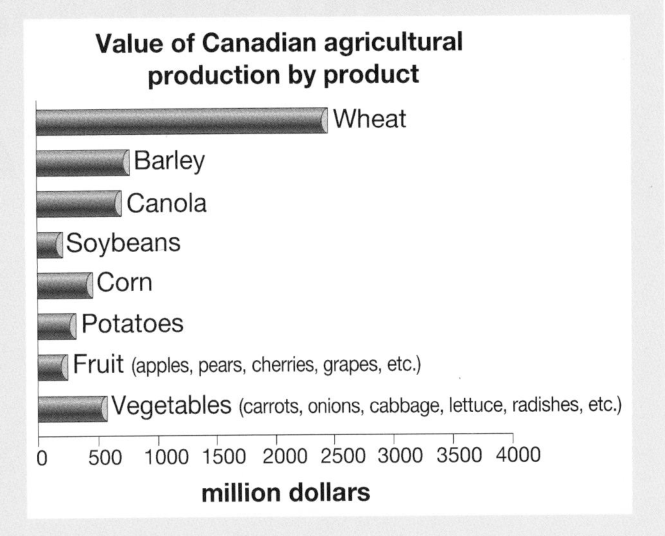

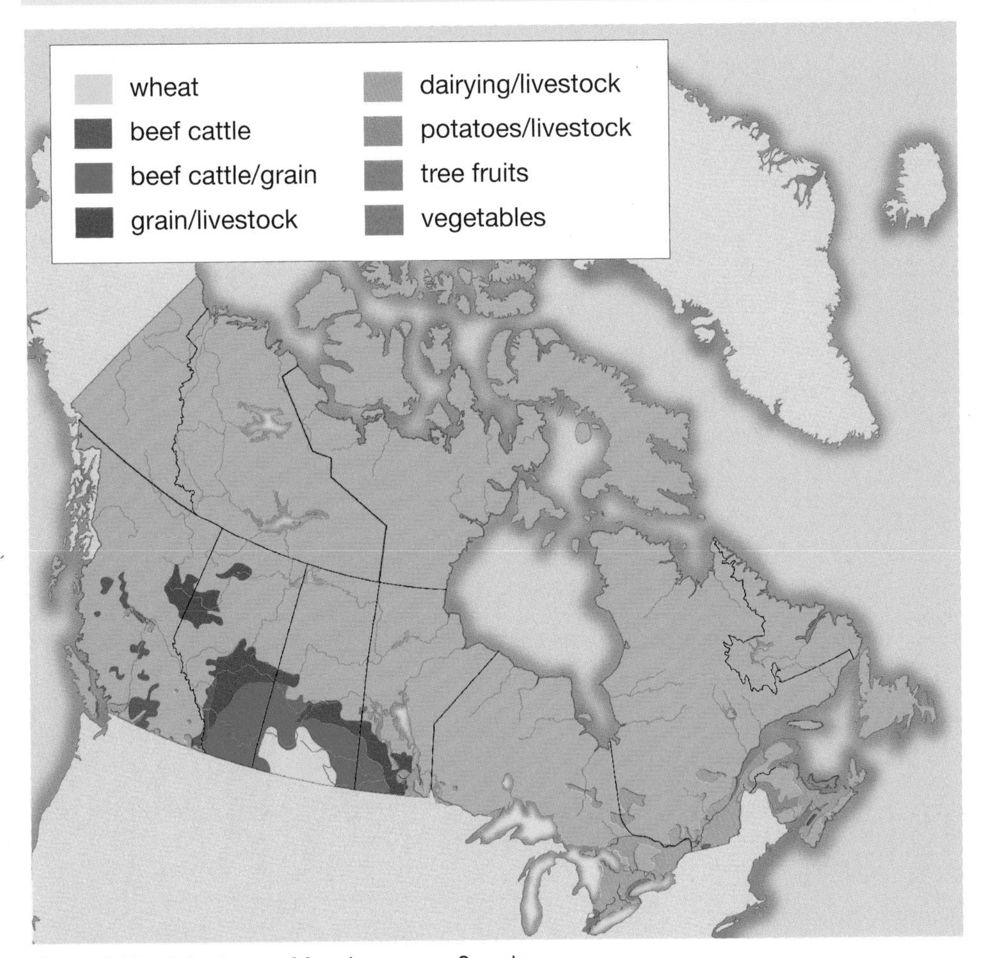

Figure 1.19 Major types of farming across Canada

Pause & Reflect

Compare Figure 1.19 with Figure 1.15 on page 21. In what biomes do the various agricultural activities take place? In your Science Log, make a chart, line graph, or other representation to answer this question.

Using Plants

Besides using plants in agriculture and forestry, Canadians also raise plants to sell as cut flowers. For example, roses are grown for this purpose in the Niagara region of Ontario. Canadians also use plants to make diet supplements and medicines. Plants that are used in this way include ginseng, evening primrose, borage, comfrey, mint, and echinacea.

Plants are important to our daily lives as Canadians. They are especially important to Canadians who work with them, depending on them to make a living. In the next section, as well as in Chapter 3, you will explore more issues relating to the use of plants in Canada.

Pause& Reflect

In the 1930s, irrigation turned parts of the dry but sunny Okanagan Valley into one of Canada's main fruit-growing areas. Today the area produces many of the country's apricots, plums, pears, cherries, peaches, and apples. What kinds of climatic conditions did the area experience before irrigation? After irrigation, what were the climatic conditions that made it ideal for fruit growing? In your Science Log, write what you think the positive and negative consequences of large-scale irrigation might be.

Check Your Understanding

1. Distinguish clearly between weather and climate. Give examples of each.
2. Prepare an illustration, flowchart, or other representation to show how climate, soil, and organisms are related to each other.
3. Solve the following "Who Am I?" puzzle. Write the letters (a) through (e) in your notebook. Beside each letter, write the name or names of the organisms that fit. Then write the name of the biome that fits.
   (a) I am related to the grizzly. My white fur is an adaptation to my snowy environment. Who am I? In which biome do I live?
   (b) We are cone-bearing organisms. We stay green through cold winters and get 35 to 40 cm of precipitation a year. Who are we? In which biome do we live?
   (c) I am a small bush that lives in unirrigated parts of one of Canada's driest areas. Who am I? In which biome do I live?
   (d) We are leaf-dropping organisms that receive 75 to 150 cm of precipitation a year. Who are we? In which biome do we live?
   (e) I am a meat-eating animal that lives in the same biome as ground squirrels, which I sometimes eat for dinner. Who am I? In which biome do I live?
4. In which biome do you live? Prepare a poem, song, painting, chart, or other representation to show what life is like there.
5. In what major plant-related activities do Canadians take part? Why are these activities important to Canadians?

1.5 The Biosphere: The Really Big Picture

All of the places on Earth where life can exist and interact with the physical environment, taken together, form the **biosphere**. The biosphere, shown in Figure 1.20, is made up of three parts, as shown in Figure 1.21.

1. The **hydrosphere** is all the water on Earth, including the water in the oceans, rivers, and lakes.
2. The **lithosphere** is the solid mineral material that covers Earth, including the soil.
3. The **atmosphere** is the blanket of air that surrounds both the hydrosphere and the lithosphere.

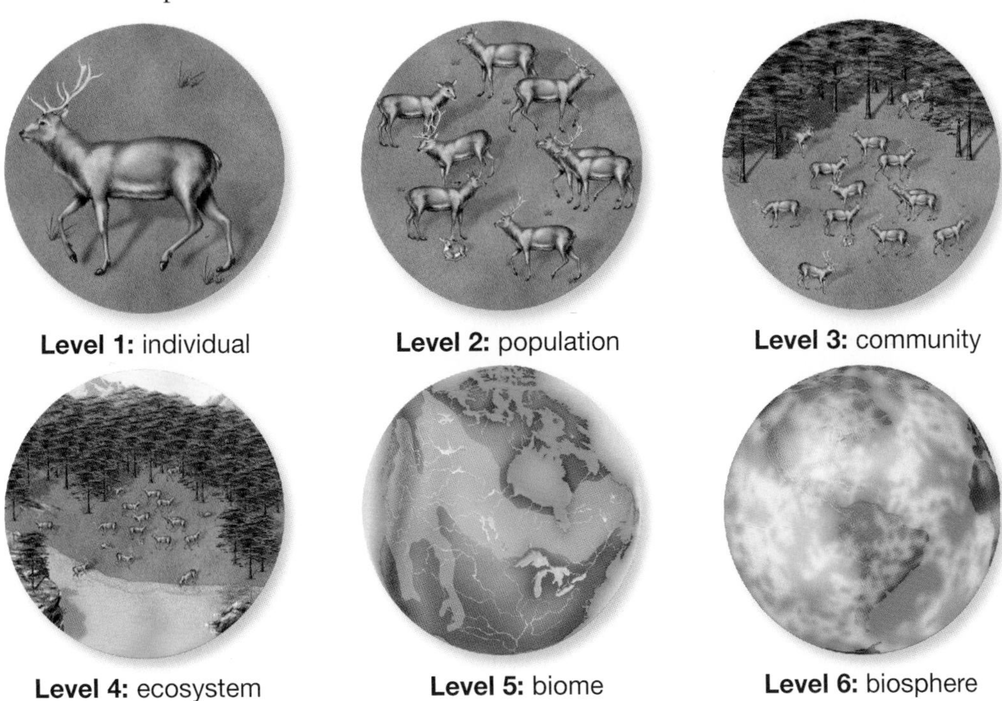

Figure 1.20 The levels of biological organization, from individual organisms to the biosphere

Word CONNECT

Look up the prefixes "hydro," "atmos," and "lithos." What do they mean? Think about their connection to the suffix "sphere," and write your thoughts in your Science Log.

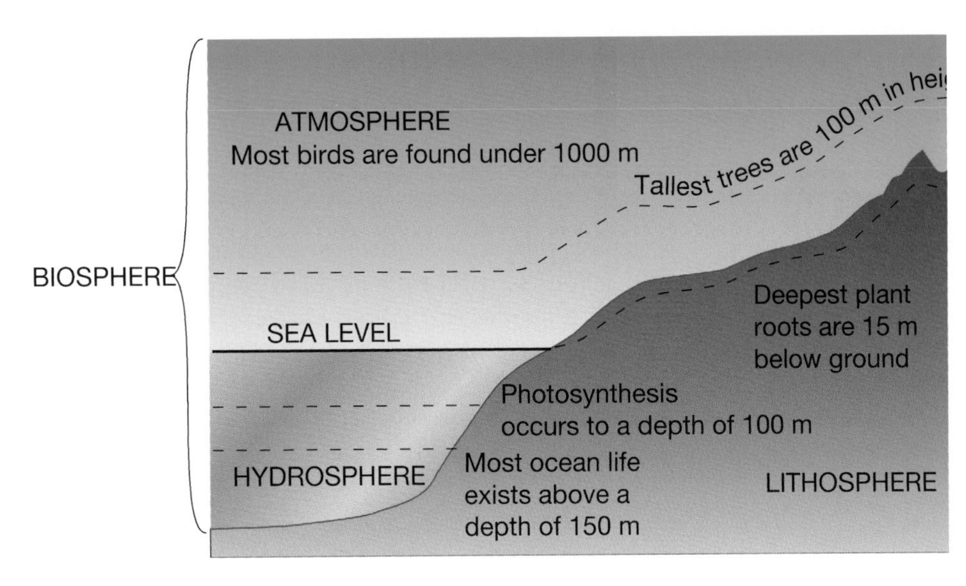

Figure 1.21 The biosphere includes every zone where life can exist.

DidYouKnow?

When the *Titanic* sank in April 1912, it is likely that the contents of its cold storage containers spilled out. All of the food that was intended for passengers probably became a feast for bacteria. Just 48 h after the sinking, the *Titanic* would have been totally covered with bacteria.

In 1960, when space flights began, humans first saw their planet as a whole, from space. Some people thought a better name might be "planet Water" rather than "planet Earth," since 70 percent of its surface is covered with the hydrosphere (water). As far as scientists know at this time, the three parts of the biosphere are the only places in the universe where life, as we understand it, can exist naturally. Within the biosphere, forms of life survive in every possible region. For example, many creatures live in total darkness deep below the ocean surface, as shown in Figure 1.22. Bacteria and the spores of many fungi have been found floating high in the atmosphere.

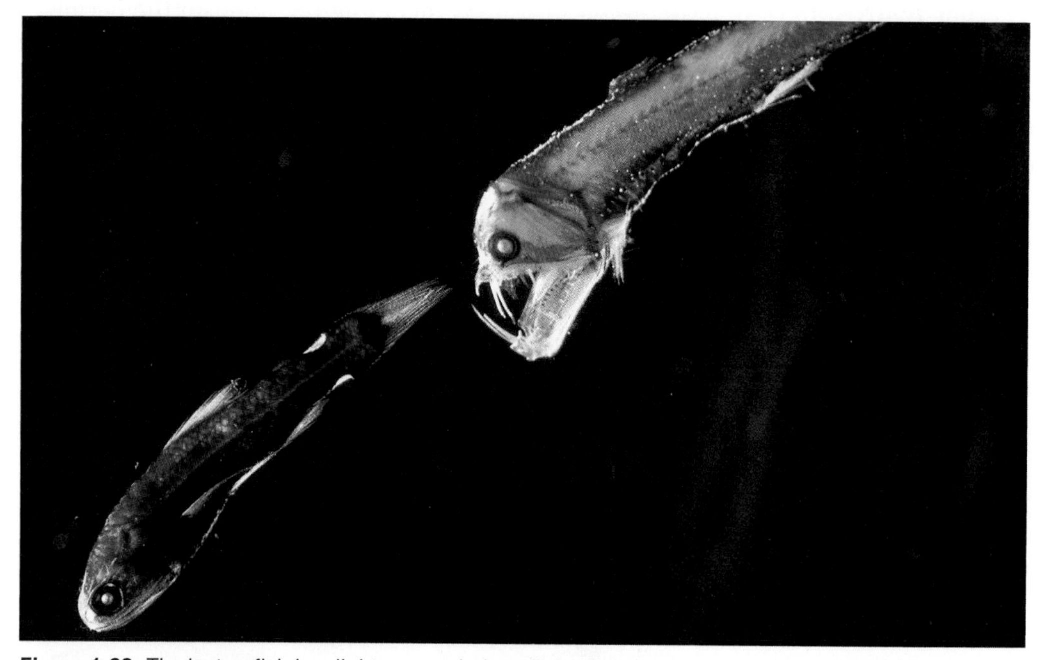

Figure 1.22 The lanternfish has light organs in its tail. In the murky ocean depths, its light helps others of its species to find it. As you can see, a viperfish has also detected it.

Word CONNECT

What is the difference between a threatened species and an endangered species? Research this question, and write your answer in your Science Log. Include some examples.

The biosphere is a giant ecosystem. All the abiotic and biotic parts of it are related — each part affects the others. The loss of a single species has some effect on other species. (Think back to the poem you read at the beginning of the chapter.) Many species have become **extinct**, meaning they have completely died out. The passenger pigeon, sea mink, Dawson's caribou, and blue walleye all used to live in Canada, and are now extinct because of human actions. As well, there is a long list of species that are endangered. They include the eastern cougar, bowhead whale, right whale, spotted owl, grizzly bear, and sea otter.

A scientist on the island of Mauritius recently observed that the few Calvaria trees that remained on the island were all about 300 years old. He wondered why no new trees were growing. He knew that the dodo bird had become extinct about 300 years before, and he hypothesized that the two events were connected. Historical research showed that dodo birds ate Calvaria seeds and that the seeds grew only after they had passed through a dodo's digestive system. Now scientists are experimenting with turkeys as a replacement for the dodo birds, hoping to ensure that the Calvaria tree does not become extinct.

Keep Them Safe

How can you help to keep endangered species from going extinct?

What You Need

research resources
markers
glue
poster paper
pieces of fabric
other materials for decorating

What to Do

1. Your teacher will provide you with a list of Canadian species that are endangered or threatened. Choose one species and do research to find out which ecosystem it lives in and why it is in danger. Try to find information about what is being done to prevent its extinction. If you cannot find any information, develop a plan to prevent its extinction yourself.

2. Create a poster, or start an advertising campaign to save your endangered species. Be creative, and be as persuasive as you can. Encourage people so that they will want to take part in saving the species.

3. Display your work to your class.

Extension

4. Many conservation and nature organizations have programs to help save a threatened or endangered species. As a class, choose a species from one of these programs. Develop a strategy to protect and promote awareness of the species.

Settlement and farming disturbed the grizzlies' grassland to such an extent that none were left on the Canadian prairies by about 1900. Grizzlies still do live in the Canadian North, as well as in Alberta and British Columbia, but they are classified as a threatened species. What effects would their extinction have on the environment?

Sea otters, like the one in the picture, use rocks to smash clams to eat. By 1900 these North Pacific mammals had been hunted almost to extinction because of their valuable fur. A few survived in Alaska and California, from where they have been successfully re-introduced into British Columbia waters.

Career CONNECT

The scientist in the photograph at the beginning of this unit (page 3) is Leah De Forest. She is placing bands on the legs of the young chicks that have hatched in nests on the cliffs, in order to keep track of their movements. Like Leah, other wildlife biologists carry out studies of animals such as grizzlies, sea otters, cougars, whales, owls, wolves, and caribou in their natural habitats. These biologists study such things as nutrition, parasites, disease, and migration patterns. They observe the animals carefully, and ask questions about what they observe. Then they find ways to try to answer their questions.

Develop your own question about the murre chicks that Leah De Forest is banding. See if you can find an answer to your question by looking on the Internet or visiting a library.

Skill POWER

For tips on how to use resources and the Internet effectively, turn to page 497.

Count the Caribou

Peary caribou are an endangered species in Canada's Far North. Imagine that a new pipeline is being planned for an Arctic area, to bring oil to more populated areas in the south. Many people question how the pipeline may affect wildlife. Your task is to find out how many caribou live in the area.

Problem

How can you estimate the number of caribou in a particular area?

Apparatus

2 dice

Materials

notebook

Sample	Row	Column	Number of caribou
1			
2			
3			
4			

Procedure

1. In your notebook, make a table like the one shown above. Give your table a title.
2. The graph on the next page is made from an aerial photograph of a herd of caribou in a 1 km^2 area. Each mark represents one caribou.
   (a) To make the sampling easier, the whole area has been divided by a grid into a number of squares. Note that the caribou are not evenly scattered. Some squares have no caribou, while other squares have several.
   (b) To estimate accurately the numbers of caribou, you will need to take samples at random.
3. Throw dice to determine which squares to select. Roll the first die to select a row. Roll the second die to select a column. Record the row and column numbers in your table. The row and column will indicate which square to use for your first sample. For example, if you rolled a 2 and 4, you would find the square to sample by selecting row 2 in the graph then moving along row 2 to column 4.
4. Count and record the number of caribou in the square you are sampling.
5. Repeat steps 3 and 4 for three more samples.
6. Find the average number of caribou per sample and then multiply this average by the number of squares in the grid. Refer to the Math Connect on page 17 if you need to review averages.

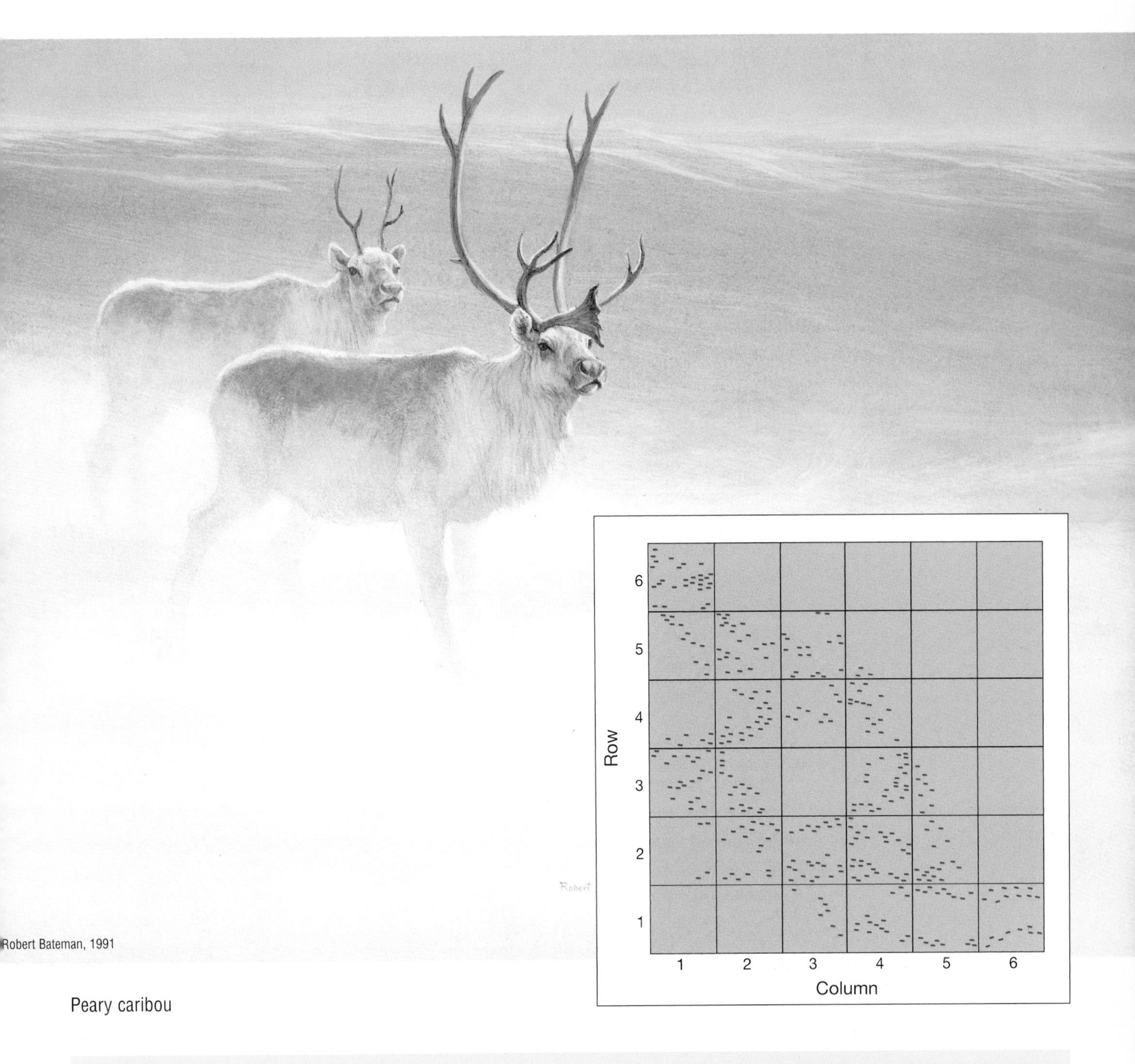

Peary caribou

Analyze

1. What is the average number of caribou in your study area?

Conclude and Apply

2. You found the average for a 1 km^2 area. Suppose that this area and your average are representative of a 500 km^2 area. How many caribou would you estimate are in the whole area?

3. Suppose that the oil pipeline will be built through the middle of the area and the caribou will not be able to go under it or jump over it. Predict what might happen to the caribou. Why?

Check Your Understanding

1. Write the letters (a) through (f) in your notebook. Use the clues in parentheses ( ) to unscramble the letters and write the scientific terms.
   (a) s h p y h d e r r o e (water)
   (b) e o r h e t h i p l s (Earth's crust)
   (c) p t h o s e m e a r (air)
   (d) s p b o i r e e h (layer around Earth in which life can exist)
   (e) c e t x i n t (no longer existing)
   (f) i a a h b t t (location where an organism lives)

2. Imagine that you are preparing to travel into space. Make a tape recording, write one or two paragraphs, or use another way to represent how you feel about the biosphere. Include some or all of the following:
   (a) how you feel as you leave the biosphere in your spacecraft
   (b) how the biosphere looks from your spacecraft
   (c) why the biosphere is important to organisms on Earth
   (d) why the biosphere is important to the members of your family
   (e) how you feel as you return to the biosphere

3. Work in a group. Imagine you are wildlife biologists. Prepare a short article, a series of labelled illustrations, a panel discussion, or another representation in which you explain some or all of the following:
   (a) reasons why some species are endangered
   (b) aspects of animals' lives that are studied by wildlife biologists
   (c) ways that biologists study animals
   (d) actions that people can take to preserve and protect wildlife

   To help you prepare your representation, you may wish to do some research. Among the organizations you could contact are the Canadian Wildlife Federation, the Canadian Nature Federation, World Wildlife Fund Canada, the Alberta Wilderness Association, Nature Conservancy organizations, or the Bruce Trail Association.

INTERNET CONNECT

www.school.mcgrawhill.ca/resources/

You are doing research on the conservation of wildlife and their habitats. Find out about conservation organizations and how you can help protect wild animals and spaces by going to the web site above. Go to **Science Resources**, then to **SCIENCEPOWER 7** to know where to go next. Make notes about your findings.

CHAPTER at a glance

Now that you have completed this chapter, try to do the following. If you cannot, go back to the sections indicated.

Explain the relationships between individuals, biological populations, and communities. (1.1)

What is the key difference between a community and an ecosystem? Give examples of a community and an ecosystem to explain the difference. (1.2)

Choose an organism, and explain how it interacts with sunlight, air, water, and soil. (1.2)

Choose an organism, and describe at least one adaptation it possesses that helps it survive in its environment. (1.2)

Give examples of different areas that could be sampled. Explain why sampling is necessary. (1.3)

Explain the relationship between biomes, climates, and weather. (1.4)

What are the names of the six land biomes? What are the characteristics of each of these biomes? (1.4)

Give four examples that illustrate the importance of plants in Canada's economy. (1.4)

List and define the three parts of the biosphere. (1.5)

Explain why some species in the biosphere are endangered. (1.5)

Prepare Your Own Summary

Summarize this chapter by doing one of the following. Use a graphic organizer (such as a concept map), produce a poster, or write the summary to include the key chapter ideas. Here are a few ideas to use as a guide:

- Create a series of pictures to illustrate the levels of biological organization. (You could use Figure 1.20 as a model.) Write two or three sentences to explain each of your pictures.
- Copy the illustration shown here into your notebook, and label the levels of the biosphere. Use illustrations and words to show how biotic and abiotic parts of the biosphere interact.
- Choose key scientific terms from this chapter. Write a summary to make their meanings clear.

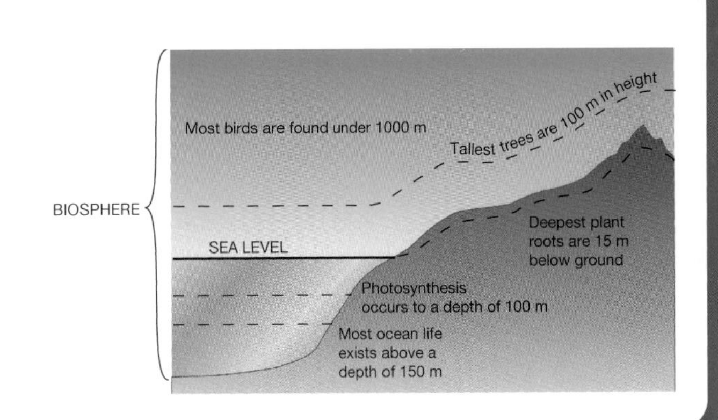

CHAPTER 1 Review

Key Terms

organisms
biotic
abiotic
ecologist
individual
biological population
species
biological community
habitat
adaptation
ecosystem
sampling
quadrat
weather
climate
biome
biosphere
hydrosphere
lithosphere
atmosphere
extinct

Reviewing Key Terms

If you need to review, the section numbers show you where these terms were introduced.

1. List in order, starting with the largest, the six levels of biological organization. (1.1, 1.2, 1.3, 1.4, 1.5)
2. For each of the following, explain the difference between the two terms.
   (a) biotic and abiotic (1.1)
   (b) abiotic and extinct (1.1, 1.5)
   (c) individual and population (1.1)
   (d) weather and climate (1.4)
3. Prepare a diagram or another representation to show the meanings of the following scientific terms: biosphere, hydrosphere, lithosphere, atmosphere. (1.5)

Understanding Key Ideas

Section numbers are provided if you need to review.

4. What was the first way scientists divided and classified objects found in the environment? (1.1)
5. Give examples of natural non-living things and non-living things created by people. (1.1)
6. Give two examples of living things that interact. (1.1)
7. Give two examples of biotic-abiotic interactions. (1.1)
8. How is the robin adapted for life as a tree-dweller? Give another example of how an organism is adapted for life in its environment. (1.1)
9. Explain the difference between a biological population and a biological community. Give examples. (1.1)
10. What is an ecosystem? Give an example. (1.2)

Developing Skills

11. Make a table like the following in your notebook, leaving more space to write. Complete the table, and give it a title.

	Land biome	Two plants that live in this biome	Two animals that live in this biome
1			
2			
3			
4			
5			
6			

12. Draw or find pictures of at least seven organisms that you might find in a wetland ecosystem. Label them.
13. Describe your own community, and tell why you would call it a community.
14. Imagine that you are teaching the topic of adaptation to a class of younger students. Devise two questions you could ask to assess the students' understanding of adaptation. Develop an example you might use to illustrate the differences between learning and adaptation.

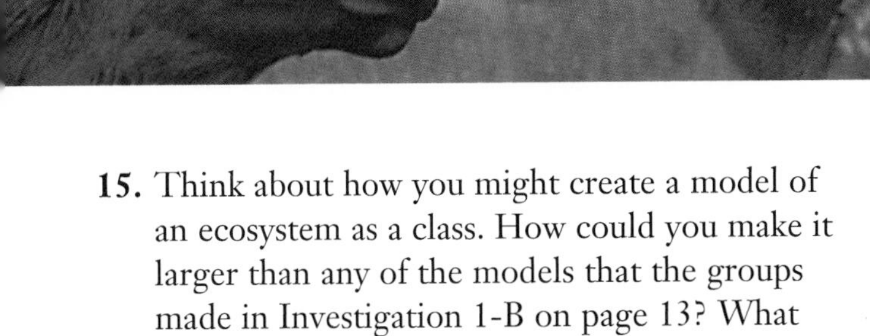

15. Think about how you might create a model of an ecosystem as a class. How could you make it larger than any of the models that the groups made in Investigation 1-B on page 13? What could you include in a larger ecosystem? Why?

16. Copy the following illustration of the biomes of Canada into your notebook, and label it. Colour your illustration, and give it a suitable title.

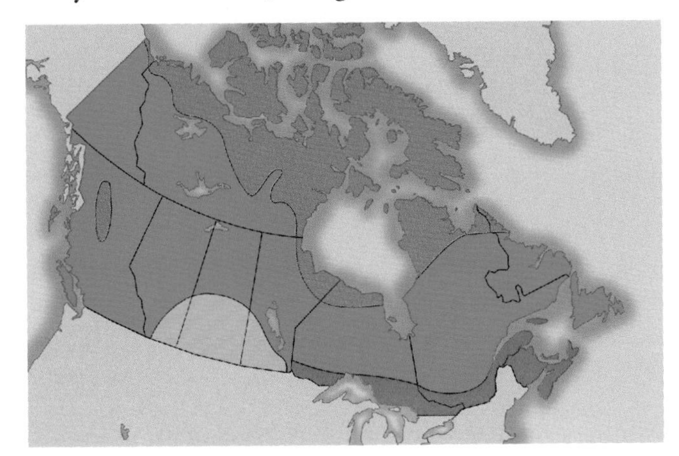

Problem Solving/Applying

17. Why do human beings use running shoes, winter boots, snowshoes, skis, or flippers? Explain why these are not biological adaptations.

18. Do animals use technology? What about the sea otter shown on page 29?

19. How does the climate of British Columbia's Okanagan Valley differ from the climate of British Columbia's coast? Prepare a flowchart, numbered list of points, labelled map, model, or other representation to account for the differences.

20. Imagine you make a living working with plants. You are having trouble growing plants in the biome in which you live. Suggest possible reasons for this. What steps could you take to solve the problem?

21. Imagine you are a scientist charged with the job of developing a plant that can grow in a spacecraft and be used as food for the crew.
    (a) What questions would you want to ask before you began your research?
    (b) What characteristics would you try to develop into your plant?

Critical Thinking

22. (a) If a rock rolls down a hill, has it moved by itself? Why or why not?
    (b) If a rock rolls down a hill and breaks into pieces, has it reproduced? Why or why not?

23. If you were an ecologist, how would you decide if an object in the evironment was living or non-living?

24. If you could live in any biome, which would you choose? Why?

25. Why is it important to use correct scientific terms when discussing scientific subjects?

Pause& Reflect

1. Describe three human activities that can have a negative effect on the biosphere. What steps can be taken to remedy such situations?
2. Now that you have been introduced to ecosystems, list three questions that you would like to ask about them. Watch for answers as you read through the next two chapters.
3. Go back to the beginning of this chapter on page 4, and check your original answers to the Getting Ready questions. How has your thinking changed? How would you answer those questions now that you have investigated the topics in this chapter?

CHAPTER

2 How Organisms

Getting Ready...

- Why do you hardly ever see the carcasses of dead animals in the environment?
- Why do termites eat wood when they are unable to digest it?
- Why do "vacant" lots never remain vacant?

Science Log

With a partner, think about the questions above. Use what you already know and make notes in your Science Log in answer to the questions. Look for the answers as you read the chapter.

Imagine that you are a hunter and trapper during the early days of Canada's history. You exchange lynx pelts for items that you and your family need, so your survival depends on knowing where to find lynxes. You learn as much as you can about these animals and their habits, and you observe them very carefully.

Last year, there were many lynxes around so you were able to obtain lynx easily. You were anticipating the same kind of year again, but strangely, lynxes seem hard to find. You notice, too, that snowshoe hares are scarce, and there were many of those last year. You wonder if there is any connection between lynxes and hares.

What you are participating in is a stage in the great cycle of life and death. This is the stage in which organisms interact in their quest for food. There *is* a connection and an interaction between lynxes and hares. The rise and fall of populations of species have been observed and recorded by ecologists who ask questions about why such changes occur.

In this chapter, you will have an opportunity to observe the interactions that occur in your environment and ask some questions of your own!

Interact

Spotlight On Key Ideas

In this chapter, you will discover

- how organisms are connected to each other in food chains and food webs
- how organisms form long-term relationships that may help or harm them
- how energy flows through food chains
- how ecological succession takes place

Spotlight On Key Skills

In this chapter, you will

- analyze food chains and food webs
- build a simple composter to study decomposers in action
- use a pyramid of numbers to show energy flow
- develop a model to illustrate succession in an ecosystem
- design an experiment to see which variables change rates of decomposition

Starting Point ACTIVITY

Ant Alert

Ants are found all over the world and are a common part of our environment. What can you find out about how they live and how they obtain food?

What to Do

1. With a partner, look around outside for some ants on the sidewalk or pavement. If you watch them for a little while, you may see a pattern in their activity. Do they ever bump into each other? What happens when they meet another kind of insect? Do any of them appear to be carrying food? Scatter a few grains of sugar in front of the ants, and watch what they do.

2. See if you can find the ants' nest. It usually looks like a small mound of sand with an opening on top. Place a few grains of sugar close to the opening of the nest, and see what happens. Now try a few grains of grass seed and a tiny bit of bacon. What happens?

What Did You Find Out?

1. Make some notes about your observations. Compare your observations with those made by others in your class.

Extension

2. How could you use a large jar with a mesh cover, soil, a moist sponge, honey, sugar, and breadcrumbs to construct a "society" for some ants? With your teacher's permission, design and construct the ant town using these or other materials. Observe your ant town to see how the ants interact. Observe which features of ants help them live in their environment.

2.1 Food Chains, Food Webs, and Energy Flow

Skill POWER

To find out how to use models in science, turn to page 493.

Did you ride your bike to school today? Did you play a sport in gym class? Have you ever mowed the lawn or shovelled the snow? Do you sleep, breathe, grow? All of these activities require energy, and to get energy you must take in food. Food contains stored energy, but how did the energy get there in the first place? Grass and other plants grow by using energy from the Sun and nutrients in the soil as sources of food. The energy of the Sun is then "stored" in plants. When an animal, such as a cow eats a plant, it obtains the Sun's energy indirectly in a useful form. When a meat-eating animal — perhaps a person sitting down to a steak dinner — later consumes the cow, the energy is passed on to the consumer.

Food Chains

A **food chain** is a model which shows how energy stored in food passes from organism to organism. Figure 2.1 shows some examples of food chains, in a lawn, a forest, and a pond. In a food chain, arrows show the direction in which energy flows through the chain.

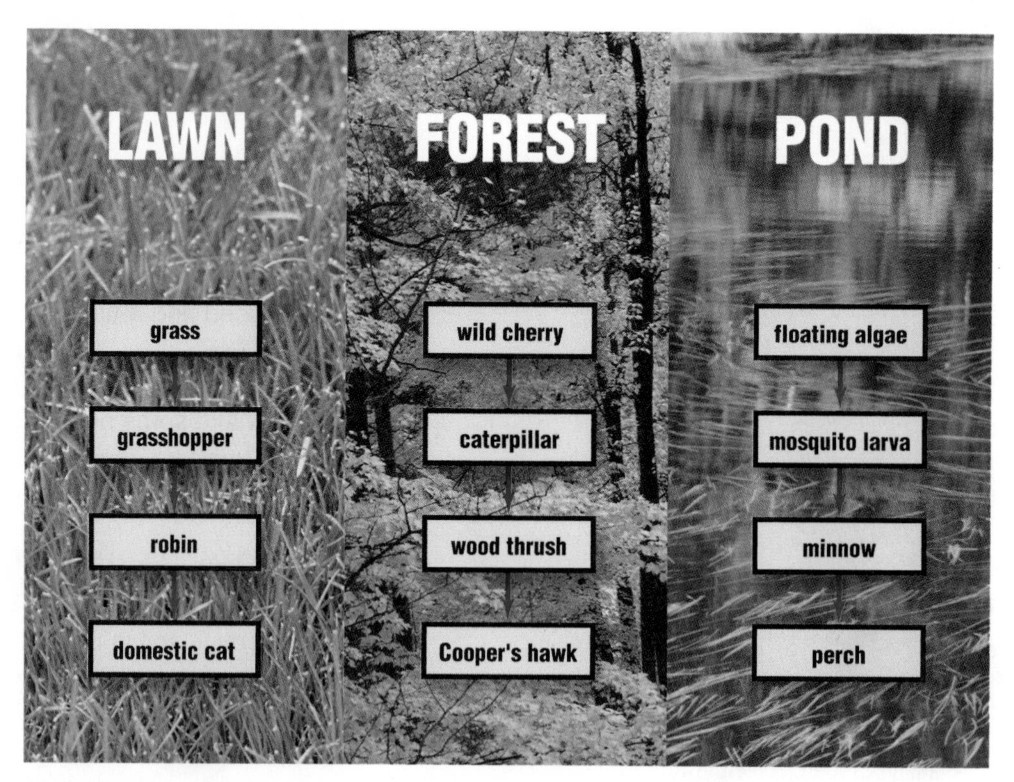

Figure 2.1 Food chains in a lawn, a forest, and a pond

An average adult human weighs as much as 1 000 000 ants. An ant is so strong that it can lift 50 times its own body weight. If adult humans were equally strong for their size, they could easily raise an elephant into the air.

Energy Flow

How does energy move through a food chain? At each step along the chain, energy is taken in by an organism. Some of this energy fuels the organism, and it is burned up and released as heat. Some energy is stored in the organism's body tissues, while some energy cannot be used and passes out of the animal as waste.

For example, when a grazing cow eats several kilograms of grass in one day, it does not gain mass equivalent to the mass of the grass it eats. Why not? Figure 2.2 shows a cow digesting the grass. About 4 percent of the stored energy in the grass goes to build and repair the cow's body tissues. A little over 30 percent fuels the cow's normal activities, such as breathing, mooing, and pumping blood through its body. Much of the grass — over 60 percent — cannot be used by the cow and passes out of its body as waste. Only the 4 percent that is used to build and repair the cow's body stays in the cow's body tissues. This is the "stored" energy, and it becomes available to the organism that eats the cow. You can see most of the energy in the grass eaten by the cow is not passed along the food chain.

Energy flow is the movement of energy, starting from the Sun, and passing from one organism to the next. In a food chain, as you saw with the cow, very little energy that is stored in one organism is passed on to the next organism.

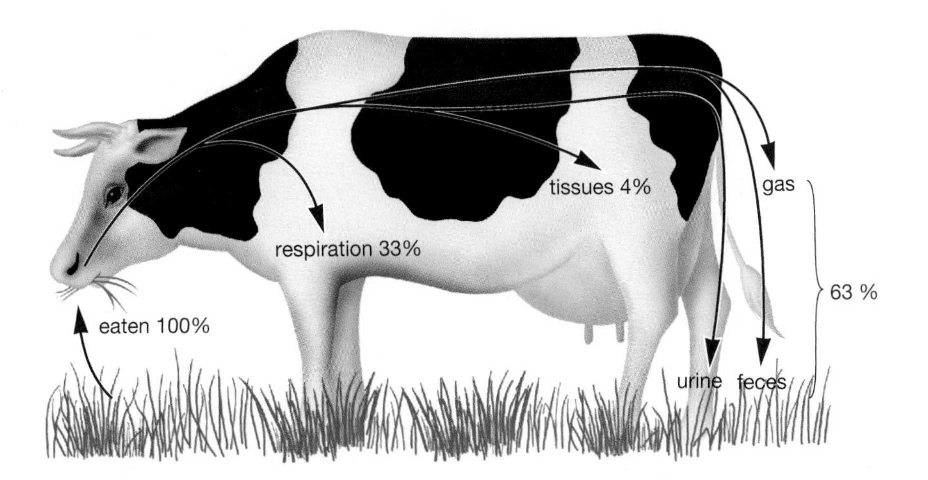

Figure 2.2 How a cow uses food energy

Pause& Reflect

Can you find out if food energy is used at every stage in a food chain? How will you determine how much energy stays in the food at each stage? Write the steps of your procedure in your Science Log.

The Role of Organisms in Food Chains

Look at the plants in the food chains in Figure 2.1. Green plants and algae contain a chemical called chlorophyll. Chlorophyll traps the energy of the Sun, in order to create food through photosynthesis. You will learn more about the process of photosynthesis in chapter 3. Plants and algae are called **producers** because they produce food for themselves and others, using the Sun's energy and nutrients in the soil and air. At the beginning of every food chain, there must be a producer. Producers make life possible for all other organisms on Earth.

All other organisms are called **consumers**, because they consume (eat) the food made by the producers. Consumers come in all sizes and shapes and include grazing animals, fish, and animals that eat other animals. Look again at the food chains in Figure 2.1. The organism next to the producer always feeds on plants or algae. Plant-eating animals are called **herbivores**. Cows, snowshoe hares, deer, herring, and tadpoles are examples of this group of consumers. Figure 2.3 on the next page shows Arctic herbivores.

Word CONNECT

The word "herbivore" comes from the Latin words *herba* (herb or plant) and *vorare* (to devour). Using this knowledge, write what you think the Latin words *carnis* and *omnis* mean.

The next organism in the food chain eats the herbivore. These meat-eaters are called **carnivores** (see Figure 2.4). Canada lynxes, cod, minnows, and dragonflies are examples of carnivores. The carnivore at the end of the food chain is known as the top carnivore.

Sometimes the top carnivore feeds on other carnivores: for example, a hawk preys on smaller, insect-eating birds. In some cases, the top carnivore may also feed on herbivores: the lynx eats snowshoe hares, and the wolf eats moose. Figure 2.5 shows a complex ecosystem.

Figure 2.3 Musk oxen are Arctic herbivores. In winter, they scrape through the snow to find lichen, grass, and moss to eat.

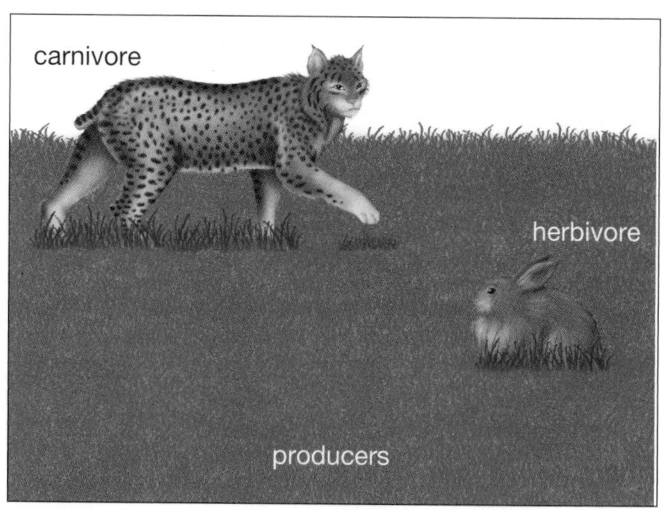

Figure 2.4 These organisms are typical producers and consumers.

Figure 2.5 A pond ecosystem. Producers in this ecosystem include algae and water lilies. Herbivores that feed on the producers include snails. Carnivores include frogs and turtles. Big fish, such as bass, are the top carnivores.

Predators are animals that kill and eat other animals called **prey**. Hawks and wolves are examples of predators. Predators usually eat only meat; however, some will eat almost anything if food is in short supply. Such animals, including bears and dogs, are called **omnivores**. How would you describe your place in the food chain?

INVESTIGATION 2-A

What Goes Up Must Come Down

Think About It

As you saw at the beginning of this chapter, the Canada lynx feeds mainly on snowshoe hares. Snowshoe hares are herbivores. When there are lots of plants for snowshoe hares to eat, more of them survive and reproduce. This means that the lynxes that feed on snowshoe hares have more food. Therefore, more lynxes survive and reproduce.

However, after several years there are so many lynxes killing snowshoe hares that the hare population starts to decline. Then the lynxes do not have enough food, and *their* numbers decline. Plants are able to grow because there are fewer snowshoe hares around to eat them. As new generations of snowshoe hares are born, there is plenty of food for them. Since there are fewer lynxes to hunt them, the hare population begins to increase. There is more food for the lynxes, so *their* numbers increase, too. So the whole cycle, which lasts about ten years, begins again.

The graph below shows how the number of lynxes and hares harvested by trappers changed over a period of 90 years.

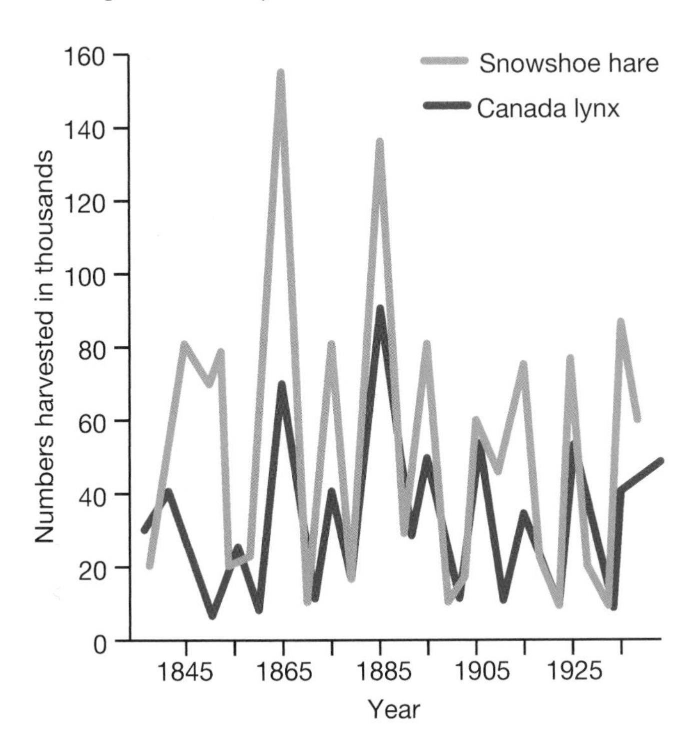

What to Do

Use the data in the graph to answer the following questions.

(a) In 1845, approximately how many lynxes were harvested by trappers?

(b) In 1845, how many hares were harvested by trappers?

(c) How many of each were harvested in 1855?

(d) In 1865, two years before Canada's Confederation, how did the two populations compare? What led to this change in the relative numbers of the two populations? What food that affects both hares and lynxes does not appear on this graph?

Analyze

1. Use the graph to explain how changes in the lynx population appear to follow changes in the hare population.

2. How can prey be said to control a predator's population? How can predators be said to control a prey's population?

3. The data in the graph are incomplete after the year 1935. Based on the data in the rest of the graph, estimate the populations of harvested lynxes and hares in 1940. Hypothesize about what might happen with these populations in 1945.

4. The last few years shown in the graph are the years of the Great Depression (1929–1939), a time of mass unemployment. How might this unemployment have affected populations such as these?

At Home **ACTIVITY**

The Masked Bandit

Raccoons are ever-present omnivores in many Canadian communities. They make their nightly journeys through neighbourhoods, rummaging in garbage containers, rattling lids, and disturbing sleep. How can you describe their ecosystem and the way they fit into it?

What to Do

1. Research and answer the following questions.
   - **(a)** Why are there are so many raccoons in cities?
   - **(b)** What do raccoons eat in the "wild"?
   - **(c)** What enemies, if any, do raccoons have?
   - **(d)** How are raccoons suited to live in different habitats?
2. People have varying responses when confronted with raccoons on their property. Some people feed them. Some people try to ignore them. Some people arrange to have the raccoons trapped so that they can be released outside of the community. What are the advantages and disadvantages of each response for the community and for the raccoons?
3. Check with your local Humane Society to find out how captured or injured raccoons are handled. If raccoons are not common in your community, ask the same questions about another animal considered to be a nuisance.

What Did You Find Out?

1. Compare your findings with what others learned. Work with a partner to write some "guidelines" for city dwellers on the best way to deal with raccoons.
2. With a partner, develop a question of your own about raccoons, then find a way to answer it, either by direct observation or through research.

The Clean-Up Squads: Scavengers and Decomposers

Have you ever wondered why you seldom see a dead carcass in a natural environment? If dead organisms stayed whole, the earth would soon be covered in bodies! In every community, there must be "clean-up squads" that get rid of garbage and sewage. In a biological community, the clean-up squads are consumers called **scavengers** and **decomposers**.

Figure 2.6 Carrion beetles at work

Scavengers

Many different organisms help to remove carcasses. Scavengers are organisms that eat dead or decaying plant or animal matter. For example, carrion beetles dig under the body of a dead bird or mammal, causing it to sink into the ground, as shown in Figure 2.6. The carrion beetles then lay their eggs in the carcass. What a legacy these beetles leave for their young — a huge supply of meat! Similarly, snails act as scavengers when they eat dead fish in an aquarium. (Snails can also act as herbivores when they eat living plants and algae.) Gulls act as scavengers when they eat dead material washed ashore along a beach. (Gulls are also carnivores when they eat earthworms in a freshly ploughed field.)

Decomposers

Have you ever seen old food in your refrigerator go mouldy? If so, you have witnessed decomposers at work. Decomposers differ from scavengers in that they do not actually eat dead material. Instead, they break down dead or waste material, such as rotting wood or manure. Decomposers like the ones in Figure 2.7, absorb some of the nutrients from the broken-down material into their own cells. The remaining nutrients recycle back into the ecosystem. Many bacteria and fungi are decomposers. Although bacteria are micro-organisms (too small to see without a microscope), some fungi are quite large and visible. In fact, you can see common fungi called mushrooms in any grocery store or vegetable market.

Though many decomposers are micro-organisms, they play major roles in everyday life. A scientist named Alexander Fleming observed that bacteria did not grow near moulds. His observations were used to develop penicillin, an antibiotic that fights disease-causing bacteria. People use their knowledge of how micro-organisms work in other beneficial ways, too. For example, moulds are used to ferment blue cheese, while bacteria are used to make sauerkraut and are added to milk to make yogurt. Yeast is another type of mico-organism, used to make bread dough rise as shown in Figure 2.8. Do you think people who were first offered blue cheese or sauerkraut knew what they were eating?

Decomposers also play a key role in breaking down much of our kitchen waste. We can assist this process by composting lettuce leaves, apple cores, carrot peelings, and other kitchen wastes in a composter like the one shown in Figure 2.9. When we compost, we let nature's decomposers turn our kitchen wastes into rich soil we can use for fertilizing the garden. In the next investigation, you will experiment with composters.

Figure 2.7 Bracket fungus digests the dead cells of tree bark.

Figure 2.8 Yeast is used in bread making to cause the dough to rise.

Figure 2.9 Kitchen wastes can be composted in a backyard composter.

You may have seen a robin eat a worm, but have you ever seen a worm eat a robin? Try to explain how this might happen.

DidYouKnow?

The wreck of the *Titanic* could disappear completely from the ocean floor by 2030. Bacteria are removing the iron from its hull at a rate of one tenth of a tonne a day.

INTERNET CONNECT

www.school.mcgrawhill.ca/resources/

Find out more about bracket fungi by researching them on the Internet. Go to the web site above, then to **Science Resources**, and then to **SCIENCEPOWER 7** to know where to go next. Find answers to these questions: How do bracket fungi attach themselves to a tree? Under what conditions do they grow most successfully? If you were to scrape a bracket fungus off a tree, would it regrow in the same spot? Make up at least one question of your own.

For tips on using the Internet effectively, turn to page 497.

Don’t Waste It!

Imagine cooking dinner at your home. You may peel potatoes, chop lettuce, and use some eggs. Each of these activities leaves you with material to throw away. Kitchen “wastes” are not necessarily garbage, however. Under the right conditions, they can be composted: broken down so the nutrients that are trapped in them are released. The composted material can then be recycled, for instance, as fertilizer in your garden. What kinds of materials break down well? What kinds of materials never break down at all? This investigation will allow you to explore the process of composting.

Problem

How can you find out which materials will decompose, and how long it takes decomposition to occur?

Safety Precautions

Apparatus

large clay pots with drainage holes (1 per test material)

labels for the pots

saucers to go under the pots

pieces of window screen or a similar material

magnifying glass

Materials

garden soil (not sterilized)

small stones

water

some or all of the following test materials: banana peels, paper, cabbage leaves, grass clippings, aluminum foil, orange peels, plastic, glass, potato peels, carrot peels, egg shells

Procedure

1. Before starting this investigation, develop hypotheses about what will happen to each of the materials you are going to test. Explain why you think as you do.
2. Set each clay pot on a saucer.
   - (a) Put a few small stones in the bottom of each pot, over the drainage hole.
   - (b) Add garden soil to each pot until the pot is about half full.
3. Put one test material in each pot. Label the pot to show what material is in it.
   - (a) Cover the materials in the pots with an equal amount of soil.
   - (b) Water the soil in each pot until a little water comes out the bottom into the saucer.
   - (c) Cover the open top of each pot with a piece of window screen.
   - (d) Put your composters in a permanent location for several months. Every few days, moisten the soil.

Analyze

1. Which test materials decomposed rapidly?
2. Which test materials decomposed slowly?
3. Which materials did not decompose over the course of the investigation?
4. Were your hypotheses from step 1 correct?

Conclude and Apply

5. Considering the health of the environment, what should be done to dispose of the materials you listed for question 2 and question 3?
6. What factors might speed up the decomposition of materials listed in question 2?

Extend Your Skills

7. **Design Your Own** Design an experiment to determine what effect, if any, temperature would have on the rate of decomposition. Make sure that your experiment is a fair test, and have your teacher approve your procedure before you carry out your experiment. (Hint: Your hypothesis should state whether you expect one variable, such as the rate of decomposition, to change when you alter another variable, such as temperature.)
8. **Design Your Own** Design an experiment that will test what effect, if any, using sterilized soil would have on the rate of decomposition. Your experiment should be a fair test, and your teacher must approve your procedure before you carry out your experiment. (Hint: Your hypothesis should state whether you expect one variable, such as the rate of decomposition, to change when you alter another variable, such as the type of soil.)
9. Find out about "red wrigglers." What are they? How can they be used? With your teacher's permission, obtain some to observe and use in your classroom.

4 After a week, remove the upper layer of soil and check that it is moist.

Check the decomposition. You could use the magnifying glass to examine the test materials. Replace the soil and continue the process until you can detect a difference in the condition of the materials.

Math CONNECT

For each test material, estimate and record the percentage of decomposition every week. For example, strawberries may be half of their original size, so they have decomposed about 50 percent. A plastic bag may have not have changed, so it has decomposed 0 percent. At the end of your investigation, graph your results. Finally, interpret your graphed data.

Skill POWER

For more practice organizing and communicating scientific results, turn to page 486.

Niches in a Community

You, like all other members of human communities, play several different roles in your daily life. When you are at school, you are a student. On the weekend, you might be a member of a sports team, or a volunteer at a food bank. The organisms in a community of plants and animals play different roles, too. Each of these roles is known as a **niche**.

One organism usually fills several niches. For instance, snails can act as both scavengers and herbivores, and gulls can be both scavengers and carnivores. To understand an organism's niche, you must look at what it eats, where it lives, and how it interacts with other organisms in its ecosystem.

Pause& Reflect

Think about the raccoon, the lynx, and the snowshoe hare that were discussed earlier in the chapter. How would you describe the niche each animal occupies? Do any of these animals occupy more than one niche in their community? Write answers to these questions in your Science Log.

Food Webs

Food chains are rarely as simple as the models you saw in the early pages of this chapter. Producers are usually eaten by many different consumers, and most consumers are eaten by more than one kind of predator. For example, a mouse, which may have eaten several kinds of plants and seeds, may be eaten by a hawk, a raccoon, or a snake. The raccoon may also eat berries, frogs, and birds' eggs. Figure 2.10 shows a typical food web. A network of inter-connected food chains is called a **food web**.

Figure 2.10 Food webs are a combination of several food chains. They show the connections among the food chains.

To draw a food web for a community, you can begin by drawing the food chains in the community. Next, look for organisms that are common to more than one of these food chains. Perhaps grass, a producer, is found in several chains. Grass can then be used to link the two chains together. Food webs, understandably, can quickly become very large and very complex.

Across Canada

"Curiosity is probably the most important characteristic that leads someone into a career in science," says Dr. Kevin Vessey, a scientist from the University of Manitoba. Kevin grew up in Prince Edward Island. As a teenager, he enjoyed watching marine biology shows, such as "The Undersea World of Jacques Cousteau", on television. Kevin's curiosity inspired him to study biology at Dalhousie University in Halifax, Nova Scotia, and then at Queen's University in Kingston, Ontario.

Dr. Vessey is interested in the many types of good bacteria in soil that help plants grow. The most common type of helpful bacteria are called rhizobia and they convert nitrogen from the air into ammonium, the mineral form of nitrogen. Plants cannot use nitrogen in the air, but they can use this mineral form of nitrogen to make protein. Rhizobia live in tumor-like growths on the roots of plants. When rhizobia attach themselves to a plant, they "infect" it. Dr. Vessey studies the development of infection by rhizobia in peas and soybeans (legumes). He says that this process of "nitrogen fixation" is similar to plants having their own fertilizer factory in their roots!

Kevin Vessey

The research Dr. Vessey is doing will help farmers use helpful bacteria and fewer pesticides. You might have a chance to hear Dr. Vessey if you listen to "The Science Quiz" portion of "Quirks and Quarks" on CBC radio. He has taken part in this show in the past and plans to contribute more science questions for curious minds in the future.

Check Your Understanding

1. Define the following terms in your own words, and give an example of each.
   (a) producer
   (b) consumer
   (c) omnivore
   (d) predator
   (e) decomposer
   (f) scavenger
2. Use arrows and words to draw five food chains based on the food web shown in Figure 2.10. For example, one food chain would be flower seeds → seed-eating birds → hawks.
3. Choose three of your food chains for question 3. Write down the niches that are occupied by the different organisms in them.
4. Show how an organism such as a horse or a seal obtains energy, and compare it with the way a person obtains energy. Try to show how much food energy each organism requires. (You may wish to do some research in order to answer this question fully.)
5. **Apply** Imagine that you are an ecologist. A group of people in your community wants to introduce an organism into the local ecosystem that will get rid of the mosquito population. Identify the mosquito's place in the food chain and explain to the group why introducing a new organism would not be a good idea.
6. **Thinking Critically** Choose and observe an ecosystem in your community. For example, you could choose a local park, a ravine, or your own backyard. Based on the organisms present, draw a diagram of a food chain. Indicate the role played by each organism in the chain. Does your food chain seem complete? Explain.

2.2 Welcome Partners and Unwanted Guests

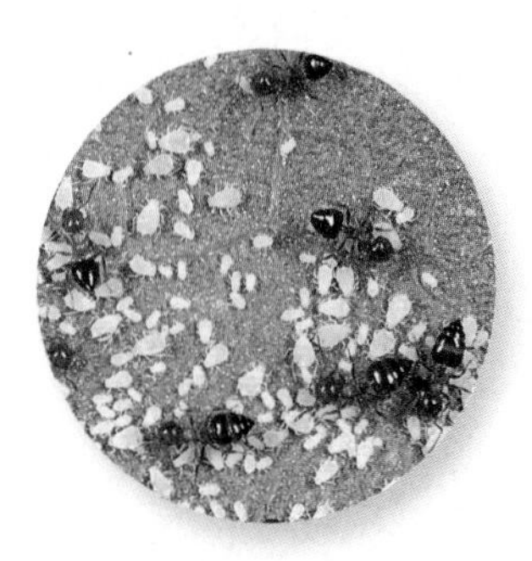

Figure 2.11 Ants and aphids have a symbiotic relationship.

Imagine a great white shark cruising toward you through tropical waters. As a human, your only thought would be to get away. Yet one small fish, called a remora, cannot get close enough! It uses suckers on its head to attach itself firmly to the shark's skin and then dines on bacteria and micro-organisms that are unhealthy for the shark. **Symbiosis** is when two species live closely together in a relationship that lasts over time. The odd association between the large, fearsome shark and the little remora is an example of a symbiotic relationship.

Symbiotic relationships are common in the natural world although some associations may seem unusual. For example aphids on a rose bush have a symbiotic relationship with the rose bush as they feed on it. Ants and aphids have a symbiotic relationship as shown in Figure 2.11. The ants protect the aphids from predators, and in return they drink the sweet liquid aphids excrete.

There are three main types of symbiotic relationships. **Parasitism** is a symbiotic relationship in which one partner benefits and the other partner is harmed. Typically, one of the partners lives on or in the other organism and feeds on it. External parasites, such as the louse shown in Figure 2.12, are often very small and so they are able to hide in hair, fur, or feathers. Some internal parasites are microscopic. The malaria-causing parasite *Plasmodium* lives and reproduces inside the human red blood cell! Other internal parasites, such as the tapeworm shown in figure 2.13, grow to be very large.

Figure 2.12 Lice are parasites that feed on the blood of mammals. This photograph of a louse has been magnified 55 times.

Figure 2.13 Tapeworms are common parasites that live inside other animal's intestines.

In 1986 scientist-filmmaker Greg Marshall watched a shark with a remora clinging to its side. He realized that if a camera could be attached to the side of the shark in a similar way, it would give an amazing close-up view of the shark's movements and behaviour. Thus was born a device called the "crittercam." It is a small battery-operated video camera that can be attached to the side of a shark by a small metal dart. The dart pierces the outer layer of the shark's hide without harming the shark. Shark food is thrown into the water to attract the shark close enough to a boat so that the crittercam can be attached. After a time, the crittercam is automatically released from the shark, tracked by radio signals, and retrieved.

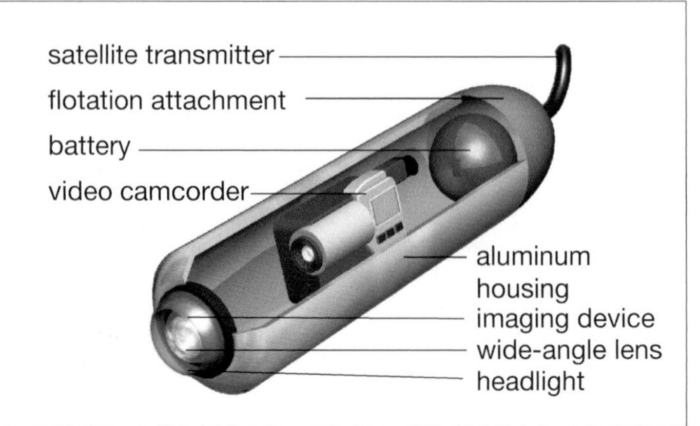

The organism that provides food for a parasite is called the **host**. Some parasites stay with the host only for a brief time while they reproduce. Other parasites, such as the tapeworm, may stay with one host for the host's lifetime. Typically a parasite does not kill its host (if it did, it would die as well), but it often weakens its host by taking nutrients away, and thus shorten its host's life.

Mutualism is a relationship between two different organisms, in which each partner benefits from the relationship. Termites, shown in Figure 2.14, are insects that live in large colonies and feed on wood, cardboard, and paper. They are decomposers, but they are incapable of breaking down the wood to get the nutrients on their own. They have a mutualistic relationship with single-celled micro-organisms, called protozoa, that live in their digestive tract. The protozoa, like the ones shown in Figure 2.15, digest the wood. The termites live on the protozoa's waste products. Their relationship is mutually beneficial — although it is not beneficial to the humans whose homes termites eat!

Figure 2.14 Caution! Termites at work.

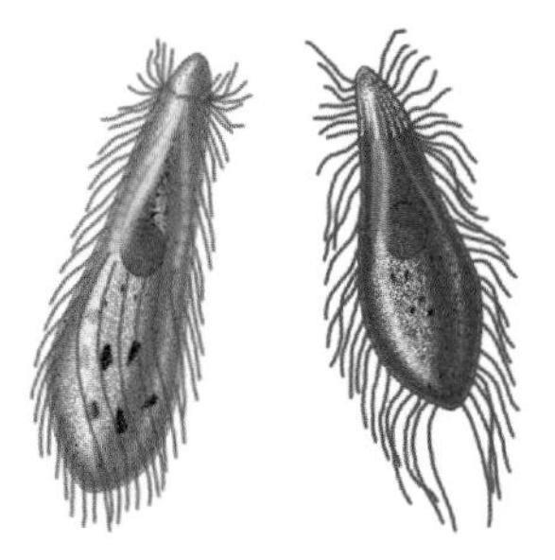

Figure 2.15 Two kinds of protozoa that live in termites.

Lichens, shown in Figure 2.16, appear to be single organisms but are, in fact, made up of two different partners. Figure 2.17 shows the partners — an alga (plural: algae) and a fungus living together. In partnership, they are able to live where neither could survive alone, such as on a bare rock or a tree trunk. The alga makes food for both, while the fungus forms a sponge-like body that protects, anchors, and holds the water they both need.

Mutualistic relationships may be critical to the survival of certain organisms. For instance, sometimes one organism cannot reproduce unless it interacts with an organism of another species. This is the case with the dodo bird and the calvaria tree described in Chapter 1 on page 28.

Figure 2.16 Leafy lichen (on the left) and crusty lichen (on the right) are both examples of mutualism.

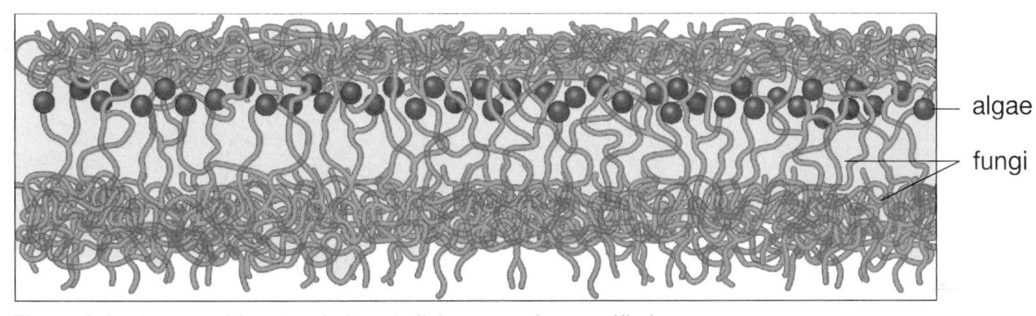

Figure 2.17 Layers of fungi and algae in lichen, greatly magnified

DidYouKnow?

Tapeworms that can live in the small intestines of human beings can grow as long as 10 m. While in the egg stage of their life cycle, these parasites enter the body in meat or fish that has not been inspected and properly or sufficiently cooked.

Pause& Reflect

Symbiotic relationships are very common in the natural world. Think of a symbiotic relationship, either commensal, parasitic, or mutualistic. Draw a picture or write a poem in your Science Log describing the relationship.

Commensalism is a symbiotic relationship in which one partner benefits and the other partner does not appear either to lose or to gain from the relationship. For example, many species of flowering orchid, like the one in Figure 2.18A, live high up, attached to the trunks of trees. The orchids benefit by having a safe place to live and a constant source of water from rain dripping down the tree trunks. The trees do not seem either to benefit or to lose from the presence of the orchids. Clownfish and sea anemones also have a commensal relationship. Figure 2.18B shows the clownfish living among the poisonous tentacles of the hosting anemone. The fish gains safety and eats the scraps left over from the anemone's dinner. There is no obvious benefit or cost to the anemone.

Figure 2.18A An orchid plant attached high up on a tree trunk.

Figure 2.18B A clownfish living among the tentacles of a poisonous sea anemone.

Check Your Understanding

1. Define the following terms in your own words, and give an example of each.
   (a) symbiosis
   (b) parasitism
   (c) mutualism
   (d) parasite
   (e) host
   (f) commensalism
2. Compare the effects of parasites and predators on a community.
3. **Apply** Brainstorm in a group and come up with examples of relationships that are similar to mutualism in a human community. Describe them.
4. **Thinking Critically** Think about each of the following pairs of organisms, and name the type of symbiotic relationship the partners might have. Indicate what the gains and/or losses might be for each partner:
   (a) a flowering plant and an insect
   (b) a whale and a barnacle living on the whale's back
   (c) a dog and a flea
   (d) a nectar-eating bat and a flowering cactus
   (e) a bird and a water buffalo

2.3 Energy Flow and Pyramids of Numbers

Imagine walking through the forest. As you look closely at the ecosystem, you notice many different types of plants, insects, and birds, and even small herbivores and carnivores. You may catch sight of larger carnivores or see their tracks. The relationships between these organisms are difficult to understand. Food chains and food webs help organize the information about relationships. They also show how energy is transferred through an ecosystem. These models do not always give enough information because they do not show how many organisms are involved in the energy transfer. For instance, looking back to Figure 2.1, we do not know how many minnows the perch eat, only that the perch eat minnows.

Pause& Reflect

Think about the local ecosystem you observed at the end of section 2.1. You probably noticed that there were more plants, such as grasses and weeds, than there were animals eating them. In your Science Log, explain why this would be so.

Figure 2.19 A pyramid of numbers is a model of an ecosystem that represents the number of organisms consumed at each level. Producers always form the broad base.

To solve this problem, ecologists build a **pyramid of numbers**. Figure 2.19 shows a typical pyramid of numbers. It includes the same organisms as you see in a food chain, but the size of each level represents the number of organisms involved at this level. There are always more animals being eaten than there are animals eating. For example, there may be one hawk eating three woodpeckers, but not three hawks eating one woodpecker. The producers always form a broad base, with all of the consumers in different levels above.

A pyramid of numbers can only indicate how many organisms at one level are eating how many organisms from a lower level. They do not indicate exactly *how much* energy is consumed. We can find this out by looking at how much each level of the pyramid weighs — how many kilograms of grasshoppers are needed to feed one kilogram of woodpeckers. **Biomass** is the total mass of all the organisms in an ecosystem. Just as each level in the pyramid of numbers has fewer organisms than the level below it, it also has less biomass. In any pyramid of numbers, the most biomass is in the base formed by the producers.

Skill POWER

To review what mass is and how it is measured, turn to page 484.

As you go up the pyramid, less and less food energy is available at each level. Figure 2.20 shows another pyramid. Each time a caterpillar eats a maple leaf, energy is lost as heat. Only a small amount of energy is stored in the tissues of the caterpillar to be passed on to the robin. This is why the 10 000 caterpillars store only enough energy to feed six robins! If no energy were lost between the levels, each level would be the same size as the one above and below it, and the pyramid would be a rectangle of numbers!

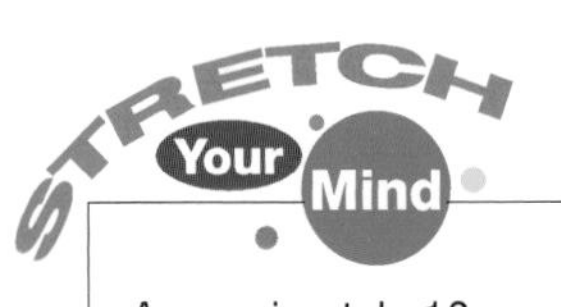

Approximately 10 percent of available food energy is passed to the next level of a food chain. If tuna fish are four steps up a food chain, what mass of phytoplankton (the producers in the ocean) would be required to provide a human with 125 g of tuna for a sandwich?

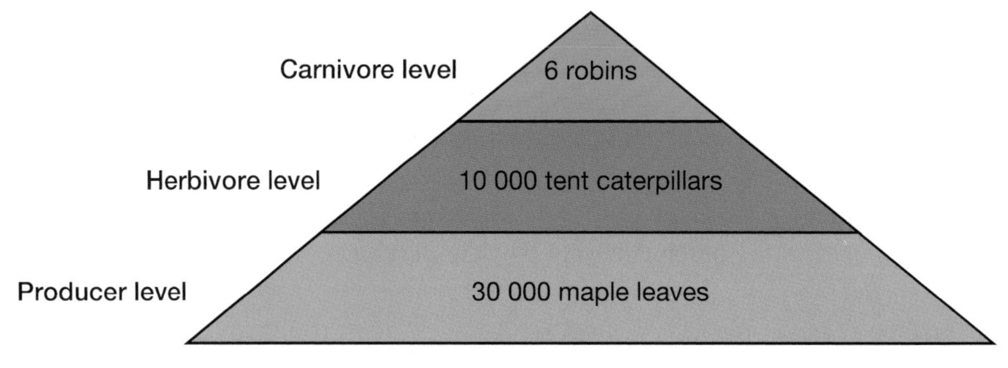

Figure 2.20 As you move up the pyramid, the levels get smaller to show less stored energy is available.

Humans, who are usually at the top of the pyramid of numbers, are able to eat many different types of food from all levels of the food chain. Other organisms may be limited in what they can and will eat. If an organism only eats one type of food, the disappearance of that food can have serious effects on the entire pyramid of numbers. For example, what would happen to the pyramid that includes the lynxes and the hares, if the plants that the hares eat were no longer available?

Find Out **ACTIVITY**

No Fishing Allowed

As you are probably aware, the cod-fishing industry in the Maritimes is in serious trouble. The cod stock has been reduced to dangerously low levels.

What to Do

1. Do some research on this topic. Look in newspapers and magazines for current articles. Check the Internet.
2. Write a brief report on your findings.

For some tips on using resources and the Internet effectively, turn to page 497.

What Did You Find Out?

1. What other organisms are in the food chain? Are the cod at the top or the bottom of the pyramid?
2. What has caused the reduction in cod supply?
3. If the cod became extinct, what would you predict would happen to the food pyramid in which the cod are found?

Extension

4. Find out about the salmon-fishing industry on Canada's west coast. How do its problems compare with those of the cod-fishing industry?

Check Your Understanding

1. In what ways is a pyramid of numbers like and unlike a food chain? Construct a Venn diagram to show your answer.

2. Humans usually eat foods from all levels of a food chain, like those shown in the photograph below. Construct two different food chains based on food you typically eat. Make one food chain long (at least four levels), and the other food chain short (two levels). You may want to illustrate your food chains, or simply use words and numbers.

Typical foods eaten by humans

3. Explain why all of the energy in one level of a food pyramid is not available to organisms higher in the pyramid. Prepare a drawing, collage, skit, paragraph, or other representation to explain your answer. If necessary, do further research. Share your representation with the class.

4. Why are shorter food chains more efficient (transfer more energy) than longer ones? How could this information influence some people to become vegetarians? You may wish to research this question. As a first step, you could try contacting an association of vegetarians.

5. Think about a food chain that includes grass → field mice → snakes → owls. Describe what would happen if many mice died as a result of disease. What would happen to the owls? What would happen to the snakes? What would be the probable result in the ecosystem?

6. **Apply** Why do some people have limited food choices? Discuss this question in a group or as a class. What might happen if their food supplies were damaged by, for example, a hurricane or a war?

2.4 A First Look at Succession

Have you ever noticed grasses or plants growing in a vacant lot near your home? If you wait long enough, bushes and trees will grow, and animals will make their homes in the lot. Can you imagine how this process happens?

Over time, changes take place in an ecosystem. Some changes are rapid. For example, a forest fire or a landslide might completely destroy an existing ecosystem. Other changes are slow. For example, seeds carried by wind or water might take root in a vacant lot or sidewalk cracks and result in a new population of plants. If conditions are good, the new plants might become established and even replace plants that were already growing.

The gradual process by which certain species replace other species in an ecosystem is called **succession**. In the process of succession, organisms that are present at one stage alter the environment in some way. This change makes it possible for some other species to move in. If the process continues naturally, the final result is a stable ecosystem in which there are few further changes. Look at Figure 2.21 to see how succession proceeds to change bare rock to forest. First, lichens, mosses, and ferns grow, preparing the area for varied plants and animals. Finally the area becomes a woodland. Succession can be a very slow process, taking hundreds, even thousands, of years.

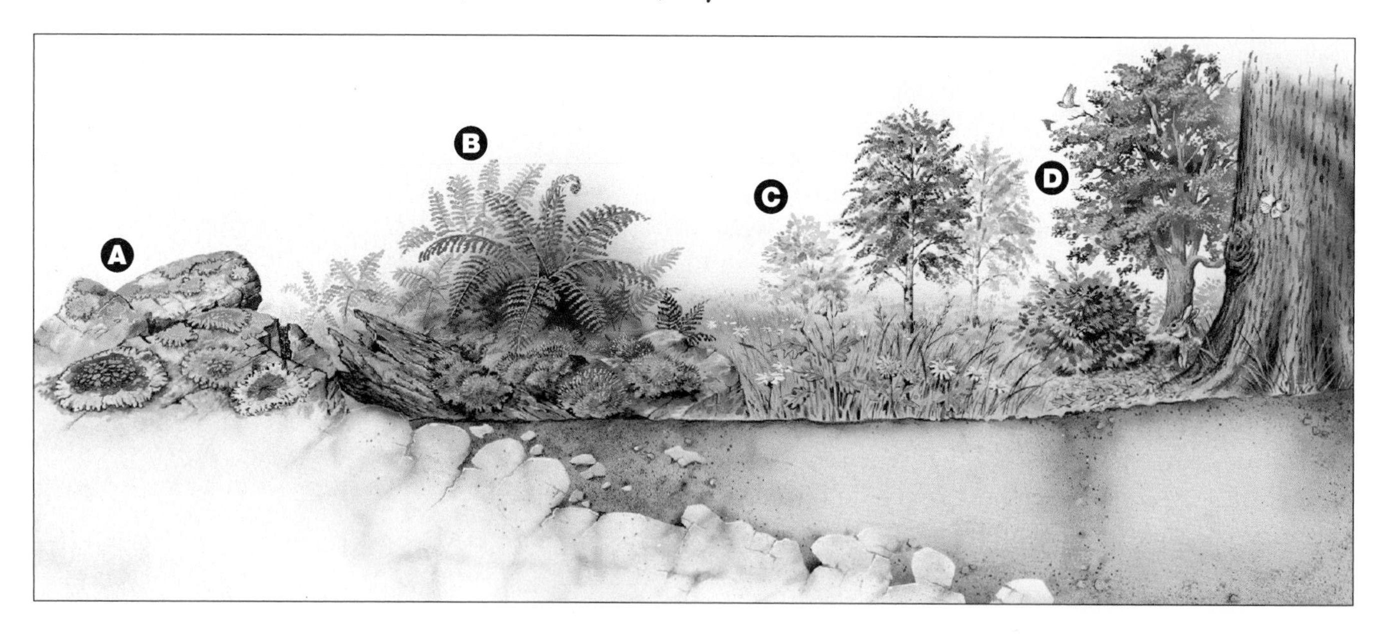

A Lichens produce acids that help to break down the rock. The broken-down rock and the decomposing bodies of dead lichens contribute to soil formation.

B The resulting soil is poor and thin. However, mosses and ferns grow and slowly replace the lichens.

C The soil layer thickens, which means it can hold more water. Plants that need more soil and moisture, such as grasses and flowering weeds, take root and grow. They attract insects, such as bees and butterflies.

D Since the soil is now thicker and richer, bushes and trees take root. They provide shelter and food for birds, mammals, and other organisms, which now start moving in.

Figure 2.21 Succession is a long, slow process in which a stable natural ecosystem gradually develops over time.

Succession occurs wherever plants can become established in all ecosystems (including open water). Succession works because as some organisms become established and grow in an area, they change the conditions and allow other plants and animals to grow and live there. The process of succession is a natural one, and it usually has positive effects. For example, a forest will regrow into a healthy ecosystem after a natural disaster has damaged the area. However, it is possible for succession to have negative effects. This is true, for instance, if succession occurs too quickly. Usually succession takes place slowly, but can speed up if extra nutrients and pollutants are present. Farmers' fields contain fertilizers, so if run-off from a field enters a stream or a lake, plants may grow too quickly and start to fill in the open water.

There is a delicate balance between organisms in a succeeding ecosystem. Some foreign species, such as purple loosestrife shown in Figure 2.22, do so well that other species, which normally develop at the next stage of succession, cannot become established. The natural balance can also be disturbed by foreign animals. Zebra mussels, shown below, are very successful in the Canadian Great Lakes. Zebra mussels eat much of the food which naturally occurring organisms usually eat, so the organisms in the next stages of succession find it difficult to get enough food.

Figure 2.22 Purple loosestrife is an unwelcome visitor in wetland areas, where it is choking out native plants.

DidYouKnow?

The shallowest of the Great Lakes, Lake Erie, is being filled in. As succession moves out from the shoreline, it is claiming what used to be open water.

"Musselling" In!

The zebra mussel has travelled the world. It now makes the Canadian Great Lakes one of its homes.

What to Do

With the aid of resources, including the Internet, answer the following questions.

(a) What are zebra mussels?
(b) Where did they come from?
(c) How do they affect plants, animals, and humans?
(d) How are they being controlled?
(e) Can you think of any benefits produced by zebra mussels?

What Did You Find Out?

Write a brief illustrated report of your findings.

CONDUCT AN

INVESTIGATION 2-C

Nothing Succeeds Like Succession

Succession can take place in any kind of area, large or small, in a short period of time or over many years. For instance, weeds quickly grow in an untended patch of soil, but trees take many years to grow back in an area cleared by forest fire or logging.

Problem

How does succession take place in an ecosystem?

Safety Precautions

Materials

2 L clear plastic soda bottle with the top cut off or large-mouthed jar

potting soil

water

a small aquatic plant (maybe from an aquarium supply store)

wild bird seed

Procedure

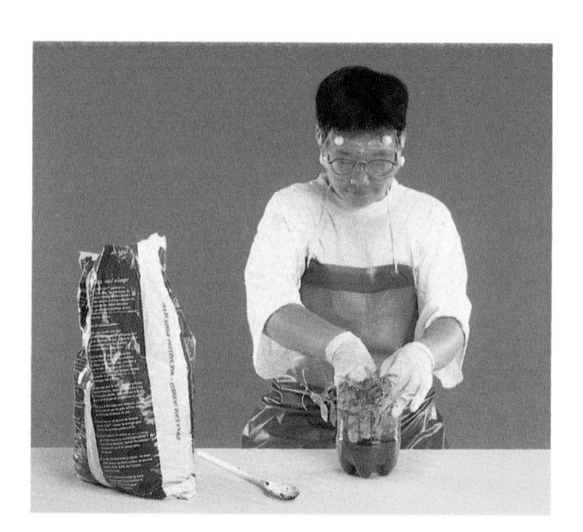

1. Put soil in the bottom of the container, to a depth of 5 cm.
   (a) Fill the container with water, to a depth of 7.5 cm, thus covering the soil.
   (b) Place the container, uncovered, on a window sill, and allow the contents to settle overnight.

2. The next day, plant an aquatic plant in the container. Although the water will evaporate over time, do *not* add more water to the container.

3. Once a week, add 3 or 4 seeds from the wild bird seed mix to the container. Note any observations.
   - (a) Continue adding seeds weekly, even though the water evaporates. What, if anything, happens to these seeds?
   - (b) After a few weeks, gradually start to add water again, as you would when watering a plant. Continue to record your observations.

Analyze

1. Keep a careful record of your observations. Describe what is occurring in your container.
2. Describe your observations during step 3. What was the significance of not adding water to your ecosystem? What happened to the aquatic plant?

Conclude and Apply

3. Compare your ecosystem at the beginning of this investigation with your ecosystem at the end. How did they differ?
4. To what extent does this investigation demonstrate succession?
5. Using your knowledge of succession, describe what you would expect to happen if a fire burned through a forest, destroying most of the mature trees and vegetation.

Extend Your Skills

6. **Design Your Own** Design an experiment to determine what effects different environmental factors would have on your successional ecosystem. Make sure that your experiment is a fair test, and have your teacher approve your procedure before you carry out your experiment. (Hint: Your hypothesis should state what effect changing one variable, for example, an environmental condition of your choice, will have on another variable, for instance, the growth of any new plants.)

Pause& Reflect

Make a list of different places succession could happen. Choose a place from your list and draw a flowchart to show a possible sequence of succession. Record your flowchart in your Science Log.

Career CONNECT

Eco-Careers

There are many careers that tie in with ecosystems. A few are noted below. How is each of these careers related to ecosystems? What others can you think of?

crop farmer | zoo worker | nursery or lawncare worker | fishing guide

Choose one of these careers or another ecosystem-related career that interests you. What qualities or experience do you already have that might help you in this career? Imagine that you are applying for a related summer volunteer position. What work experience can you list? Do you belong to a club? Have you studied ecosystems on your own, perhaps in your own yard or while on vacation? Perhaps you have items that record your interest or experience: a letter of recommendation, a badge of achievement, a ribbon for growing a prize-winning vegetable, a photograph of yourself handling an animal or landing a large fish. Write a letter outlining why you would be the best candidate for the position.

Check Your Understanding

1. Examine the local ecosystem you observed for question 6 at the end of section 2.1 (page 47), or find another one to observe. What signs can you find to show that the process of succession is taking place there?
2. In nature, succession occurs naturally and slowly. Humans, however, sometimes make drastic changes that affect how succession works in an ecosystem. Explain, using either purple loosestrife or zebra mussels as an example.
3. Should humans simply leave all natural ecosystems exactly as they are, or is it acceptable to make changes? What kinds of changes are acceptable? Use these questions as the basis for a report, essay, article, or letter to a newspaper editor. Work alone or with a group.

Humans affect natural ecosystems. What kinds of changes are acceptable? What kinds are not?

CHAPTER at a glance

Now that you have completed this chapter, try to do the following.
If you cannot, go back to the sections indicated.

Define the terms "food chain" and "food web." (2.1)

Explain how energy is lost in a food chain. (2.1)

Give two examples of organisms and describe the niche they fill. (2.1)

Explain what energy flow is and why it is important. (2.1)

Explain the difference between scavengers and decomposers. What different roles do decomposers play in everyday life? (2.1)

Explain how the relationship between a human and a pet could be described as symbiotic. (2.2)

Explain why termites cannot function alone as decomposers. (2.2)

Explain what a pyramid of numbers is. What information does it provide that a food chain or a food web does not provide? (2.3)

Explain what succession is and how it occurs. (2.4)

Prepare Your Own Summary

Summarize this chapter by doing one of the following. Use a graphic organizer (such as a concept map), produce a poster, or write the summary to include the key chapter ideas. Here are a few ideas to use as a guide:

- Draw a diagram that illustrates energy flow. In your diagram, show the source of the energy flow and include at least five organisms. Indicate what happens to the energy as it moves from one organism to the next.
- Look at the illustration. The dashed line at the top of the graph indicates that above that line the environment in which the fish live cannot support any more fish. (There will not be enough food.) What do you think will happen to the line representing the number of fish if the present level of food remains available? Design an experiment to investigate what will happen to the number of fish if more food is available. (Alternatively, select a different variable to investigate that you think would affect the number of fish.)
- Ladybugs are sometimes imported by gardeners and released in their gardens to eat aphids. Create a poster, and show at least two other organisms that could be affected by the ladybugs. Explain why this effect might happen.
- Why do most humans have more food choices than many animals do? Brainstorm answers to this question, alone or in a group. Summarize your answers in a written report.

- Look again at the ecosystem you began to observe for question 6, at the end of section 2.1 (page 47). What effects might changes in this ecosystem have on food pyramids within the ecosystem? For example, what might happen if a forest is cleared for farming or if a grassy schoolyard is paved? Make predictions based on observing your ecosystem. You could check your predictions by researching similar cases elsewhere.

CHAPTER

2 Review

Key Terms

food chain
energy flow
producers
consumers
herbivores
carnivores
predators
prey
omnivores
scavengers
decomposers
niche
food web
symbiosis
parasitism
host
mutualism
commensalism
pyramid of numbers
biomass
succession

Reviewing Key Terms

If you need to review, the section numbers show you where these terms were introduced.

1. Complete the following table in your notebook.

Type of organism	Examples of organism
producer	1. 2.
consumer: herbivore	1. 2.
consumer: carnivore	1. 2.
scavenger	1. 2.
decomposer	1. 2.
parasite	1. 2.
omnivore	1. 2.

2. Define the term "niche." List, with examples, five different niches in an ecosystem. (2.1)

Understanding Key Ideas

Section numbers are provided if you need to review.

3. If you found hawks, field mice, and corn in the same ecosystem, what role would each probably be playing in a food chain? (2.1)

4. What can happen to an ecosystem if one or more of the organisms are removed? Give a specific example. (2.1)

5. Why are scavengers and decomposers important in an ecosystem? How do they differ? (2.1)

6. Cattle egrets are birds that follow herds of large animals, such as African Cape buffaloes. Can you explain what kind of relationship this is? What might the birds and the buffaloes get out of the relationship? (2.2)

7. In the ecosystem described in question 3, which would have the greatest biomass: the corn, the field mice, or the hawks? Which would have the least biomass? Explain. (2.3)

8. A family cultivated part of their lawn and turned it into a vegetable garden. The family then moved away. The house remained empty, and nobody looked after the garden. Ten years later, the family came back for a visit. Their lawn looked similar, though much weedier. They were surprised, however, to see wildflowers, shrubs, and small trees growing in their deserted garden. Explain why this is an example of succession. (2.4)

Developing Skills

9. Explain the food web below. Draw three food chains based on it.

10. Select a newspaper or magazine picture that includes living organisms in an ecosystem. List all the organisms in the picture. Now trace each back to its original niche in an ecosystem.

11. Why are longer food chains less efficient (transfer less energy) than shorter ones? Draw a pyramid of numbers to help explain your answer.

Problem Solving/Applying

12. A sailor survived a shipwreck. She managed to save several hens and a bag of grain from the cargo. She is now on an island far from land, in an area where there are no other people. It may be months before she is rescued. To survive as long as possible, what should she do?
    (a) Feed the grain to the hens, and eat the eggs they lay.
    (b) Eat the grain, and then eat the hens.
    (c) Eat the hens, and then eat the grain.

    Explain why you think the option you chose is best. If you do not agree with any of the options listed above, what other solution would you propose?

13. Imagine that you have just eaten a meal consisting of a salmon sandwich (with lettuce and tomato), a cup of mushroom soup, an almond cookie, and a glass of apple juice. All of these foods once occupied a niche in the food chain.
    (a) Classify each food as having been a producer, consumer, or decomposer in the food chain.
    (b) Sketch a food chain showing possible relationships and the organisms that now make up your meal.

Critical Thinking

14. In an ecosystem, what does a consumer lack that a producer possesses?

15. Consider the local ecosystem you have been observing. Predict what would happen if each of the following major changes occurred.
    (a) People begin to use chemical fertilizers.
    (b) The rain becomes acidic.
    (c) A hydro line is built through the area.
    (d) The area is hit by a severe hailstorm.
    (e) The butterflies do not appear one spring.
    (f) The earthworms are attacked by a parasite, and their numbers are severely reduced.
    (g) An oil spill occurs.

16. Think again about the changes you considered in question 15. What roles might people play to reduce or reverse their impact on the ecosystem?

Pause& Reflect

1. In a number of cases, humans have introduced different organisms or, in other ways, deliberately changed an eco-system. To what extent can we or should we change natural ecosystems? What are some advantages and disadvantages of changing ecosystems?
2. Go back to the beginning of this chapter on page 36, and check your original answers to the Getting Ready questions. How has your thinking changed? How would you answer those questions now that you have investigated the topics in this chapter?

CHAPTER

3 Keeping the Systems

Getting Ready...

- What prevents dandelions from covering the surface of Earth?
- How can plants warm buildings?
- If water is constantly cycled and recycled in the water cycle, how can it be "wasted"?

Try to answer the questions above in your Science Log. Think about what you know, and use your imagination. Look for the answers as you read this chapter.

At the time of writing, the ferris wheel shown in the photograph above was the largest in the world. It was being built in London, England, to welcome the twenty-first century. This amazing structure is just one of many examples of technology in the modern world.

Look at the photograph of the athlete. Wheelchair athletes do not have complete use of their legs, but technology enables them to participate fully in a wide range of athletic events.

What does technology have to do with the interactions among living things? Think of the positive and negative effects of the ferris wheel. It provided work for its designers and builders, and it continues to provide work for its operators. On the negative side, it required non-renewable energy and Earth's resources to manufacture, and it now requires more energy to operate. Perhaps an ecosystem was disrupted or destroyed as the metals were mined for its construction.

How can we find a balance between our technological needs and Earth's requirements? How can we help Earth to flourish and to continue to support life — including our own? These are topics you will investigate in this chapter.

Going

Spotlight On Key Ideas

In this chapter, you will discover

- how the biosphere cycles water, carbon dioxide and oxygen, and carbon
- how technology affects water, soil, and air
- how natural factors in ecosystems help to keep them in balance

Spotlight On Key Skills

In this chapter, you will

- model the water cycle
- demonstrate the greenhouse effect
- demonstrate how limiting factors work
- learn various ways to reduce, re-use, and recycle products
- design your own experiment to show that humans give off carbon dioxide

Starting Point ACTIVITY

Protecting Your Planet

Are you familiar with the word "fragile"? It refers to something that is easily broken or damaged. An eggshell is a fragile object. Can you protect your eggshell from any harm for a whole day?

What You Need

egg
straight pin
small bowl
coloured markers
globe of Earth (optional)

Safety Precaution 

Be careful when using sharp objects such as a straight pin.

What to Do

1. Write your name on the eggshell using one of the markers. You may want to decorate your egg to look like Earth.
2. Carefully poke holes in each end of the egg using the straight pin. Blow the contents of the egg out one end into a bowl.
3. This is your fragile "planet Earth". Your job is to carry it with you for one day, and to try to keep it safe. Watch out for the actions of other students in your school. They may not care about your "Earth" as much as you do.

What Did You Find Out?

1. Were you able to keep your "Earth" safe all day? Why or why not?
2. What kinds of dangers to your "Earth" did you expect to encounter? What were the unexpected dangers?
3. In what way is our Earth fragile?
4. What steps do you already know you can take to help protect Earth? Start a list in your notebook, and add to it as you study this chapter.

3.1 Cycles in the Biosphere

Think about what balance means to you. Does it mean equality? Does it mean an even distribution of weight, or stability, or counteracting something? The idea of balance was mentioned in the introduction to this chapter. Balance is an important idea to keep in mind as you begin to explore the cycles that constantly occur throughout the biosphere.

Have you ever been given a book that someone finished reading, and enjoyed reading it yourself? Have you ever given a toy that you had outgrown to a younger child who could enjoy playing with it? If so, you were **recycling** — using the same item over and over. The biosphere is excellent at recycling.

The Water Cycle

As Figure 3.1 shows, an apple is 84 percent water, a carrot is 88 percent water, and a tomato is 94 percent water. The human body is 60 to 70 percent water. All living things require water. Water is used for life processes such as supplying food throughout an organism's body in a form it can use in its cells, and carrying away wastes from those cells.

Figure 3.1 The percentages of water in some living things

DidYouKnow?

Does it surprise you to know that the amount of water in the biosphere always stays more or less the same? The same water particles that were present hundreds of years ago are still around today. Hundreds of years ago, your great-great-great-great-great-grand-parents drank water. Perhaps you drank some of the same water particles today!

As you may have learned in earlier studies, the **water cycle** is the continuous movement of water through the biosphere. There are four main processes in the water cycle:

1. **Evaporation** is the process of changing a liquid into a vapour (see Figure 3.2). Liquid water evaporates to form invisible water vapour. For example, when laundry is hung in the sunlight, the water evaporates until the laundry is dry. As well, solid water (ice) can change directly into water vapour without going through the liquid stage. This process is called **sublimation.** For example, have you ever noticed that ice cubes shrink in size if the ice-cube tray is left in the freezer for too long? The solid ice cube shrinks as it sublimates into invisible water vapour.

Figure 3.2 Water evaporates from streams and other bodies of water. It forms invisible water vapour.

2. **Transpiration** is the process in which water that is taken in through a plant's roots evaporates from the plant's leaves, stem, and flowers.

3. **Condensation** is the process of changing a vapour to a liquid. Warm air contains water vapour. As air cools, however, it is able to hold less and less water. Condensation happens when air becomes so cool that it can no longer hold as much water vapour, and liquid water is released. The liquid forms clouds, fog, or dew. For example, water droplets form on the outside of a cold glass of juice. As the glass cools from contact with the juice, the air surrounding the glass cools. The cooled air can no longer hold as much water vapour, so the water is released as droplets on the outside of the glass.

4. **Precipitation** is the process in which liquid water forms from condensation occurring inside clouds, and then falls as rain, sleet, snow, and hail.

Figure 3.3 illustrates the water cycle. The first two processes — evaporation and transpiration — move water up from Earth into the atmosphere. The second two — condensation and precipitation — return water to Earth. **Ground water** is water in the soil. Plant roots can grow down to reach ground water. People can reach ground water by digging wells. **Run-off** is water that runs off the ground into lakes, rivers, or streams.

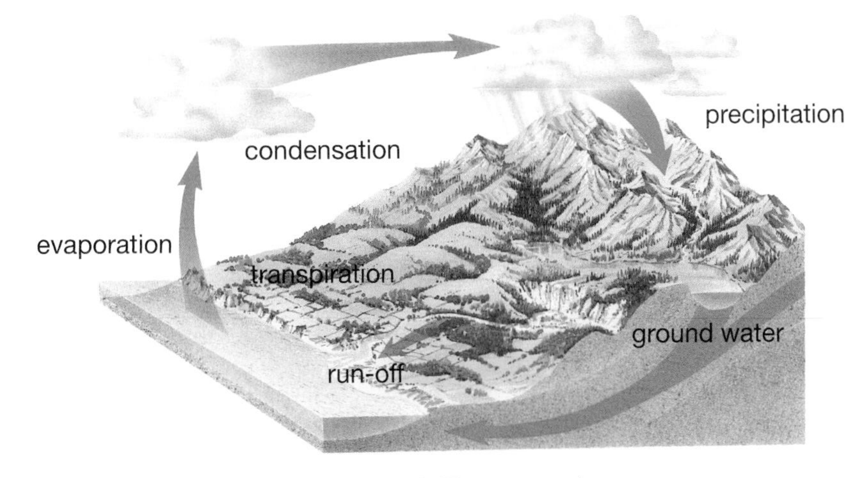

Figure 3.3 The water cycle

INVESTIGATION 3-A

Water, Water Everywhere!

The water cycle is a very important process, and it occurs constantly on Earth. Modelling the water cycle will help you to understand how it works and how important it is.

Challenge

Water can exist as a liquid, a gas, or a solid. Design a model to show water changing from a liquid to a gas and then back again. Then, change the water to a solid and back again to a liquid.

Safety Precautions

- Be careful — you will be working with hot water and steam.
- Wear your safety glasses, apron, and gloves.

Materials

everyday items of your choice: for example, electric kettle, large shallow bowl, oven mitts or gloves, ice, hot plate, modelling clay, sand, soil, water, small bowl, refrigerator

Design Criteria

A. Your model must demonstrate how water can exist as a liquid, as a gas, and as a solid.

B. Your model must demonstrate how water can change from a liquid to a gas, from a gas to a liquid, from a liquid to a solid, and from a solid to a liquid.

C. Your model does not need to be all in one location.

Plan and Construct 

1. With your group, plan how you will cause the changes in the state of water to occur.
2. Draw a labelled sketch of your model, indicating what materials you will use.
3. Obtain your teacher's approval, and then build your model.
4. Demonstrate your model.
5. Wash your hands after completing this project.

Evaluate

1. **(a)** Did your model work as expected?
   (b) What adjustments were needed in order to make it work, or work better?
   (c) What scientific knowledge helped you to develop your model?
2. How did the models constructed by the other groups work? Did the other groups have ideas that you would like to use? Did your group have ideas that others wanted to use?
3. If you could carry out this investigation again, what would you do differently?
4. What part did temperature play in this investigation? What part do you think it plays in the water cycle?

Skill POWER

To learn more about how to use models in science, turn to page 493.

Skill POWER

For more information on the safety symbols used in this textbook, turn to page 492.

The Carbon Dioxide and Oxygen Cycle

In the late eighteenth century, scientists discovered that a mouse could not survive in a closed container in which a candle had been burned. However, when a plant was left to grow in the container for eight or more days, a mouse could survive.

We now know that living things, including you, need oxygen to survive (see Figure 3.4). The burning candle used up the oxygen in the container, so the oxygen was not available for the mouse to breathe. Why do you think the plant made it possible for the mouse to survive in the container? Another major cycle constantly occurs on Earth, and it will help you answer this question. The **carbon dioxide and oxygen cycle** is the process by which carbon dioxide and oxygen are cycled and recycled in the biosphere.

When you breathe, you — like the mouse — take in oxygen. Your body cells use this oxygen, in a process called respiration, to release energy from food. **Respiration** is the oxygen-using process, shown in Figure 3.5, that takes place in the cells of living things to get the energy out of food. All living things — both animals and plants — must respire all the time. Fish, algae, and plankton in the water respire. So do the decomposers — bacteria, fungi, yeasts, and moulds — that you explored in Chapter 2.

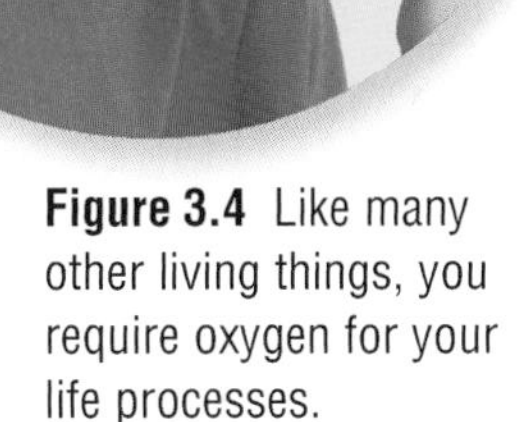

Figure 3.4 Like many other living things, you require oxygen for your life processes.

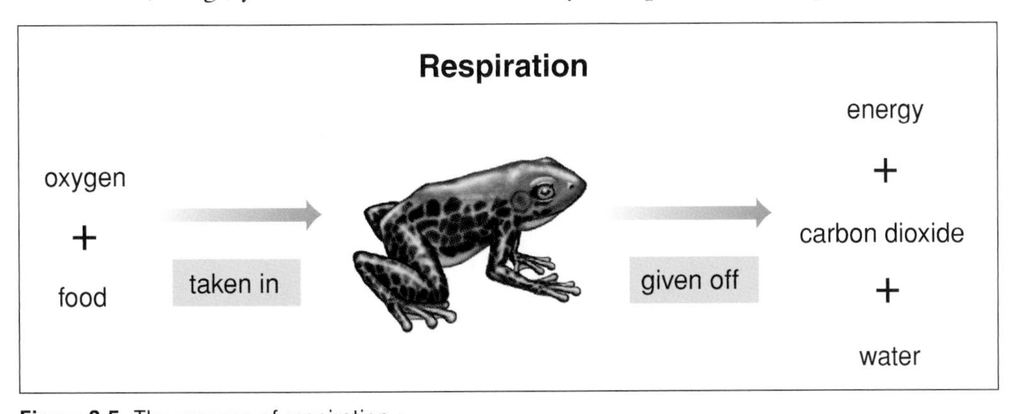

Figure 3.5 The process of respiration

The process of respiration is much like the process of burning a candle. Wax is the fuel for the burning candle shown in Figure 3.6. What do both burning and respiration take out of the air? Do Figures 3.5 and 3.6 help to explain what happened to the mouse when it was placed in a container after a candle was burned?

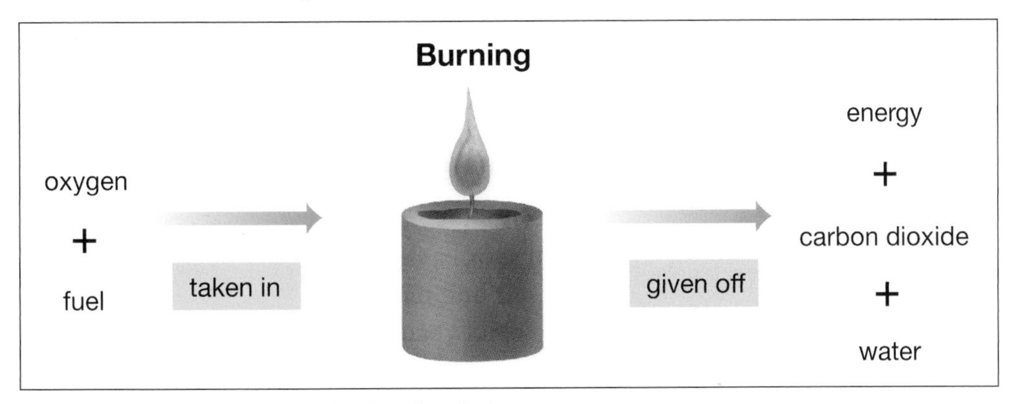

Figure 3.6 The process of combustion (burning)

Pause& Reflect

Think about what you have learned about the water cycle and the carbon dioxide and oxygen cycle. Summarize what you know about these cycles by drawing a diagram or making a flowchart in your Science Log.

In addition to energy, respiration has two other products: carbon dioxide and water (refer back to Figure 3.5). Carbon dioxide is a gas made up of carbon and oxygen. Plants need carbon dioxide, along with other substances, to make their own food in a process called **photosynthesis**. In some ways, photosynthesis (shown in Figure 3.7) is the reverse of respiration because the plant takes in carbon dioxide and releases oxygen. Does Figure 3.7 help explain what happened to the mouse in the container after a growing plant had been left in the container for eight or more days?

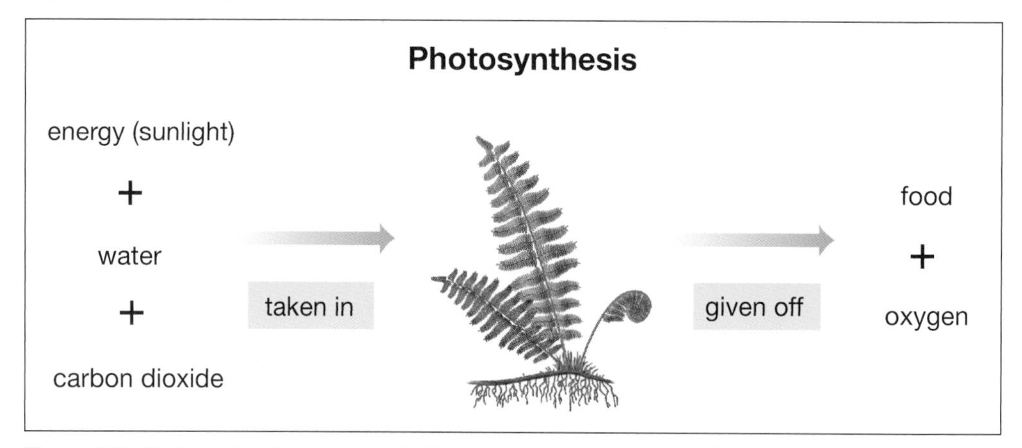

Figure 3.7 Photosynthesis occurs in plants, algae, and some bacteria that contain green chlorophyll.

Imagine that you are preparing a labelled diagram as part of a project. You really need a green pencil, but you do not have one. A classmate has a green pencil that she is not using. She needs a brown pencil, which you have and do not need. You and your classmate exchange pencils.

Plants and animals have a similar useful relationship. Plants give off the oxygen that animals require, and animals give off the carbon dioxide that plants require. Figure 3.8 shows the entire carbon dioxide and oxygen cycle. In nature respiration and photosynthesis are more or less balanced. When the Sun is shining, plants can use the carbon dioxide about as quickly as it is produced.

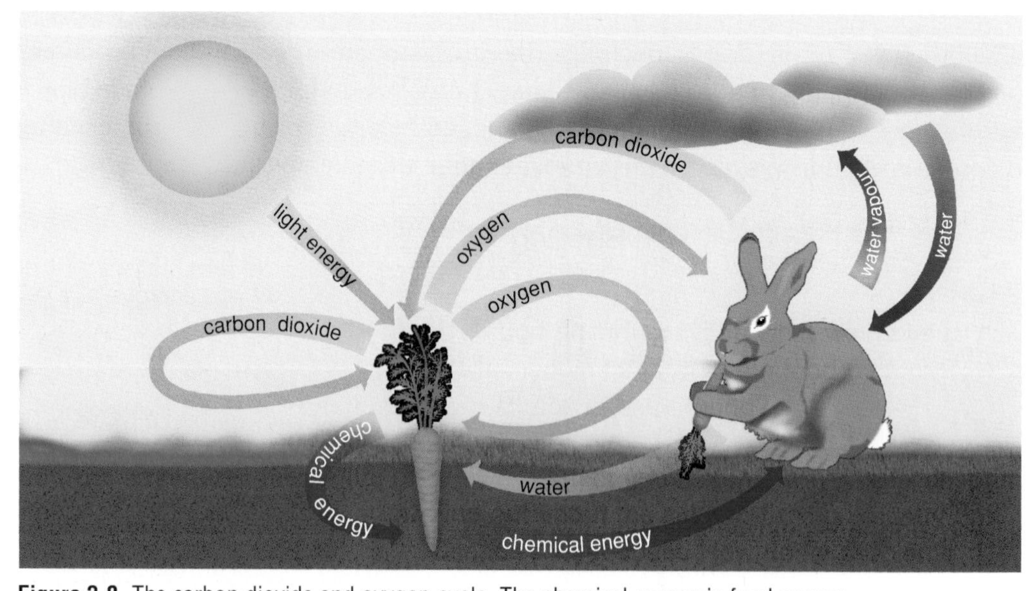

Figure 3.8 The carbon dioxide and oxygen cycle. The chemical energy is food energy.

INVESTIGATION 3-B

Tell-Tale Snails

Bromthymol blue is a liquid that changes colour from blue to green to yellow when carbon dioxide levels increase. How could you use it to demonstrate that snails give off carbon dioxide?

Challenge

Show that snails give off carbon dioxide.

Safety Precautions

- Wear an apron, safety glasses, and gloves.
- Handle chemicals with care. Bromthymol blue may stain clothing.
- Follow your teacher's directions for disposing of materials safely.
- Always handle live creatures with care and respect, and return them to their habitat after you have studied them.

Materials

2 Erlenmeyer flasks
2 small, clear cups or similar containers
modelling clay
flexible drinking straws
bromthymol blue indicator
distilled water
small, live aquarium snails

Design Criteria

A. Your demonstration must cause a change in the colour of the bromthymol blue.

B. You must be able to show that the change could have been caused only by the presence of the snails.

Plan and Construct

1. With your group, think about the materials, and decide how you might use them.
2. Decide how you can demonstrate that any observed carbon dioxide was given off by the snails.
3. Draw a labelled diagram of your set-up for the demonstration. On your diagram, indicate how long you think your set-up should be in place.
4. Show your diagram to your teacher for approval, and then set up your demonstration.
5. Record your observations.

Analyze

1. Did the snails produce carbon dioxide? Explain how you know.

Evaluate

2. Was it clear from your observations that the snails gave off carbon dioxide? If not, how could you make it clear if you repeated the demonstration?

3. Did any other group come up with a better way to carry out the demonstration? What did other groups do that was different from what your group did?

4. Predict what might happen if you put water plants in with the snails. With your teacher's permission, try it.

Extend Your Skills

5. Design Your Own Design an experiment to show that humans give off carbon dioxide in their breath. Make sure your experiment is a fair test, and have your teacher approve your procedure before you carry out your experiment.

Going in Cycles

The water cycle and the oxygen and carbon dioxide cycle are never-ending processes in our environment. Without them, life on Earth could not exist. These cycles may become unbalanced or even cease to operate as a result of human activity. Can we use technology to restore the balance of these crucial cycles instead of damaging them? In the next section, you will explore the effect of human activity on nature's cycles.

Check Your Understanding

1. Why is water important to all living things?
2. Water is not always a liquid. What other forms does it take? How does it change during the water cycle?
3. "Plants and animals have a useful relationship." Explain this statement in terms of the carbon dioxide and oxygen cycle.
4. In what way is respiration like burning?
5. Write the letters (a) through (e) in your notebook. Use the clues in parentheses ( ) to solve the x puzzle. Write the scientific terms, replacing each x with the letter that fits.
   - (a) xvxxoxation (the process in which a liquid changes into a vapour)
   - (b) xxxxxpixation (something that plants do; it involves evaporation)
   - (c) cxxxxxxation (the process in which clouds are formed in the water cycle)
   - (d) xxxcxxxxation (water that falls to Earth)
   - (e) rexxxxation (something we need to do all the time to get energy from food; it happens in our cells)
6. Choose three or more key terms from this section. Make up an x puzzle, like the one in question 5, and see if a classmate can solve it.
7. **Apply** Imagine that you are a water particle. Trace a possible path that you might take as you move through one complete water cycle. Begin the cycle as a snowflake falling in the winter.

3.2 Redirecting Water Flow

Humans often use technologies to change ecosystems in order to meet their own needs. For example, the people in a community might benefit from damming a nearby river. What might be the positive effects for the community? Figures 3.9 and 3.10 show the effects of damming a river.

Figure 3.9 These diagrams show a river before and after it is dammed.

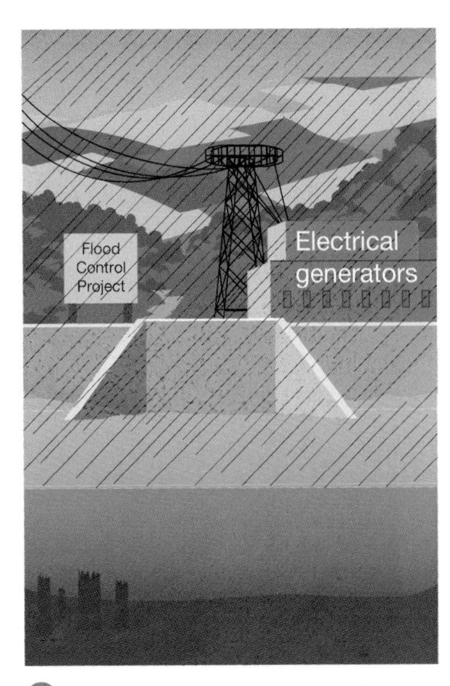

A The forested hills around the dam site were bulldozed to make room for farming. Birds and large animals were forced to flee. Insects and small animals fled or were destroyed.

B Machines destroyed some plants in the flooded area. Other plants "drowned" in the water.

C Wind and water eroded the soil on the newly bare hills around the reservoir. The soil slowly washed down into the reservoir, making it muddy.

Figure 3.10 Long-term effects of a dam

DidYouKnow?

Schistosoma is an internal parasite that afflicts 300 million people worldwide. The eggs of this parasite mature inside snails, hatch, and then burrow into humans. The occurrence of this very painful disease has increased in tropical countries because irrigation and damming have created more habitats for the snails.

How might a dam affect an area's ecosystems? Redirecting water flow can have major effects not only on a moving fresh-water ecosystem and the ecosystems immediately surrounding it, but also on ecosystems that are farther away. For example, when birds, animals, and other organisms are forced to leave an area because a dam is built, they must try to survive elsewhere. When the animals leave, they move to an ecosystem where food and other resources may be scarce. They must compete with the birds, animals, and other organisms that are already living there.

When areas that were previously covered by trees are cleared, the soil can be washed away by run-off. This process of moving soil and rock from one place to another is called **erosion**. Wind and water both cause erosion (you will learn more about these processes in Unit 4). If the reservoir that is formed by damming a river is large, an area downwind could receive more rain, snow, and other precipitation. This occurs because there is a larger accumulation of water to evaporate, condense, and fall.

Find Out ACTIVITY

Save the Soil

If erosion results in the loss of some topsoil, how much does that matter? In this activity, you will use an apple to represent Earth.

What You Need

cutting board
knife
apple

Safety Precautions

The knife is sharp and may cause cuts if not handled properly.

What to Do

1. Carefully cut the apple into four quarters.
2. Set aside three quarters. Cut the fourth quarter lengthwise into two equal pieces (each piece is one eighth of the apple.)
3. Set aside a one-eighth piece. Carefully cut the other one-eighth piece into four equal pieces. (Each piece is one thirty-second of the apple.)
4. Set aside three thirty-second pieces. Carefully peel the skin from the last thirty-second piece.
5. Each piece of apple represents the following:
   - The three quarters represent the part of Earth covered by oceans.
   - The one-eighth piece represents land that is unsuitable for human life.
   - The three thirty-second pieces represent land that is unsuitable for food crops.
   - The peel from the last piece represents the layer of topsoil that can support life.

What Did You Find Out?

1. Suppose a crop of food that could only grow in fairly limited regions was affected by loss of topsoil. What might be the effect on this food supply? What might be the effect on other species?
2. Aside from the loss of food crops, what other long-term effects can be caused by loss of topsoil?

Fossil Fuels, Carbon, and Air

Canadians use plants in agriculture, forestry, and other plant-based economic activities (see Figure 3.11). We use plant products when we burn coal to make electricity, and we use natural gas or oil to heat our homes. We also use plant products when we drive cars, trucks, or farm machines, because gasoline comes from oil. Coal, oil (petroleum), and natural gas are fossil fuels. **Fossil fuels** are fuels that originated from plants and other organisms that died and decomposed millions of years ago and were preserved deep under the ground (see Figure 3.12).

Fossil fuels are products of plankton that lived millions of years ago. **Plankton** are microscopic plants, algae, and other organisms that float in seas and other bodies of water. **Phytoplankton** are plankton, such as algae, that use sunlight to make their own food through photosynthesis. Millions of years ago, decomposing plankton were repeatedly buried under layers of mud and silt that, with time and pressure, became rock. You will explore this topic in more detail in Unit 4.

Figure 3.11 In what two ways is the farmer in the photograph using plants?

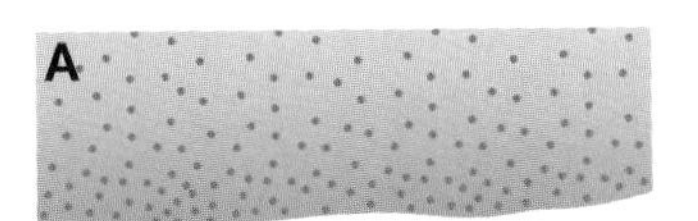

Long ago plankton in the seas trapped energy from sunlight.

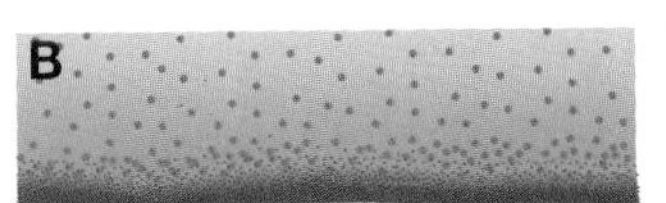

Over time, layers of thick ooze formed on the bottom of the sea from the dead cells of the plankton.

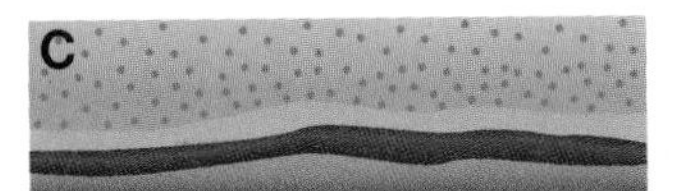

The ooze became trapped by layers of mud and silt. The sea and these layers put pressure on the ooze.

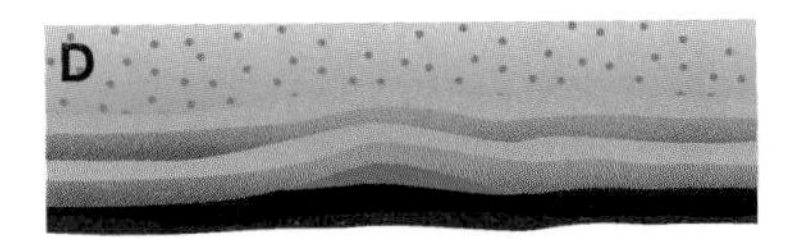

In time, fatty substances in the ooze changed into natural gas and petroleum oils.

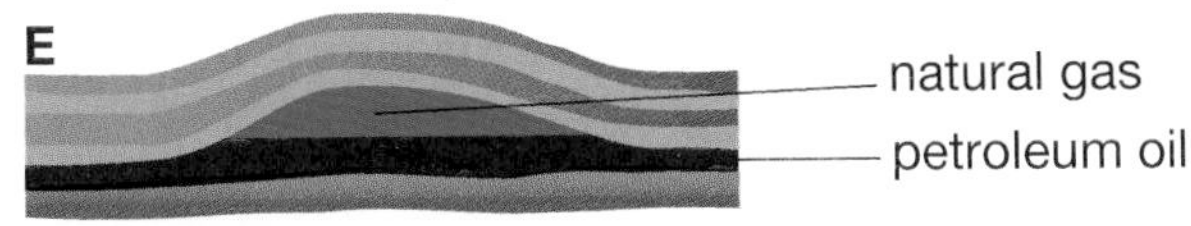

The sea drained away. The materials were trapped underground.

Figure 3.12 Scientists hypothesize that oil and natural gas are formed as shown in this diagram.

The Carbon Cycle

Carbon is necessary for all life to exist. Plants use carbon dioxide from the air for photosynthesis, and moose, mice, and other organisms eat the plants. They respire, releasing carbon dioxide into the air. Wolves, foxes, and other organisms eat the moose and mice, and they also respire, releasing carbon dioxide into the air. In this way, carbon cycles around and around in an ecosystem. Figure 3.13, on the next page, shows the carbon cycle.

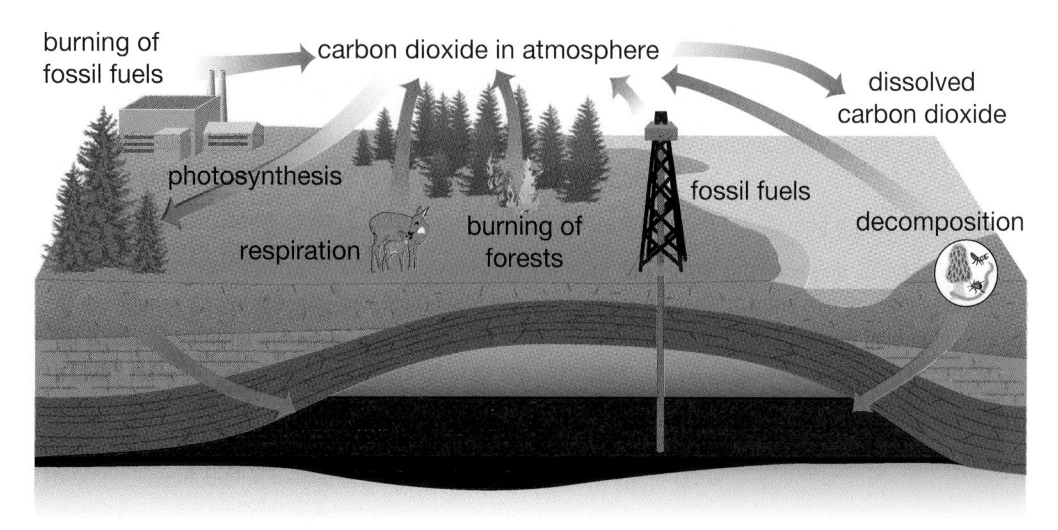

Figure 3.13 This diagram shows the carbon cycle. The oil derrick is pumping oil from deep under the ground.

Figure 3.14 shows the carbon cycle in balance. Many things might upset this balance. Fossil fuels burn, and this releases carbon dioxide into the air. The amount of carbon dioxide released into the atmosphere has greatly increased over the past 100 years. This is mostly because people are burning more and more fossil fuels. As well, people have produced large amounts of carbon dioxide by the widespread burning of tropical rain forests to clear areas for farming. The destruction of rain forests also reduces the number of photosynthesizing organisms, making the cycle more unbalanced.

Figure 3.14 As long as there is not more carbon than plants can use, the carbon cycle works well and the system stays in balance.

Pause & Reflect

As you have learned, fossil fuels are non-renewable resources. If the carbon cycle is really a cycle, like the water cycle or the oxygen and carbon dioxide cycle, why is there concern about using up all of the fossil fuels? Write your answers in your Science Log.

Greenhouse Gases

Imagine stepping into a greenhouse on a sunny but cool day. The air inside the greenhouse is much warmer than outside because most parts of the Sun's rays pass through the glass walls of the greenhouse. These rays are taken in by the soil and plants. Energy released by the soil (called heat radiation) travels back up to the top of the greenhouse, but this type of energy cannot easily pass through the glass. The trapped energy warms the air in the greenhouse. (You will find out more about radiation in Chapter 9.)

The atmosphere surrounding Earth acts like the glass walls of a greenhouse. It allows sunlight to pass through to Earth's surface, and prevents much of the returning heat radiation from passing back out into space. This is one reason why Earth is warm enough for life to exist. Without the atmosphere, Earth would be too cold to support most species.

The glass of a greenhouse prevents some of the heat radiation from escaping, so the air in the greenhouse warms up. If another covering were added to prevent the loss of more heat radiation, the air in the greenhouse would get even warmer. This same effect is occurring in our atmosphere (see Figure 3.15). **Greenhouse gases** are gases, such as carbon dioxide, that result from the burning of fossil fuels as well as other fuels, such as wood. These gases add to the heat-trapping effect of the atmosphere, forming a "covering." This "covering" causes the atmosphere to hold in more heat radiation than it would naturally. Many scientists think that the result over time is **global warming**: a gradual warming of Earth's atmosphere. Scientists still disagree about many aspects of global warming, including how much Earth's atmosphere will warm.

Skill POWER

For tips on using the Internet effectively, see page 497.

INTERNET CONNECT

www.school.mcgrawhill.ca/resources/

Some scientists consider greenhouse gases to be a real threat to life on Earth; others maintain that the current warming trend is simply part of a much larger pattern. Record-keeping of Earth's temperature is fairly recent and does not allow scientists to observe long-term trends very readily. Do some research on this issue by going to the web site above. Go to **Science Resources**, then to **SCIENCEPOWER 7** to learn where to go next.

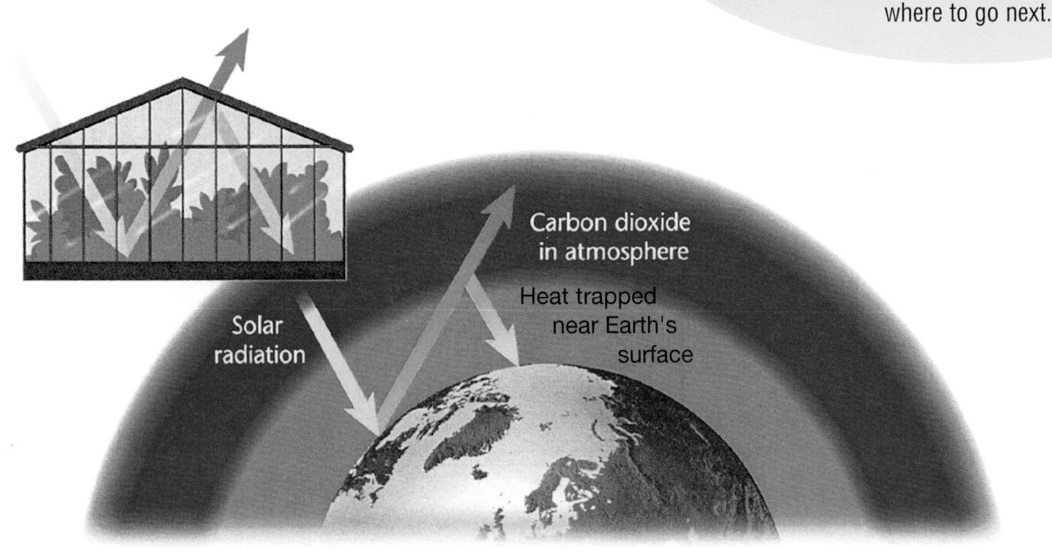

Figure 3.15 This is a model of how many scientists believe the greenhouse effect takes place.

Check Your Understanding

1. Why do people sometimes redirect rivers or other waterways? Is it acceptable for humans to change natural ecosystems like this? In what situations might it be all right? In what situations might it not? Discuss your opinions with a group or the class.

2. Deer live in a river valley where a dam is going to be built. What might result from each of the following situations?
   (a) When water floods the valley, the deer move to nearby sand dunes. There they eat the few green plants that are struggling to become established in the windblown sand.
   (b) When water floods the valley, the deer move to nearby pastures. There they compete with farmers' cattle and horses for grasses and other food plants.
   (c) Before water floods the valley, the deer are rounded up and moved to a national park where there is lots of food for them. The park already has a population of deer, and the new deer cause an overpopulation.

   Which situation do you think is best in terms of preserving ecosystems? Which might cost more money in the short term? Which might cost more money in the long term?

3. How do you use fossil fuels? Why are you using plants when you do so?

4. What is the carbon cycle? What role do each of the following play in it?
   (a) fossil fuels burning
   (b) forests burning
   (c) animals respiring
   (d) plants carrying out the process of photosynthesis

5. The carbon cycle is balanced in stable ecosystems. What events might occur to upset this balance?

6. **Apply** Why might a gardener cover an area of the garden with plastic during a cool spell in spring?

7. **Thinking Critically** What is global warming? List at least four ways that it might affect your life in the future. List two ways that people can work towards reducing global warming.

INTERNET CONNECT

www.school.mcgrawhill.ca/resources/

What role do the rain forests play in global warming? What part is played by algae in the world's oceans? Find information about these topics by going to the web site above. Go to **Science Resources**, then to **SCIENCEPOWER 7** to find out where to go next. Make a poster to illustrate your findings.

3.3 Ecosystems Regulate Themselves

Have you ever blown the parachute-like seeds off a dandelion seed head or scattered seeds from a milkweed pod? If so, you have some idea of how many seeds one plant can produce and how easily they can spread. Why, then, do these plants not take over a large part of Earth?

Figure 3.16 What kinds of factors limit the spread of dandelions?

Limiting Factors

Like dandelions and milkweed, many species have an amazing potential to spread. Rabbits, for example, are very efficient at reproducing. They do not spread over large areas, however, because of limiting factors that regulate population size. A **limiting factor** is an abiotic or biotic factor that restricts the number of individuals in a population.

There are many different types of limiting factors. The following are some of the many natural (not caused by humans) limiting factors in ecosystems. All of these factors control population growth. There may be more than one limiting factor at work in an ecosystem.

Pause& Reflect

As a group project, design a board game about limiting factors and their effects on a population of Canada lynx and a population of snowshoe hares. Write the rules of the game in your Science Log. With your teacher's permission, make and play the game.

Predator-Prey Populations

Remember the hare-lynx cycle at the beginning of Chapter 2. Interactions between predators and their prey control the population size of both predators and prey. When prey populations increase, there is more food available for predators, so the predator population increases. More predators means that more prey are killed and the prey population stops getting bigger. As the prey population stops growing, there is less food for the predators and this population stops growing, as well. Predator-prey populations often increase and decrease in cycles.

Competition for Resources

Only a certain amount of food and living space are available in any ecosystem. For example, brook trout, perch, and sunfish compete for food and living space, as shown in Figure 3.17. This means that these species cannot multiply as they would with unlimited resources.

Figure 3.17 This brook trout is defending its home territory.

Diseases and Parasites

Diseases and parasites are common population controls. Foxes, for example, can get rabies, which can seriously reduce their population in an ecosystem. What might be the effect of a reduced fox population on the mice that foxes catch and eat?

Diseases affect not only natural ecosystems, but also areas of agriculture and forestry. A number of plant diseases are caused by micro-organisms known as fungi. (You were introduced to fungi in Chapter 2.) The fungus called wheat stem rust, for example, has severely reduced the Canadian wheat crop in some growing seasons. Corn smut damages crops and lowers corn yields (see Figure 3.18). White pine blister rust has nearly eliminated the eastern white pine.

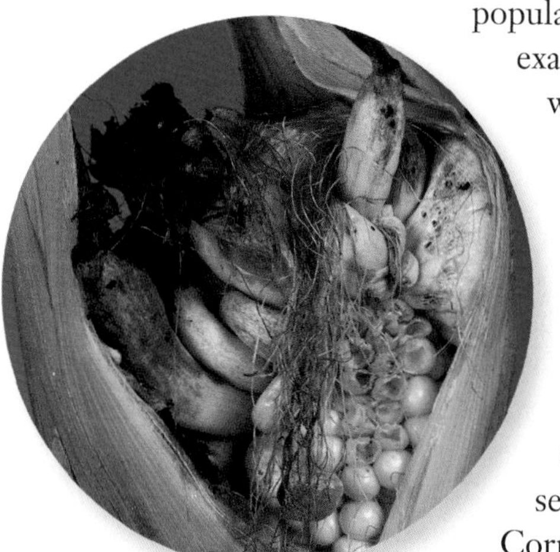

Figure 3.18 Corn smut can destroy much of a corn crop.

Climate Changes and Weather Patterns

Climate changes can result from natural events, such as volcanic eruptions and changing ocean temperatures. Suppose that an unusual amount of snow falls in a northern ecosystem where caribou live. The thick snow cover makes it difficult for the caribou to find the grasses and other plants on which they feed. The lack of food causes some caribou to die. The result is a reduced caribou population in this ecosystem. What might happen to the population of wolves that feeds on the caribou?

Climate changes affect not only natural ecosystems but also economic activities, such as agriculture and forestry. Unusually early winters, for example, can shorten the growing season so that crops such as sunflowers, canola, and wheat do not have time to ripen. Unusual buildups of ice can break tree branches. As Figure 3.19 suggests, a drought (lack of rainfall) can have a major impact on dairy farming.

Did You Know?

Natural climate changes can be major and long term. For example, about 18 000 years ago, large parts of Earth, including what is now Canada, were covered with a thick layer of ice. The climate was much colder than today's climate. By about 8000 years ago, the climate was warmer than our climate today, and major global warming had melted the ice all over Canada and high into the Arctic.

Figure 3.19 How might a severe drought affect the plants that these dairy cattle eat?

Forest Fires

Lightning strikes have caused forest fires ever since forests existed. Despite what you might expect, deer, bears, blue jays, and many other animals and birds survive a forest fire since they are quick enough to escape. In fact, some species need occasional forest fires to survive. For example, the Kirtland's warbler, an endangered black and yellow bird, builds its nest in stands of young jack pines. Fire burns away older jack pines and stimulates growth of young trees.

Many bushes, trees, and other plants are burned and destroyed in a forest fire as shown in Figure 3.20. Forest fires naturally occur in some ecosystems. In most cases, burned forest areas quickly renew themselves. In Chapter 2, you were introduced to succession, which is a gradual change in the structure of a community over time. **Primary succession** is the gradual growth of organisms in an area that was previously bare, such as rock. **Secondary succession** is the gradual growth of organisms in an area that previously had a number of organisms. The regeneration of a burned forest area is an example of secondary succession.

DidYouKnow?

The cones from jack pine and black spruce trees lie closed for years on the forest floor, until the heat of a fire pops them open, releasing the seeds that will grow into new trees. When the branches of these trees are burned, their root systems send out new shoots, which will grow into new trees.

Figure 3.20 Forest fires often burn off the foliage but leave the tree trunks standing.

Figure 3.21 These wildflowers are thriving in soil that has been burned by fire.

In the secondary succession of burned forest, wildflowers and other plants that grow best in strong sunlight are among the first to spring up (see Figure 3.21). Blueberry bushes also thrive since they are adapted to grow better in soil containing ash. Like blueberry bushes, a number of tree species actually grow better in soil that has been burned.

The more diverse an ecosystem, the better able it is to regenerate. **Diversity** is a measure of the number of different species living in an ecosystem. The greater the number of species, the more diverse the area is. Forests, for example, have burrowing animals that dig homes in the ground. In the process, the burrowers bring to the surface soil that mixes with nutrient-rich ash from the fire. When seeds of wildflowers and other species blow in, they can take root. Among the burrowing species that are found in forests are moles, eastern chipmunks, and wolverines. If a forest ecosystem had only one burrowing species, and the whole population died of a disease, forest regeneration would be slower. This is because other burrowing species would not be present to turn over soil, and make it ready for new growth.

Find Out **ACTIVITY**

That's the Limit!

You can observe limiting factors at work if you carefully control some of the abiotic factors that plants require in order to grow.

What You Need

4 small seedlings (about the same size)
water
masking tape
marker
cardboard container

What to Do

1. Make "DO NOT WATER" labels for two of the plants, using the marker and masking tape.
2. Put the masking tape labels on two of the plants. Place one of these plants on a windowsill. Place the other plant in a dark place, such as in a closet or under an overturned cardboard box.
3. Place one of the unlabelled plants on the windowsill. Place the other unlabelled plant beside the labelled plant in the closet or under the cardboard box.
4. Leave the plants for several days, remembering to add water to the two unlabelled plants (do not over-water). Write a hypothesis about what you think will happen to each plant and why you think it will happen.

What Did You Find Out?

1. What happened to each of the four plants? Did your observations agree with your hypothesis?
2. What are some of the abiotic factors that plants require in order to live? How do you know?
3. Why was it necessary to place a watered and an unwatered plant in each location?

Check Your Understanding

1. In China, a certain species of wasp preys on a certain species of destructive tree beetle. Some of these tree beetles have come into Canada along with wood-based imports from China. The beetles could cause a lot of damage to Canadian trees, especially since their natural enemy — a species of wasp — does not live in Canada. Explain how this situation relates to the predator-prey method by which populations regulate themselves.
2. Identify and explain at least two methods by which ecosystems regulate themselves.
3. For each of the following, explain the difference between the two terms.
   (a) primary succession and secondary succession
   (b) natural forest fires and forest fires that result from human carelessness
4. Explain how the extinction of various species results in a less diverse ecosystem. How can this lead to the extinction of even more species?
5. **Thinking Critically** If the beetles mentioned in question 1 were to multiply in Canada, how would you try to control them? Why would you have to be extremely careful?

3.4 Technology and Nature's Regulators

Human technologies, such as bicycles, telephones, computers, tractors, and medicines, are important in our lives. Technologies can, however, affect nature's limiting factors and change the balance in ecosystems.

Forestry and Agriculture

The technologies associated with forestry and agriculture affect wildlife habitats. For example, the brown-headed cowbird, shown in Figure 3.22, is a bird that lays its eggs in the nests of other species of birds, such as warblers and vireos. When the cowbird chicks hatch, they loudly demand food. Cowbird chicks also grow quickly, often squeezing the smaller warbler and vireo chicks out of the nest. As a result, the warblers and vireos do not reproduce as successfully as they would without the cowbirds.

Figure 3.22 Brown-headed cowbirds have spread across Canada.

The natural habitat of brown-headed cowbirds is the grassland biome of the Canadian prairies. Since brown-headed cowbirds are adapted to living in open country, they quickly spread into newly cleared farmland. A lot of forest east of the prairie grassland, across central Canada, and in the Atlantic provinces has been cleared to prepare the soil for growing oats, barley, potatoes, beans, and other crops. Brown-headed cowbirds have been quick to take advantage of this new habitat and are now found across Canada.

Pesticides and Pests

The spruce budworm is a moth larva that eats spruce and fir needles, killing the trees. A pesticide was developed that was effective against spruce budworms, but when New Brunswick sprayed its forests with this pesticide, bees were also killed. As a result, there were not enough bees to pollinate blueberry plants. This severely reduced the production of blueberries, which are an important crop in the province.

Medicines and Micro-Organisms

Your family doctor has probably prescribed an antibiotic for you when you have a sore throat or an ear infection. Penicillin and other antibiotics are technologies developed to treat medical problems (see Figure 3.23). Unfortunately, many species of disease-causing micro-organisms have become resistant to antibiotics. This means that the more resistant individuals in a micro-organism population survive an antibiotic attack, and continue to multiply. The next time the same antibiotic is used, it is not as effective in killing the micro-organism population. Researchers constantly search for new antibiotics to combat disease-causing species of micro-organisms that have become resistant. In time, these new antibiotics may also become less effective if micro-organisms develop resistance to them.

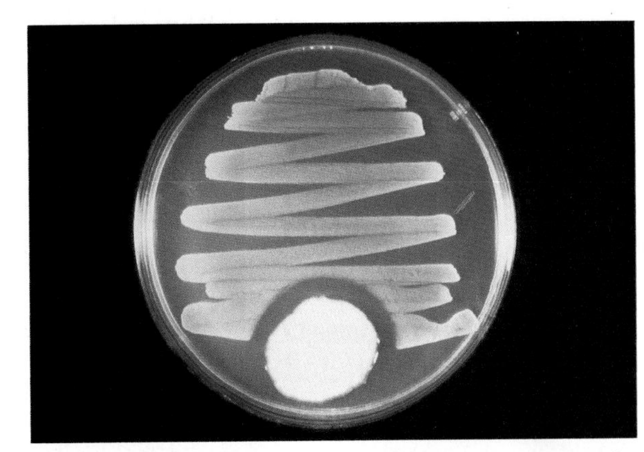

Figure 3.23 No bacteria can grow around the round patch of penicillin in this petri dish. Early researchers realized that penicillin might have the same effect on disease-causing bacteria, leading to the extensive use of antibiotics today.

Helping the Disease-Fighters

How can you help medical technology fight disease-causing micro-organisms?

What to Do

1. Do not take antibiotics unless you really need them.
2. Take antibiotics only when a doctor prescribes them.
3. Do not take antibiotics that have been prescribed for someone else.
4. Follow the directions on the container. Even if you feel better after taking only half of the antibiotics, finish taking the prescription. Why is this very important?
5. Share what you have learned about antibiotics with your family and friends.

Not all technologies affect the balance of an ecosystem in a negative way. Sometimes ecosystems are out of balance and technologies can help bring them into balance. In the 1930s, for example, rabbits were introduced into Australia and New Zealand to provide a convenient source of meat. As mentioned at the beginning of this section, rabbits are highly efficient at reproducing, and the populations soon got out of control. The rabbits did enormous damage to agricultural crops in these countries, and there seemed to be no way to control the population growth. Finally scientists introduced a virus, called myxomatosis, that affected only rabbits and not any other organisms. Most of the rabbits soon died, but some rabbits were resistant to the virus. The virus still exists, and it affects rabbits that have no resistance. There will always be rabbits, and there will always be the virus to control them. A balance has been reached.

Pause& Reflect

You have seen that limiting factors can occur naturally or as a result of human technologies. Imagine you have to prepare an explanation of limiting factors for a class of younger students. Think of two examples you could use to show the difference between these two types of limiting factors. Write your ideas in your Science Log.

Check Your Understanding

1. How have farming and forestry affected brown-headed cowbird populations in Canada? Give your answer in the form of a numbered list of steps, a labelled diagram, or a flowchart.
2. List two other ways that farming and forestry have affected ecosystems in Canada?
3. Imagine that you are a researcher working on methods of controlling insect pests. Identify some concerns you might have relating to preserving ecosystems? As well, what are some concerns relating to economic activities, such as forestry and farming?
4. Imagine that you are a researcher working with antibiotics. You are concerned about how people use them. Do one of the following.
   (a) Make a brochure or poster about antibiotic use for the general public.
   (b) Make a brochure or poster about antibiotic use for doctors.

   You may wish to illustrate your brochure or poster with cartoons.

3.5 Making a Difference

There are many other greenhouse gases besides carbon dioxide, and one of these is methane. Methane is produced when garbage decomposes in a landfill. A **landfill** is a site where garbage is disposed of by being buried under a shallow layer of soil. In Canada, much of our garbage is taken to landfills or dumps. There, the garbage not only produces greenhouse gases that affect the atmosphere, but it can also contribute to water and soil pollution. Water from precipitation and run-off sometimes soaks through a landfill or dump, carrying polluting chemicals and other contaminants into the soil and ground water (see Figure 3.24).

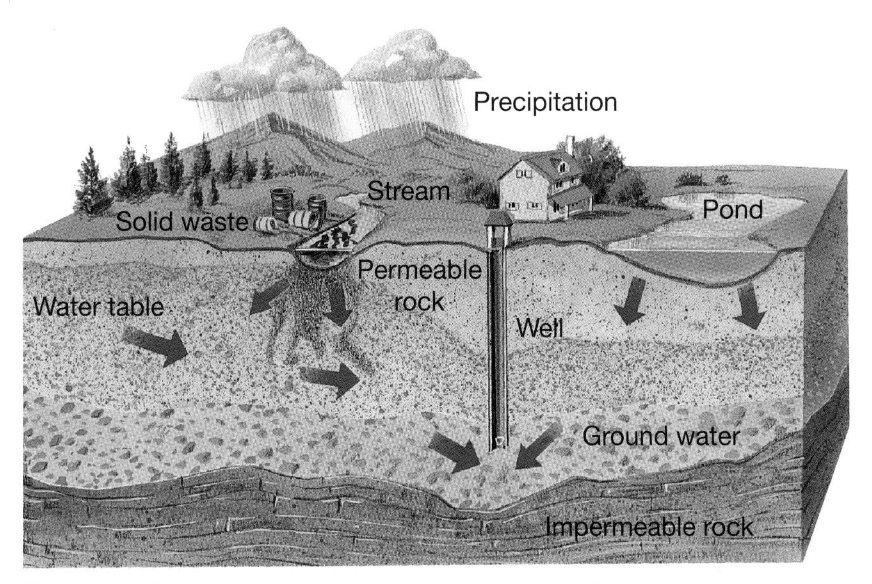

Figure 3.24 Solid waste from oil barrels in a dump is leaking through the ground into the ground water below.

DidYouKnow?

The disappearance of the ant population from Earth would be a major disaster. Ants turn the soil, eat other insects, are scavengers of dead animals, and pollinate plants.

Reduce, Re-Use, Recycle!

There are many ways in which you can help protect the environment. One key way is to reduce your use of fossil fuels. Figure 3.25 shows one way to do this. Another key way is to cut down on the amount of garbage you add to your community's waste disposal system. The 3 R's — reduce, re-use, and recycle — will help you.

Figure 3.25 This technology can help to reduce your use of fossil fuels.

1. Reduce the amount of garbage you produce. For example, try to buy products that have little or no packaging for you to throw away. Try to avoid buying individually wrapped items and small CD-ROMs in big boxes, so that manufacturers may be encouraged to use less packaging.
2. Re-use products rather than throwing them away. For example, buy products that you can use again and again, rather than disposable ones. Use plastic yogurt containers to store leftovers. Hand down used clothing to other family members, sell it to a consignment store, or give it to charity.
3. If you cannot re-use materials in their present form, look for ways to recycle them — turn them into something else. For example, compost kitchen scraps and yard wastes. Place newspapers, egg cartons, jars, cans, and other recyclable materials into recycling boxes, and set the boxes at the curb for pickup. Whenever you have a choice, buy products made of recycled materials.

DidYouKnow?

Technology has caused waste management problems, but it can also help to solve them. For example, one creative company has recently found a way to turn plastics back into the material they were made from — refined oil. Researchers have also developed a process for making sweaters out of recycled soda bottles like the one shown here.

This sweater was made from recycled soda bottles.

Find Out **ACTIVITY**

Going Green

You have been thinking about global warming, the greenhouse effect, and other environmental issues. There are a number of things that all of us can do to help make our planet a "greener" and healthier place.

What to Do

In your notebook, write the names of people (classmates, family members, or friends) who do the following.

(a) try to reduce the amount of paper they use

(b) recycle paper, cans, and bottles

(c) ride a bicycle to school

(d) take public transit to school rather than travelling by car

(e) have recently supported an environmental issue in the media

(f) have participated in some kind of environmental clean-up

(g) would buy a product with less, rather than more, packaging

(h) wear warm clothing rather than turning up the thermostat in the winter, and encourage family members to do the same

(i) compost kitchen wastes

(j) would choose a toy or game that does not require batteries over one that does

(k) take empty paint cans, used paint thinner, used batteries, used motor oil, bleach containers, and other hazardous wastes to a hazardous waste disposal site

(l) sell unwanted clothes to a secondhand store, or donate them to charity

(m) wrap gifts in previously used wrapping paper or re-usable cloth bags

(n) use washable dishes and cutlery instead of disposable ones

What Did You Find Out?

1. Are most people you know involved in reducing, re-using, and recycling? How can you encourage the people on your list to become more involved?
2. What are the most common ways that people reduce, re-use, and recycle?

Extension

3. What other "environmentally friendly" activities can you add to the list?

CONDUCT AN

INVESTIGATION 3-C

Waste Reduction Diary

Many companies conduct "waste audits" to keep track of what they throw away because everything they throw away costs them money and puts extra stress on the environment. With family members, classmates, or others, try a similar investigation yourself. You could conduct this investigation at home, at school, or perhaps at a workplace. Becoming aware of the garbage you make will help you think of new ways to reduce, re-use, and recycle.

Problem

How can you keep track of all the garbage thrown away during a week?

Materials

waste reduction diary

Procedure

1. Use the sample waste reduction diary shown to you by your teacher to make your own diary. Use different columns for the categories of waste: paper, metal, plastic, organic, and so on.
2. Place your waste reduction diary on a bulletin board, refrigerator, or other place where all the participants in your investigation can easily use it.
3. Discuss the contents of your diary with the other participants. You will need each of them to be willing to record each item discarded.
4. Each day for a week, have all the participants keep track of what they put into the garbage. For example, if you throw away a magazine on Day 1, put a check mark in the paper column under Day 1. If another participant also throws away a magazine on Day 1, he or she puts another check mark in the paper column under Day 1.

Packaging and human carelessness contributed to the pollution you see here.

Analyze

1. At the end of the week, total the number of times the participants threw out the various items. Did some participants create more waste than others?
2. Compare how much waste was thrown away in the different categories. For example, how much paper was thrown out compared to organic waste?

Conclude and Apply

3. Use your diary as the basis for a waste reduction meeting. Along with the other participants, plan how you can reduce, re-use, and recycle. Co-operate to carry out your plans.

Extend Your Skills

4. Just as a company can save money by reducing waste, so can you. You could make a quilt out of patches cut from old clothes, or build a doghouse out of scrap lumber. Be creative and come up with your own ideas.

Pause& Reflect

The 3 R's are necessary if we want to protect our environment for the future. Imagine you are writing a letter to a friend who lives far away in another country. Explain the 3 R's to your friend, and be sure to show how they are helping in your community. Write the letter in your Science Log.

Career CONNECT

Reducing Waste

Waste management consultants are people who study and give advice on garbage disposal matters, including Blue Box programs. Mayors, councillors, and other members of community governments are also involved with Blue Box programs and waste management. Executives of manufacturing companies and executives of companies who turn recyclables into resaleable goods need to work closely with waste management specialists.

Conduct research to find out what kind of education is needed to become a waste management consultant. What kind of background would you need in order to work in waste management? What skills are most important to be effective at this job? Find out if there are any waste management consultants in your area. What types of companies do they work for?

Check Your Understanding

1. What is a landfill site, and how can it contribute to pollution?
2. How can technology be used to help reduce the harmful effects of other types of technology?
3. What is a waste audit? How might it help the person who conducts it to "go green"?
4. **Apply** During one month, 17 Canadian universities reported that the following amounts of materials, in tonnes (t), were sent to recycling depots: 49, 39, 62, 51, 73, 40, 82, 94, 44, 49, 64, 88, 82, 75, 90, 64, 73. Make a stem-and-leaf plot showing amounts of recyclable materials sent to depots. Remember to give your stem-and-leaf plot a suitable title.
5. In question 4, you organized your data in a stem-and-leaf plot. When data are organized in this way, it is easier to make decisions about how to represent your results in a graph. Analyze your stem-and-leaf plot, then select and draw an appropriate type of graph to represent the data.

Skill POWER

For tips on how to make stem-and-leaf plots and on graphing data, turn to page 486.

CHAPTER at a glance

Now that you have completed this chapter, try to do the following. If you cannot, go back to the sections indicated.

Demonstrate that the biosphere is an excellent recycler. (3.1, 3.2)

Describe the four main processes in the water cycle. (3.1)

Explain what part plants play in the following cycles: the water cycle (3.1), the carbon dioxide and oxygen cycle (3.1), the carbon cycle (3.2).

What forms can water take? Explain. (3.1)

Identify the benefits and problems relating to redirecting natural water flow. (3.2)

Describe what fossil fuels are and how they are made. (3.2)

Use the example of a greenhouse to explain the process of global warming. (3.2)

Explain why species of plants and animals that reproduce quickly do not spread and take over Earth. (3.3)

Describe the different kinds of limiting factors found in nature. (3.3)

Explain the forest fire regeneration cycle. (3.3)

Identify effects of human technologies on natural regulators in ecosystems. (3.4)

Explain benefits and problems relating to the 3 R's: reduce, re-use, and recycle. (3.5)

Prepare Your Own Summary

Summarize this chapter by doing one of the following. Use a graphic organizer (such as a concept map), produce a poster, or write the summary to include the key chapter ideas. Here are a few ideas to use as a guide:

- Find or draw examples of technologies mentioned in this chapter. Use them in a presentation to explain how human technologies relate to ecosystems.
- Design a symbol that represents the idea of cycles and recycling. Use it in a poster to explore cycles and recycling as they relate to this chapter.
- Use the balance scale below to explain the concept of balance as it relates to ideas and issues in this chapter.

CHAPTER

3 Review

Key Terms

recycling
water cycle
evaporation
sublimation
transpiration
condensation
precipitation
ground water
run-off
carbon dioxide and oxygen cycle
respiration
photosynthesis
erosion
fossil fuels
plankton
phytoplankton
greenhouse gases
global warming
limiting factor
primary succession
secondary succession
diversity
landfill

Reviewing Key Terms

If you need to review, the section numbers show you where these terms were introduced.

1. In your notebook, match the description in column A with the correct term in column B. Use each description only once.

A

- antibiotics fight them
- a change from a solid to a gas
- made up mainly of ancient decomposed organisms
- relates to the number of species in an ecosystem
- a change from a solid to a liquid
- forest renewal is an example
- a change from a gas to a liquid
- evaporation from a plant
- humans dig wells to reach it
- include algae and plants floating in seas

B

- condensation (3.1)
- melting (3.1)
- sublimation (3.1)
- transpiration (3.1)
- micro-organisms (3.4)
- phytoplankton (3.2)
- fossil fuels (3.2)
- diversity (3.3)
- secondary succession (3.3)
- ground water (3.1)
- primary succession (3.3)

2. Choose five or more key terms from the list above. Use them to create a matching puzzle like the one in question 1. Ask a classmate to solve your puzzle.

Understanding Key Ideas

Section numbers are provided if you need to review.

3. Photosynthesis and respiration are two processes that contribute to the cycling of carbon dioxide and oxygen. Make a diagram to show how the two processes are related. (3.2)

4. Based on your knowledge of cycles, explain the slogan "Have you thanked a plant today?" (3.1, 3.2)

5. Based on what you have learned in this chapter, add five "environmentally friendly" activities to the list in the Find Out Activity, Going Green (page 84). (3.1, 3.2, 3.3, 3.4, 3.5)

6. Show how scientists think coal was formed. (If you prefer, show how scientists think oil and natural gas were formed.) Give your answer as a numbered list of steps, a poem, or a flowchart. (3.2)

Developing Skills

7. Design a model to show what happens to an ecosystem when water is diverted from its natural course. Be creative and show your model to your class.

8. Draw a map or illustration of your community. Show what it might look like if the following technologies had never been invented: coal-mining machines, oil-pumping machines, agricultural machines, forestry machines. How would the ecosystems be different? How would human life be different?

9. In what ways do we overuse technologies in our modern Canadian society? Draw a map or illustration to show how we might arrive at a better balance between the use of technologies and the preservation of ecosystems.

10. In your notebook, copy and complete the following spider map.

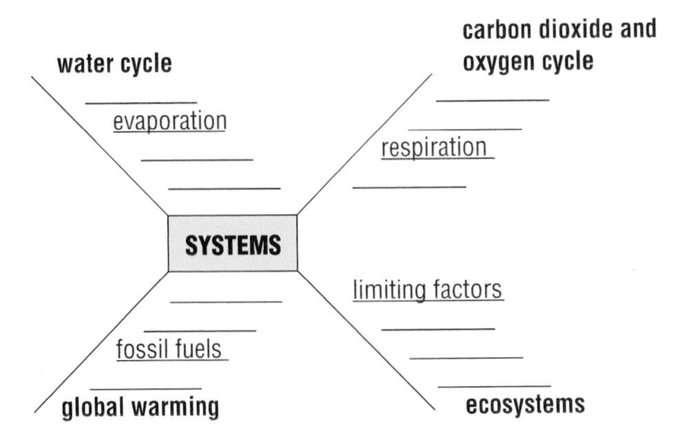

Problem Solving/Applying

11. Based on your knowledge of cycles, what are two advantages of keeping plants in your home?

12. Design and carry out an activity to increase the diversity of species where you live. What role might be played by a bird feeder or herbs growing in pots? What are some advantages of species diversity?

Critical Thinking

13. Imagine that you are a planning consultant hired by a community to help plan a dam. What might you tell the people about each of the following?
    (a) benefits the community could receive from the dam
    (b) possible impacts on ecosystems where the dam is going to be built
    (c) possible impacts on ecosystems in the surrounding area
    (d) ways to minimize impacts on ecosystems

    Prepare an oral or written presentation, and present it to the class. If you wish, you could include visuals such as overhead transparencies, before and after models, or before and after drawings.

Pause& Reflect

1. Identify a specific issue raised by this chapter. Tell why it is important to you. Plan and carry out positive actions by which you personally improve a situation related to the issue.
2. Go back to the beginning of this chapter on page 62, and check your original answers to the Getting Ready questions. How has your thinking changed? How would you answer those questions now that you have investigated the topics in this chapter?

UNIT 1 Ask an Expert

Have you ever heard of parasites? They are small organisms that live outside or inside other animals and obtain their food from these animal hosts. Hilda Ching can tell you all about parasites. She is a parasitologist. She studies parasitic worms and the way they use food chains to get to their hosts.

Q How did you become interested in parasites?

A Growing up in Hawaii, I was surrounded by coral reefs, tropical fish, and colourful plants. They inspired my interest in nature. Then when I was in Grade 11, we learned about a parasite called the cattle liver fluke. I was fascinated that what looked like a tiny speck on a watercress leaf could develop into a huge worm inside the liver of a cow or a person. I knew, then, that I wanted to study parasites.

Q A tiny speck can grow into a huge worm? How does this happen?

A Parasites can take dramatically different forms at different stages of their lives. It's a bit like a caterpillar turning into a butterfly. In order to develop, however, some parasites have to "hitchhike" their way up the food chain. Unless they find their way into the right host animal, they can't develop to the next stage of life.

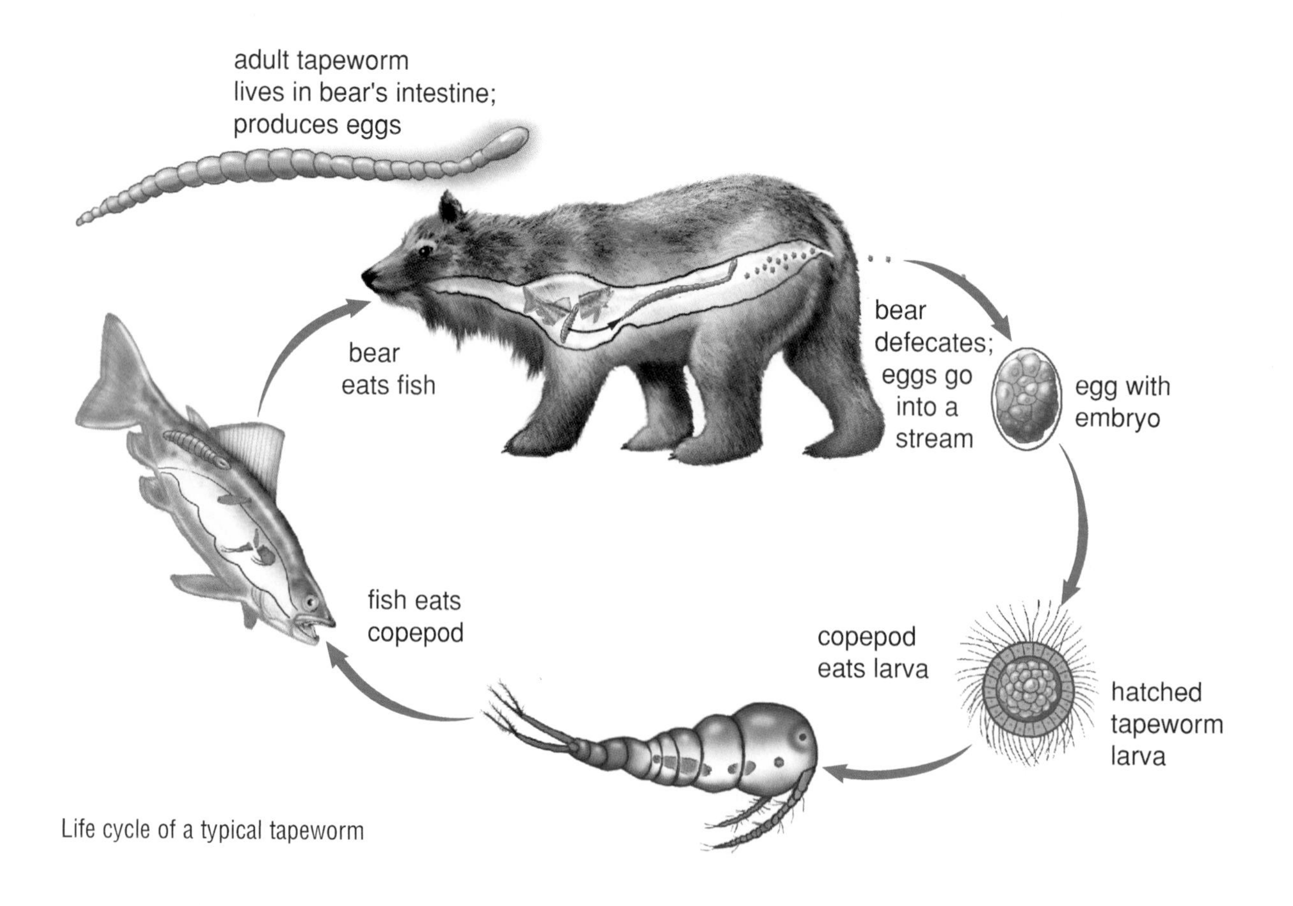

Life cycle of a typical tapeworm

Q **How can something so tiny find its way into the right animal? Are they intelligent?**

A They are, in a way, by being at the right place at the right time. It's a natural process. Recently I've been studying a type of tapeworm, called *Diphyllobothrium dendriticum* (die-FI-lo-BAW-three-um den-DRI-ti-COME), here in British Columbia. Its life cycle begins in fresh water when its eggs hatch into larvae and tiny crustaceans called copepods (KOPE-uh-pods) eat them.

The copepods, with the tapeworm larvae inside them, are eaten by salmon or trout. Inside a fish, the larvae grow a little more, then remain in the fish for the rest of its life, up to about four years. When the infected fish gets eaten by a larger animal, say a bear or a gull, the larvae develop into mature tapeworms in their host's intestines.

Q **That's quite a journey!**

A The journey's not over yet! The tapeworms grow as much as 30 cm a week inside their final host. As they grow they release millions of eggs. The bear's or the gull's feces, with the eggs inside them, get washed into the river. In the water, the eggs hatch into larvae which get eaten by copepods, and the cycle begins again.

Q **People eat salmon and trout, too. Can these tapeworms affect us?**

A Yes. This particular tapeworm can infect people, but the salmon and trout we buy are pretty clean of tapeworms. When you catch a salmon or trout yourself, make sure that you clean it promptly and cook it thoroughly to kill any parasites that might be lurking. If you want to eat raw salmon, such as in sushi or sashimi, make sure that the fish has been previously frozen.

Q **Who benefits from the work that you do?**

A Doctors often use the information that parasitologists provide to find out what parasite might be causing a health problem in a patient. I am often asked to identify parasites from fish or other animals. Also, companies and the government ask me to assess changes in the environment that may cause diseases in fish. One recent concern that we have is the introduction of new, possibly harmful, parasites into our marine and fresh-water environment.

The work I did, studying *Diphyllobothrium dendriticum* with colleagues in other parts of the world, showed that it is a common tapeworm in North America, Europe, and Japan. Like many other kinds of zoologists, parasitologists try to learn more about the creatures with which we share the planet Earth.

EXPLORING Further

Parasites and You

Although most people in Canada are fairly safe from parasites, there are still some, such as tapeworms, that pose a risk. There are others, such as heartworms and roundworms, that are a serious threat to our pets. Contact a family doctor or a veterinarian in your area, and get some information about these parasites. Consider asking questions like the ones on the right:

- How does a person or the pet get a parasite?
- How can people avoid contracting parasites?
- How do you treat patients to get rid of a parasite?

An *Issue* to Analyze

A DEBATE

Beyond the Curb: Is Recycling Really Reducing Garbage?

Think About It

In this unit you have learned about landfills, which produce greenhouse gases and add toxins to the ground water and run-off. Presently, landfills and garbage dumps are being filled faster than ever. The main way that you can help to reduce waste in landfills is to follow the 3 R's: reduce, re-use, and recycle.

Assume that you follow the 3 R's to the best of your ability — you are confident that you separate and sort all the recyclable items in your garbage. How can you be sure that other people are as committed as you are, so that, together, you can make a difference in the garbage crisis?

Rising municipal taxes have caused some people to complain about the cost of the Blue Box Program. These people argue that recycling is not helping the environment, but instead it is using up a lot of time, energy, and money. They would prefer to see the Blue Box Program eliminated.

Resolution

Be it resolved that recycling programs, such as the Blue Box Program, have not been effective in reducing the impact of garbage on the environment.

What to Do

1. Read the In Favour and Against points listed on the next page, and begin to think about other points that could be made in favour of and against the resolution.
2. Four students will debate the resolution. Two students will speak in support of the resolution and two will speak against it. **Note:** No matter what view you actually hold, you must try your best to convince the jury, or debate listeners, of the point your side is defending.

❸ To aid the two teams, two other students will work with them to help gather background information. This is needed to put forward a strong case for the point that side is defending.

❹ The rest of the class will act as the jury in hearing the debate. In preparation for the debate, they should do their own research in order to understand the science and technology behind the issues raised.

❺ Your teacher will provide you with the proper *Debating Procedures* to follow.

In Favour

- Garbage is only one of many environmental concerns we face today. The rising level of energy consumption is perhaps of greater concern. The full recycling process uses a great amount of energy — much more than is required simply to dump garbage at a landfill.
- The quality of products made from recycled materials is generally lower than the quality of products made from new materials. In addition, most plastics and papers can be recycled only once, or maybe twice, before the product quality is so low that the products are unusable.
- Many people are still unwilling to make the effort required to recycle properly. The end result is that landfills are still overflowing with both recyclable and non-recyclable garbage.
- Recycling costs money and unless companies can make a profit by manufacturing products from recycled materials, they cannot afford to do it. Consumers pay for the cost of recycling by paying more for products made with recycled materials. Some consumers are unwilling to do this.

Against

- Recycling programs have made people much more aware of the garbage they produce.
- Much less energy is needed to recycle some materials than to process them from raw materials because of refined processing technologies.
- Although early technologies for cleaning, purifying, and reusing recycled materials were very limited, present technologies have allowed a wide range of plastics and papers to be added to the list of recyclable materials.
- Many resources are non-renewable and limited in quantity. If they are not recycled, they end up in dumps and landfills where they are inaccessible. It is important to continue recycling before our resources become depleted.

Analyze

❶ Which team won the debate, based on a class vote?

❷ Did the winning team produce better research or make a better presentation? Explain.

❸ Did you find any studies or any other reliable information that contradicts the arguments that were presented? If so, explain.

❹ Did your initial viewpoint change as a result of the points presented in the debate? Explain why or why not.

Extension

❺ One proposal to motivate people to recycle more, and more effectively, is to charge individuals and businesses for each bag of garbage that is picked up. What do you think? Prepare arguments in favour of and against this idea to help you determine whether this proposal might, in fact, work.

UNIT 1 PROJECT

Natural Habitats in Your Own Schoolyard

Imagine spending your lunch break eating under a tree at the edge of a forest, the warm air filled with butterflies and the sweet fragrance of wildflowers. As you have learned, soil, water, and air come together in a variety of different ways to provide healthy, life-giving habitats for plants and animals. Developing and growing a patchwork of Canadian natural habitats in your schoolyard will take several years, but classes can contribute to the project each year.

Challenge

In a group, design, plant, and maintain a model habitat of native Canadian plants in your schoolyard.

Materials

Bristol board
coloured pencils, or markers, or paints
ruler
eraser
glue (optional)
variety of soils
peat moss and compost
variety of gardening tools, such as a trowel and shovel
watering can or hose

Design Criteria

A. Naturalization projects are intended to include native plants. Find out which habitats and which indigenous (native) plants occur in the region where your school is located.

B. Make a large sketch of your habitat on a single piece of Bristol board. Indicate each plant species using a unique symbol and include these symbols in a legend. Write the species' niche in the habitat on the legend as well.

C. Write a paragraph describing your habitat on the Bristol board. Include what you have learned about the habitat and list the basic requirements needed to keep it healthy.

Plan and Construct

1. Work in a group of 4 to 6 students. Study your schoolyard, noting soil drainage and sunny and shady areas to determine which habitats you could develop.

2. Choose a habitat to develop, based on suitable areas in your schoolyard. For example, you could

develop an herb habitat in a dry, sunny location, a forest-edge habitat of low bushes and wild-flowers beside a group of shady trees, or a prairie meadow habitat in a space with lots of sunlight. If space in your schoolyard is limited, it may be necessary to make habitats in small-scale plots or raised beds.

3. Decide how large your habitat will be. Choose three or four main plants, then decide how many of each to get. The larger your habitat is, the more plants you will need and the more work you will have to do.

4. Decide where your plants will go. How do the different plants grow together in their natural habitats? Are they mixed together, or do they grow in clusters? Draft several designs of their arrangement before you choose your final design. You may wish to include some non-living elements, such as rocks, bark-nugget pathways, or logs for seating.

5. Have your design approved by your teacher. There may be safety or comfort considerations. For example, some plants may aggravate allergy sufferers, or logs may need to be secured to the ground.

6. Make a list of the jobs that need to be done. Decide which group member will be responsible for each job. For example, who will dig and prepare the soil, buy the plants, weed, water, trim? Also, decide how you will protect your new habitat from schoolyard traffic.

7. Make your habitat.

8. Keep a log of all activities related to your habitat. Each member of your group should record in the log what she or he did to the habitat and when. This will be especially important for regular maintenance.

9. As a class, prepare a one-page information sheet that can be posted near your habitat. This will encourage all the students at your school to learn about your project.

Evaluate

1. Submit a one page report each week, detailing the progress of your habitat. Comment on the overall health of the plants. Include information on their growth, flowering, and any signs of disease or pests.
2. What parts of your habitat might you change if you did the project again?
3. Choose one problem that has occurred and write a proposal explaining how you could solve the problem. Explain why you think your solution would work. For example, if you are having trouble controlling weeds, you might propose laying mulch or bark-nuggets on the exposed soil.

1. Hold an "odds and ends drive" at your school. Have people and businesses in the community donate items to the school for teachers to use in their classrooms. As a class, design a campaign and advertise when and where items can be dropped off.
2. Set up a "Green Council." Divide your class into groups that will be responsible for various eco-projects. Each group will develop and implement plans to increase public and community awareness of their projects. Make a difference at your school!

UNIT 2

Pure Substances and Mixtures

Battered gold nuggets twinkle in the bright light. A carved and polished work of art gleams from a display case. Tiny veins of gold glitter from rough silver in which they are trapped.

The distinctive colour and properties of gold catch the eye and captivate the imagination. From earliest times this material has been a favourite of artists because it is easy to carve. Also, it has symbolized purity because gold coins and jewellery do not erode or tarnish, even in salt water.

The gold nuggets shown here were probably found in the late 1800s, washed downstream from a mountain slope. The gold ring is ancient. Its design is based on a pattern thousands of years old. The gold in the ring may have come from a nugget, but it was more likely obtained by extraction from gold-bearing rock. The veins of gold were embedded in the silver millions of years B.C.E.

Examine the title of this unit. Which of the objects shown here seems most like a mixture? Which seems more like a pure substance? In the next three chapters, you will explore the scientific meaning of these terms and their practical impact on your life. Along the way, you will find out how gold is extracted from rock and purified.

Unit Contents

CHAPTER

4 Mixtures or Pure

Getting Ready...

- What substances make up the foam and the clouds in the photograph?
- What common liquids are really made up of two or more different materials?
- Are there some basic substances from which everything is made?

Science Log

In your Science Log, write what you think are the answers to these questions. You will find the answers as you study this chapter.

Have you ever spent time at a beach? Perhaps you have visited a beach like the one shown here. If you take a closer look at the photograph, you can see that there is more to the beach than just water and sand. How many different materials can you find? You can see that some of them, such as the sand and the gravel, are made up of smaller bits and pieces of different-looking materials. What about the air and the water? Do they seem to be made up of different-looking bits and pieces, or are they much the same everywhere in the photograph? Are they the same everywhere in the world? In this chapter, you will learn about many everyday materials that are really mixtures, and you will find out why a pure substance is the same everywhere it occurs.

Skill POWER

For tips on how to make and keep a Science Log, turn to page 476.

Substances?

Spotlight On Key Ideas

In this chapter, you will discover

- how to use properties to classify a variety of materials as heterogeneous or homogeneous
- how to find out the difference between mixtures and pure substances
- how to use the particle theory to interpret observations of matter

Spotlight On Key Skills

In this chapter, you will

- distinguish between a mechanical mixture and a solution
- distinguish between a solution and a pure substance
- Design your own methods for separating mixtures

Starting Point ACTIVITY

Mixed or Pure?

Suppose that you pick up a rock on the beach. You see that some parts of it are grey, some parts are white, and some parts are silver. You conclude that the different-coloured parts of the rock must be different materials.

There are many observable characteristics that can help us tell one material from another. What are some of the differences between glass and steel, water and milk, and gasoline and oil?

What to Do

1. With a partner, select one of the pairs of materials mentioned above. Write down as many differences as you can. Hint: Think of colour, transparency, "heaviness," hardness, and strength.
2. Repeat step 1 for the other pairs of materials.

What Did You Find Out?

Be ready to explain your ideas to the rest of the class.

4.1 Matter in Your World

DidYouKnow?

Dry cleaning might better be called "wet" cleaning. Because water can damage some fabrics, dry cleaners use different liquids, such as perchloroethylene. The clothing is still put into a washing machine full of liquid, but because the liquid is not water, we call the process "dry" cleaning.

What is matter? **Matter** is any material that occupies space and has mass. Matter includes all solids, liquids and gases. These forms of matter have different characteristics.

How would you describe the substances in Figure 4.1A? The characteristics that are used to describe matter are called **properties**. Every material has its own set of properties. For example, the water at the beach shown on page 98 is a clear, colourless, odourless liquid. Under normal conditions, water freezes at 0°C and boils at 100°C. These are all properties of water.

Several other materials share some of water's properties. The three liquids shown in Figure 4.1B look like water because they are clear and colourless. However, they do not have the same properties as water. The liquids in Figure 4.1B have an odour, and they freeze and boil at different temperatures than water. No other material has the same properties as water.

Figure 4.1A Are these substances homogeneous or heterogeneous?

Figure 4.1B The WHMIS safety symbol indicates that these are poisons. Carbon tetrachloride was once used as a dry cleaning fluid but was banned because it is such a dangerous substance.

Skill POWER

For more information on the safety symbols used in this book, turn to page 492.

Some materials cannot be described by a single set of properties. Are the rocks in the photograph of the beach white, grey, pink, or other colours? They have bits of different materials in them, and the different materials have their own sets of properties.

Whenever you see a material that has more than one set of properties, you know that it is a **mixture**. A mixture contains more than one kind of matter combined in such a way that each keeps its own properties. Mixtures that are made up of parts that can be seen are called heterogeneous. **Heterogeneous** means made up of parts, or mixed.

Materials that have only one set of properties are called homogeneous. **Homogeneous** means that every part of the material is the same.

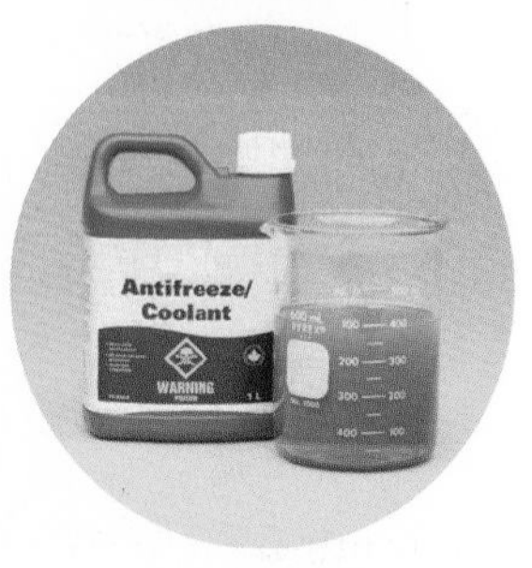

Figure 4.2 The salt and pepper pictured here can easily be identified as two separate parts — a mixture. The antifreeze looks the same throughout; it is homogeneous.

Find Out **ACTIVITY**

A Classified Photograph

Study a photograph to determine if the materials shown are heterogeneous or homogeneous.

What to Do

1. In your notebook, make a data table like the one below. Give your table a title.

Material	Classification (heterogeneous or homogeneous)	Why I think so

2. In the first column, list all the materials you see in the beach photograph on page 98.
3. Decide whether each material is heterogeneous or homogeneous. Use personal experience to help you decide, not just what you see in the photograph. Record your decision in the second column.
4. In the third column, identify the observations that helped you decide.

What Did You Find Out?

1. When you have completed your table, your teacher will ask you to pair up with another student. Compare your table with your partner's. If your partner included any materials that you did not, add them to your table.
2. Were there any differences of opinion about any of the materials? Make a note of your reasons for disagreeing, and discuss them with the class.

Heterogeneous or Homogeneous?

Would you like to breathe the air shown in Figure 4.3A? Like most people, you would probably prefer the air shown in Figure 4.3B. What is the difference? Is air the same everywhere you go? No. In fact, the "air" in Figure 4.3A is really smog. Smog is a mixture of air plus a lot of pollutants. If you were standing in the middle of the smog, you might not be able to see any of these pollutants. When you look at smog from a distance, however, there is no question that it contains more than just air. So smog is heterogeneous. The air in Figure 4.3B looks perfectly clear. Is it homogeneous?

Figure 4.3A Is the substance pictured here homogeneous or heterogeneous?

Figure 4.3B Is the air in this photograph homogeneous?

Figure 4.4 The water in the Thompson River in British Columbia has two colours. Does the colour indicate whether it is homogeneous or not?

Figure 4.4 shows the Thompson River in British Columbia. Notice that the colour of the water is not the same throughout. What do you think makes the water shown at the bottom of the photograph brown? Is it really just water? Do you think the blue water shown at the top of the photograph is homogeneous? What reasons can you give to support your answer? Compare the appearance of the river water with the appearance of the water in Figure 4.5.

Figure 4.5 Why might you describe the water in this photograph as homogeneous?

Figure 4.6 This concrete wall is clearly not homogeneous.

In the concrete wall in Figure 4.6, you can easily see stones of several sizes, shapes, and colours. If you watch workers mix concrete by hand, you can see that there are many bits of even smaller rocks and sand in the concrete, too. The workers also mix in powdered cement before adding water and mixing for the final time. Obviously concrete is a heterogeneous material. Now look at the photograph of the steel beam. Unlike the concrete, it does not *appear* to have any different bits in it. Is it homogeneous?

The words homogeneous and heterogeneous can be used to describe gases, liquids, or solids. As you saw in the photographs, air, water, and building materials may be heterogeneous mixtures or homogeneous. As you read through this chapter, you will better understand the meanings of the words homogeneous and heterogeneous and how to classify materials correctly.

Figure 4.7 How many bits of different materials can you see in this steel beam?

Taking a Closer Look

Heterogeneous or homogeneous? For some materials, it can be difficult to decide, and different observers may disagree. For example, you may feel quite certain that peanut butter is heterogeneous, but someone else may insist that it is homogeneous. Maybe you interpret the label on the jar differently: the label says that the peanut butter is "homogenized." Maybe the other person has never heard of "chunky" peanut butter. Perhaps you have used different methods of observation: you rub it between your fingers; the other person looks at it in the jar. What can ensure that everyone uses the same standards when classifying materials as heterogeneous or homogeneous?

At Home

Treasure Hunt

Your task in this treasure hunt is to examine five materials from the kitchen in your home and five materials from the bathroom or laundry room. From the kitchen, you could select ketchup, mustard, spices, soft drinks, dishwashing liquid, cereal, bouillon cubes, jam, molasses, or bread. From the bathroom or laundry room, you could select shampoo, conditioner, soap, toothpaste, shaving cream, a pumice stone, an emery board, hand or body lotion, hair gel, or detergent.

CAUTION: In this activity, you will be choosing a variety of household materials to examine. The suggested materials have been carefully chosen as safe for you to examine. Do not touch any materials other than those listed here.

What to Do

1. Prepare a data table similar to the table you made on page 101. Give your table a title.
2. In the first column, list each material you examine. In the second column, indicate whether it is heterogeneous or homogeneous, and, in the third column, give your reasons.
3. Put away all materials, and wash your hands.

What Did You Find Out?

After all the students in your class have completed their hunt, find out the total number of different materials that were heterogeneous and the total number that were homogeneous. Write these class totals underneath your table.

Examining Three Common Beverages

On a picnic, what do you like to drink? Is your favourite beverage heterogeneous or homogeneous? What properties of this beverage make it your favourite? In this investigation, you will examine three common beverages — milk, orange juice, and soda water — to find out in which category each beverage belongs.

Problem

Are milk, orange juice, and soda water heterogeneous or homogeneous?

Safety Precautions

In science class, it is not safe to taste samples. Do not even taste fluids you drink at home every day. Laboratory containers that appear clean may still contain invisible traces of harmful materials left over from a previous activity.

Apparatus

3 clean test tubes

marking pen and masking tape (for labels)

test tube rack

hand lens

medicine dropper

watch glass or Petri dish

Materials

homogenized milk

orange juice, fresh or from concentrate

soda water

Procedure

Beverage	Method of Observation	Observations	Inference	Reasons
Milk	Unaided eye			
	Hand lens			
	Microscope			
Orange Juice	Unaided eye			
	Hand lens			
	Microscope			
Soda Water	Unaided eye			
	Hand lens			
	Microscope			

1. On a full sheet of paper turned sideways, prepare a data table like the one above. Give your table a title.

2. Label the three test tubes: M (for milk), O (for orange juice), and S (for soda water). Half fill each test tube with the correct beverage, and place it in the test tube rack.

3. Using only your unaided eye, examine the beverage in each test tube. Can you see any bits that are different from the rest of the beverage? Record your observations for each beverage in your table.

4. Based only on what you see for yourself, infer whether the beverage is heterogeneous or homogeneous. Record your inference in the correct row of your table, and include the reasons for your inference.

5. Place a small amount of one of the beverages on a watch glass or Petri dish. Using the hand lens, examine the beverage again. Record your observations in your table. Repeat for the other two beverages.

6. Infer whether each beverage is heterogeneous or homogeneous. Remember to include your reasons.

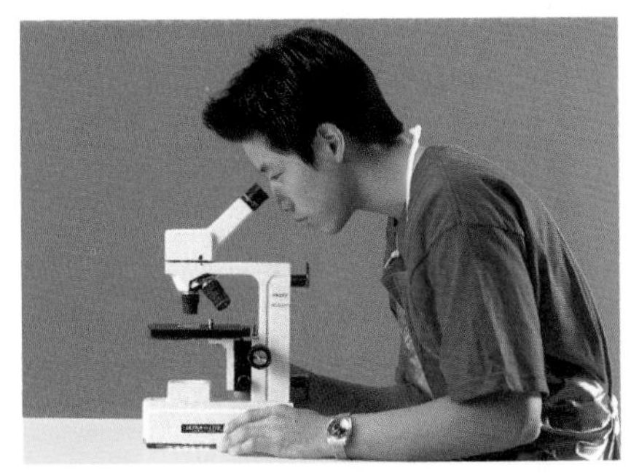

7. Your teacher will set up three compound light microscopes with each of the beverages. Observe the beverages using the low power magnifying objective of the microscope.

8. What new observations might lead you to think some of your previous inferences may not be accurate? Record your new observations in your table.

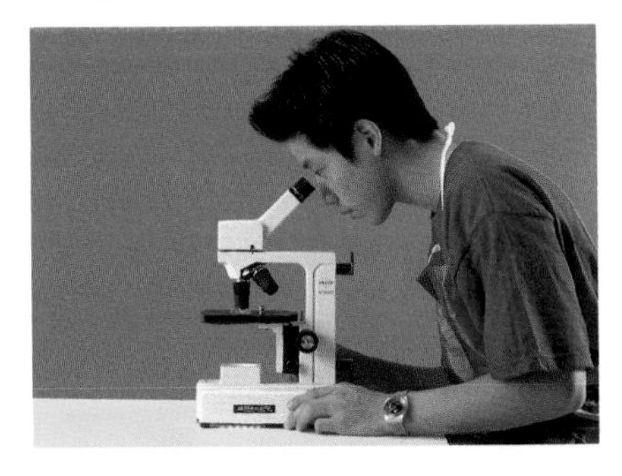

9. Next, observe what each beverage looks like under the microscope at high power. Note the sizes and shapes of any bits of different materials that you see. Also note whether or not the bits are evenly spaced through the liquid.

10. Make your final inferences, based on the microscopic evidence, and record them in your table.

11. Wash your hands after completing this investigation.

Analyze

1. When you examined the beverages with your unaided eye, you looked for bits of different materials with different properties within each liquid.
   (a) Which bits were big, small, or microscopic?
   (b) Were the bits all the same shape?
   (c) Explain which bits were solids, liquids, or gases.
2. Describe what, if anything, you were able to see with the hand lens that you could not see with your unaided eye
   (a) in the milk **(b)** in the orange juice **(c)** in the soda water
3. **(a)** Which beverage appeared to be homogeneous until you saw what it looked like under the microscope?
   (b) Describe the new evidence you could see in the microscope photographs.
   (c) Under high power, you could see bits of different materials with different properties, but the bits were not all exactly the same size and shape. What reasons can you give to explain the differences?
4. Summarize your findings. Write one or two sentences about each beverage. Be sure to state whether it is heterogeneous or homogeneous. Give reasons for your inference.

Conclude and Apply

5. Materials that are homogeneous can be tricky to identify. Which heterogeneous beverage in this investigation did you think was homogeneous at first? Explain how you found out it was not.
6. What would you have to observe before you would be willing to call a material homogeneous?

Bubbles and Blobs

The fizz in pop is bubbles of carbon dioxide gas. The pop in an unopened bottle appears to be all liquid. What makes it start to fizz? Where do the bubbles come from? Where do you think the carbon dioxide was before the bottle was opened?

Did you guess that the tiny blobs you saw in the milk under the microscope in Conduct an Investigation 4-A were fat globules? Many years ago, milk was not homogenized. Instead of staying mixed with the milk, the fat globules floated to the top and clustered together, forming a noticeable layer of cream. Today milk is homogenized — specially prepared so that the fat globules remain mixed with the rest of the liquid. One single drop of milk contains approximately 100 million fat globules. They are so tiny that the milk appears homogeneous when observed with the unaided eye and even with a hand lens.

Do you like a lot of pulp in your orange juice, or none at all? You can probably tell if your juice is just right without tasting it first. Even with the unaided eye, you can usually see the bits of pulp mixed in with the juice. You could certainly see them with a hand lens or with a microscope. The pulp in orange juice is made up of the remains of the cells of the orange.

DidYouKnow?

In the times of the ancient Greeks and Romans, soda water came from springs — it bubbled right up out of the ground. People could just scoop it into a cup and drink it. Centuries later, some people began bottling and selling this bubbly spring water. Scientists later learned how to make soda water by adding carbon dioxide gas to ordinary water. The great variety of soft drinks we now have are made by adding flavouring and colouring to the plain soda water.

Check Your Understanding

1. List five properties of the following:
   (a) water,
   (b) glass,
   (b) steel.
2. Explain how you can tell the difference between a homogeneous material and a heterogeneous material.
3. (a) List five homogeneous and five heterogeneous materials in your classroom that have not already been discussed in this chapter.
   (b) Identify two materials in your classroom that you are not sure how to classify.
4. **Apply** The water in some rivers appears blue. The water in other rivers may appear green or brown.
   (a) How many sets of properties does the water in one river have?
   (b) Is river water heterogeneous or homogeneous? Why do you think so?
5. **Thinking Critically** Based on your experience so far in this chapter, do you think most materials on Earth are homogeneous or heterogeneous? Why do you think so?

Pause& Reflect

1. Of all the materials you have classified so far, how many are heterogeneous? How many are homogeneous? Remember to include your class totals from the Treasure Hunt. List examples of each.
2. How many materials are you unable to classify for sure? In your Science Log, list these materials.

4.2 What Is a Mixture?

Anything with two distinct sets of properties must be a mixture of at least two materials. For example, it is easy to decide that sand, pop, and smog are mixtures. All are heterogeneous. When a mixture's different parts can be identified this easily, the mixture is usually called a **mechanical mixture**.

Even so-called "homogenized" milk turned out to be a heterogeneous mixture. It *looks* homogeneous. The unaided eye can detect no bubbles, blobs, or bits with a different set of properties. Examination under a microscope reveals that milk contains many tiny globules of a second liquid — fat. These fat globules have different properties that can easily be observed with ordinary laboratory equipment. No matter how smooth milk appears, it is actually a heterogeneous mixture.

By now you may be wondering …

"Isn't anything pure?"

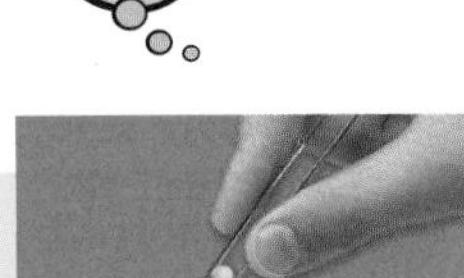

Find Out **ACTIVITY**

Homogeneous Mixtures

What you already know about familiar materials, such as sugar and water, can help you find out more about them.

What to Do

1. In your notebook, make a data table like the one below. Give your table a title. Using what you already know, fill in rows 1, 2, 3, and 4.

Observable properties	Sugar	Water	Sugar-and-water mixture
1. Colour			
2. State			
3. Taste			
4. Transparency			
Inference (heterogeneous or homogeneous)			

2. Explain whether you would expect a spoonful of sugar from the bottom of a bag
   - to taste the same as one from the top
   - to look the same
   - to be in the same state (solid, liquid, or gas)
3. Explain whether you would expect a sip of water from the bottom of a glass
   - to taste the same as a sip from the top
   - to look the same
   - to be in the same state
4. What would a mixture of distilled water and the purest possible sugar look like? Would it be heterogeneous or homogeneous?

What Did You Find Out?

1. How did you classify the imaginary sugar-and-water mixture?
2. How do the properties of a sugar-and-water mixture compare with the separate properties of sugar and water? Choose one of the following statements. Record it, and give reasons for your choice.
   - The mixture has all the properties of water and *only* those properties.
   - The mixture has all the properties of sugar and *only* those properties.
   - The mixture has two distinct sets of properties.
   - The mixture has a blend of sugar's properties and water's properties.

A Homogeneous Mixture?

Drink powders mix readily with water. The colour will help you see whether or not the mixture is homogeneous. (*If* a homogeneous mixture is possible, *then* you should be able to see the same colour in every sample.)

Problem

Is a homogeneous mixture possible?

Safety Precautions

Do not taste anything in science class.

Apparatus

250 mL beaker
4 glass slides
hand lens
medicine dropper
marking pen

Materials

paper towels
coloured drink powder
water
250 mL clear plastic cup
5 mL plastic spoon
drinking straw

Procedure

1. Half fill the beaker with water. Note the observable properties of the water. Set the beaker aside.
   (a) Cover your work area with a clean paper towel.
   (b) Mark the 4 glass slides T, M, B, and A. From your teacher, collect a paper container of drink powder.
   (c) Open the container over the paper towel. Use the plastic spoon to isolate bits of the drink powder. Note their colour and shape. Try crushing them with the back of the spoon. Decide if they are hard or soft. Record their properties.

2. Add the drink powder to the water in the beaker. Stir with the spoon until the two materials are thoroughly mixed. To make sure that they are thoroughly mixed, stop stirring. When the water stops swirling around, make sure that nothing settles to the bottom of the beaker. Record the properties of the mixture.

3. Use the straw to take a sample of the mixture from somewhere near the bottom of the container. Drop the sample onto slide marked B.

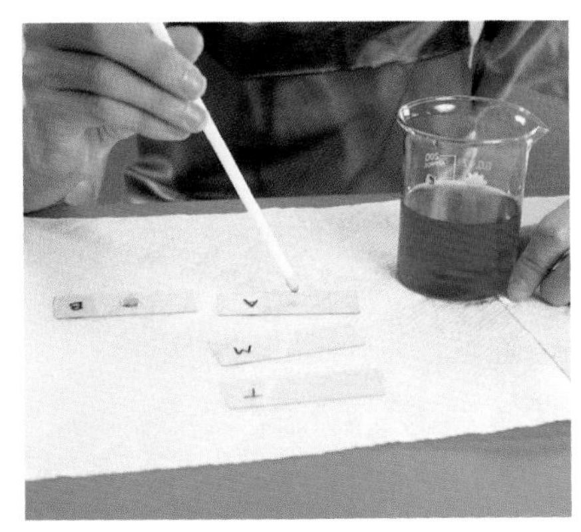

4. Repeat step 3, taking three more samples from different locations: one from the middle of the mixture (M), one from the top (T), and one from anywhere you choose (A).
   (a) Drop each sample into its own glass slide.
   (b) Observe and record the properties of the four drops. Use the hand lens. Note especially any differences in colour.

5. Pour some of the drink mixture into the empty cup, to a depth of about 1 cm. Do not cover the top of the cup. Your teacher will tell you where to store this sample so that it will be undisturbed for several days.

6. Your teacher will tell you where to dispose of the rest of your drink mixture. Do *not* drink it. Wash your hands after completing this investigation.

Analyze

1. To the unaided eye, in what way did the drink mixture differ from a sugar-and-water mixture?
2. (a) Were both the drink powder and the water still present after you mixed them?
   (b) Was the drink powder evenly distributed throughout the mixture?
3. What properties of the mixture support your answers to question 2?
4. (a) Using the hand lens, could you see bits of the drink powder or water in the mixture?
   (b) Based on your answer to part (a), what can you say about the bits of the drink powder and water?

Conclude and Apply

5. Because you made the drink mixture yourself, you know that it was made up of two different materials. Give two reasons why this mixture should be classified as homogeneous.
6. At the end of this investigation, you set aside a cup containing 1 cm of your drink mixture. Predict what you expect to find after a few days. Check the cup after it has sat undisturbed for three weeks.
7. Based on your findings in this investigation, is a homogeneous mixture possible? Explain how you know. Is a sugar-and-water mixture homogeneous? Explain your answer.

Homogeneous Mixtures

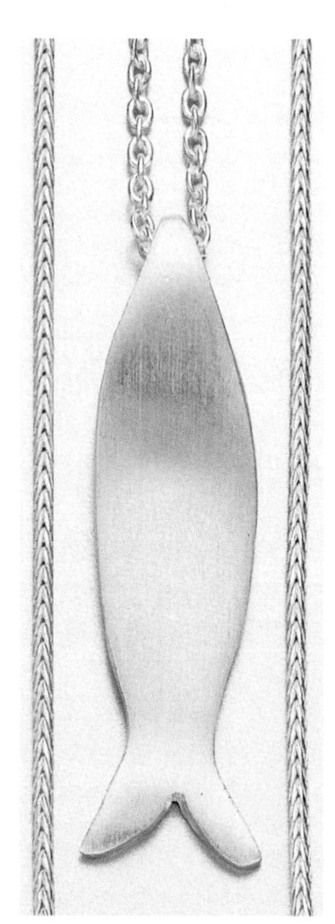

Figure 4.8 Sterling silver is a solution of silver and copper.

A homogeneous mixture is called a **solution**. Solutions are everywhere. Two thirds of Earth is covered with a liquid solution — the salty water of the oceans. Earth is surrounded by a gas solution that we call air — a homogeneous mixture of oxygen, nitrogen, and other gases. There are even solid solutions, as you can see in Figure 4.8.

Where Do the Bits Go?

When sugar is mixed with salt, the mixture is heterogeneous. The bits of sugar can be detected with a hand lens or even a sharp eye, as you can see in Figure 4.9. Salt crystals are shaped like cubes. Sugar crystals have a different shape, so they are easy to spot. We can see where the sugar and salt "bits" have gone.

When sugar is mixed with water, however, the sugar crystals disappear from view. The sugar-and-water solution has the sweet taste of sugar, but no sugar bits can be detected. Where do the bits go in a solution? You will consider this question in the next section.

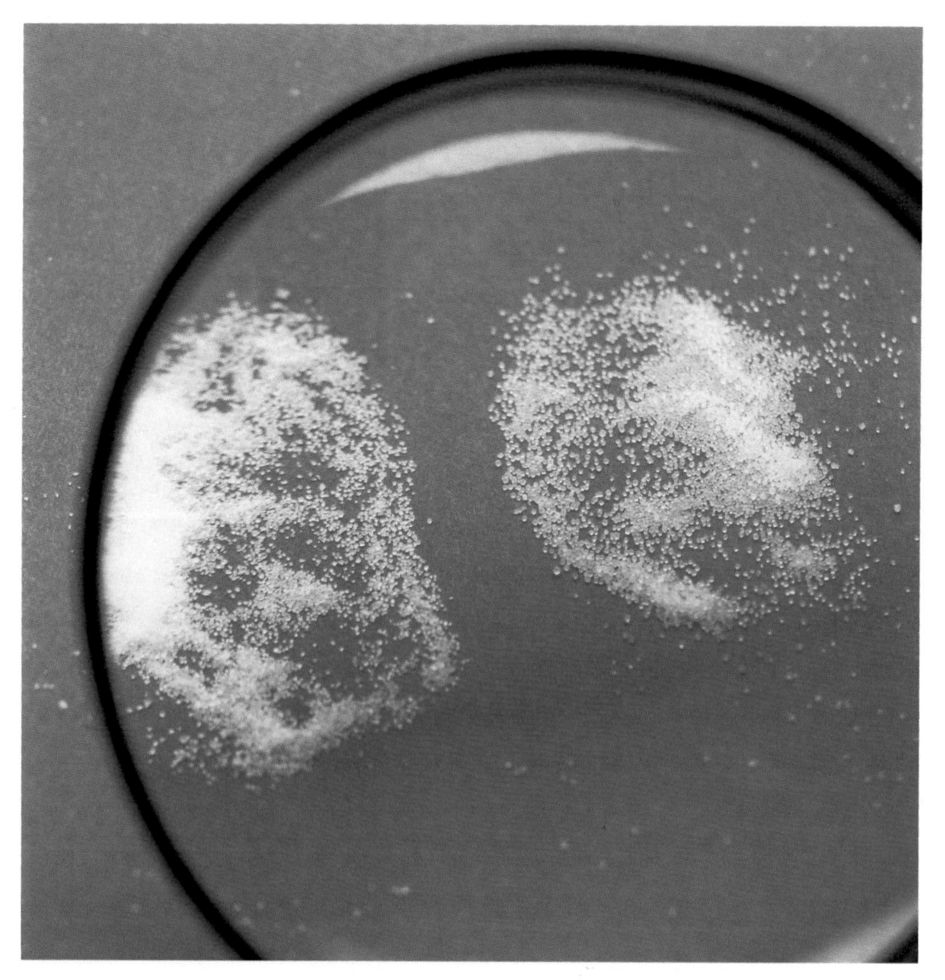

Figure 4.9 It is easy to distinguish between salt and sugar when you observe them through a hand lens.

Across Canada

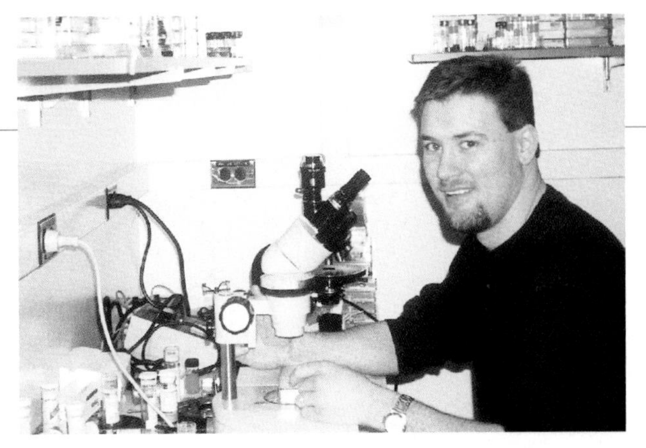

Gerald Audette

To treat everything from cold sores and chicken pox to cancer, researchers are always looking for better medicines. Gerald Audette, a young University of Saskatchewan biochemist, examines in the smallest detail substances that could be medically useful. How do you study something so small that you cannot see it with a microscope? "If you have a pure sample," Gerald explains, "you can use X-ray crystallography to find out what it looks like."

When X-rays pass through a crystal, they produce a dot pattern on a photographic plate. Gerald uses a computer to analyze this pattern. He studies mixtures as well as pure substances. He is currently trying to find out how the interaction takes place in a special kind of protein-sugar mixture found in type O blood.

Born in Edmonton, Alberta, Gerald enjoyed soccer in school. An asthma sufferer, he refused to let his condition interfere with his enjoyment of the game. "I always knew when I was getting tight in the chest," he says. "I would stop, get my breath and take medication if I needed it, and go back out — I kept at it." Keeping at it has been a big part of Gerald's life. He still plays soccer, as well as coaching soccer and practising kendo, the Japanese art of fencing. His love for kendo took on new meaning recently when he visited Japan to collect data for his ongoing study of important biochemical processes.

Check Your Understanding

1. How many sets of properties do you need to observe in order to classify a material as a heterogeneous mixture?
2. (a) What is a solution?
   (b) What evidence do you need in order to classify a material as a solution?
3. You are examining a glass that contains a liquid. You think the glass contains a mixture of water and salt. Your lab partner thinks the glass contains pure water. How can you find out who is correct, without tasting the liquid?
4. (a) Does ocean water have all of the properties of water? Support your answer.
   (b) Does ocean water have all of the properties of salt? Support your answer.
5. **Apply** You describe the appearance of some blobs and bits you saw in a homogeneous mixture. Your lab partner says that what you are describing was not a homogeneous mixture.
   (a) How can you support your point of view?
   (b) What evidence would you need to observe before you would be convinced of your lab partner's interpretation?
6. **Thinking Critically** A cloud is a mixture of water droplets suspended in air. Is a cloud homogeneous or heterogeneous? Give reasons to support your answer.
7. Based on the information in question 6, what substances do you think make up the foam on top of waves? Give reasons to support your answer.

Pause& Reflect

Give another example of a homogeneous mixture from your own experiences. (Go to your Science Log and look at your answer to the "Pause and Reflect" on page 106.)

4.3 Mixtures and Pure Substances

You have seen that heterogeneous mixtures have two sets of properties. There are also homogeneous mixtures, which have properties that are a blend. The blend of properties depends on how much of one material and how much of the other material are in the mixture.

Are there any pure substances — materials whose properties are not a blend and are always the same? How can you find them? How do you know when you have a pure substance? These sound like simple questions, but scientists took hundreds of years and thousands of investigations to come up with the answers.

Yes, there are pure substances. Some familiar examples are water, gold, oxygen, copper, silver, sugar, and salt.

How can you find a pure substance? First you look to see if the material is homogeneous. A pure substance must be homogeneous.

How do you know whether the homogeneous material is a pure substance? Investigate the properties of the material to find out if they are always the same. If they are always exactly the same in all parts of the material, no matter what part of the world the material comes from, then you may infer that you have a pure substance.

Look again at the photograph of the gold ring shown on page 97. If you turned it over, you would expect to see the same properties on the reverse — the same colour, texture, and shininess. When scientists test gold's other properties, they find that those properties, as well, are always the same. They can therefore conclude that gold is a pure substance. A thorough investigation is necessary before anything can be classified as a pure substance.

Figure 4.10 summarizes what you have learned so far about matter. All materials can be classified as heterogeneous or homogeneous. Homogeneous materials can be further classified as solutions or pure substances.

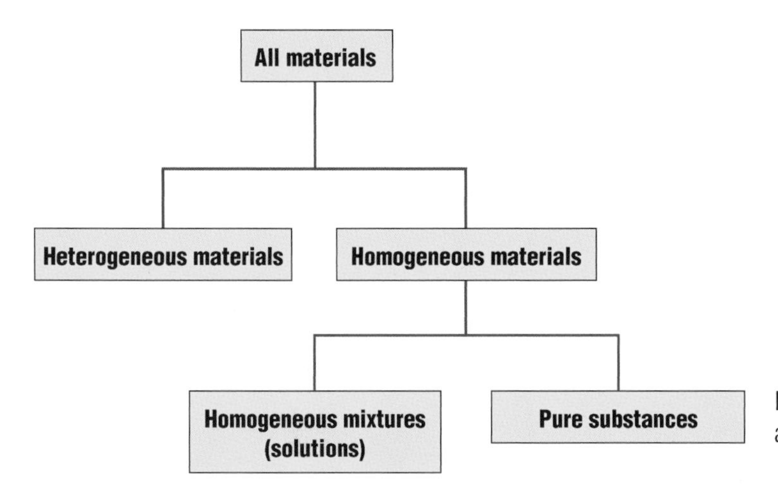

Figure 4.10 All matter can be classified as heterogeneous or homogeneous.

Find Out ACTIVITY

Investigating Money

What are the properties of a penny?

What You Need

new penny
short piece of thick copper wire
hammer
nail

What to Do

1. Compare the shininess and colour of the penny with the shininess and colour of the pure copper. (For properties of pure copper, refer to the table shown.)
2. Compare how easily the penny and the copper wire bend.
3. Strike the sample of pure copper 10 times with the hammer. Repeat with the penny. Notice which one flattens out more.
4. Scratch the penny with the nail 10 times, as deeply as you can. Repeat with the copper wire.

Some Properties of Pure Copper

shiny solid
reddish-brown colour
bendable
malleable (can be hammered into a sheet)

What Did You Find Out?

1. Make an inference to explain how you know, *from your own observations* and from the information in the table,
   (a) that a penny is not made of pure copper
   (b) that a penny probably contains copper
2. Examine the information about the composition of Canadian pennies, which your teacher will give you.
   (a) Does the information support your answers to the previous question?
   (b) Explain how you know the penny is a mechanical mixture rather than
   - a solution of the two metals
   - a simple heterogeneous mixture

   (c) Is a mechanical mixture more suitable for coins than pure copper would be? Give reasons for your opinion.

A Theory to Explain Mixtures and Pure Substances

How can pure substances always be the same? How can solutions have blended properties but still be homogeneous? To answer these questions, scientists need a theory.

What is a theory? Is it just an off-the-wall idea? Can anyone make up a theory about anything? Well, anyone can, but no one will take it seriously unless it is useful in explaining observations. As more and more observations are made, a theory has to be useful in explaining those new observations, as well. A good theory can also help to predict what will be observed even before an investigation is done. For more information about theories, see "Science Inquiry," pages IS-9–IS-12.

The theory that scientist use to explain the properties of various mixtures and substances is called the **particle theory of matter**. Two important points of this theory state that

- All matter is made up of extremely tiny particles.
- Each pure substance has its own kind of particle, different from the particles of other pure substances.

DidYouKnow?

The idea that all matter is made of particles was first proposed about 2400 years ago in ancient Greece. Modern scientists have been gathering evidence to test the theory for more than 200 years. They have used the theory to explain their observations and to predict results of their investigations. The particle theory of matter has been so useful that it is now universally accepted.

Based on the particle theory, a **pure substance** is a material made up of only one kind of extremely small particle.

Before going on, you need to know just what the words "extremely small" mean in relation to the particle theory. Imagine a drop of water balanced on your fingertip. Can you guess how many individual water particles are clinging together to create the drop? The answer is about 1 700 000 000 000 000 000 000 — one thousand seven hundred million million million! No wonder you cannot see the particles with your unaided eye.

According to the particle theory, all water particles are exactly the same. Pure water, no matter how or where it is obtained, always contains identical particles. The properties of water are the way they are because of the particles. The particles do not change, so the properties of water do not change.

Similarly, all particles of table sugar are exactly the same. They are different from water particles, different from salt particles, different from the particles of any other pure substance. All particles of table sugar are the same, however.

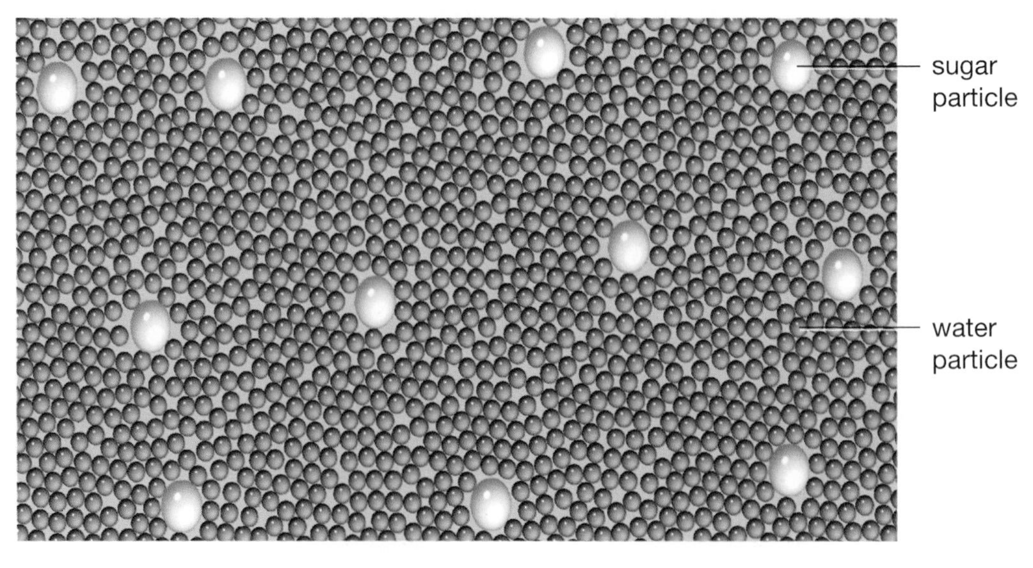

Figure 4.11 This is a solution, as explained by the particle theory. Every sugar particle has all the properties of sugar. Every water particle has all the properties of water. The solution blends the properties of both.

A solution of water and sugar contains particles of both water and sugar. There may be more or less water, more or less sugar. Whatever the proportion, the sugar particles remain sugar and the water particles remain water. The solution has both sets of properties. It is homogeneous because the sugar particles are evenly scattered among the water particles, as shown in Figure 4.11.

You will be using the particle theory often, to explain what you find out about solutions and mixtures. Try using the theory to explain what you already know about solids, liquids, and gases in the following activity.

Pouring? Shaping? Filling?

According to the particle theory, everything is made up of particles. Scientists use the idea of particles to explain the properties that are common to all solids, all liquids, and all gases. The following diagram shows how a solid, liquid, and gas are explained by the particle theory.

What to Do

Examine the diagram to find the answers to the following questions.

1. Name the state(s) in which a material
   - **(a)** has a fixed shape
   - **(b)** takes the shape of its container
   - **(c)** always fills whatever container it is in
2. Name the state(s) in which the particles are
   - **(a)** far apart from each other
   - **(b)** relatively close together
   - **(c)** free to move around
   - **(d)** held in fixed positions

What Did You Find Out?

Use the particle theory to explain your answers to questions 1 and 2.

A solid

B liquid

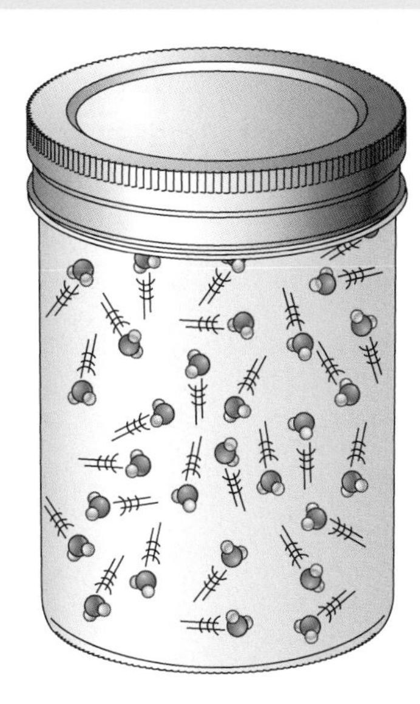

C gas

Career CONNECT

Don't Spoil It!

What if someone found a way to prevent all food from spoiling? Imagine how far food could be transported and the numbers of hungry people it could feed throughout the world.

Figuring out how to keep foods from spoiling is part of the varied job of a food scientist. From the moment it is harvested, food begins to deteriorate. Food scientists have to understand the changes that occur and find ways to control them. Other tasks of food scientists are related to determining shelf life, quality, safety, and nutritional content of existing food. Food scientists are also involved in the development of new food products that must be carefully analyzed and tested before they go on the market.

A food scientist needs to understand how one ingredient will combine with other ingredients and what the effects will be on taste, texture, and nutrition. Food scientists can work in areas as diverse as quality assurance, product development, food industry management, marketing, consumer relations, and food policy and regulation. A knowledge of the substances and mixtures that make up food and how they interact is essential in food science.

As a class, write a letter to a local person who works in the food industry, such as someone in a lab, or someone in a hotel or restaurant. Invite the person to speak to your class about his or her background and about the challenges of working in the food industry. Take careful notes and then see if you can find out more about the area of the food industry that most interests you.

Check Your Understanding

1. List the two points of the particle theory that you learned in this chapter.
2. According to the particle theory, how does a pure substance differ from a mixture?
3. Is the difference between a pure substance and a mixture always easy to observe? Explain your answer.
4. **Thinking Critically** A few coloured solid crystals of iodine dissolved in colourless liquid alcohol produce a coloured liquid. Every sample taken from the liquid has the same colour. No solid pieces can be seen.
   (a) In your notebook, draw a simple labelled diagram to represent the mixture.
   (b) Use this example to explain the meaning of "blended properties."
   (c) Use the particle theory to explain how the properties became blended.

CHAPTER at a glance

Now that you have completed this chapter, try to do the following. If you cannot, go back to the sections indicated.

In your classroom, identify something made of a material that is heterogeneous. To show you can be *sure* that the material is heterogeneous, list a set of properties for each different kind of matter in the material. (4.1)

In your classroom, identify something made of a material that *might* be homogeneous. List its properties. Explain why further investigation would be needed to be sure that the material really is homogeneous. (4.1)

Are heterogeneous materials more common in the natural environment than homogeneous materials? Why? (4.1)

Are heterogeneous materials more common in the human-made environment than homogeneous materials? Why? (4.1)

Explain how an understanding of mixtures and pure substances can help people make decisions about what to do when

- handling materials in the school laboratory
- handling materials, such as paint thinner, at home
- hearing about an "air quality advisory" on the news (4.1, 4.3)

Use the particle theory to explain why a mixture can be either homogeneous or heterogeneous. (4.2, 4.3)

Are the particles in each of the following identical or not identical? Give reasons for your answers. (4.2, 4.3)

- the bubbles of soda water
- the blobs of milk
- the pulp bits of orange juice

Prepare Your Own Summary

Summarize this chapter by doing one of the following. Use a graphic organizer (such as a concept map), produce a poster, or write the summary to include the key chapter ideas. Here are a few ideas to use as a guide:

- What is a sure way to know whether or not a substance is a mixture?
- What kind of observations are necessary to find out if a material is heterogeneous or homogeneous?
- Why does the identification of pure substances demand thorough investigation?
- How does the particle theory explain why the different parts of milk have different properties?
- What do smog and clear air have to do with the subject of this chapter?

CHAPTER

4 Review

Key Terms

matter
properties
mixture
heterogeneous
homogeneous
mechanical mixture
solution
particle theory of matter
pure substance

Reviewing Key Terms

If you need to review, the section numbers show you where these terms were introduced.

1. Divide a clean page so that you have eight roughly equal blocks. Write one key term at the top of each block. Now write the three most important facts or ideas about each key term in the blocks. Use a different-coloured pen to see how many ways the ideas are connected. (4.1, 4.2, 4.3)

2. What is the relationship between the following pairs of terms?
   (a) mixture, solution (4.2)
   (b) solution, homogenous (4.2)
   (c) pure substance, homogeneous (4.2, 4.3)
   (d) mixture, heterogeneous (4.2)
   (e) mechanical mixture, properties (4.2)
   (f) particle, pure substance (4.3)

Understanding Key Ideas

Section numbers are provided if you need to review.

3. Properties of a material can often help you to classify it as homogeneous or heterogeneous. (4.1)
   (a) Give an example of a material that is definitely heterogeneous.
   (b) Give an example of a material that is definitely homogeneous.
   (c) Easily observed properties can sometimes mislead you. Give an example of a material that looks homogeneous but is actually heterogeneous.

4. Define the terms "mixture" and "pure substance." Give two examples of each. (4.1, 4.3)

5. What scientific theory explains the make-up of matter? State the two main ideas of the theory that you learned in this chapter. (4.3)

6. Identify which of the following diagrams represents
   (a) a homogeneous mixture (4.2)
   (b) a mechanical mixture (4.2)
   (c) a pure substance (4.3)
   (d) a different pure substance (4.3)

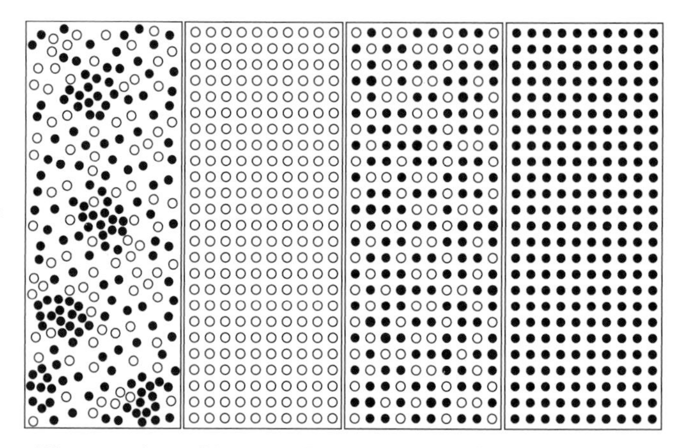

Developing Skills

7. **(a)** In your notebook, draw a concept map using the key terms listed in this chapter.
   (b) Where in your concept map could you put the following terms: gas, liquid, and solid? How could you show their relationship to the other terms in your map?

8. Use Styrofoam™ peanuts in two colours to make a model of
   (a) two pure substances
   (b) a heterogeneous mixture of two pure substances
   (c) a homogeneous mixture of two pure substances

9. **Design Your Own** Develop procedures to separate each mixture below. List the equipment you would use and outline your methods in a flowchart.
   (a) oil and water
   (b) paper clips and pennies
   (c) sawdust and sugar

10. A class of students decided to compare how many heterogeneous materials they had each identified as they studied this chapter. Here are 25 students' results: 10, 35, 30, 15, 14, 15, 20, 22, 26, 17, 15, 31, 26, 15, 23, 21, 12, 10, 18, 20, 30, 14, 16, 17, 18. Organize their data using a stem-and-leaf plot.

Problem Solving/Applying

11. In this chapter, you classified several mixtures as either heterogeneous or homogeneous.
    (a) How would you classify a soft drink when the sealed bottle is on the store shelf? Use the particle theory to support your answer.
    (b) How would you classify the soft drink when it is poured into a glass? Use the particle theory to support your answer.

12. Compare the properties and kind of particles in a crystal of sugar with the properties and kind of particles in a cube of sugar.

13. Using a hand lens and a pair of rubber gloves, decide if each of the following is heterogeneous or homogeneous.
    (a) potting soil for house plants
    (b) fertilizer for house plants
    (c) pesticide pellets for house plants

Skill POWER

For tips on making a stem-and-leaf plot, turn to page 486.

Critical Thinking

14. In which cases in question 13 was the hand lens necessary in helping you make your decisions? Explain why.

15. Callia says that salt and water can make a heterogeneous mixture. Mike says that it cannot. Decide who is right, explain why, and give evidence to support your answer.

16. In section 4.3, you learned that water is a pure substance. How is it possible that water from the tap in one location tastes different from water from the tap in another location?

Pause& Reflect

1. Pharmacists used to mix or "compound" prescription pills by hand, right in the drug store. Visit your local pharmacy, and find out what goes on behind the counter today. Do pharmacists still need to know about the properties of mixtures and pure substances? What is your personal opinion? Ask your pharmacist for a professional opinion.
2. Scientists use theories to explain what they observe. List two or three things that you took for granted before you started this chapter and that you can now explain. What theory has improved your understanding? How did it improve your understanding?
3. Go back to the beginning of this chapter on page 98 and check your original answers to the Getting Ready questions. How has your thinking changed? How would you answer those questions now that you have investigated the topics in this chapter?

CHAPTER

5 Solutions

Getting Ready...

- Why do some substances dissolve in water while others do not?
- Why does water not taste the same everywhere?
- Why does cold water contain more dissolved oxygen than warm water?

Science Log

Write what you think are the answers to these questions in your Science Log. Exchange answers with a partner and compare your ideas. You will find the answers as you study this chapter.

Solutions are everywhere you look. In the photograph, you can see maple sap being collected to make maple syrup. Both sap and syrup are solutions, but how are they different? A number of processes are used to change the watery maple sap into thick, delicious maple syrup.

Water in the environment is also a solution. It may contain several dissolved substances, including oxygen, carbon dioxide, minerals, and salts. Other substances found in water are the result of human activities. Knowing which substances are which is an important step in understanding our natural environment.

In this chapter, you will find out about solutions in general and then concentrate on water, maple syrup, and other water-based solutions.

Spotlight On Key Ideas

In this chapter, you will discover

- why some materials dissolve while others do not
- the many different kinds of substances that are dissolved in water in the natural environment
- what modern industrial process is based on a technique first carried out by the Aboriginal people in Canada

Spotlight On Key Skills

In this chapter, you will

- learn how the particle theory helps to explain dissolving
- carry out a distillation to separate the parts of a solution
- follow the steps in a diagram of an industrial process
- design your own experiment to test unknown solutions

Starting Point ACTIVITY

Cold Tea?

Most directions for making tea stress the need to use boiling water. What if you want a cold beverage, such as iced tea? Can you start with cold water?

What You Need

cold tap water

a transparent container, such as a glass mug or clean jam jar

tea bag

What to Do

1. Examine the dry tea bag. Smell it. Shake it.
2. Half fill the container with cold water. Place the tea bag on the water.
3. Observe for a few minutes. Sketch what is happening. Use arrows to indicate motion and labels to indicate colour changes. Let the water and tea bag sit undisturbed for several hours.

What Did You Find Out?

1. Is the dry tea a pure substance or a mixture? Give reasons for your answer.
2. Why does the dry tea stay inside the tea bag?
3. How do you know that something escaped from the tea bag?
4. What do you think escaped from the tea bag?

Extension

5. Repeat the activity using warm water. What differences did you notice? How would you explain your findings?

5.1 What Makes Materials Dissolve?

When you stir sugar into a glass of water, it forms a homogeneous mixture — a solution of sugar and water. Forming a solution by mixing two or more materials is called **dissolving**. We say that sugar dissolves in water. Mixing materials together does not always make a solution, however. Neither orange juice nor milk is a solution. For some reason, the pulp of the orange does not dissolve, and neither does the milk fat. Why do these materials not dissolve? What determines whether or not materials dissolve? To help answer this question, think about the following scenario.

On a late August afternoon, a group of students are sitting on the grass in a park, talking about the great times they had during the summer. As the afternoon creeps on, the occasional rumbling of an empty stomach is heard. During the next hour, the students get up one by one and drift off home for dinner.

Why did the students stay together in the park? Probably they were attracted to the friendship of the other members of the group. How long did this attraction keep them together? Only until something more attractive came along — dinner!

Think of the group of students as a group of particles. The particle theory of matter explains that materials are made from extremely tiny particles. Like the students in the park, these particles stay together because they are attracted to each other.

What happens to the attraction among sugar particles when you place sugar crystals in water? Why does each sugar crystal break up and dissolve? There must be other attractions. The sugar particles are also attracted to the water particles, so they mix with the water particles.

A group of water particles can attract a sugar particle more strongly than the other sugar particles around it can. Figure 5.1 shows what happens to sugar particles on the edge of a crystal. First the water particles pull a sugar particle away from the other particles in the crystal. Then the motion of the water particles carries it away. This makes room for more water particles to move in and attract another sugar particle. This process continues until all of the sugar is dissolved. Particles of sugar gradually move around and mix evenly throughout the water.

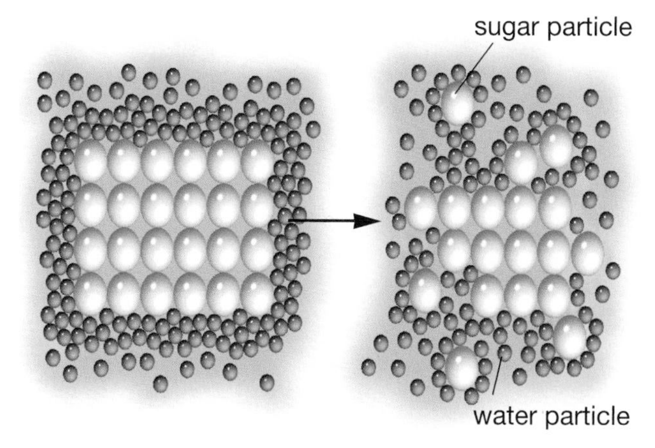

Figure 5.1 The particle theory can be used to explain dissolving.

This explanation uses two new ideas:

- Particles attract each other.
- Particles are always moving.

These ideas are useful additions to the particle theory that was introduced in Chapter 4. They can also help to explain why, for example, a drop of water, left on a table, seems to disappear. In liquid water, the water particles are attracted to each other. The particles are always moving about, however, and some are always on the outside of the drop. The particles on the outside of the drop occasionally "jump off" into the air. Over time, all of the particles jump off. They still exist, but they are independent and free to move about — that is, the water is now a gas, usually called water vapour. The change from a liquid to a gas, illustrated in Figure 5.2, is called evaporation.

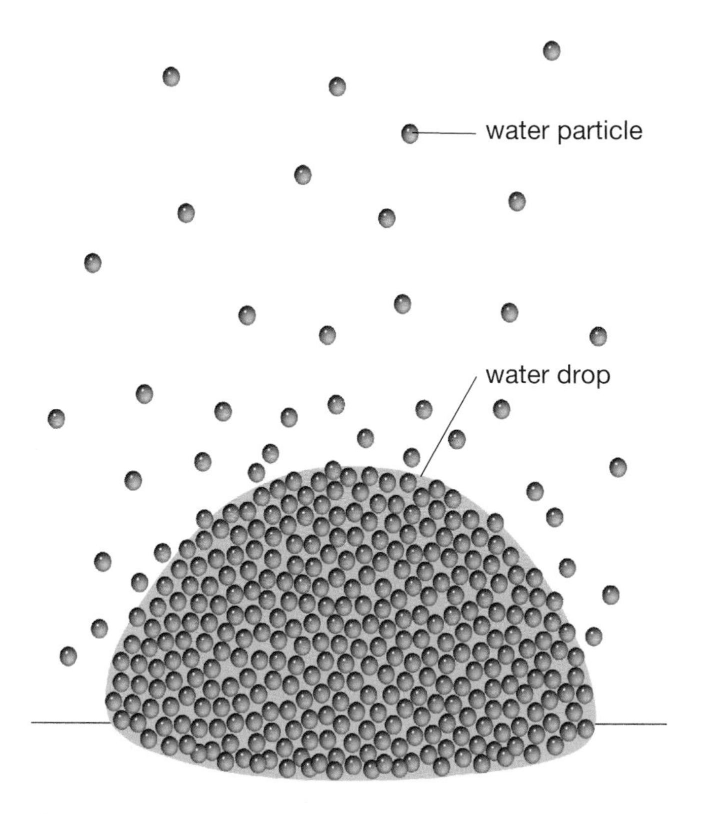

Figure 5.2 Evaporation occurs when a liquid changes to a gas.

At Home ACTIVITY

Here Today, Gone Tomorrow

How can you use the particle theory to explain an everyday event?

What to Do

1. Fill a glass with cold water, and let it sit for a few seconds until the water is still. Gently pour one level teaspoonful of sugar into the bottom of the glass. Set the glass aside, where it will not be moved.

2. Check the glass the next day. Write a brief description of the changes you observe, and explain them using the particle theory.

Why Do Some Materials Not Dissolve?

Figure 5.3 Nail polish remover can be used as a solvent for some materials that are insoluble in water. Because it is such an effective solvent, however, it may also dissolve plastic, making marks on counters and tabletops.

When sugar dissolves in water, we call the water the solvent and the sugar the solute. The **solute** is the substance that dissolves in a solvent to form a solution. There is usually less solute than solvent in a solution. The **solvent** is the substance that dissolves a solute to form a solution. There is usually more solvent than solute in a solution.

Another way to say "sugar dissolves in water" is to say "sugar is soluble in water." **Soluble** means able to be dissolved in a particular solvent. For example, sugar is soluble in water. The pulp in orange juice is not soluble, however, and neither are the small droplets of fat in milk; these are insoluble. **Insoluble** means not able to be dissolved in a particular solvent.

Simply mixing materials together does not always result in a solution. Why are some materials insoluble? This question can be answered by considering the attractions among particles. For the particles of milk fat to dissolve, they would have to be more attracted to the water particles than to the other particles of milk fat. However, they are not more attracted to the water particles. That is why particles of milk fat stay together and form insoluble globules.

The idea of particle attraction applies to many other combinations of substances. For example, it can help to explain why grass stains are difficult to get out of clothing. Grass stains are caused by the chlorophyll found in grass. The particles of chlorophyll are more attracted to each other than they are to water particles. To remove grass stains, you need to use a different solvent — one whose particles attract the particles of chlorophyll. Rubbing alcohol can be used to remove grass stains and other spots, as can the other solvents shown in Figure 4.1B on page 100. Chlorophyll is insoluble in water, but it is soluble in other solvents. Both solutes and solvents may be solids, liquids, or gases. Table 5.1 shows some examples.

Table 5.1 Examples of Solutions in Solid, Liquid, and Gaseous States

Example	Made up of	States of matter	
		Solute	Solvent
air	oxygen, other gases in nitrogen	gas	gas
humid air	water vapour in air	liquid	gas
mothballs	naphthalene in air	solid	gas
soda water	carbon dioxide in water	gas	liquid
vinegar	acetic acid in water	liquid	liquid
ocean water	various salts in water	solid	liquid
amalgam	mercury in silver, tin, zinc	liquid	solids
brass	zinc in copper	solid	solid

A beam of light can reveal whether a liquid that seems to be a solution really is one. The liquid on the left is a solution. Because it is truly homogeneous, there are no pieces of solid to reflect and scatter the light. In the liquid on the right, the undissolved pieces reflect and scatter the light. You may have seen a similar effect if you have watched dust "dancing" in a beam of sunlight or if you have seen fog through a car's headlights.

Changes in Mixtures and Solutions

In the rest of this chapter and in Chapter 6, you will learn about changes in mixtures and solutions. You will find out about mixtures and solutions in the environment, about techniques for making and separating mixtures, and about industrial processes that use these techniques.

One type of change that can occur in mixtures and solutions causes one material to *change into* another! For example, when you burn toast, part of the bread changes into a black material. You can scrape the black material off, but you cannot change it back into bread. In this kind of change, called a *chemical change*, it is not always possible to go back to the starting materials.

Check Your Understanding

1. (a) Define solution.
   (b) Give two examples of a solution, one from this book and one from your own experience.
2. (a) Identify a solute in ocean water.
   (b) Identify the solvent in ocean water.
3. In Table 5.1 on page 124, find
   (a) two solutions that are gases
   (b) two solutions that are solids
   (c) a solution in which both solute and solvent are liquids
   (d) a solution of a gas in a liquid
4. **Apply** Is fog a solution or a heterogeneous mixture? Explain.
5. **Thinking Critically** You have read that grass stains are soluble in alcohol. Suppose that you sponge alcohol onto your grass-stained jeans. When your jeans dry, however, the grass stains remain and have taken the shape of the areas you wet with alcohol. Now what will you do to remove the grass stains? (Hint: What will dissolve alcohol?)

At Home ACTIVITY

Same or Different?

A special kind of change occurs when you form a mixture in which one material changes into another and cannot be changed back.

What You Need

plaster of Paris, water, spoon, empty milk carton (1L), mixing bowl

What to Do

1. Examine the plaster of Paris, and make notes, describing its texture. (**Note:** Follow the safety precautions printed on the bag of plaster of Paris, as well as your teacher's instructions.)
2. Pour the plaster of Paris into the bowl and add some water. Stir the contents until its texture is smooth and creamy. Add more water, if necessary, to get the desired consistency.
3. Pour the mixture into the empty milk carton, which serves as a mold. Set the mold aside for several hours or overnight.
4. Tear away the milk carton from the plaster of Paris. Describe the texture of the plaster of Paris.
5. Use the spoon to carve a sculpture from the plaster of Paris. (Be as creative as you like!)

What Did You Find Out?

1. What was the texture of the plaster of Paris initially?
2. What was the texture after you had set it aside for a few hours?

3. What kind of change has taken place? How do you know?

5.2 Water in the Environment

Water has been called the "universal solvent" because it can dissolve so many materials. Calling it "universal" is exaggerating, of course, but no other common substance can match this property of water.

Water is a very important solvent. For example, about half of your blood is made up of water. It carries dissolved food materials and other essential substances all around your body. Water carries dissolved wastes out of your body, too. Because water is such a good solvent, it is used to wash people, clothes, food, cars, floors, and countless other things. How fortunate we are to have such a plentiful supply!

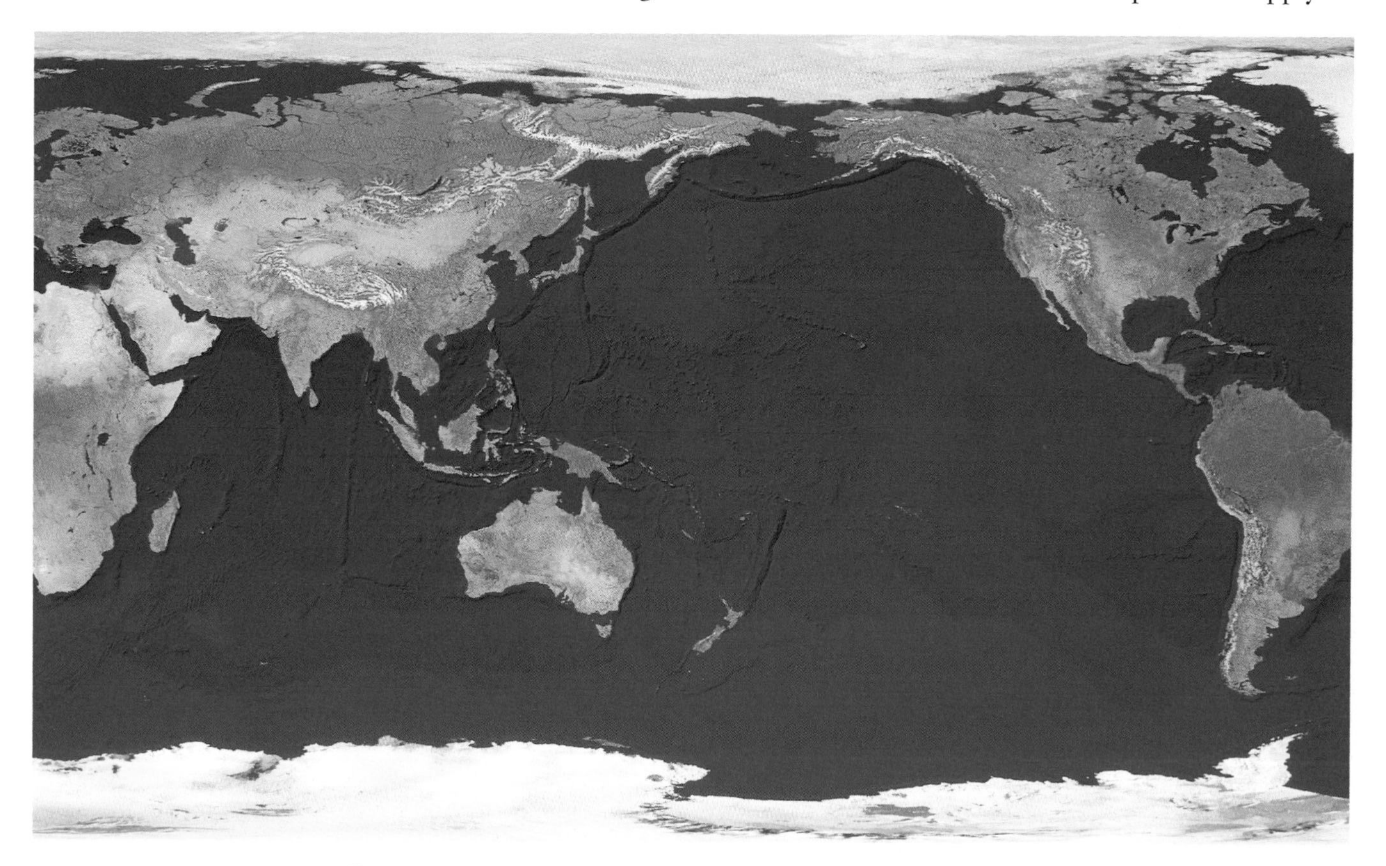

Figure 5.4 Two thirds of Earth's surface is covered with water.

Wait — is there really such a plentiful supply of water? The oceans hold 97 percent of the water on Earth, but this water contains a lot of dissolved materials. Ocean water contains so many solutes and is so salty that it is not drinkable. Perhaps it should *not* count as part of the supply of water we can use. Of the remaining 3 percent of Earth's water, approximately 2 percent is locked up as ice in the Arctic and the Antarctic. That leaves only about 1 percent of Earth's water that is usable "fresh" water.

Is this 1 percent safe to use, however? Because water is a good solvent, it may contain many solutes. Some of these may be harmless, but some are not. As well, water mixes easily with solid materials and may appear discoloured or cloudy. People prefer to use pure, clean water for drinking and for other purposes, such as cooking and washing, and sometimes even for irrigation and industrial purposes.

What affects the usability of water? Its usability depends on what is mixed with it. In this section, you will examine four types of materials that may be mixed with water in the environment:

- dissolved salts
- other dissolved solids
- undissolved solids
- dissolved gases

DidYouKnow?

Canada's Great Lakes are known as fresh-water lakes, but *all* fresh-water lakes and rivers are salty to a very small degree.

Dissolved Salts

All seawater contains salt, but the percentage can vary from one body of water to another (see Table 5.2). Also, many inland locations have pockets of "brackish" water — not as salty as seawater, but still too salty to drink or to use for irrigation. How can the dissolved salt be removed, so the water will meet people's needs?

It is relatively easy to recover the solute from seawater. Leave the seawater in the sunlight, and let the water evaporate as shown in Figure 5.5. Eventually you will be left with solid salt. Recovering the solvent from the solution is more complicated, as you will learn in the next investigation.

Table 5.2 Percentage of Salt in Some Bodies of Water

Body of water	Salt (%)
Arabian Sea	3.7
Atlantic and Pacific Ocean	3.2-3.7
Baltic Sea (some areas)	1 (or less)
Dead Sea	27
Great Salt Lake, Utah	5-27
Red Sea	4.1

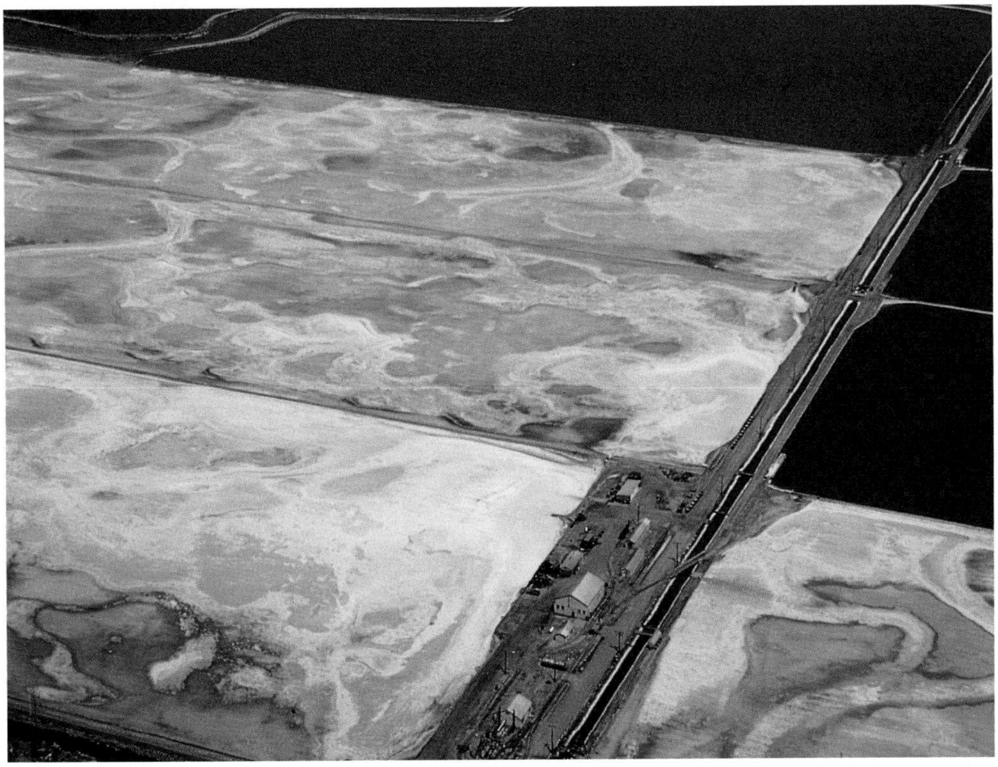

DidYouKnow?

There are many different kinds of salt. The scientific name for ordinary table salt is sodium chloride. Other salts include potassium chloride, magnesium sulfate, calcium nitrate, and ammonium carbonate. In the ocean, there are many dissolved salts, but the most common is ordinary table salt.

Figure 5.5 These shallow basins are being used to recover salt from seawater. The water evaporates from the basins, leaving behind crystals of salt.

CONDUCT AN

INVESTIGATION 5-A

Distillation

You just read that it is fairly easy to recover the solute from seawater. In this investigation, you will try out a process to separate water from a salt water solution.

Problem

How can the parts of a salt water solution be separated?

Safety Precautions

You will be working with a hot plate, boiling water, and steam. Use care around hot products, especially steam.

Apparatus

glass plate
marker
2 beakers (250 mL)
graduated cylinder
medicine dropper
hot plate
flask (500 mL)
stopper with glass tubing already inserted
50 cm rubber or plastic tubing
tongs for handling beaker or flask
measuring spoon (5 mL)

Materials

2 to 3 mL distilled water
salt

Skill POWER

To review units of measurement, turn to page 479.

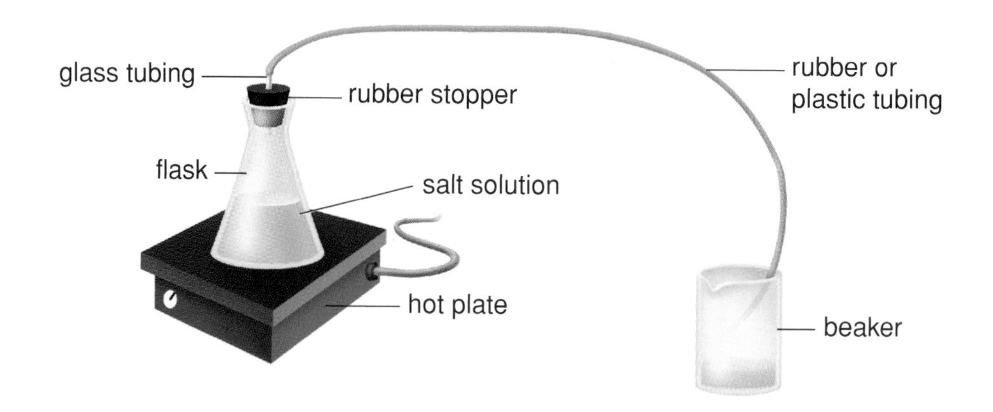

Procedure

1. Measure 100 mL of tap water in a graduated cylinder. Pour the water into a 250 mL beaker.

2. Add 10 mL of salt, and stir until it is dissolved.

3. Pour almost all of this solution into a 500 mL flask. (Save a few drops in the beaker to use for testing in step 5.)

4. Using the other beaker, set up the apparatus as shown in the illustration above.
   - **(a)** Turn on the hot plate, and boil the mixture until about half of the liquid has boiled away. Use the second beaker to collect any drops that come out of the tubing. **CAUTION:** Turn off the hot plate when half of the liquid is gone, and remove the end of the tubing from the beaker. If you heat the solution too long, the flask will break. **CAUTION:** The tubing will be hot!
   - **(b)** Allow the apparatus to cool thoroughly.

5 To test for the presence of dissolved salt, you will allow three different samples of liquid to evaporate. Be sure to use the same amount of liquid for all three samples.

(a) With the marker, divide the glass plate into three parts, marked A, B, and C.

(b) Use the dropper to put two or three drops of the salt solution you saved in step 3 on the glass plate (sample A).

(c) Clean the dropper, then use it to get some of the liquid you collected from the end of the tube. Put two or three drops on the glass plate (sample B).

(d) Clean the dropper, and do the same with the distilled water (sample C).

(e) Set the glass plate aside, where it will not be disturbed. After the liquid has evaporated, examine the plate to see what is left on the plate. Record your observations for each sample.

Skill POWER

For tips on measuring liquids, turn to page 481.

Analyze

1. **(a)** What did you see in the top half of the flask after the water began to boil?

(b) What change of state must have occurred inside the flask?

(c) How do you know?

2. **(a)** Describe what you saw at the end of the rubber tubing in the beaker.

(b) What change of state must have occurred inside the tubing?

3. After the drops of liquid evaporated, what remained on the glass plate for

(a) the salt solution?

(b) the liquid collected from the end of the tube?

(c) the distilled water?

Conclude and Apply

4. **(a)** Which substance, solute or solvent, did you collect in the beaker in step 4? Explain how you know.

(b) What happened to the substance that you did not collect in this beaker?

5. According to the particle theory, solute particles and solvent particles attract each other. In this investigation, what do you think overcame this attraction? That is, what do you think caused the salt particles and the water particles to separate?

6. In your opinion, is distillation a suitable method for producing large amounts of pure water (for example, to provide water for a large city)? Give reasons to support your answer.

Distillation and the Particle Theory

As you saw in Conduct an Investigation 5-A, **distillation** is a method for separating the parts of a liquid solution. In distillation, the solvent is heated to change it to a gas, then condensed to a liquid again, as shown in Figure 5.6. Condensation is the change from a gas to a liquid. Because the solutes do not change state, they remain behind. Distillation can remove other dissolved solids as well as salt by leaving them behind as the pure water evaporates.

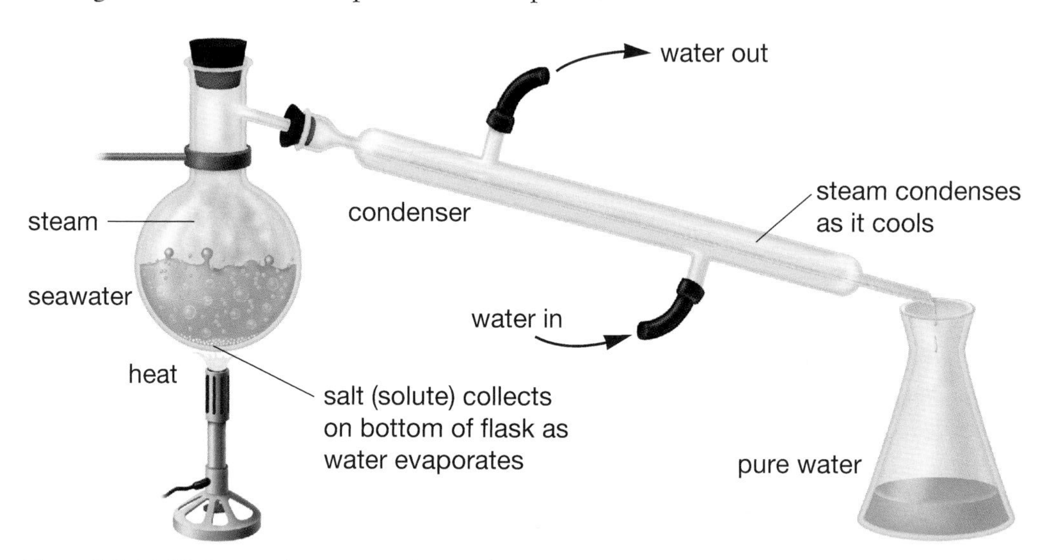

Figure 5.6 Distillation involves evaporating a solvent to separate it from the solute and then condensing it to a liquid. Water circulating in the condenser helps cool the steam as it passes through the tube.

How can distillation be explained by the particle theory? According to the particle theory, solute particles and solvent particles must attract each other in order to form a solution. What overcomes the attraction and allows them to separate? The particle theory also states that particles are constantly moving. This idea can be used to explain why water evaporates: occasionally some particles move away from the others and go off into the air. Now we can add one more idea to the theory:

- Particles at a higher temperature are moving faster than particles at a lower temperature.

When water is heated, for example, its particles move faster, so more of them escape. Water changes to a gas at a lower temperature than salt. Does this piece of the theory fit with what you already know? When you want to dry your hair quickly, does it help if you use a hair dryer set to "hot"? Of course it does. When the particles in the liquid water are heated, they move faster, so more of them escape the attractions of the other particles, and they move into the air.

In distillation, heating the solution causes all of the particles — salt and water alike — to move more quickly. The salt and water particles have different properties, however. Although both kinds of particles move more quickly, the water particles escape and the salt particles do not. The water evaporates, and the salt does not. In the tubing, as the hot gas cools, the water particles move more slowly again. The gas changes back into a liquid, and water drips out the end of the tube. The salt is left behind. The product is pure water, or distilled water.

Pause& Reflect

You have just considered the fifth (and last) idea in the particle theory, as it is discussed in this textbook. In your Science Log, list the five ideas with page references, so you can refer to them again.

Desalinating Water

Imagine being trapped on a desert island in the middle of the ocean. Is there any way you could produce drinking water? You could make the seawater drinkable if you could desalinate it. **Desalination** means removing the salt from salty water. In many parts of the world, fresh water is scarce, and seawater must be desalinated to provide drinking water. The desert tent method shown in Figure 5.7 is not expensive, but it is very slow. Also, it is practical only in areas that receive a lot of bright sunlight.

Along the Red Sea, where people live between salt water and the desert, huge desalination plants provide drinking water for thousands of people. These plants are very expensive to run, and use enormous amounts of energy.

INTERNET CONNECT

www.school.mcgrawhill.ca/resources/

Learn more about desalination by going to the web site above. Go to **Science Resources**, then to **SCIENCEPOWER 7** to find out where to go next.

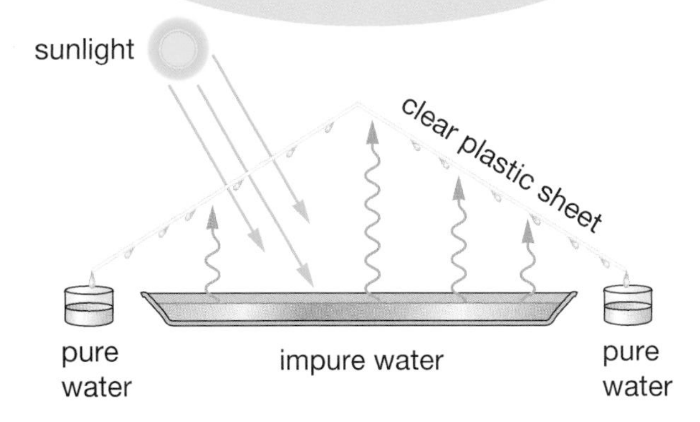

Figure 5.7 This "desert tent" apparatus uses a process very much like distillation. The water in the pans does not boil, but energy from the Sun causes it to evaporate. When the rising water vapour comes in contact with the cooler plastic sheets, it condenses. The drops of water run down the plastic into containers.

Other Dissolved Solids

In the natural environment, many other substances besides salt are dissolved in water. As rainwater runs over rocks and soil, minerals are dissolved and carried away in solution. Water that contains a lot of dissolved minerals is called **hard water**. Water with few dissolved minerals is called **soft water**. Rainwater, if it is caught before it comes in contact with the ground, is "soft." Distilled water is also "soft," since it contains no dissolved substances at all.

Hard water is perfectly natural. Hardness is not caused by human activity, nor is it harmful to humans. Hard water is safe to drink and usually safe to use for irrigation. Hardness does affect the usefulness of water for people, however. The dissolved minerals in hard water may interfere with the cleaning action of soap, so much more soap is needed to form the same amount of lather as in "soft" water. As well, hard water may cause a soap scum to form in a sink or bathtub. Hard water may also cause a solid material to be deposited on the inside of pipes and kettles. This deposit is caused by the dissolved materials separating from the solution, becoming "undissolved" (see Figure 5.8).

Figure 5.8 From time to time, electric kettles must be "descaled" so that the mineral deposits left by water evaporation do not interfere with the operation of the kettle's electrical element shown here.

Undissolved Solids

Would you want to drink the water shown in Figure 5.9? You can see that it contains undissolved materials. Tiny grains of soil and clay get stirred by the action of the river's currents against its banks. How could you obtain drinking water from this water? Distillation would certainly do the job, since it removes everything — dissolved and undissolved — from the water. Distillation is usually very costly, however, so it is used only when less expensive methods will not work.

Figure 5.9 Grains of clay and soil that wash into a river can give the water a muddy appearance.

Testing Water for Hardness

Does all hard water have the same hardness? How could you find out?

What to Do

Design the following tests. Ask your teacher's permission to perform the tests at school, or perform them at home with the help of an adult.

(a) a test for several water samples, to show which are "hard" and which are "soft"

(b) a test to determine the relative hardness of two water samples

Undissolved solids can often be removed from water using a much less expensive method: settling. If you have ever collected tadpoles or minnows for observation, you have probably noticed settling as the muddy water in your collecting jar gradually cleared. **Settling** means that undissolved matter will, over time, settle to the bottom of the container. If this method is too slow, the water can be filtered. Water purification plants in towns and cities filter water through sand and other substances. More natural methods to purify water are also being developed, as shown in Figure 5.10.

Figure 5.10 This tooth-shaped "sculpture" is located near the Ontario Science Centre in Toronto, Ontario. It is part of a search for methods to clean up the polluted water in the nearby Don River. The river water is pumped to the top of the "tooth," and the soil there filters out undissolved solids. The plants use many of the dissolved substances. The water that is returned to the river is relatively clear and pure.

Find the Undissolved Solids

Knowing if water contains undissolved solids can help you to decide whether a particular water supply can be used for a specific purpose without being treated (purified). How can you tell when a liquid contains undissolved solid(s)?

What to Do

1. Explain how you could decide, without a formal test, whether a water sample contained undissolved solids.
2. Design a test to compare the amounts of undissolved solids in several water samples. Ask your teacher's permission to perform the test at school, or perform it at home.
3. How could you treat water to remove undissolved solids? Design a process you could use, and test your process at school or at home.

Dissolved Gases

In the natural environment, water contains several dissolved gases from the air. Oxygen and carbon dioxide are necessary for the animals and plants that live in lakes and rivers. Some species of fish, such as trout, require large amounts of dissolved oxygen in the water in which they live. This means they can live only in cold water, because only cold water can contain sufficient oxygen.

Some Canadian waterways have naturally cold water, and some have warmer water. Human activities can cause the water in lakes and rivers to become warm, reducing the amounts of dissolved gases they contain. Figure 5.11 shows the holding pond of a dam. The surface layer of the water is heated by the Sun. It gets warmer than it would if the dam were not present. As the water warms, dissolved gases become "undissolved" and escape. The water that spills into the river contains less carbon dioxide, which is needed by water plants. It also contains less oxygen, which is needed by aquatic animals such as fish. Industries often have the same effect. They use water as a coolant, then release it back into the environment. Even if the water appears clean and pure, it may have been warmed. The warmer water contains less dissolved oxygen and carbon dioxide.

Figure 5.11 The water behind this dam does not contain harmful dissolved substances. However, the sunlight has increased the temperature of the water, which causes essential gases that were dissolved in it to escape.

Find Out ACTIVITY

Controlling Water Temperature

The water that industries use often becomes warm. If it is cooled before it is released, however, it will not affect the dissolved gases.

What You Need

2 identical bowls

thermometer

measuring cup (250 mL)

warm tap water

What to Do

1. Pour 750 mL of warm tap water into each bowl.
2. Check the temperature of the water in each bowl. If necessary, add hot tap water to one bowl until the water in both bowls gives the same reading. Then measure the amount of water in the bowls again so that both bowls contain the same amount.
3. Leave one bowl of water to cool by itself.
4. Use the other bowl for the procedure below.
   - **(a)** Scoop out a cupful of water.
   - **(b)** Raise the cup about 30 cm above the bowl.
   - **(c)** Allow the water to trickle slowly back into the bowl.
5. Repeat step 4 ten times.
6. Check the temperature of the water in each bowl.

What Did You Find Out?

1. Which bowl of water became cooler?
2. Explain how the trickling procedure affects
   - **(a)** the temperature
   - **(b)** the amount of dissolved oxygen in the water
3. **(a)** Why should an industry cool water before releasing it to the environment?
   - **(b)** How could this be done without using costly refrigeration?

Our Use of Water in the Environment

People everywhere must have a supply of clean drinking water. In many places, they must use water for irrigation in order to grow food. Where water is scarce, it is recognized as a precious resource.

Canadians, and citizens of other industrialized countries, use water for more than drinking and irrigation. Think about the many ways that you use water and observe it being used every day. Because we have so much water available, we seem to find ways to use it. Water's ability to dissolve many other substances makes it helpful in thousands of different ways in homes and industries. In the investigation on the following page, you will work in a group to explore some of the ways in which we use water. How does our use of water affect its quality?

Pause& Reflect

How do you affect the water in your environment? With a partner, think of some changes that you could make to reduce the amount of water you use. In your Science Log, write notes about these changes and how you might encourage others to make them.

INVESTIGATION 5-B

Using Water

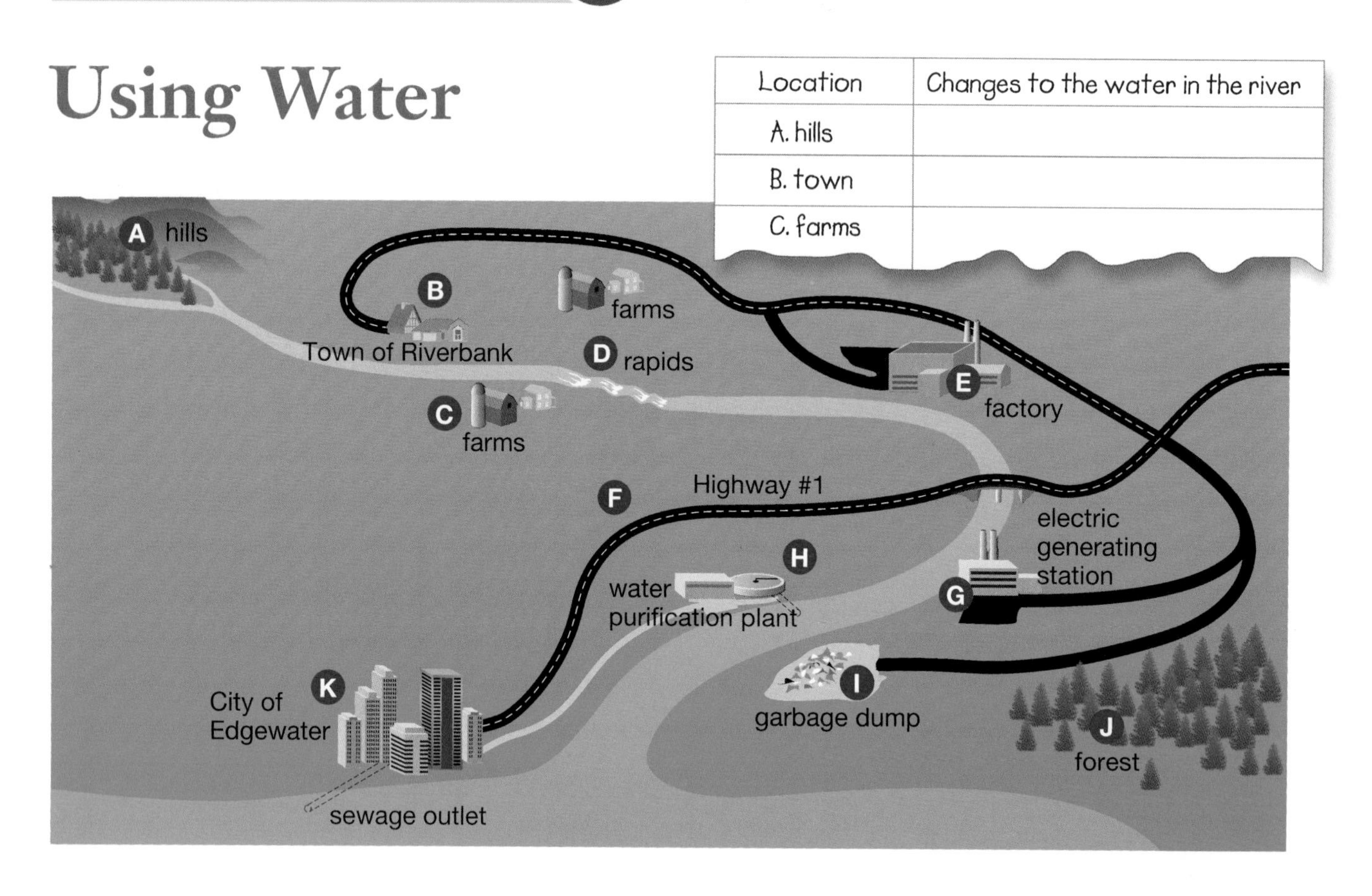

Location	Changes to the water in the river
A. hills	
B. town	
C. farms	

Think About It

You have been thinking about some of the ways that water is used in our society and about human activities that affect the quality of water. This activity will help you to examine some of these ideas in more detail.

What to Do

1. With your group, examine the illustration. For each marked location, discuss how the water in the river might be affected.

2. Make a table like the one above, including all the locations. Give your table a title.

3. For each location, list materials that might be added to or removed from the water. (Hint: Think of solids, liquids, and gases, both dissolved and undissolved, that occur naturally and that result from human activities.)

Analyze

1. (a) How many human activities that might improve the quality of the water did your group identify?
   (b) How many human activities that might reduce the quality of the water did your group identify?

2. Select one of the human activities identified by your group. In your group, discuss the benefits and the drawbacks of this activity for the community.

Extend Your Skills

3. Have each person in your group select one potentially hazardous material you identified in this investigation. Find out about any government regulations that control its use.

Check Your Understanding

1. (a) What is the purpose of distillation? Use the terms "solute" and "solvent" in your answer.
   (b) Use the particle theory of matter to explain how distillation works.
2. Is the water that is found in nature always pure? Give reasons and examples to support your answer.
3. Compare the distillation method shown in Figure 5.6 with the desert tent evaporator in Figure 5.7. What is similar? What is different? Refer to the particle theory of matter in your answer.
4. **Apply** A solution can contain more than one solute. Identify several solutes that are dissolved in water to make a cola drink.
5. In the past, people collected rainwater from their roof in a barrel. Water from the rain barrel was used for washing hair and clothes. Why do you think people did this, rather than using water from a river or a well?
6. **Thinking Critically**
   (a) How does river water become "hard"?
   (b) How does hardness affect the safety of water for drinking?
   (c) How does hardness affect the usefulness of water?
7. Canada has the largest supply of fresh water on Earth. Describe at least five ways in which human activities affect our freshwater supply.
8. In what kind of environment is distillation likely to be used to produce water for drinking?

INTERNET CONNECT

www.school.mcgrawhill.ca/resources/

Fresh water occurs in many lakes and rivers. However, most of Earth's fresh water is found underground, within the soil and rocks. How did water get there and where is it going? What substances are dissolved in it? How might human activities affect groundwater? To learn more about groundwater, visit the above web site. Go to **Science Resources**, then to **SCIENCEPOWER 7** to find out where to go next. Prepare a brief report on your findings.

5.3 Sugar in Solutions and Mixtures

Figure 5.12 The process used in making maple syrup has changed very little since early European settlers were taught by Aboriginal people how to process the sap from maple trees.

In North America, long before European settlers arrived, Aboriginal people were processing the sap from maple trees. From the sap, which is mostly water, they produced maple syrup and maple sugar (Figure 5.12). Although the equipment that is used today for producing maple syrup is different, the basic process has changed very little.

To make maple syrup from sap, most of the water has to be removed from the solution. The starting solution is very **dilute**. A dilute solution is a solution that contains relatively little solute. When most of the solvent (water) has been removed from the sap, the solution of sugar and other materials is much more **concentrated**. A concentrated solution is a solution that contains a lot of solute for the amount of solvent. In fact, it takes 30 to 40 L of sap to make 1 L of maple syrup.

To get rid of the excess water, many pails of sap are poured into a large boiler and heated over an open fire for several hours. The boiling process makes the sweet sugar solution more concentrated and also changes some of the sugar into caramel, giving the syrup its golden colour.

DidYouKnow?

Sugar maples grow in eastern Canada, but other maples can be found all the way to the Pacific Coast. The sap that rises in the trunk of a maple tree in the spring carries food and water that are needed for the buds to develop into leaves. The water comes from the soil around the tree's roots. The sugary food was made the previous year by the tree itself and has been stored in the tree's roots all winter. When the new leaves emerge, they will make food for the rest of the season. Some of this food will be stored in the roots for the following spring.

Processing Maple Sap

The sap of a sugar maple tree carries nutrients to all parts of the tree. The sap is mostly a dilute solution of sugar in water, but there are also small amounts of many other substances needed by the tree. As the water is boiled off, all of the solutes become more concentrated. The combined flavours of these solutes give maple syrup its distinctive taste. Figure 5.13A shows the stages in the commercial production of maple syrup.

Figure 5.13A Stages in the commercial production of maple syrup

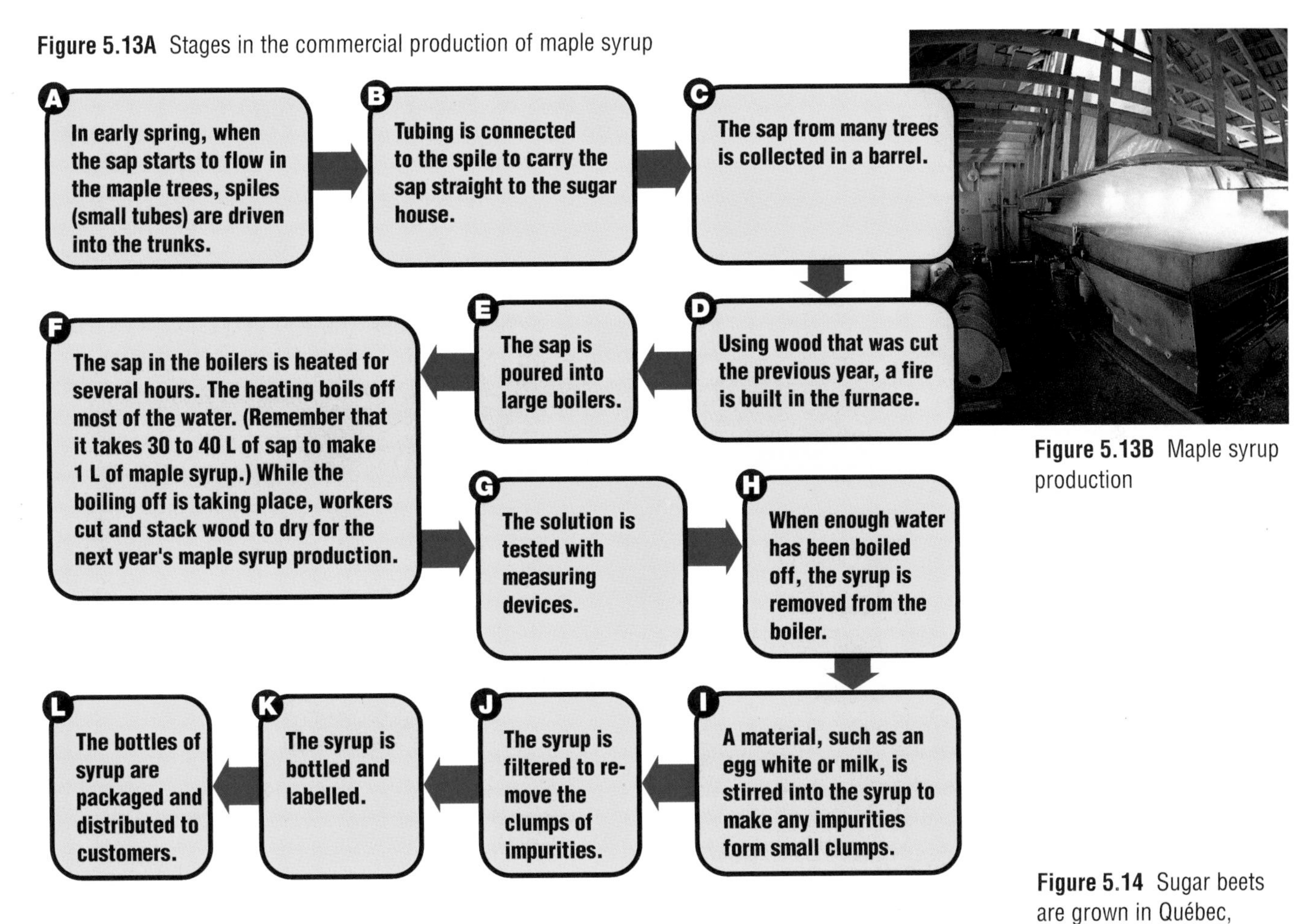

Figure 5.13B Maple syrup production

Making Sugar

In early days, the maple sap was usually boiled until almost all of the water was removed. This process produced solid maple sugar. Except for the slightly darker colour and different taste, maple sugar is very similar to the table sugar we use today. In fact, many of today's sugar bush operations produce small amounts of maple sugar to be sold as candy. Maple sugar is expensive to produce, however. Almost all of the sugar we buy today is produced from sugar cane and sugar beets. Most sugar refineries in eastern Canada use imported sugar cane as the raw ingredient. Most refineries in western Canada use domestically grown sugar beets (see Figure 5.14). Regardless of which raw material is used, the end products are very similar.

Figure 5.14 Sugar beets are grown in Québec, Manitoba, and Alberta.

INVESTIGATION 5-C

A Sweet Process

Think About It

White sugar, golden brown sugar, dark brown sugar, and molasses are all produced from either sugar beets or sugar cane. You might wonder how one raw material can be turned into so many different products. The answer lies in the processing.

What to Do

With a partner, examine the flowchart below that shows how various sugar products are made.

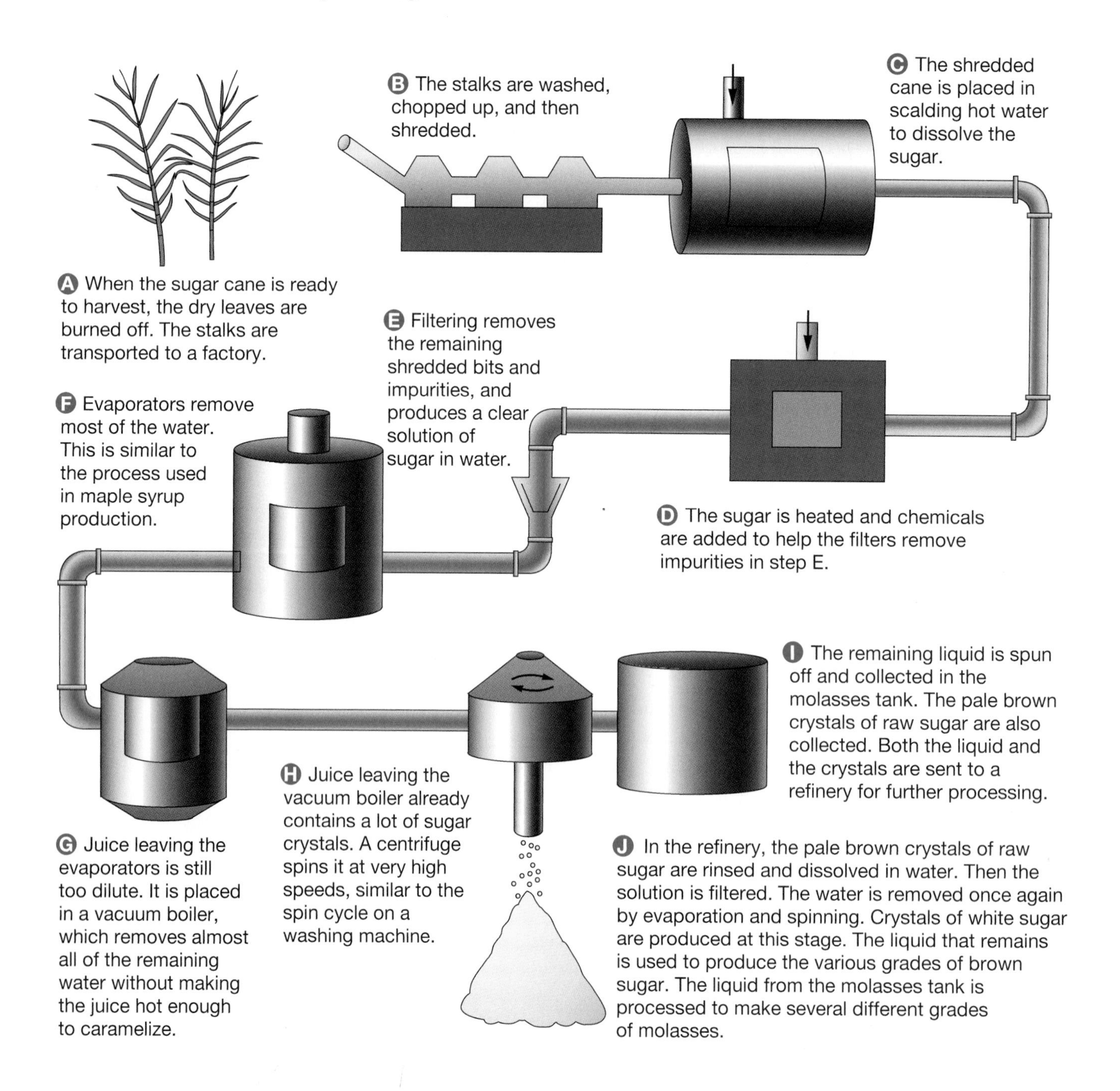

Sugar cane is an important economic crop in many countries. This photograph shows a sugar cane field in Cuba.

Analyze

1. What is the purpose of step B?
2. Does step C produce a solution or a mechanical mixture?
3. Filtration (step E) is a method for separating mechanical mixtures. Draw a sketch to show how you think filtration works. Use the particle theory to create labels that explain how filtration works.
4. Evaporation (steps F and G) is a method for recovering a dissolved solute from a solution. Draw a sketch to show how you think evaporation works. Use the particle theory to create labels that explain how evaporation works.
5. **(a)** In what ways is the processing of cane sugar the same as the processing of maple sap?
   (b) What extra steps are involved at the beginning of sugar cane processing? Explain why the extra steps are needed.
6. **(a)** What are the first two products that come from a sugar factory?
   (b) What happens to each of these products?
   (c) What is the waste product that is discarded in this process?

Career CONNECT

A Seasonal Industry

Centuries ago, Canada's Aboriginal people collected "sweetwater" from maple trees and poured it into hollowed-out logs. They turned the clear, colourless sap into a sweet, amber liquid that we now call maple syrup, adding heated rocks to speed up the evaporation of the sap.

Some of Canada's 13 000 maple syrup producers in Ontario, Québec, and the Maritimes still use the traditional method to collect and process the sap. They carry it to the sugar shack on sleds or in wagons. There it is boiled to remove the extra water and reduce the syrup so that it is the right flavour and thickness.

Other producers use more modern methods: kilometres of plastic tubing connect thousands of trees, carrying the sap to the sugar shack. Filtering systems and pasteurization processes are then used to prepare the syrup for sale all over the world. The basic process is still the same, however — many hours of boiling large quantities of sap down to a small quantity of delicious concentrate.

Since the sap runs for only four to six weeks in the early spring, maple syrup farmers usually conduct other kinds of farming operations during the rest of the year. Employees who help these farmers during the sap season are called seasonal workers. Are there farms in your part of Canada that require "hands" or helpers during planting or harvesting times? Contact an employment centre to find out what seasonal jobs exist in your area and how people are hired for this type of work. If someone in your class has an older sister or brother who has done seasonal work on a farm, invite the person to visit your class to talk about the experience.

Check Your Understanding

1. (a) Find any products around your home that are sold in a concentrated form, to be diluted before use.
   (b) Why might such a product be labelled as a "concentrate"?
2. Compare the raw materials that are used to make maple sugar with those that are used to make ordinary white sugar.
3. (a) Name two end products that are made from maple sap.
   (b) Name four end products that are made from sugar cane or sugar beets.
4. Sugar can be produced from three different raw materials: cane, beets, and maple sap. Regardless of the raw material, a lot of fuel is needed. Explain why.
5. (a) Write your own definition of filtration.
   (b) What is the role of filtration in maple syrup production? What is its role in cane or beet sugar production?
   (c) What other separation method is used to make these products? Write your own definition for this other separation method.
6. **Apply** Which requires more steps to produce, cane sugar or maple syrup? Explain why.
7. (a) Describe the role of filtration in making coffee.
   (b) What enables the coffee filter to do its job?
   (c) "A tea bag is a kind of filter." Explain.

CHAPTER at a glance

Now that you have completed this chapter, try to do the following. If you cannot, go back to the sections indicated.

Use the particle theory of matter to explain how sugar dissolves in water. (5.1)

A drop of water falls on a tabletop, and evaporates after a period of time. How does the particle theory explain this process? (5.1)

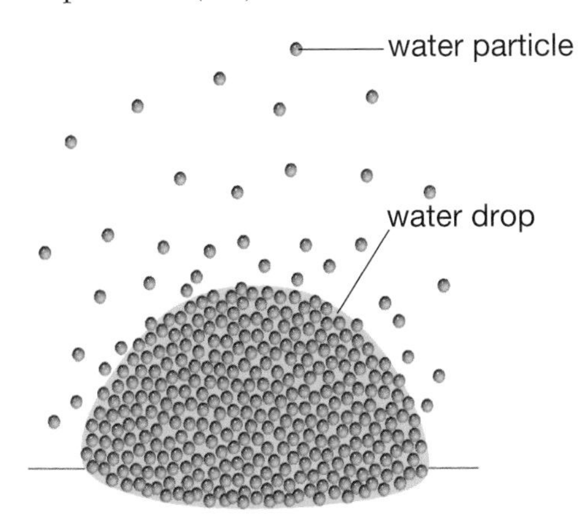

A solute and a solvent combine to form a solution. Explain the difference between the two components. (5.1)

When oil and water are combined, the two materials do not mix. Use the particle theory to explain why this happens. (5.1)

Why is water called the "universal solvent"? Give some examples to help explain your answer. (5.2)

Describe what happens when a liquid solution is distilled. How is this process explained by the particle theory? (5.2)

Compare the processes of distillation and desalination. Describe how desalination works. (5.2)

Describe alternatives to distillation for removing "undissolved" solids from a water supply. (5.2)

Explain how warmer water can affect aquatic life. (5.2)

Explain the difference between a dilute and a concentrated solution. (5.3)

Describe the process used to make maple syrup from sap. (5.3)

Compare traditional and modern methods of collecting maple sap. (5.3)

Prepare Your Own Summary

Summarize this chapter by doing one of the following. Use a graphic organizer (such as a concept map), produce a poster, or write the summary to include the key chapter ideas. Here are a few ideas to use as a guide:

- Use the particle theory of matter to explain why some materials dissolve while others do not.
- Identify some of the substances that are dissolved in water in the environment.
- How does the particle theory explain the evaporation of liquids?
- Explain the terms "hard water" and "soft water."
- List the five key points in the particle theory, as discussed in this text.
- Copy the following diagram into your notebook and label it.

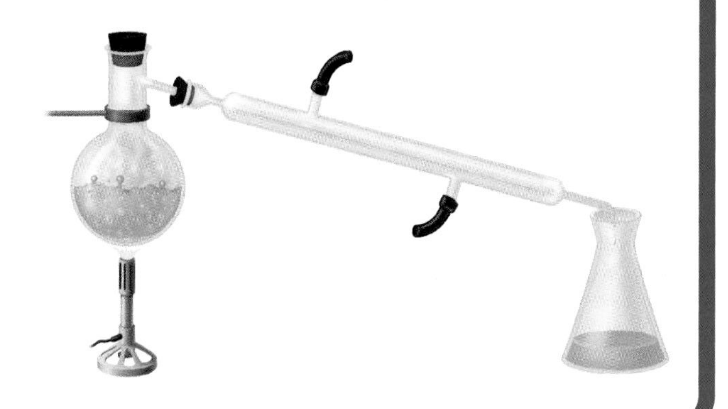

CHAPTER

5 Review

Key Terms

dissolving
solute
solvent
soluble
insoluble
distillation
desalination
hard water
soft water
settling
dilute
concentrated

Reviewing Key Terms

If you need to review, the section numbers show you where these terms were introduced.

1. In your notebook, copy and complete the following statements.
   (a) In a sugar and water solution, the sugar is the ▇▇▇▇▇ and the water is the ▇▇▇▇▇. (5.1)
   (b) Substances that ▇▇▇▇▇ in a liquid are ▇▇▇▇▇ in that liquid. (5.1)
   (c) Iron is ▇▇▇▇▇ in water. (5.1)
   (d) ▇▇▇▇▇ is a way of recovering the solute and the solvent from a solution. (5.2)
   (e) When steam is cooled, it changes back into water. This process is called ▇▇▇▇▇. (5.2)
   (f) Both ▇▇▇▇▇ and ▇▇▇▇▇ take place during the process of distillation. (5.2)
   (g) You need to ▇▇▇▇▇ seawater to make it drinkable. (5.2)
   (h) If a solution contains a lot of solute in a small amount of solvent, it is ▇▇▇▇▇. (5.3)
   (i) Weak solutions are also called ▇▇▇▇▇. (5.3)

Understanding Key Ideas

Section numbers are provided if you need to review.

2. Give examples of three common solutions. For each example, name the solute and the solvent. (5.1)

3. Compare the attractive force of the particles in a solvent and a solute when
   (a) the solute will dissolve in the solvent
   (b) the solute will not dissolve in the solvent (5.1)

4. Draw and label a diagram to show a particle model of
   (a) a dilute solution
   (b) concentrated solution. (5.1)

5. A single grain of house plant fertilizer is a mixture made up of many millions of particles of different substances. Use the particle theory to answer the following questions:
   (a) Why do the particles of house plant fertilizer stay together and form a grain?
   (b) When the grain of fertilizer is put into water, what causes the particles to dissolve?
   (c) House plant fertilizer will not dissolve in oil. Explain why. (5.1)

6. Identify the solute and the solvent in each of the following solutions: (5.1)
   (a) instant coffee
   (b) soda water
   (c) ocean water
   (d) air
   (e) brass (5.1)

7. When sugar dissolves in water, the water particles exert a force of attraction on the sugar particles. Compare the action of this force between a dilute and a concentrated solution. (5.1)

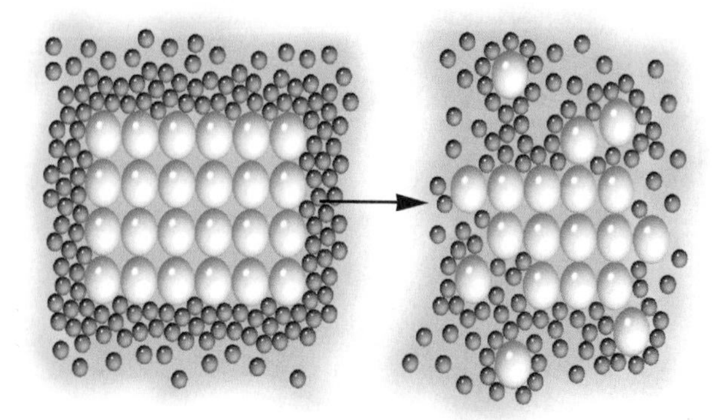

Developing Skills

8. (a) Describe the steps you would take to recover pure water from dirty dishwater. What is the name of this process?
   (b) Would this process be practical for conserving water in your home? Explain your answer.

9. (a) Human activities add harmful materials to our water. Name five such activities and explain why water is used in each.
   (b) Choose one of these activities, and suggest how the same result could be accomplished without polluting our water supply.

10. In your notebook, copy and complete the following concept map. Use the following words as needed (you may want to add more words to your concept map): solvent, distillation, dissolves, soda water, ocean water, mothballs.

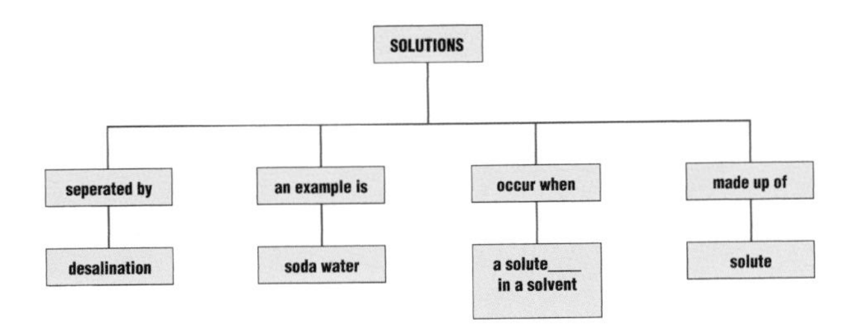

11. **Design Your Own** You are given two samples of water, one from Lake Ontario and one from the Atlantic Ocean. Somehow the labels have fallen off. Design an experiment, other than tasting, to determine which is which.

Problem Solving/Applying

12. Suppose you have been given a jar of water that contains a pinch of salt and sand. Explain the steps you would use to remove the sand. Then explain how you would remove both materials.

13. Suppose you work for a company that produces distilled water. The director of marketing has asked you to design an advertising campaign to show that your company's product is the purest one on the market. Outline your design.

Critical Thinking

14. The Dead Sea is located in the Middle East between Jordan and Israel. Its surface is about 400 m below sea level. The water in the sea has a salinity level almost six times that of the ocean. Explain why the Dead Sea is so salty.

15. Fats and oils are not soluble in water. Why do greasy dishes become clean when you wash them?

Pause& Reflect

1. You have now examined five ideas from the particle theory:
   - All matter is made up of extremely tiny particles.
   - Each pure substance has its own kind of particle, different from the particles of other pure substances.
   - Particles attract each other.
   - Particles are always moving.
   - Particles at a higher temperature are moving faster than particles at a lower temperature.

   In your Science Log, state which of these ideas you think are most useful to explain
   (a) why water is different from sugar
   (b) why nail polish is insoluble in water
   (c) why water evaporates

2. Go back to the beginning of this chapter on page 120 and check your original answers to the Getting Ready questions. How has your thinking changed? How would you answer those questions now that you have investigated the topics in this chapter?

CHAPTER

6 Working with

Getting Ready...

- What do a diver and a carbonated liquid have in common?
- Why do some materials dissolve quickly and some dissolve slowly?
- How is gold mined?

Write your ideas about these questions in your Science Log. You will find the answers as you read through this chapter.

Where are the solutions in the large photograph shown here? They are everywhere! By now, you are familiar with solids, liquids, and gases in solution. Did you know, however, that there is a life-threatening solution in this photograph, inside the diver's body?

The diver was participating in a controlled dive to examine a shipwreck. The increased atmospheric pressure beneath the sea caused higher than normal amounts of dissolved gases to be present in the diver's blood. If a diver ascends to the surface too quickly, nitrogen bubbles will be released from the blood into the body tissues, with painful and possibly even fatal results. This condition is often referred to as "the bends." To avoid it, a diver must resurface gradually from any depth greater than 9 m. This helps the nitrogen to be released slowly from the blood (into the lungs) so that it can be breathed out safely.

In this chapter, you will find out more about solutions and mixtures as you investigate how they are made and how they are separated. You will see how knowledge of solutions and mixtures is applied in industries, and how the processes result in products we use every day.

Solutions and Mixtures

Starting Point

Separating Mixtures

How many ways can you find to separate a mixture? What effect will different items in the mixture have on the method(s) you choose?

What You Need

paper clips
cotton balls
plastic pen caps
pennies
marbles
large bowl
water

What to Do

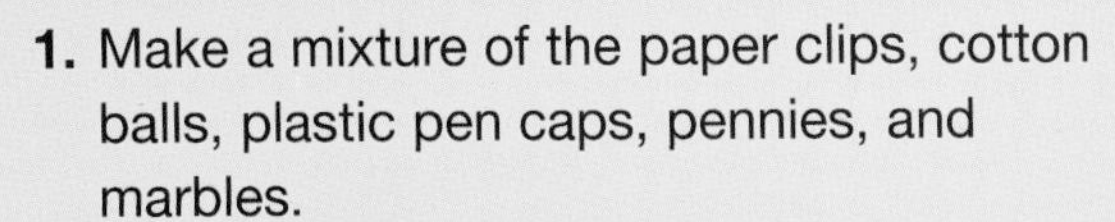

1. Make a mixture of the paper clips, cotton balls, plastic pen caps, pennies, and marbles.
2. Suppose that you have a huge amount of the mixture. In a group, discuss methods you could use to separate the parts.
3. Place the mixture in the bowl, and pour enough water in the bowl to cover it. What method (not discussed in step 2) could you use to separate the water?

What Did You Find Out?

1. What simple device lets you pull one part of the mixture away?
2. What parts are more difficult to separate?
3. In your notebook, write each word from the illustration below, and explain how the word relates to this activity. (Look up unfamiliar words in a dictionary.)

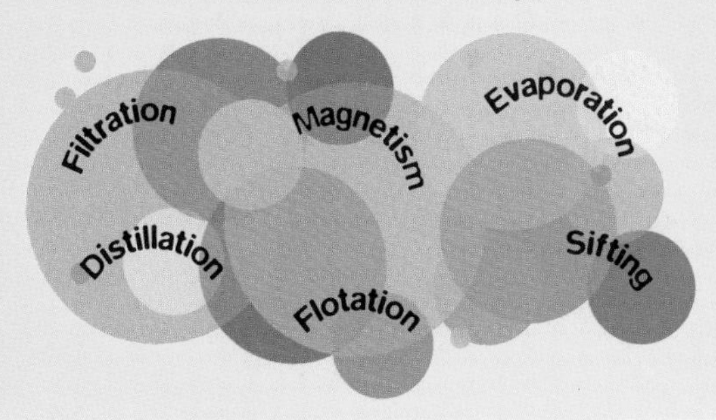

Spotlight On Key Ideas

In this chapter, you will discover

- why there are differences in the rate of dissolving and concentration of solutions
- how the particle theory can explain observations about solutions and mixtures
- why two mixtures in particular are valuable natural resources in Canada

Spotlight On Key Skills

In this chapter, you will

- use scientific equipment to make accurate measurements
- use numerical data to communicate your findings
- design experiments to find out how the rate of dissolving is affected by different variables

6.1 How Much Can Be Dissolved?

Skill POWER

For information on how to use a balance, turn to page 481.

You have seen that water can dissolve many different materials. In fact, more substances are soluble in water than in almost any other solvent. It is for this reason, as you learned in Chapter 5, that water is often called the universal solvent. In the next few pages, you will find out about a few of the substances that dissolve in water and discover how much of each can dissolve in a given amount of water.

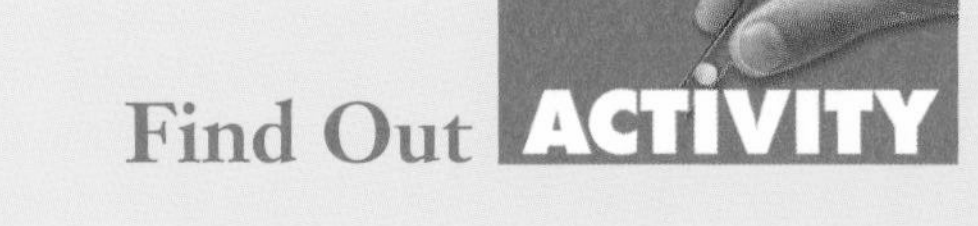

Find Out ACTIVITY

How Much Is Too Much?

According to the particle theory, solutions form because particles of different substances are attracted to each other. Because of the attractions, they mingle, forming a homogeneous mixture. This leads to two questions:

- Is there a limit to the amount of solute that can dissolve? In other words, can a solution be made more and more concentrated, indefinitely?
- If there is a limit to the amount of solute, is the limit the same for different solutes and solvents?

What You Need

graduated cylinder
250 mL beaker
2.5 mL measuring spoon
salt
water
stirring rod
balance

What to Do

1. Using the graduated cylinder, measure 100 mL of cold tap water and pour the water into the beaker. Add 2.5 mL of salt to the water. Use a stirring rod to stir until all of the salt is dissolved.
2. Now add another 2.5 mL of salt, and stir. Repeat until some salt will not dissolve, even after prolonged stirring.
3. Record the quantity of salt that you were able to dissolve in 100 mL of water.
4. Using a balance, determine the mass (in grams) of this quantity of salt.
5. Record your result on the class chalkboard along with the results of the rest of the class.

What Did You Find Out?

1. **(a)** How many grams of water were in your beaker at the beginning? (**Note:** 1 mL of water has a mass of 1 g.)
   (b) How much salt (how many grams) can be dissolved in this mass of water?
2. **(a)** Did everyone in your class obtain a similar result?
   (b) Use your knowledge of the particle theory to explain why you might expect everyone's results to be similar.

Homogeneous or heterogeneous?

Saturated Solutions

There is a limit to just how concentrated most solutions can become. When you prepare a solution, you reach this limit when no amount of stirring can make more solute dissolve in the solvent. At this point, the solution is said to be saturated. A **saturated solution** is one in which no more solute will dissolve at a specific temperature. An **unsaturated solution** is one in which more of the solute could dissolve at the same temperature.

How can the particle theory explain saturated solutions? Look at the photograph and the diagram in Figure 6.1. The photograph shows crystals of bluestone in a saturated bluestone solution. According to the particle theory, no more solid dissolves because all of the water particles are already attracted to as many bluestone particles as they can be. No additional water particles are available to attract more bluestone particles away from the bluestone crystals, even though individual particles are moving.

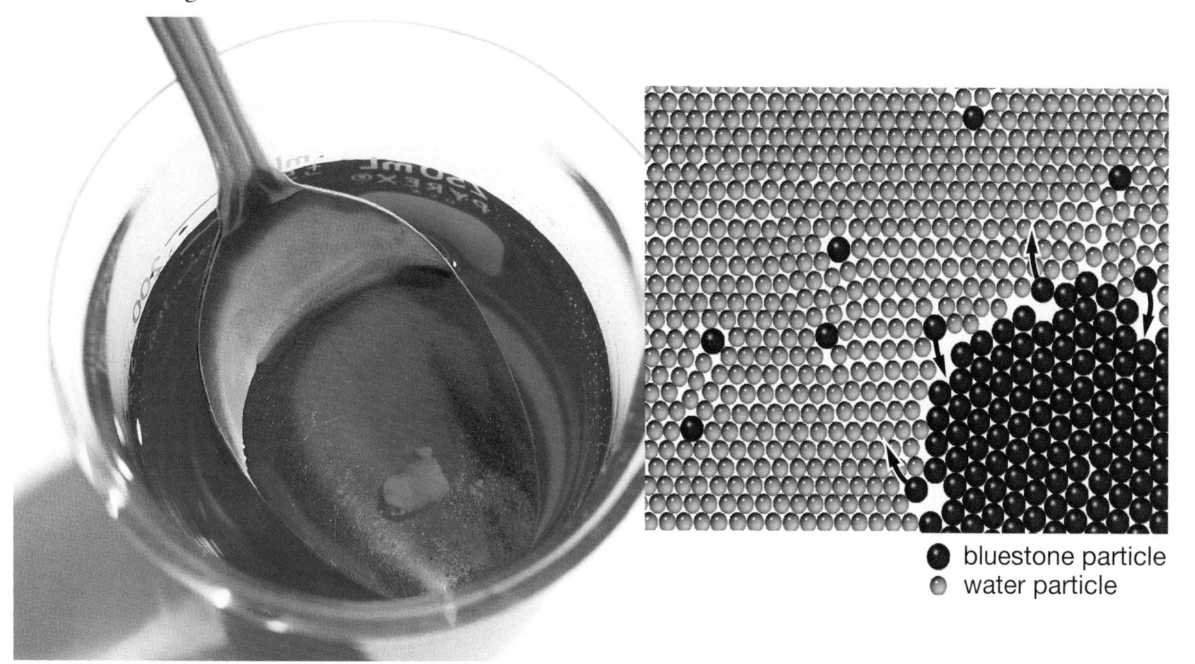

Figure 6.1 The particle theory can be used to explain how a substance, such as bluestone, reaches a point where no more of the substance will dissolve.

The particles of one pure substance are not the same as those of another, so the degree of attraction is different for different substances. The limit to how concentrated a solution can become is called solubility. **Solubility** refers to the mass of a solute that can dissolve in a given amount of solvent to form a saturated solution (at a given temperature). For example, scientists have determined that no more than 35.7 g of salt will dissolve in 100 g of ice-cold water (at 0°C). So the solubility of salt is 35.7 g/100 g of water (we say "35.7 grams of salt per 100 grams of water"). Note that solubility is stated in grams. Figure 6.2, on the next page, shows the solubility of several common substances.

Math CONNECT

Suppose that you are dissolving a solute into 550 g of water at 0°C. You find that no more than 495 g of solute will dissolve. Calculate the solubility of the solute.

Figure 6.2 Solubility of some common substances. How many grams of each of these substances will dissolve in 100 g of ice-cold water?

Substance	State	Solubility (g/100 g of water)
alum	solid	11.4
baking soda	solid	6.9
bluestone	solid	31.6
canola oil	liquid	insoluble
carbon dioxide	gas	0.34
Epsom salts	solid	70.0
ethyl alcohol	liquid	unlimited
limestone	solid	0.0007
nitrogen	gas	0.003
oxygen	gas	0.007
salt (sodium chloride)	solid	35.7
sugar (sucrose)	solid	179.2

Beyond the Limit: Supersaturated Solutions

It is possible to pass the saturation limit in some solutions. A solution that contains more solute than would normally dissolve at a certain temperature is called a **supersaturated solution**. You can prepare a supersaturated solution from some solutes by making a saturated solution, then cooling it without stirring. The solute stays dissolved for a short time. When a small crystal of solute is added, the extra solute quickly becomes crystals, as shown in Figure 6.3.

Pause& Reflect

Imagine that you are conducting a test for saturation. Add a small amount of solute to a solution. In your Science Log, explain what the following results would tell you, using the words "saturated," "unsaturated," and "supersaturated."

- The crystal of solute dissolves.
- The crystal of solute does not dissolve.
- Many more crystals form.

Figure 6.3 The solution of sodium acetate on the left is still homogeneous. No solid crystals have come out of the solution — at least, not yet. When a single crystal enters the supersaturated solution shown on the right, the excess solute crystallizes almost instantly.

DidYouKnow?

Have you ever wondered what "dew point" means when you hear it on a weather report? Dew point is related to saturation. All air contains some dissolved water vapour. The dew point is the temperature at which the air is saturated with water vapour. If the air gets colder than the dew point, it will not be able to keep as much water vapour in solution. The water that cannot stay dissolved will turn into droplets of rain or flakes of snow.

Check Your Understanding

1. (a) In your own words, explain what solubility means.
   (b) In an investigation, how can you determine the solubility of a solid in water?
2. Refer to Figure 6.2 on page 150 to answer the following questions.
   (a) Which solid is the most soluble in water?
   (b) Which gas is the most soluble in water?
   (c) Use your knowledge of the particle theory to compare the solid and gas you named in parts (a) and (b).
3. Refer to Figure 6.2 again, to answer these questions.
   (a) Which substance is the most soluble in water?
   (b) Which substance is insoluble in water?
   (c) Use your knowledge of the particle theory to compare the substances you named in parts (a) and (b).
4. List five materials in your home that are almost completely insoluble in water. For each material, state what evidence tells you that it is insoluble.
5. (a) **Thinking Critically** Many of the nutrients required by plants are soluble in water. How does this help plants to survive?
   (b) **Thinking Critically** What change would you find in rivers, lakes, and oceans if oxygen were not soluble in water? Give reasons for your answer.

6. **Apply** After completing an activity to measure the solubility of an unknown substance in water, Joanne found that she could dissolve only 15 g of the substance per 450 g of water. Raylene tried the same activity at home to see if she could confirm Joanne's results. Raylene found that she could dissolve 40 g per bowl.
   (a) Is it possible that both girls' results are correct?
   (b) Why do the different numbers create a problem?
   (c) How could this problem be avoided?
7. Refer to Figure 6.2. Which combination of solute and solvent particles has the strongest forces of attraction? Explain why you think so.

Computer CONNECT

Suppose that you work in the school laboratory, making solutions for teachers to use in their classes. Use the information in Figure 6.2 to prepare a spreadsheet or database showing the number of grams of solute that are needed to make 10 mL, 50 mL, 100 mL, 500 mL, and 1000 mL of the following saturated solutions at 0°C: alum, bluestone, Epsom salts, salt, sugar.

6.2 Rate of Dissolving

DidYouKnow?

We stir or shake mixtures that are not solutions. You may have seen the clerk in a paint store put a newly purchased can of paint in a machine that shakes it to make sure that the colour pigments are thoroughly mixed. Paint is not a solution. If it were, the colour pigments would need to be mixed only once. After that, they would stay evenly spaced. They do not stay evenly mixed, however. In fact, this is one clue that will help you to know whether or not a mixture is a solution. As long as their containers are covered so that none of the solvent can evaporate, solutions do not separate when they are left standing.

When we study how quickly something dissolves, we are observing how fast a solute dissolves in a solvent. The speed at which something happens is often called a rate. The **rate of dissolving** is the measure of how fast a solute dissolves in a solvent. What factors do you think might affect the rate of dissolving? The factors that affect this rate are conditions that make the solute dissolve faster or slower. You already have had experience in making a solute dissolve faster. For example, to prepare a drink by mixing some flavour crystals with water, you would probably stir the mixture. In this case, stirring is the factor that changes the rate of dissolving.

How Does Agitation Make Solutes Dissolve Faster?

Scientists often refer to stirring or shaking as **agitation**. You agitate a mixture when you do something that makes it move around inside its container. To understand how agitation works, first consider how a solute mixes with a solvent without agitation. Figure 6.4 shows a particle-sized view of a sugar crystal being dissolved in water.

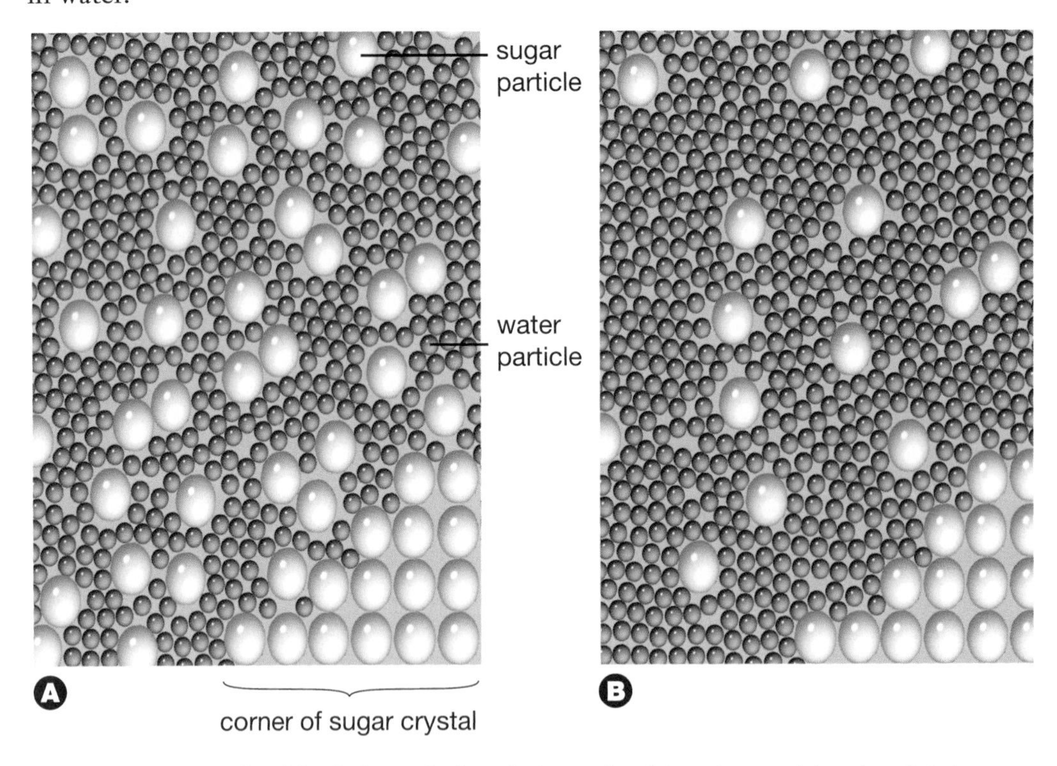

Figure 6.4 **A** shows dissolving before agitation. **B** shows dissolving when a mixture is agitated.

Without agitation, dissolving proceeds in two stages, as shown in Figure 6.4A. First, the water particles pull sugar particles loose. Then the motion of the particles carries the sugar particles away from the crystal. It takes a long time for the motion of the water particles to spread the sugar evenly through the solution. As a result, the area near the crystal becomes very concentrated. As more and more sugar particles are mixed in, the solution near the crystal comes closer and closer to being saturated. The water particles near the crystal are using all of their attractive forces on the sugar particles that are already dissolved. They do not have enough attractive force left to pull more sugar particles away from the sugar crystals.

How does this change if we agitate the mixture? In Figure 6.4B, the solution has been agitated. Now the water that was next to the crystal has been pushed far away. At the same time, water that was far away from the crystal has moved next to it. The water that is near the crystal now has a much lower concentration of sugar particles. These water particles do not have to use so much of their attractive force to keep sugar particles that are already dissolved in solution. Therefore, they can exert a much greater force of attraction on the sugar particles that are still part of the crystal.

Agitation is one way of making a solute dissolve faster. Are there other ways? In the next investigation, you will test several other factors to see if they can change the rate at which a substance dissolves.

Figure 6.5 How is this student helping the soup to dissolve?

Figure 6.6 Drink crystals have a fast rate of dissolving.

Figure 6.7 Soap has a slow rate of dissolving.

Figure 6.8 Epsom salts dissolve easily in water. They are used to increase the relaxing effects of hot tubs and whirlpool baths. They are also used to help soothe sore feet.

Figure 6.9 Medication that is used to treat cold and flu symptoms contains "time-release" capsules to allow the medication to continue dissolving in the body over a period of time.

Changing the Rate of Dissolving

In this investigation, you will test three variables to see if they can change the rate at which Epsom salts dissolve in water.

Problem

How would you design experiments to find out how the rate of dissolving is affected by three variables: changing the temperature of the water (Part 1), changing the rate of stirring (Part 2), and using small or large crystals of Epsom salts (Part 3)?

Safety Precautions

Do not drink any solution you make in your science classroom.

Apparatus

graduated cylinder or measuring cup
2 beakers (250 mL)
2 measuring spoons (10 mL)
2 stirring rods or plastic spoons
digital timer or watch with second hand
sieve
bowl

Materials

Epsom salts
water (hot and cold)

(Note: For information on planning an experiment, see "Science Inquiry," pages IS-9–IS-13.)

Skill POWER

For tips on working in a group, turn to page 478.

Procedure

1. In your notebook, make a table like the one shown below. Give your table a title.
2. In your group, discuss the variables you must keep constant. Make a list of these for each part of the investigation. (Hint: You will need to do *some* stirring in Part 1. What will you do to ensure that your results relate to water temperature rather than to stirring?)
3. Make a prediction about the effect of each variable on the rate of dissolving. Explain your prediction using the particle theory of matter.
4. For each part of this investigation, plan how you will carry out your experiments to answer the Problem question. (Hint: For all tests, use 100 mL of water in a 250 mL beaker, and add 10 mL of Epsom salts.)
5. Distribute tasks among the members of the group. You will need at least one person to do the timing and recording, and two people to do the mixing.
6. Record the time required for the solute to dissolve in Parts 1, 2, and 3. Enter the values into your table.

	Beaker 1		Beaker 2	
Variable	Description	Time to dissolve	Description	Time to dissolve
temperature				
rate of stirring				
size of crystals				

Part 1

Does temperature affect the rate of dissolving?

Part 2

How much does the rate of stirring affect the rate of dissolving?

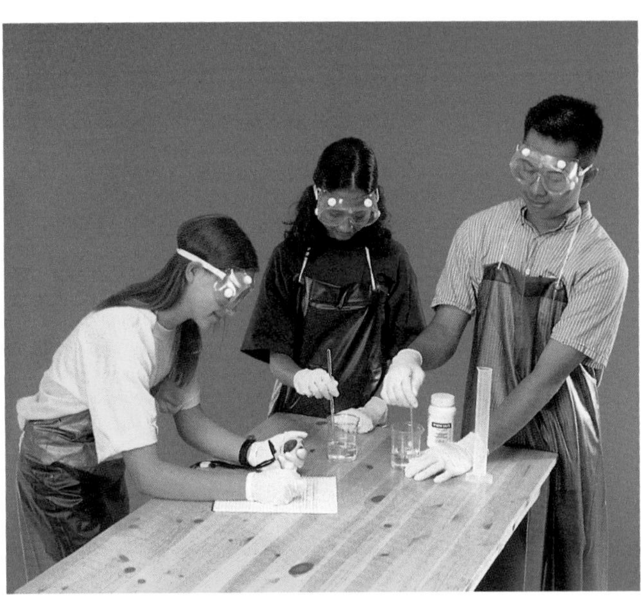

Part 3

Do smaller pieces of solute dissolve faster?

CONTINUED ▶

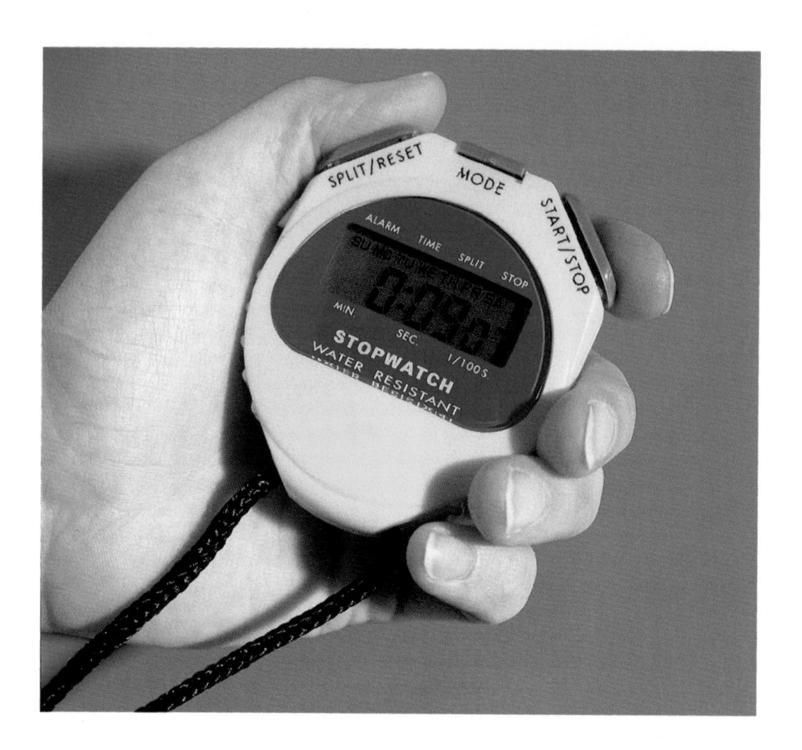

Analyze

1. (a) Was your prediction for Part 1 correct? Explain how the particle theory helped you make your prediction.
   (b) Was your prediction for Part 2 correct? Explain how the particle theory helped you make your prediction.
   (c) Why do you think the size of the pieces of solute affected the rate of dissolving?

2. (a) In your group, brainstorm to add to the list of variables that you kept constant in Parts 1, 2, and 3. List everything you can think of, even variables that seem far-fetched. Here is a start to your list: shape of beaker, amount of light, time of day. Think of at least ten more variables.
   (b) Discuss why you think each variable you listed might or might not affect the rate of dissolving.

3. (a) Based on your results for Part 3, what additional variable could you have controlled in Parts 1 and 2?
   (b) Why would this have been an improvement in the design of the investigation?

Conclude and Apply

4. Most gold that is mined today is found as tiny flakes embedded in rock. This mixture is called gold ore. Since a tonne of ore often contains only a few grams of gold, the ore is placed in a solvent that will dissolve the gold but not the rock. What could you do to the ore to make the gold dissolve as quickly as possible?

Extend Your Skills

5. There are many investigations you could develop from this one:
   (a) Use water at more than two temperatures, and measure the rate of dissolving at each temperature.
   (b) Try three or four different speeds of stirring.

The Particle Theory and Rate of Dissolving

In your hypothesis about stirring and the rate of dissolving, did you recall the explanation in Chapter 5? Figure 5.1 on page 123 shows how particles on the outer edge of a piece of solute are attacked and carried away by solvent particles. Even when there is no stirring, the particles are always moving. Stirring moves water particles next to the sugar crystal. The water has a lower concentration of sugar, so the water particles can attract more sugar.

For your explanation of the effect of temperature on the rate of dissolving, did you recall the fifth point in the particle theory (page 130)? The fifth point states that particles at a higher temperature are moving more quickly. So, again, the sugar-saturated water particles move away from the sugar, allowing pure water particles to take their place near the sugar.

You have seen that when a solid solute is dissolving in a solvent, the dissolving occurs on the surface of the pieces of solute. When the solute is in smaller pieces, there is more surface area where dissolving can occur (see Figure 6.10).

Figure 6.10 Both sugar cubes have the same mass of sugar, and both were completely coloured on the outside. When the sugar cube on the right was broken up, a lot of white surface appeared. Which would dissolve faster — the whole sugar cube or the one that has been broken into pieces?

Pause& Reflect

After you read Conduct an Investigation 6-B but before you begin it, look back to the Pause & Reflect on page 145, where you noted all five points of the particle theory. Make a hypothesis, based on the particle theory, for Part 1 and Part 2 of the investigation.

Can the Solubility of Alum Be Changed?

According to Figure 6.2 on page 150, the solubility of sugar in ice-cold water is 179.2 g per 100 g of water. This means that when 179.2 g of sugar is dissolved in 100 g of water, the solution is saturated. Is there a way to make more solute dissolve? In this investigation, you will work with alum, a common substance that is sold in drugstores.

Problem

Is solubility affected by temperature (Part 1)? Is solubility the same in different solvents (Part 2)?

Safety Precautions

Use hot water with care.

Apparatus

balance
graduated cylinder
thermometer
2 beakers (250 mL)
stirring rod
scoop

Materials

small paper cup
alum
canola oil
10 paper squares (about 10 cm x 10 cm)
water

Procedure

Part 1

Changing the Temperature 

1. Prepare 10 squares of paper, each containing a measured mass of alum, as follows: 6 squares with 3 g of alum each; 4 squares with 1 g of alum each. Designate one member of your group as the data recorder.

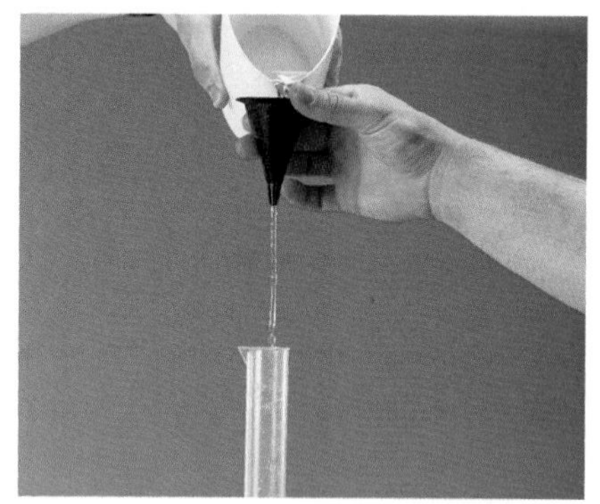

2. Use the graduated cylinder to measure 100 mL of water from the hot tap. (Recall that this is a shortcut for measuring 100 g of water.)

3. Quickly, before the water has a chance to cool, pour the water into a beaker, add one of the measured 3 g of alum, and stir. If the alum dissolves completely, add another 3 g amount of alum. Repeat until no more alum will dissolve. (Hint: To keep the water from cooling, you could place the beaker in a hot-water bath.)

4. Record how many grams of alum dissolved in 100 g of hot water.

5. Carefully place the thermometer in the beaker. Record the temperature of the water.

6. Repeat steps 1 to 5, but use cold water, and the measured 1 g amounts of alum.

Part 2

Changing the Solvent

1. Measure 100 g of canola oil in a 250 mL beaker.

2. Measure 10 g of alum, and add it to the oil in the beaker. Measure another 10 g on a paper square.

3. Stir the alum and oil mixture for at least 1 min. Set the beaker aside until any undissolved alum has settled to the bottom.

4. Compare the amount of alum that settled to the bottom of the beaker with the amount of alum on the paper square.

5. Record your estimate of the amount of alum that dissolved in the oil.

Analyze

1. **(a)** Name the solute and the solvent in Part 1.
   (b) Name the solute and the solvent in Part 2.
2. **(a)** Write a statement comparing the solubility of alum in hot water with the solubility of alum in cold water.
   (b) State the solubility of alum in hot water in grams per 100 g of water.
   (c) State the solubility of alum in cold water.
   (d) Compare your value in part (c) with the value in Figure 6.2, on page 150. Why might your results be different?
3. **(a)** Write a statement in your notebook comparing the solubility of alum in canola oil with the solubility of alum in water.
   (b) Does your result in Part 2 confirm your earlier prediction? What personal experience led you to make this prediction?
4. Use the particle theory to explain why the solubility of alum in water is different from the solubility of alum in canola oil.
5. Explain why it would be difficult to find out if any alum dissolved in the canola oil.

Conclude and Apply

6. Use the particle theory to explain what happened
   (a) to the first 3 g of alum that you added to the hot water
   (b) to the last 3 g of alum that you added to the hot water
7. Verena prepared a saturated solution of alum in water at 80°C. She then set the solution aside to cool. Predict what you think she would find the next day. Give reasons for your answer.
8. Why did you take the temperature of the solutions in step 5? How might this knowledge help you compare your results with other class members?

Extend Your Skills

9. The salad dressing shown in the photograph includes oil, sugar, spices, and vinegar. Sugar is not soluble in oil, but no sugar can be seen in this dressing. Explain why.

Cleaning Up with Solvents

A house that smells "clean" may actually smell of the solvents that were used to clean it. Television commercials remind us daily that modern window cleaners and bathroom sprays can "remove solid dirt in minutes — without scrubbing!" Now you have the knowledge you need to interpret such claims. Any solvent that can dissolve an accumulation of grime this fast and easily must have a powerful ability to attract solute particles in the dirt. If it is so powerful, however, you would have to wonder about what else the solvent can attract. Can it attract the natural oils in your skin? Specialized cleaning products are used to clean off decades, even centuries, of accumulated dirt from paintings without dissolving the paint (see Figure 6.11). As helpful as cleaning solvents are, they must be used with caution. Always read the label on the solvent container. Pay close attention to any safety warnings. This is especially important for household cleaning products, which often do not have the WHMIS symbols shown on page 492.

Figure 6.11 The cleaned painting on the right appears as the artist would have seen it.

Factors That Affect the Solubility of Gases in Liquids

So far this section has focussed on the solubility of solids in liquids. What about gases that are dissolved in liquids? You have already seen that cold water in lakes and rivers can contain more dissolved oxygen than warm water can.

You Be the Scientist: Dissolved Carbon Dioxide

Soda water is plain water with carbon dioxide gas dissolved in it. In this activity, you will make observations about solubility with your ears as well as your eyes. (**Note**: For information on planning an experiment, see "Science Inquiry," pages IS-9 – IS-13.)

What to Do

Your task is to design an experiment that will enable you to compare the solubility of carbon dioxide gas in cold water with its solubility in warm water.

You will not be able to measure the solubility in grams of solute per 100 g of solvent. You will need to find another way to compare the two mixtures. (Hint: What do you hear when you open a can of pop? What causes this noise?)

Safety Precautions

- Use sealed, plastic bottles of pop, not glass ones.
- Compare cold and warm bottles, not cold and hot ones.

What Did You Find Out?

1. According to the particle theory, in which state (solid, liquid, or gas) do particles have the most freedom to move around? (Refer to the illustration in the Find Out Activity on page 115.)
2. According to the particle theory, how are particles affected by higher temperature?
3. How can these ideas be used to explain what you observed in this activity?

Extension

Refer to Figure 6.2 on page 150. Which gas is most soluble in water? Explain why you think so.

DidYouKnow?

The diver in the photograph on page 146 has a lot in common with a sealed bottle of pop. Both are "under pressure," and both contain concentrated gas-in-liquid solutions.

Carbon dioxide is pumped under pressure into the water in the pop bottle. The pressure forces extra gas particles into the spaces between the water particles. The extra gas particles dissolve only because they were forced into the solution by the outside pressure. When the cap is removed from the bottle, the pressure drops rapidly. Some of the gas particles come out of the solution in a hurry, forming many small bubbles.

Compressed air moves from breathing tanks into the diver's lungs. The diver's body is under pressure caused by the deep water, so the diver's blood contains more gas particles than normal. These respiratory gases stay dissolved only as long as the pressure remains. If the diver resurfaces too quickly, nitrogen in the diver's blood comes out of solution rapidly, forming bubbles in the diver's blood, just like the bubbles in the bottle of pop.

Pause& Reflect

Ecologists tell us that the ice-cold waters of the Arctic Ocean contain far more fish than do warmer waters, such as the Indian Ocean or the Gulf of Mexico. How is this possible? Write an explanation in your Science Log.

"The bends" can occur high above Earth as well as at Earth's surface. They have affected people flying in airplanes at very high altitudes. Normally the air pressure inside an airplane is kept similar to the air pressure at ground level. If an accident causes decompression, however, the sudden change can cause "the bends."

Check Your Understanding

1. What does the term "rate of dissolving" mean?
2. **(a)** Name three factors that can change the rate at which a solid dissolves in a liquid.
   (b) Name one factor that you think would have no effect on the rate.
3. Use the particle theory to explain why each factor you identified in question **2. (a)** causes the rate to change.
4. **(a)** Identify two factors that affect the solubility of a solid in a liquid.
   (b) Describe the effect of each of these factors.
5. **(a)** How does heating affect the solubility of a gas dissolved in a liquid?
   (b) How does pressure affect the solubility of a gas dissolved in a liquid?
6. Explain why hot water can dissolve more sugar than cold water can.
7. **Apply** Water is often called the universal solvent because it can dissolve so many substances.
   (a) List ten materials that will dissolve in water. Include some solids, some liquids, and some gases.
   (b) List ten materials that will not dissolve in water.
8. **Thinking Critically** To keep winter roads ice-free, highway crews often spread salt on the snowy or icy surface of the roads. Putting salt on the ice or snow helps it to melt. Which form of salt would you expect to stay on the roads longer, table salt (fine grains) or rock salt (large chunks)? Give reasons for your answer.

6.3 Processing Mixtures from Underground

Our planet is rich in natural mixtures. Over thousands of years, humans have developed technology to process these mixtures to make useful products. This section will focus on two natural mixtures: one liquid (petroleum) and one solid (gold ore).

What Is Petroleum?

Petroleum is the source of most of the fuels used today — gasoline, diesel fuel, and kerosene, for example. It is found as an oily liquid mixture deep beneath Earth's surface. In Alberta, petroleum was first discovered in 1904, but the "big one" — an oil well at Leduc — did not blow sky-high until 1947 (see Figure 6.12).

Figure 6.12 This picture of the Leduc gusher was published in all the major Canadian newspapers and shown in newsreels in movie theatres across the country. Everyone was excited because, at last, Canada would probably be able to supply all of its own petroleum needs.

Extracting Petroleum from Underground

Thousands of years ago, people in the Middle East noticed an oily black liquid bubbling up through the sand. They recognized that it could be burned, but the flame was smoky and not very bright. They used it only when they had none of their usual lamp fuel. Lamps from the past used fuel that was made from vegetable or animal oils, as shown in Figures 6.13A and B.

Figure 6.13A In the Middle East, oil lamps burned plant-based fuels, such as olive oil.

Figure 6.13B In the Arctic, stone lamps burned animal fat.

Even 150 years ago, there was little interest in petroleum. The Europeans who settled in North America used other fuels: coal to run steam engines, wood for home heat, and whale oil for light. By 1850 whale oil was becoming scarce. Its price rose dramatically, and people began to search for alternative lighting fuels.

In Nova Scotia, a researcher named Abraham Gesner (1797–1864), shown in Figure 6.14A, began to experiment with natural tar found in surface deposits. This solid form of petroleum did not look much like a lighting fuel at the beginning. It did look like a mixture, however. By 1853 Gesner had perfected a process to separate the tar mixture. This process produced a new liquid fuel, which was ideal for lighting. It was easy to pour; it burned with a clear, steady, bright yellow flame, and the flame was nearly smoke-free. Gesner named the new fuel kerosene. Kerosene would not burn in lamps designed to burn whale oil, so Gesner also invented a new kind of lamp to burn it (see Figure 6.14B). Today we think of kerosene lamps as old-fashioned, but they are still the main source of lighting in many parts of the world.

Figure 6.14A Dr. Abraham Gesner patented many inventions, but none was more important than the technology for kerosene lighting.

Figure 6.14B The kerosene lamp represented an important technological advance. As one tribute stated, Gesner truly did "give the world a better light."

Suddenly, underground petroleum was seen as a valuable natural resource, and active exploration began in North America. Petroleum was discovered near Sarnia, Ontario, in 1857. It served an important purpose as a fuel for lighting, but once cars became common, there was a great demand for fuel made from petroleum. Alberta oil wells have satisfied this demand in Canada. More recently, large reserves of petroleum have been discovered offshore near Newfoundland (see Figure 6.15).

Figure 6.15 In the Hibernia oil field, large oil rigs are used to extract petroleum from beneath the ocean floor.

Because our modern world runs on the energy of petroleum, oil exploration companies now spend millions of dollars drilling test holes to locate new underground deposits of petroleum. Petroleum products, such as kerosene, gasoline, and diesel oil, are burned to produce electricity, move vehicles of all kinds, and do many other kinds of work.

INTERNET CONNECT

www.school.mcgrawhill.ca/resources/

Where is petroleum found on Earth? Why is it sometimes found under the ocean? Learn more about this energy resource by going to the web site above. Go to **Science Resources**, then to **SCIENCEPOWER 7** to find out where to go next. Prepare a brief report of your findings.

INVESTIGATION 6-C

Pump Up the Volume

Think About It

Like water wells, oil wells use pumps to bring the underground liquid up to the surface. Pumping petroleum is not as easy as pumping water because the oil is a very thick liquid. It is found in small, sponge-like pores of underground rock. The diagram below shows how petroleum is brought up to the surface.

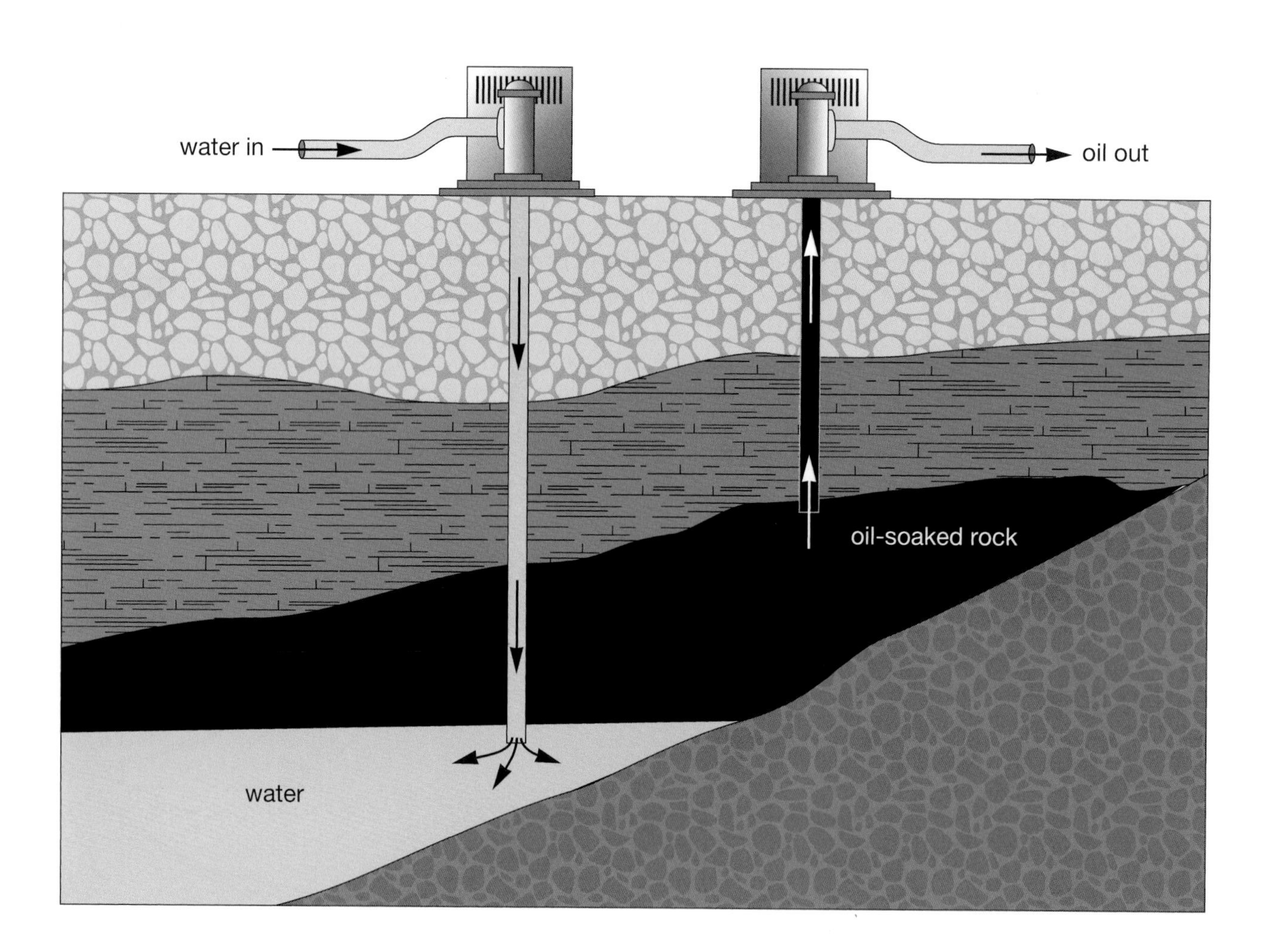

What to Do

Examine the illustration and answer the following questions.

1. How many pumps are involved?
2. Which pump is lifting oil up to the surface?
3. What is the other pump doing? Explain how it helps to separate the oil from the rock.

Analyze

Refer to the photograph of the salad dressing on page 159. What are the two liquid layers in the photograph? Explain the connection between these two layers and this method of pumping petroleum.

Processing Petroleum

Pumping petroleum to the surface is only the first step. What comes out of the pump is crude petroleum, a raw material. As you can see in Figure 6.12 on page 163, burning crude petroleum produces a lot of black smoke. To make usable products, petroleum must be processed. The factories that process petroleum are called oil refineries. Refining is the processing of petroleum to separate it into its parts.

In 1857 the refineries in Ontario made just two petroleum products. One was buggy wheel grease, which is thick and gooey; the other was kerosene lamp fuel, which is thin and runny. Why are their properties so different? The main reason is particle size. When petroleum comes straight from the ground, the particles vary in size. Some are nearly as small as water particles; others are bigger than sugar particles. The mixture of particles is not very useful in its natural state. Before petroleum can be used in a vehicle, for example, it must be processed to make specialized fuels, such as kerosene, gasoline, and diesel fuel.

Did You Know?

In the term "fractional distillation," the meaning of "distillation" is the same as the meaning on page 130. The distillation part of the term indicates that petroleum is heated until it is vaporized, and then cooled so the vapours condense. What does the other part of the term mean? "Fractional" comes from the same root word as "fracture," meaning "break up." The fractional distillation of petroleum "breaks up" the mixture into several separate products, or "fractions."

Fractional Distillation

The process that yields all of these different petroleum products is known as **fractional distillation**. Fractional distillation is done in a two-tower structure, as shown in Figure 6.16. In the shorter tower, the petroleum is heated strongly enough to vaporize every part of the mixture. Then the mixture of hot vapours is pumped into the bottom of the taller tower.

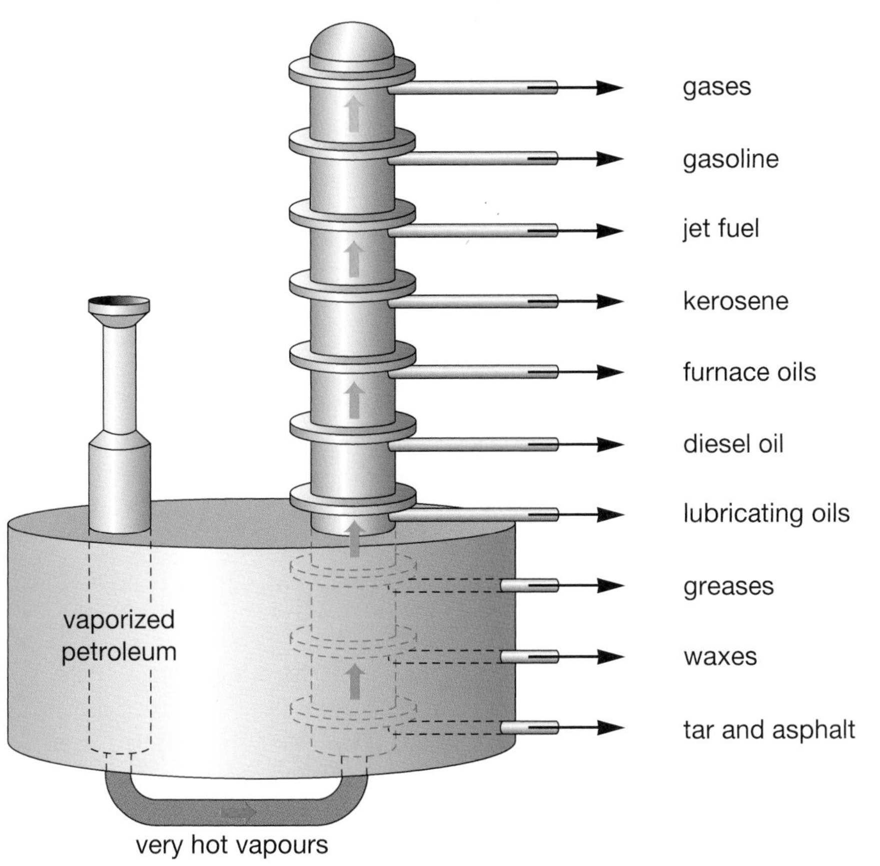

Figure 6.16 Fractionating towers are a common sight in the oil-producing regions of Canada. What raw materials enter the short tower? What change takes place there? What processed materials leave the tall tower?

Word CONNECT

The word "vaporized" was used in the "Did You Know?" above. In your notebook, write how the meanings of "vaporize" and "evaporate" are similar and how they are different.

Inside the tall tower, the hot vapours rise. As they rise, they cool. Remember that these are different pure substances, so they have different properties. This means that some of them condense and form a liquid while they are still very hot, near the bottom of the tower. As the remaining vapours continue to rise, different ones condense at different levels in the tower. Near the top (the coolest part) of the tower, a few remain as a gas.

Each fraction is drawn off by the collecting pipes at its own level and is sent to a different part of the refinery for further processing. There each material may be converted into petrochemicals. **Petrochemicals** are entirely new products made from the same raw material — petroleum. Scientists have developed and produced over 500 000 different petrochemicals.

Career CONNECT

Checking Mixtures

Have you ever wondered just what is in your milkshake or soft drink or in the cleaner you use to wash the sink? You can be sure that a chemical research analyst knows. Chemical research analysts examine mixtures. They look at the smallest particles to find out what is in a mixture and to make sure that it contains no harmful substances, such as pesticides.

John Persaud is a chemical research analyst. He works for a petrochemical company that makes products from crude oil. John's job is to check the contents of these products before they are sold to consumers. He makes sure that the crude oil is good quality and safe for the environment. In the laboratory, he tests a sample of products made from the crude oil, such as natural gas, motor oil, gasoline, and diesel fuel. If he finds unwanted particles, he develops ways to remove them from the whole batch.

John could not possibly tell what is in the petrochemical products he tests just by looking at them. Specialized equipment and computers in the laboratory give him a close-up view of the molecules and measure the amounts for him.

Computers are a very important part of many types of work. With the help of an adult, telephone three people and ask them about how they use computers. To get a range of responses, try to interview three people who use computers for different tasks. Here are some ideas of possible people to interview:

- dentist
- water treatment plant employee
- newspaper editor/reporter
- architect/draftsperson
- order desk clerk

After the interviews, prepare a report of your findings. Put your findings in a class book, entitled "Computers in the Workplace."

Pause& Reflect

You have learned that the products of oil refining can be "converted into" petrochemicals. First, petroleum is "separated into" fractions, then the fractions are "converted into" something else. You saw this type of change in Chapter 5, when you did the At Home Activity involving the hardening of plaster of Paris and read the example of burnt toast (page 125). The example here — fractions from an oil refinery changing into new materials with different properties — is also a chemical change.

In your Science Log, write the heading "Chemical Changes" and list these three examples. If you can think of other examples, list them as well. (Hint: Think of changes that cannot easily be reversed, such as baking cookies or burning gasoline in a car engine.)

Solid Mixtures From Underground

Most underground materials are solids — solid rocks. Most rocks are mixtures. For example, the rock shown in Figure 6.17 is a mixture of two pure substances: white quartzite and yellow gold. This rock is called gold ore because it can be processed to extract gold. An **ore** is a mineral (or a group of minerals) that contains a valuable substance. Another example of an ore is iron ore.

The discovery of a large deposit of gold ore is exciting. Gold is valuable because of its beauty and its scarcity. Gold is also valuable because of its practical properties: it resists corrosion, and it is soft enough to carve easily.

Gold is so valuable that people are willing to go to a lot of trouble to get it. Twice in the nineteenth century, reports of newly discovered gold deposits led to a gold rush. Thousands of people travelled to northern California and to Canada's Yukon Territory to "try their luck."

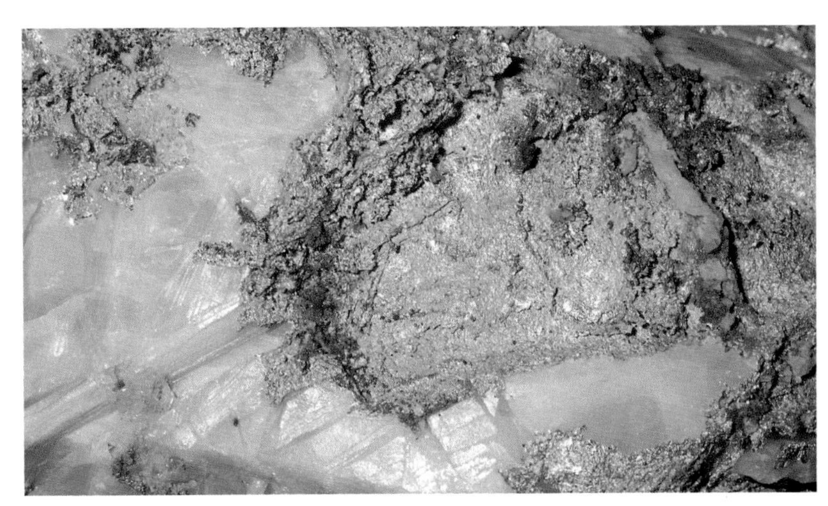

Figure 6.17 The white part of this rock is quartzite. The yellow parts are nearly pure gold.

Pause & Reflect

1. **(a)** On a full page in your Science Log, draw circles like those shown here. Print the title "RAW MATERIAL" in the innermost circle. Underneath it, write "crude oil."
   (b) Inside the larger circle, print the title "PROCESSED MATERIALS." Then, around the circle, write the names of the products listed in Figure 6.16.
   (c) Outside the circle, print "PRODUCTS OF FURTHER PROCESSING." Then write the names of as many petrochemical products as you know. Share information with your group or your classmates.

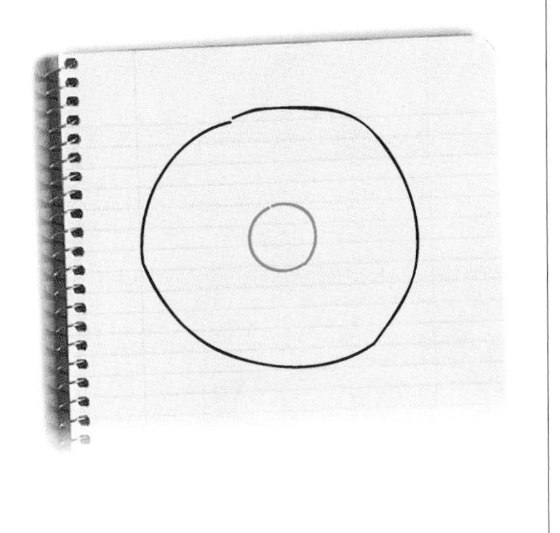

2. What does the diagram show about the link between raw materials and processed end products?

INTERNET CONNECT

www.school.mcgrawhill.ca/resources/

Examples of products that are made from petrochemicals include Aspirin™, basketballs, chewing gum, duct tape, eyeglasses, and fertilizer. Continue the alphabetical list of petrochemicals by going to the web site above. Go to **Science Resources**, then to **SCIENCEPOWER 7** to find out where to go next. Try to find a petrochemical product for every letter of the alphabet.

Figure 6.18 During the 1800s, for permission to travel beyond the Chilcoot Pass into Yukon Territory, would-be gold miners had to carry a year's worth of supplies with them. Those without sufficient supplies were turned back by the police, who feared that they would starve.

The amateur miners in Figure 6.18 did not have to dig to find gold. They looked in stream beds for chunks of gold that had been washed out of gold ore by the running water. These chunks of gold, mixed with sand or gravel, could be washed a long way downstream from the original deposit.

Separating these chunks or nuggets from the gravel required only settling and sifting. Material from the stream bed was dug out and swirled in a pan with plenty of water. Gold nuggets, no matter how small, have much more mass than gravel or sand grains of the same size. As the mixture was swirled, the lighter pieces of sand, gravel, and mud were washed away, leaving the heavier nuggets of pure gold behind.

Find Out **ACTIVITY**

Panning for "Gold"

Make a working model to show panning for gold — separating "gold" from "gravel."

What You Need

about 1 L Styrofoam™ peanuts
about 1 L marbles
large paper bag
large basin
hair dryer

What to Do

1. Pour the "peanuts" and marbles into the large paper bag, and mix thoroughly. (The marbles represent the gold nuggets.)
2. Place the mixture in a large basin. Use a circular motion to swish and swirl the basin steadily. Have a partner aim a hair dryer across the top of the basin.

What Did You Find Out?

1. In the model, what represents each of the following?
   (a) gold nuggets
   (b) gravel
   (c) running water
2. How is your model similar to the method used to pan for gold?

INVESTIGATION 6-D

Mining for Gold

Think About It

Compared to other kinds of mining, gold panning was easy, even for amateurs. In fact, it was so easy that most stream-bed deposits have now been "worked out." Very little gold is found this way today. Instead, gold ore must be brought up from underground and then processed to extract the gold from the rock.

What to Do

Use the flowchart to answer the questions below.

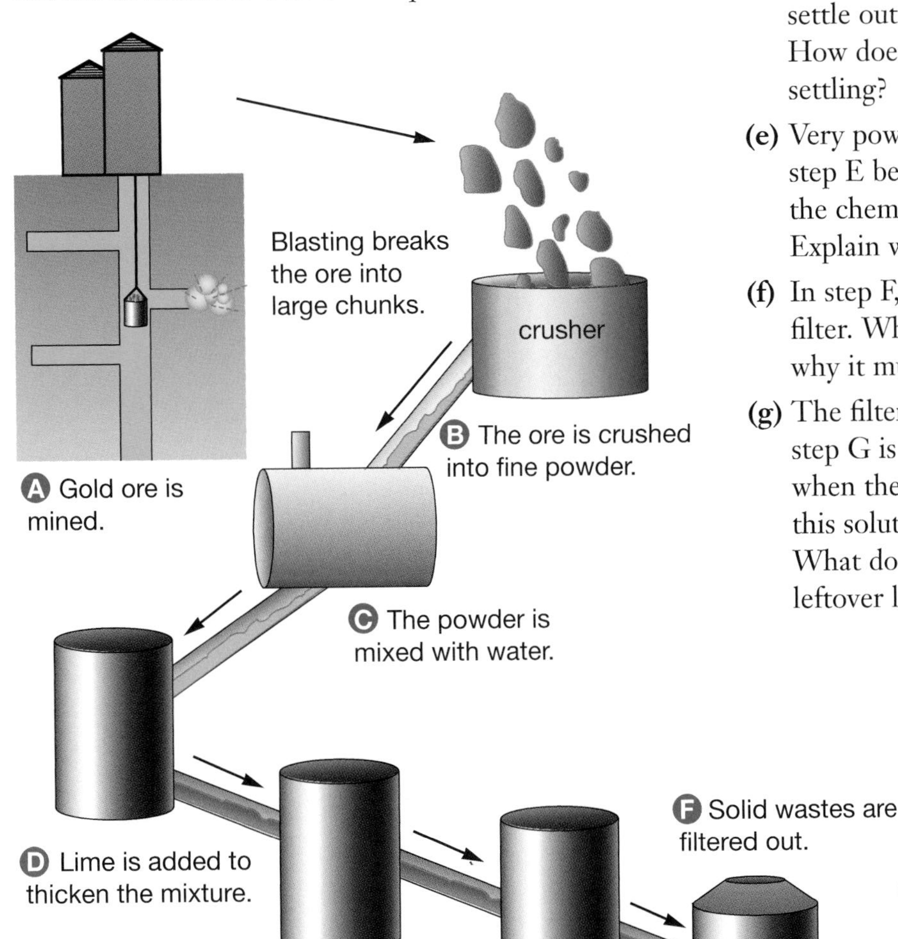

(a) The rock that is being blasted out in step A is the same kind of gold ore that is shown in Figure 6.17. What are the main parts of the ore? Do you think blasting separates the parts? Explain.

(b) What is the purpose of step B? Explain this purpose in relation to the overall process.

(c) How does step C make it easier to pump the ore through the system?

(d) The liquid mixture that is produced in step C is very runny. The powdered ore will settle out rapidly unless something is done. How does step D prevent the powder from settling?

(e) Very powerful chemicals must be used in step E because gold resists dissolving, but the chemicals must not dissolve the rock. Explain why.

(f) In step F, solid waste material collects on the filter. What is in this waste material? Explain why it must be treated before disposal.

(g) The filtered liquid that runs into the vat in step G is a solution. What is the main solute when the liquid enters? What happens to this solute when powdered zinc is added? What do you think must be done with the leftover liquid?

Across Canada

"Believe in your dream and make it happen," says Dusanka Filipovic, President of Energy and Environment Research (EER) in Markham, Ontario. She was one of the first women engineers in Canada and became a professional chemical engineer in 1974. She likes to be involved in turning technologies that she has developed into products that can be used by the general public.

As a student, she was very self-motivated. She was encouraged to pursue chemistry, physics, and math — her strongest subjects. She recalls, "I was always interested in understanding how something was made."

Dr. Filipovic is now working on ways to reduce pollutants in the troposphere, the lowest layer of Earth's atmosphere, which is about 10 to 16 km above Earth's surface. The ozone in this layer is actually smog, produced by motor vehicles and industrial emissions from pulp and paper mills, cement industries, mines, chemical refineries, and other manufacturing industries. She is trying to change the emissions chemically so that they do not harm the environment. She is also trying to stop their formation completely.

Dusanka Filipovic

Dr. Filipovic previously studied the stratosphere, which is 30 km above Earth. She developed a method to prevent substances like chlorinated fluorocarbons (CFCs) from destroying the ozone, the protective layer that keeps harmful rays from reaching Earth's surface.

After working on methods to improve the environment in both layers of Earth's atmosphere, Dr. Filipovic says with a laugh, "I am a real expert on ozone, I guess."

Check Your Understanding

1. Give four examples of natural mixtures, two from this chapter and two from your own experience.
2. (a) What is petroleum?
   (b) What makes petroleum difficult to remove from the rock underground?
3. (a) What is a petroleum refinery?
   (b) Name the process that is used to refine petroleum. Explain how it works by drawing a simple labelled sketch.
4. Gold nuggets are nearly pure gold, but collecting them involves separating a solid mixture.
   (a) What else is mixed in with the nuggets?
   (b) Where is such a mixture likely to be found?
   (c) How are the components of this mixture separated? (Be sure to explain what makes them separate.)
5. Pure gold can also be obtained from gold ore.
   (a) How does gold ore differ from the solid mixture referred to in question 4?
   (b) How are the components separated? (Be sure to explain what makes them separate.)
6. (a) Why is petroleum valuable?
   (b) Why is gold valuable?
   (c) Which is more important to our society? Give reasons for your opinion.
7. **Apply** Describe how settling is used to separate gold nuggets from gravel. Could settling also be used to separate the components of petroleum? Explain why or why not.
8. **Thinking Critically** Does your province produce oil? If not, where does your petroleum-based fuel come from?

CHAPTER at a glance

Now that you have completed this chapter, try to do the following. If you cannot, go back to the sections indicated.

List six ways that the components of a mixture can be separated. (6.1)

Use the particle theory to explain why solutions are formed. (6.1)

Explain the difference between saturated and unsaturated solutions. (6.1)

Use the particle theory to explain why a solution reaches a saturation point. (6.1)

In your own words, explain the meaning of the term "solubility." (6.1)

Explain how a supersaturated solution is made. (6.1)

What three things could you do to make a solid dissolve more quickly in a liquid? Explain how each of these factors affects the rate of dissolving. (6.2)

Describe how the solubility of a solid in a liquid can be affected by (a) temperature, (b) the solvent. (6.2)

Describe how the solubility of a gas in a liquid can be affected by (a) temperature, (b) pressure. (6.2)

Why was the discovery of kerosene a timely event? Why was this fuel suitable for lighting? (6.3)

Why is the petroleum industry so important? (6.3)

Two early products that were obtained from petroleum were buggy wheel grease and kerosene lamp fuel. Compare these two products. (6.3)

List the steps that occur during the fractional distillation of petroleum. (6.3)

Why is gold an important natural resource? What property of gold led to the introduction of "panning"? (6.3)

Prepare Your Own Summary

Summarize this chapter by doing one of the following. Use a graphic organizer (such as a concept map), produce a poster, or write the summary to include the key chapter ideas. Here are a few ideas to use as a guide:

- What are saturated and unsaturated solutions?
- How does the particle theory explain saturated solutions?
- What is the solubility of a substance?
- How do you make a supersaturated solution?
- What is the rate of dissolving?
- What factors affect the rate of dissolving?
- How does the particle theory help to explain the rate of dissolving?
- How does temperature or the type of solvent affect solubility?
- How does pressure or temperature affect the solubility of gases and liquids?
- How does fractional distillation work to produce petrochemicals?
- How is gold extracted?

CHAPTER

6 Review

Key Terms

saturated solution
unsaturated solution
solubility
supersaturated solution
rate of dissolving
agitation
petroleum
fractional distillation
petrochemicals
ore

Reviewing Key Terms

If you need to review, the section numbers show you where these terms were introduced.

1. In your notebook, match the description in column A with the correct term in column B.

A

- oily liquid mixture found underground
- solution that has all the solute it can dissolve at a given temperature
- amount of a solute that can be dissolved in a specific amount of solvent at a given temperature
- method that is used to separate solutions of two or more liquids and that depends on boiling points
- solution that can dissolve more solute at a given temperature
- moving a mixture inside its container
- minerals containing metals
- time it takes for a solute to dissolve in a solvent
- solution that contains more dissolved solute than it can usually hold at a given temperature

B

- rate of dissolving (6.2)
- petrochemicals (6.3)
- solubility (6.1)
- unsaturated (6.1)
- ores (6.3)
- petroleum (6.3)
- supersaturated (6.1)
- saturated (6.1)
- fractional distillation (6.3)
- agitation (6.2)

Understanding Key Ideas

Section numbers are provided if you need to review.

2. Why is it not possible to dissolve an unlimited amount of salt in water? Refer to the particle theory in your answer. (6.1)

3. Why is sugar more soluble in water than in canola oil? (6.1)

4. Rock salt (also called pickling salt) has very large crystals. A test tube contains a saturated solution of salt water, and a single crystal of rock salt is added. Will the crystal eventually disappear? Explain your answer. (6.1)

5. Why do you think scientists state solubility in grams of solute per 100 grams of solvent? (6.1)

6. Which would dissolve faster, powdered sugar or regular granulated sugar? Explain your answer using the particle theory. (6.2)

7. The instructions on a package of jelly powder state the following: "Dissolve package in 250 mL of boiling water." Why do you need to use boiling water? (6.2)

8. If you were given some rock salt to dissolve, what would you do to speed up the process? (6.2)

9. What effect does increasing the temperature of water have on the ability of water to dissolve a solid? What effect does this have on the ability of water to dissolve a gas? (6.2)

10. You have been given two bottles of soda water. How could you show the effect of pressure on the solubility of carbon dioxide gas in water? (6.2)

11. During the mining of gold, the rock is crushed to a fine powder. Why is this step necessary? (6.3)

12. Explain the meaning of the term "fraction" in fractional distillation. (6.3)

13. How is fractional distillation different from the process used to produce distilled water? (6.3)

Developing Skills

14. **Design Your Own** Choose one of the variables from Analyze question 2 in Conduct an Investigation 6-A that you think might affect the rate of dissolving. Design an experiment to test its effect on rate of dissolving. For example, you could test whether stirring twice as fast made the crystals of alum dissolve twice as quickly.

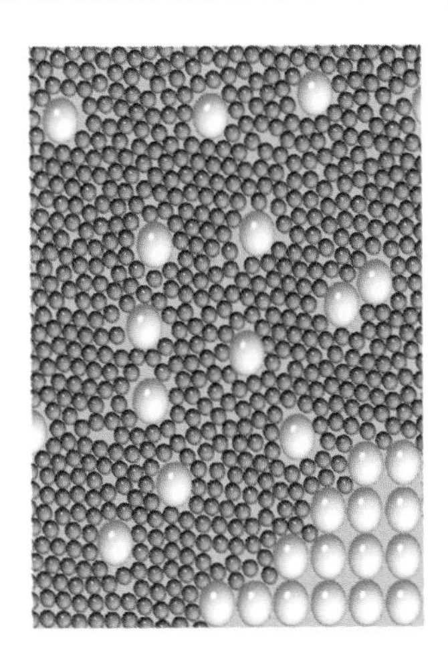

15. Formulate one or more questions of your own about solutions and mixtures, and explore resources of your choice to find possible answers.

16. **Design Your Own** As mentioned in Chapter 5, water from the Dead Sea is about eight or nine times as salty as ocean water. Design an experiment to show whether a sample of water from the Dead Sea is saturated or unsaturated.

Problem Solving/Applying

17. To make iced tea, people first make tea with boiling water. Then they put the tea in the refrigerator. What might be the advantage of this method? How would you test to see if there really is an advantage?

18. The word "volatile" is used to describe a substance that vaporizes at a low temperature. List the following products of a fractionating tower in the order of their volatility, from lowest to highest.
    - diesel oil
    - gasoline
    - kerosene
    - propane
    - aviation fuels
    - asphalt and tar
    - greases
    - lubricating oils

Critical Thinking

19. The top rock layer of Niagara Falls is made of a kind of rock called limestone. So are most of the boulders at the bottom. According to Figure 6.2 on page 150, the solubility of limestone in water is 0.0007 g of solid per 100 g of water. With 200 000 000 t of water going over the falls every day, 1400 t of limestone could be dissolved. Only a tiny fraction of this amount actually dissolves, however. Explain why.

Pause& Reflect

Go back to the beginning of this chapter on page 146, and check your original answers to the Getting Ready questions. How has your thinking changed? How would you answer those questions now that you have investigated the topics in this chapter?

Ask an Expert

Imagine looking at a handful of mixed grains and knowing what each type of grain is, approximately how much of each type there is, and whether it grew from high- or low-quality seed. About 120 Canadian Grain Commission inspectors in the Vancouver area alone complete this daunting task many times a day. Most of these inspectors learned their skills from Art Colebourne, regional staff training officer with the Canadian Grain Commission.

Q If most of Canada's grain is grown on the Prairies, what are so many grain experts doing in Vancouver?

A The grain is grown on the Prairies, but it has to come to a terminal grain elevator in Vancouver, Prince Rupert, Thunder Bay, or some other seaport in order to be shipped to our overseas customers.

Q How does the grain get to the terminal elevator?

A During harvest, a farmer or producer transports the grain by truck or trailer to a grain agent at one of the primary elevators found all across Canada's Prairies. The agent and the farmer agree on the grain's quality, then the agent pays the farmer based on the value of the crop. The crop is placed in a bin in the elevator according to its type and quality. The agent keeps track of what is in each bin and, as each additional farmer arrives, the agent adds the farmer's grain to the bin that contains the same type and quality. Then the agent loads the grain, according to type and quality, from the elevator into special railroad cars, called hoppers. The grain is then transported to a terminal elevator.

Hopper cars, emptied of their grain cargo, sit in front of one of Vancouver's five terminal elevators.

Q What happens there?

A Each hopper of grain is unloaded and officially weighed by Canadian Grain Commission staff. Automatic mechanical grain samplers take a sample and send it to the Canadian Grain Commission inspection office. An inspector determines the type and quality of the grain, and the percentage of other grains it contains. The inspector also evaluates the moisture content of the grain and the amount of dockage (small seeds, dust, and chaff) that is mixed in. Samples are sent to the entomology office to be checked for tiny insects. Factors such as sprouted seeds, frost damage, and excessive dockage degrade the sample and lower the price that the agent is paid.

Mixed in with this sample of Red Spring wheat was some grain of a different quality. The grain inspector has separated this other grain and will weigh it to figure out what percentage of it is contained in the shipment. This information helps the inspector classify the shipment according to its correct grade.

The entire shipment of grain is put through cleaning machinery to remove the dockage. If the grain is too moist, large dryers are used to dry it. Then the clean, dry grain is mixed with other grain of the same type and quality in one of the separate bins in our huge terminal elevators.

Q Is the inspector's job done at this point?

A No. Inspecting samples from the 500 hopper cars of grain that arrive each day at Vancouver's five terminal elevators is just part of the job. Canadian Grain Commission staff also monitor the shipments leaving the terminal, bound for overseas customers. Ships arrive, each with an order for a specific type, quality, and amount of grain. As the grain is loaded from the elevator, the grain sampler sends samples to an inspector who continually monitors the grain to be sure it is clean and dry, and of the proper grade and quality.

Q How much training is needed to become a grain inspector?

A New employees have at least a Grade 12 education. In their first year, we train them to identify the different grains, grade the quality of each grain, and assess the degrading factors. At the end of the year, they take final examinations. If they score 70 percent or more, they are qualified to assist a grain inspector. After three to five years of experience, an employee is eligible to compete for an inspector's position.

Q How did you become a grain inspection trainer?

A Shortly after completing high school, I was hired by the Canadian Grain Commission as an entry-level inspector and, over the years, advanced through the various training programs and promotions. Then I got the opportunity to act as Pacific Region trainer for one year. The position seemed to suit me. I have been the trainer in this region for 11 years and continue to enjoy creating new grain-related training programs for our employees.

EXPLORING Further

Full of Beans

It would be impossible for Art's team of grain inspectors to inspect every grain of each shipment. Instead, they take small samples. Test the principle that a small sample is a good representation of the whole.

In a large bowl, thoroughly mix together random amounts of three different types of dried beans. Scoop out a small cupful of the mixture, and count how many of each type of bean is in your cupful. Record your amounts, and express them as a ratio. Stir your sample back into the bowl, and repeat. Compare your ratios. Are they similar? As a class, count how many of each type of bean is in the bowl, and record the three totals as a ratio. How does this ratio compare with your sample ratios? Was a small sample representative of the whole?

UNIT 2

An *Issue* to Analyze

A SIMULATION

Dunne County Rattled By Quake

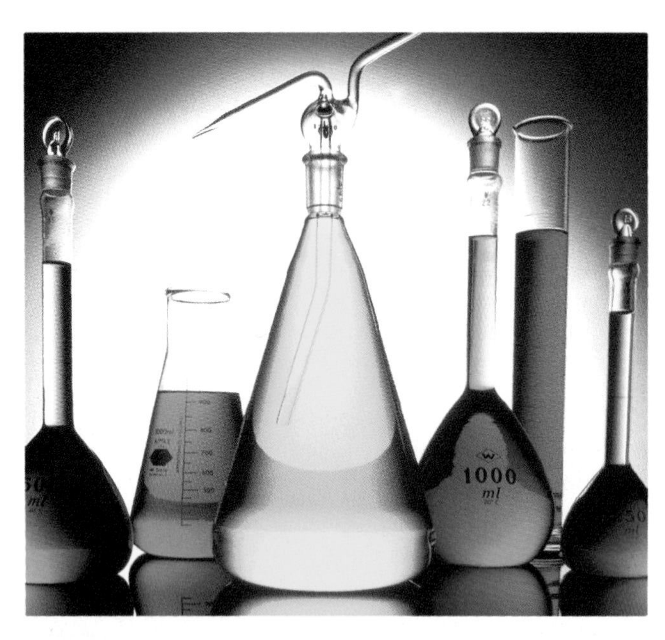

Think About It

All of these pure substances are toxic, corrosive, or flammable. Each is safe to handle with proper precautions, but what if the unexpected happens?

Imagine that you live in Dunne County. Although it lies in an area that is at low-risk for earthquakes, the rattling that woke up everyone last night was definitely an earthquake — the first this area has ever experienced! Most of the buildings are in surprisingly good shape even though they were not designed to be earthquake-resistant. Rockwell High School, however, is in ruins. At first, everyone assumes that a new school will be built on the original site. However, the Dunne County School Board soon learns that idea has its problems.

Background Information

After the earthquake occurred, the newspapers were filled with photographs and news about the event. The destruction of the local high school was one of the most serious problems because so much of the community depended on it. It was the main school in the area for most high school students. As well, it was a community centre for its neighbourhood. It held classes for adults and gatherings for senior citizens. It had the only auditorium large enough for concerts. It also was surrounded by small shops that depend on business from the teachers and students. Thus, everyone in the community was alarmed by its destruction.

Two days after the earthquake the local newspaper, the *Rockwell Examiner*, printed the following headline: TOXIC CHEMICALS LURK IN ROCKWELL HIGH RUBBLE. The article stated that many years ago, when chemistry laboratories in high schools used highly toxic chemicals in many of their experiments, an arrangement that seemed perfect—at the time—was carried out. Some local residents recalled that, in the early 1950s, the School Board had decided to install a concrete holding tank for the chemicals. A deep pit was dug on the school grounds. The custodian was simply to deposit the chemicals in the holding tank on a regular basis. The idea was that the various acids and bases in the tank would neutralize each other. Any undissolved solids would just sink to the bottom. Members of the board were proud to announce their plan for disposal of the toxic chemicals when most towns and cities were simply flushing such substances down the drain.

By the 1970s, everyone became more aware that using fewer strong chemicals in high school laboratories was a wise approach to cutting down the amount of toxic chemicals disposed of. Thus, the holding tank has not been needed or used in recent years.

More headlines appear in the *Rockwell Examiner*. The newspaper publishes a special edition, filled with facts and opinions such as the following:

- Nobody knows where the tank was installed because no records of the installation can be found.
- Modern building codes say that the soil beneath a school must be free of dangerous substances.
- Building costs will be the same no matter where a new school is built, but land costs have risen dramatically. The most affordable land is on the far edge of Dunne County.
- Some people think it would be best to build on cheap land, to avoid cleanup costs.
- The newspaper's lawyer thinks the School Board cannot legally abandon the site, and would likely be sued if anyone is harmed at the site in the future.

The *Rockwell Examiner* announces that it will sponsor an open hearing. Should the school be rebuilt on the original site or built on a new site? The newspaper's publisher will act as a facilitator.

Plan and Act

1. Plan to attend the open meeting. The following people will make a presentation:
   - School Board Treasurer
   - Ministry of Education representative
   - Ministry of Environment representative
   - Ministry of Public Works representative
   - City Planner
   - Toxic cleanup engineer

   The audience will include the following groups:
   - Taxpayers who have no children in school
   - Business owners
   - Parents of young children
2. As a group, research your role and be prepared to make a presentation or ask questions. Use the following to guide your research:
   - If your group is presenting, what kind of information should your group provide to the audience?
   - What are some questions you may expect from the different audience groups?
   - If your group is an audience representative, what questions should your group ask?
   - Think about the concerns your group might have and decide which expert your question should be directed to.
3. At the public meeting, your task as a class will be
   - to define the problem faced by the Dunne School Board
   - to identify the alternatives available to solve the problem
   - to decide what should be done

   Your teacher will provide you with a blackline master to show you the correct *Procedure* for a Public Hearing, as well as specific information to help your group prepare to play its role in the meeting.

Analyze

1. Which alternative did the class decide on as the best solution to the problem? Was there a single main factor that led to this decision?
2. Do you agree or disagree with this decision? Explain.
3. How did your study of mixtures and pure substances help you to understand the problem and the best way to solve it?

UNIT 2 PROJECT

Purifying Mixtures

From the air you breathe to the water you drink to the soil underfoot — mixtures are everywhere! You have learned that mixtures contain two or more pure substances that are combined together *physically*. One way to identify a mixture is to separate it into its components, or fractions. Some methods include magnetism, flotation, sifting, evaporating, and filtration. Many industries rely on these methods to isolate the fractions in as pure a state as is necessary. For example, our drinking water is cleaned and purified before it runs out of our taps for drinking and cooking. Tap water still contains some minerals, but these do not harm us (and are actually good for us). Therefore, it is not necessary to purify water any further. If pure water is needed, it must be distilled, which requires much more time and energy to accomplish. Each new step of a separation may cost a company more money. This is why ingredients and materials that are almost 100 percent pure can be costly to buy. As companies develop methods and technologies that can separate a mixture or purify a contaminated substance more efficiently, the cost can come down.

This giant magnet is being used to lift bits of iron. How could this method be used to purify materials in a mixture?

Challenge

Separate a dry mixture into the pure substances that make it up.

Materials

variety of mixtures prepared by your teacher
several beakers (250 mL to 600 mL), bottles, or cups
magnet
variety of sieves (fine to coarser mesh)
small funnels
filter paper
hot plate (optional)
evaporating dishes
water
paper
labels or grease pencil

Safety Precautions

- Be careful when using the hot plate. Unplug it when it is not in use.
- Be sure to wipe up any spills as wet floors are slippery.

Design Criteria

A. In a small group, separate a mixture into its component substances.

B. Include the smallest number of separation steps as is necessary to produce pure fractions.

C. Design a flowchart, similar to the one presented in Think & Link Investigation 6-D on page 171, showing the steps and separation methods that you used in your final separation procedure.

Plan and Construct

1. Obtain a dry mixture from your teacher. With your group, try to identify as many different substances you can see in the mixture. Record these in a list. Then suggest any substances that might be present but not clearly visible (for example, salt and sand can look very similar in a mixture and, therefore, may be difficult to tell apart). Record these substances in a separate list.

2. Brainstorm the types of methods you might be able to use to separate your mixture. Make a list of your ideas. Examine the list and make a group decision on which methods you will use and in which order.

3. Outline the steps in your method of separation and the pure substance isolated at each step. You may need to revise your outline a few times before you come up with the best sequence of steps.

4. Separate your mixture. Collect each fraction in a separate, labelled container.

5. If your fractions look contaminated (for example, sand stuck to larger pieces of gravel), try to purify them by using one of the separation methods.

6. After several trials, decide on the method that will give you the best results in the fewest number of steps.

7. Make a flowchart to reflect your final method.

Evaluate

1. How pure were your fractions? How did the order of steps affect how pure you were able to get them?

2. Which methods did you use to separate substances that were not clearly visible? How did you know that each method worked?

3. How could you improve your methods of separation? What other equipment or methods might have helped you to improve the quality and efficiency of your separations?

Paper chromatography is a special method that can be used to separate mixtures such as ink and various drinks made from flavour crystals. This method is particularly useful in forensic science when the components of solutions must be examined very carefully. Conduct research on the Internet and/or your library about how mixtures can be separated using the method of paper chromatography. With a partner, use this method to identify the various colours used to make blue or black ink, or your favourite unsweetened drink made with crystals. Present your results in a poster with labels and a brief explanation of the procedure and practical uses for it.

UNIT 3

Thermal Energy and Heat Technology

Imagine a world where people had not learned how to warm or cool anything. Without furnaces or air conditioners, homes and schools would get uncomfortably hot or dangerously cold. No one would ever enjoy a hot meal (no stoves) or an ice-cream treat (no freezers). Almost nothing would be made of metal or glass because these materials require intense heat for shaping. Automobiles, trucks, buses, and even bicycles would not exist. Most of the comforts and conveniences we now enjoy would not exist.

Machines that use heat and control temperature are all around us. Most of the time, these devices work so well that we do not even notice them. They were designed, however, using knowledge gained from hundreds of years of observation and experiment. Inventors developed ingenious devices to measure temperature. Scientists developed theoretical pictures of the effect of heating and cooling on the tiniest particles of materials. Engineers developed methods of calculating and predicting how materials will behave when they are heated or cooled. As a result of all this work, scientists now have a clear and useful picture of heat and temperature. By studying what these scientists and engineers have learned, you can develop a much better understanding of the heat technology in your own life.

Unit Contents

CHAPTER 7

Warming and

Getting Ready...

- What are the hottest and coldest temperatures you have ever experienced?
- Why does spraying water on most types of fires put out the fires?
- What is energy?

Science Log

Discuss the questions above with a partner. Then record your answers and ideas in your Science Log. If you and your partner disagree, summarize the reasons for each person's ideas.

Skill POWER

For tips on how to make and use a Science Log, turn to page 476.

Search for cool shade in the hot summer sun; try to warm up in the winter wind. Put on extra layers of clothing to keep warm; take off layers of clothing to cool down. Turn up the furnace to heat a building; install an air conditioner to cool it. Heat food to cook a meal; freeze the leftovers. Warming and cooling things are very common events in everyday life.

It is important to know how hot or cold things are. Is the campfire hot enough to use for cooking? Is the air cold enough to freeze the skating rink at the park? Is the weather forecaster predicting warm enough temperatures to go camping on the weekend? We use thermometers and check temperatures every day.

In fact, our lives are full of tools for measuring temperatures and for warming or cooling things. By inventing and using such technology, scientists have learned a lot about heat and temperature. In this chapter, you can start to learn about exactly what is happening when you warm or cool something, and about how hot and cold objects differ. By studying these ideas, you will be able to develop a better understanding of the thermometers, stoves, refrigerators, and other "heat machines" that we rely on every day.

Cooling

Spotlight On Key Ideas

In this chapter, you will discover

- how different types of thermometers are made and used
- what happens when solids, liquids, and gases are warmed and cooled
- the difference between temperature and thermal energy

Spotlight On Key Skills

In this chapter, you will

- design, construct, and calibrate your own thermometer
- examine and use a laboratory thermometer to measure temperatures
- compare the behaviour of different substances when they are warmed and cooled
- construct mental models to explain how particles of matter behave as temperature changes

Starting Point

Warm-Ups

Scientists often look closely at simple things to try to find clues to explain more complicated problems. What can you infer from this simple experiment?

What You Need

small, empty glass pop bottle that has been cooled in a refrigerator

a coin large enough to cover the top of the bottle

What to Do

1. Hold the palms of your hands against your forehead. Notice how warm your hands feel.
2. Rub your hands together briskly for 30 s.
   (a) Immediately touch your hands to your forehead again. Notice if they feel warmer or cooler.
   (b) Record your observations. Then suggest how moving your hands can cause what you observed.
3. Dip the top of the cold pop bottle in water.
   (a) Cover the moist top of the pop bottle with a wet coin. The water should make an air-tight seal between the coin and the bottle.
   (b) Clasp your hands around the pop bottle.
   (c) Watch the coin very carefully. Notice what it starts to do after about half a minute. Record your observations.

What Did You Find Out?

1. Give the best explanation you can for the behaviour of the coin.
2. What connection might there be between the effect of rubbing your hands together and the cause of the behaviour of the coin?

7.1 Temperature and Thermometers

"Ooh, that wind is as cold as ice. Better stir up the campfire to get those red-hot coals burning again. There, that feels a lot warmer. I hope my hot chocolate hasn't cooled down too much."

DidYouKnow?

The record Canadian low temperature of –62.8°C was recorded at Snag, in the Yukon Territory. The Canadian record high temperature of 45°C was recorded in Sweetgrass, Saskatchewan. Try to locate Snag and Sweetgrass on a map of Canada. Then try to find the record high and low temperatures in your area and the dates they were recorded.

Everyday life is full of descriptions of **temperature**: that is, how warm or cool things are. One way to estimate temperature is just to touch something. Some nerve endings in human skin are quite sensitive to different temperatures, so people can learn to recognize the feeling of particular temperatures by experience. A drop of liquid on the inner wrist is all a parent usually needs in order to tell when a bottle of juice or milk is the right temperature for a baby to drink. Health-care workers can recognize dangerous body temperatures by touching a patient's forehead with the back of a hand. People who work with very hot, glowing materials can estimate the temperature of the materials by the colour of the light they give off. Welders and glass blowers can estimate when a flame is hot enough to soften metal or glass. Astronomers judge the temperature of stars by the colour of the light they emit.

Estimating temperatures with your eyes or skin is not always safe or reliable, however. Even if glass and metal are not glowing, they can be hot enough to burn you badly. In the winter, when the air temperature rises above freezing after a cold snap, people feel warm and take off their heavy clothing. In the summer, cool winds before a thunderstorm can make people shiver and reach for sweaters, even though the temperature is still far above freezing.

At Home **ACTIVITY**

Baffle Your Skin

How hot something seems to be when you touch it depends on how warm your skin already is. You can experience this for yourself.

What You Need

3 bowls of water, large enough to dip a hand in

hot (not burning) water

room-temperature water

cold water (refrigerator temperature)

What to Do

1. Put one hand in the bowl of cold water and the other hand in the bowl of hot water. Hold them there for 1 min.
2. Quickly put both hands in the bowl of room-temperature water. Notice how each hand feels.
3. Repeat steps 1 and 2, but switch hands in step 1.
4. In clear sentences, record how warm the room-temperature water felt to each hand in step 2 and in step 3.

What Did You Find Out?

1. Was there any difference in your observations in steps 2 and 3? If there was, suggest a reason why.
2. Use your observations in this activity to explain how the same air temperature can seem warm in the winter and cool in the summer.

Extension

3. Midori and Erin go to a swimming pool on a warm summer day. Midori spreads her blanket in the Sun, and Erin lies in the shade. After 20 min, they decide to jump into the pool. Predict how warm the water will feel to each girl.

Thermometers

Your senses are easily fooled, but **thermometers** are more reliable. Thermometers are mechanical or electrical devices for measuring temperature. A thermometer similar to the one in Figure 7.1A was constructed by the Italian scientist Galileo in the early seventeenth century. One hundred years later, the design was improved, as Figure 7.1B shows. However, an important part of modern thermometers was still missing. Examine the photographs carefully to find out what it was.

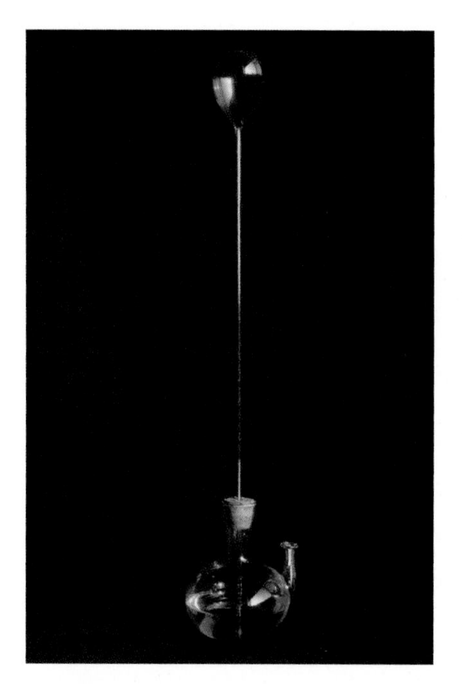

Figure 7.1A Galileo invented his air thermometer around 1600. As the air in the upper bulb cooled or warmed, liquid moved up or down in the tube.

Figure 7.1B More portable thermometers, like this liquid thermometer invented around 1700, were made by putting the liquid in the bulb and part way up the stem.

Temperature Scales

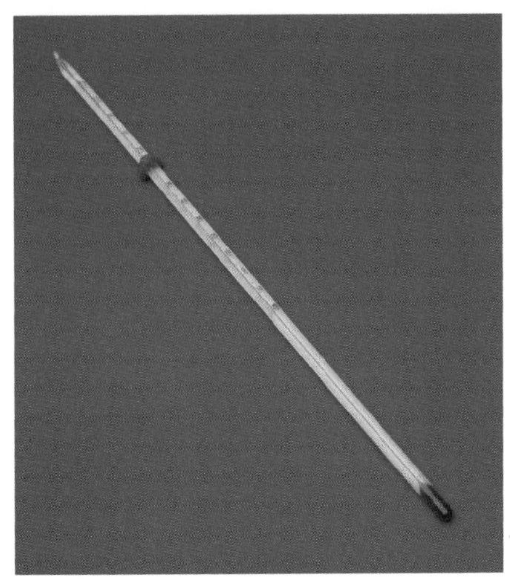

Figure 7.1C A modern laboratory thermometer has a smaller bulb and a much narrower opening in the glass stem.

When you examined Figures 7.1A and 7.1B, you probably noticed that these early thermometers do not have **scales**. That is, they have no markings with numbers to indicate a precise temperature. As scientists discovered more and more about the way that temperature affects the characteristics of materials and the rates of chemical processes, they needed to measure temperatures with precision. Modern thermometers, such as the one in Figure 7.1C, have gradations or evenly spaced lines that allow you to read exact temperatures.

Figure 7.2 Anders Celsius suggested his temperature scale in 1742.

For any form of measurement, someone has to decide on a unit and a standard for comparison. Today, the temperature scale commonly used in Canada and many other countries is called the **Celsius scale** in honor of Anders Celsius (1701-1744). He used the "degree" as the unit of temperature. He based his standards for comparison on the properties of water, the most abundant liquid on Earth. Celsius assigned zero degrees to the temperature at which ice melts at sea level. He assigned a value of one hundred degrees to the temperature at which liquid water boils at sea level. Then he separated the region between these temperatures into 100 evenly spaced units or degrees.

Do you wonder why it is important to use the freezing and boiling temperatures of water at sea level and not just any altitude? The reason for specifying sea level is that the temperature at which these changes occur is affected by the weight of the atmosphere pressing down on the water. At high altitudes, the weight of the atmosphere is smaller than it is at sea level and it takes a temperature higher than zero degrees Celsius to melt ice. When a large amount of weight is pressing on ice, it will melt at temperatures lower than zero degrees Celsius. In Figure 7.3, you can even see a difference in ice when it is under a large weight.

Figure 7.3 The ice at the bottom of this glacier looks bluish green but it really has no colour at all. The tremendous weight of the ice and snow above, pressing down on the base of the glacier, changes the nature of the ice crystals. Light shining through these unique crystals appears bluish green.

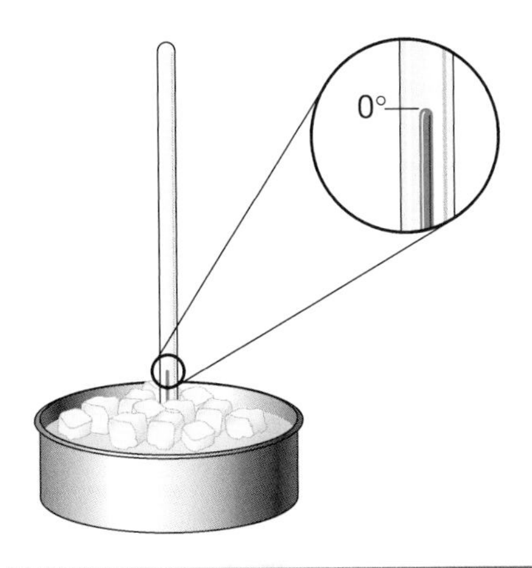

The liquid level in an ice-water bath is marked as 0°.

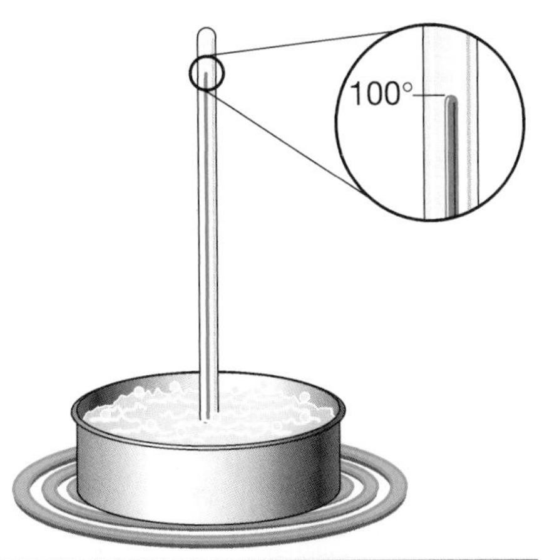

The liquid level in boiling water is marked as 100°.

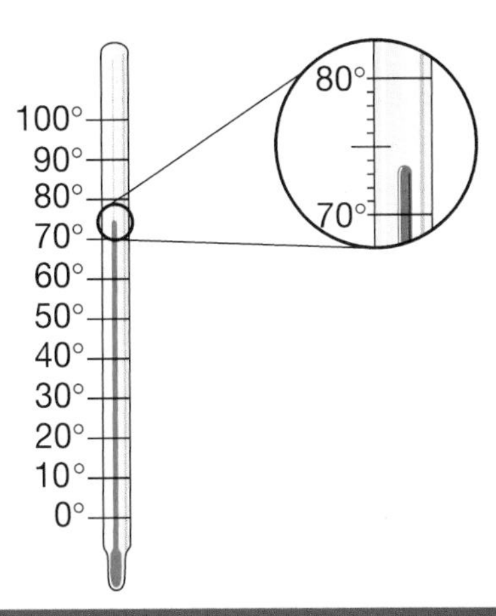

The scale is divided into 100 equal degrees and numbered.

Figure 7.4 Steps in calibrating a thermometer

When manufacturers make thermometers, they must **calibrate** them. That means that they must paint or etch the lines on the thermometer and place the numbers in the correct positions. Study Figure 7.4 to see how to calibrate a thermometer.

As scientists developed theories to explain the behaviour of gases at different temperatures, they realized that they needed a temperature scale that started at the coldest possible temperature, or "absolute zero." This new temperature scale was named the **Kelvin scale**, in honour of William Thomson (1824-1907), who was given the title, Lord Kelvin. Although no one has ever been able to cool anything down to absolute zero, scientists know that the temperature is –273.15°C.

It is easy to verify the value of absolute zero. First, consider how thermometers measure temperature. The volume of the gas or liquid in the thermometer becomes smaller as the temperature drops. What would the temperature be if the volume reached zero? You are probably thinking, "But the volume cannot go to zero because the gas or liquid has mass and takes up space." You are correct. However, a graphing technique allows you to "imagine" that the volume becomes zero millilitres and to observe the temperature on the graph. You would observe the volume of a gas at several temperatures, then plot these values on a graph. Finally, you would draw the line. If you extend the line to zero volume and then read the temperature at that point, it would be –273.15°C.

Figure 7.5 William Thomson (Lord Kelvin) developed the Kelvin scale. This scale can be used to make accurate scientific measurements.

The units of temperature on the Kelvin scale are not degrees but are simply called kelvins. For example, the freezing temperature of water at sea level is 273.15 K (read, two hundred seventy three point one five kelvins). When it is not necessary to be extremely precise, this temperature is usually rounded to 273 K.

DidYouKnow?

Another temperature scale used in Canada until the 1970's, and still used in the United States, is the Fahrenheit scale, named after Daniel Fahrenheit (1686–1736). Fahrenheit is famous for more than his temperature scale. He invented and built thermometers filled with coloured alcohol, like those used today. For more precise measurements, he made thermometers filled with mercury, a silver-coloured metal that is liquid at room temperature. (Mercury thermometers are seldom used in schools because if they break, they release mercury vapours which are poisonous.) Using his precision thermometers, Fahrenheit was the first person to observe that water does not always boil at the same temperature. As well, he discovered that water can sometimes remain liquid at temperatures below freezing!

INVESTIGATION 7-A

The Wonderful Water Thermometer

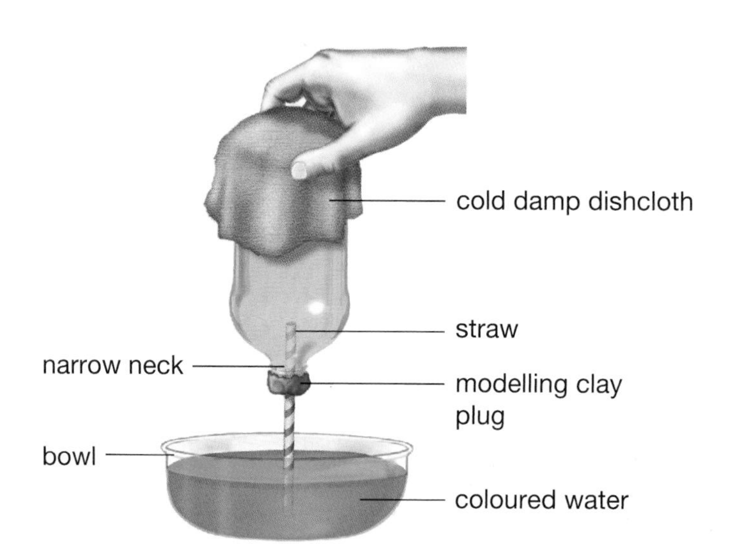

Today thermometers and other scientific instruments are mass-produced in factories. Early scientists, however, had to build their own measuring devices. Their clever designs used ordinary materials, yet produced accurate measurements. Can you create a useful measuring instrument yourself?

Challenge

Use everyday materials to build a thermometer that accurately measures temperatures in your classroom.

Materials

small glass bottle with a narrow neck (for example, a small pop bottle)
drinking straw or length of glass tubing
laboratory stand and ring clamp
dishcloth
paper
pen
ruler
calculator
bowl of water with food colouring added
modelling clay or silicone glue
ice-cold water

The class also needs two calibration devices, assembled like the apparatus illustrated but with regular laboratory thermometers instead of the straw or glass tubing, and a hair dryer for testing.

Safety Precautions

Silicone glue does not wash off hands or clothing. It irritates skin and emits fumes as it hardens. If you use it, follow your teacher's directions carefully and work in a well-ventilated area. Wear gloves, eye protection, and an apron, and work on newspaper. Use craft sticks or wide toothpicks to apply and shape the smallest possible quantity of the glue. Roll up the craft sticks in the newspaper when you are finished, and discard them in the garbage.

Design Criteria

A. Thermometers built in Part 1 should detect increases in temperature when your teacher warms them gently with a hair dryer and decreases in temperature when they are cooled with a cold washcloth.

B. At the end of Part 2, the thermometer will have a properly-constructed scale with evenly-spaced degree markings and suitable numbering.

C. The thermometer must measure the temperature of the classroom accurately. The reading should be within 2°C of the temperature measured by a standard laboratory thermometer.

Part 1

Assembling the Thermometer

Plan and Construct

1. Using the materials your teacher provides, your group will design and assemble a thermometer like the one illustrated opposite. The straw or glass tubing needs to have an air-tight seal against the bottle neck. Tape does not work very well; modelling clay or silicone glue is better. If you use silicone glue, be careful to follow the handling instructions your teacher gives and let it harden for 24 h before continuing.

2. Warm the bottle with your hands. Record what happens in the dish at the end of the straw. Troubleshooting: If nothing happens, your hands are probably about the same temperature as the bottle. Try wetting a dishcloth with warm water, wringing it out, and draping it over the top of the bottle.

3. Wet a dishcloth with cold water, wring it out and drape it over the bottle. What happens to the level of water inside the straw?

4. When you are sure that your thermometer is working correctly, have your teacher certify that it meets Design Criterion A.

Evaluate

1. Which part of your thermometer responds to changes in temperature? Describe how it responds when the air in the bottle
   (a) warms up
   (b) cools down

2. Why might you add marks and numbers to your thermometer? Where would you put them?

Part 2

Calibrating the Thermometer

Plan and Construct 

1. Read over the text section "Measuring Temperature." Then plan how to create a scale for your thermometer so that it can measure temperatures accurately. Here are some hints:
   (a) Your scale needs to be fastened to the thermometer, then taken off for measuring and marking, and then fastened back on to the thermometer in its original position.
   (b) If you put your thermometer beside the cool calibration device, you could mark the liquid level at a known cool temperature. Let your thermometer cool first.
   (c) If you put your thermometer beside a warm calibration device, you could mark the liquid level at a known high temperature.
   (d) You could measure the distance between the "cool" and "warm" marks in millimetres. You could also find the number of degrees difference between the two temperatures. Then you could calculate the number of millimetres the liquid level changed for each degree.
   (e) With the information from hint (d), you could divide up your thermometer scale evenly and number it carefully.

2. Calibrate your thermometer.

3. Show your teacher or another lab group that your calibrated thermometer meets Design Criteria B and C.

Evaluate

1. Did your thermometer meet the design criteria?

2. Describe the main problems that you had building your thermometer. How did you overcome each problem?

3. Why are thermometers designed like yours not very useful in everyday life?

CONDUCT AN

INVESTIGATION 7-B

Tracking Temperature Changes

Some traditional societies used tightly woven baskets filled with water for cooking. These baskets could not be heated directly over a cooking fire. Instead, hot stones from the fire were placed in the water-filled basket, warming the water and cooking the food. You can use a similar method to study exactly what happens when a material is warmed or cooled. How fast does its temperature change? Do different materials warm and cool at the same speed?

Problem

Do different liquids warm at the same rate?

Safety Precautions

- The metal blocks in this investigation are hot enough to burn you badly. Handle them with tongs, an oven mitt, or a hot pad.
- Use care when handling hot liquids.
- Laboratory thermometers are fragile and expensive. Handle the thermometer you are using with care.
- Work on newspaper to absorb spills.

Apparatus

teacher:

hot plate
large beaker or saucepan
hot metal block
tongs

student:

400 mL plastic beaker or tin can
laboratory thermometer
stirring rod
tongs, oven mitt, or hot pad

Materials

water
cooking oil
newspaper

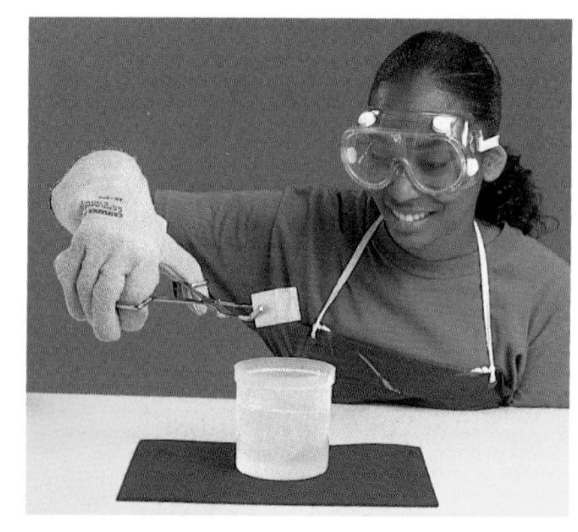

Procedure

1. Prepare a data table like the one below to record your observations. Give your table a title.
2. Measure 200 mL of water or oil into the beaker. Measure and record the temperature of the liquid. Estimate to one decimal place: for example, 21.0°C or 22.5°C.
3. Obtain a hot metal block from your teacher. Carefully lower the metal to the bottom of your beaker of liquid. **CAUTION:** Do not drop the metal into the beaker. Lower it gently to avoid splashes and broken glassware.

   Stir the liquid gently so that the hot metal warms it evenly.

	Time (min)	Temperature (°C)
water		

Skill POWER

To learn how to make a line graph, turn to page 486.

4. Every half minute, measure and record the temperature of the liquid. Continue until the temperature stops changing, or for 5 min at the longest.

5. Return the metal and oil, and clean up as directed by your teacher. Wash any apparatus that is coated with oil in hot soapy water. Rinse the apparatus, and leave it to air dry.

6. If you have time, repeat the experiment using oil (if you first observed water) or water (if you first observed oil).

Analyze

1. Draw a graph to display your first set of observations. Since the temperature probably changed smoothly, even though you only measured it every 30 s, a line graph would show the data best.

2. Add a second line to the same graph to show how quickly the other liquid warmed. If you made only one set of observations, trade results with another group.

3. Use your graph to compare how the two liquids warmed. Which liquid warmed more quickly? Which one reached a stable final temperature more quickly?

Conclude and Apply

4. Identify some features of the design of this investigation that make it a fair test of how the liquids warmed. What variables were kept the same, whether you observed water or oil?

5. Identify the independent variable and the dependent variable in this investigation.

6. Make a prediction about cooling the same two liquids. Which liquid would cool more quickly? Give a reason for your answer.

Extend Your Skills

7. **Design Your Own** Design an experiment to test the prediction you made in question 6 above. Make sure that your experiment is a fair test, and have your teacher approve your procedure before you carry out your experiment.

Extend Your Knowledge

8. At the edge of a lake or an ocean, the water and land are warmed similarly by the Sun. How will their temperatures differ at mid-afternoon? How will they differ at midnight? You can find more information under the heading "land breezes" or "sea breezes" in a reference book or on the Internet.

INVESTIGATION 7-C

Boiling Hot, Freezing Cold

Think About It

You can probably guess many familiar temperatures quite accurately. Other temperatures may surprise you! As you follow the directions, make sure that you learn the temperatures described in italics.

What to Do 

1. In your notebook, make a table with three columns labelled "Very cold," "Everyday," and "Very hot." Give your table a title.
2. Copy each description from the table on the right into the proper column in your table.
3. For each description, choose the correct temperature from the right-hand column of the table. Write the temperature beside the description. Discuss your answers with your partner until you agree on each one.
4. Check your answers against the list your teacher has. Correct any mistakes you made.
5. Have your partner quiz you to make sure that you know the common temperatures, which are printed in italics.

This "Morning Glory Pool" is heated by energy from deep within Earth. The water remains about 95°C even with snow on the ground nearby.

	Description	Temperature (°C)
1	temperature of lava from Hawaiian volcanoes	4 to 10
2	temperature of ocean currents off Canada's east coast	–5
3	temperature of ocean currents off Canada's west coast	–87
4	world record coldest air temperature	–121 to –156
5	*comfortable room temperature*	92
6	body temperature of a budgie bird	15 000 000
7	temperature where the Space Shuttle flies in orbit	–10 to –15
8	temperature of a candle flame	200
9	comfortable temperature for heat-loving bacteria	20 to 25
10	*normal human body temperature*	37
11	temperature of ice cream	40
12	oven temperature for baking bread	1
13	temperature of food in a freezer	100
14	temperature of the interior of the Sun	6000
15	temperature of hot tea or coffee	1150
16	*temperature of boiling water at sea level*	55
17	*temperature of a slush of pure water and ice*	800
18	temperature of the surface of the Sun	0

The descriptions in this table do not match the temperature in the column beside them. Your job is to work with a partner to unscramble them.

The Right Device for the Job

Could you use the same device to measure the temperature of the surface of the Sun and the body temperature of a parrot? Probably not. Thermometers have been developed to suit almost every purpose, from measuring the extreme cold of outer space to estimating the temperatures of stars. Examine the illustrations below to see how some specialized thermometers work.

The Thermocouple

In a thermocouple, wires made of two different metals are twisted together. When the twisted wire tips are heated, a small electrical current is generated. The other ends of the wires are connected to a meter or computer, which can be calibrated to show the temperature of the joined wires. The electrical current from the thermocouple can be used to turn a switch or a valve on or off if the temperature changes.

Thermocouples can be used to measure temperatures so high that ordinary laboratory thermometers fail because the liquid in them would start to boil.

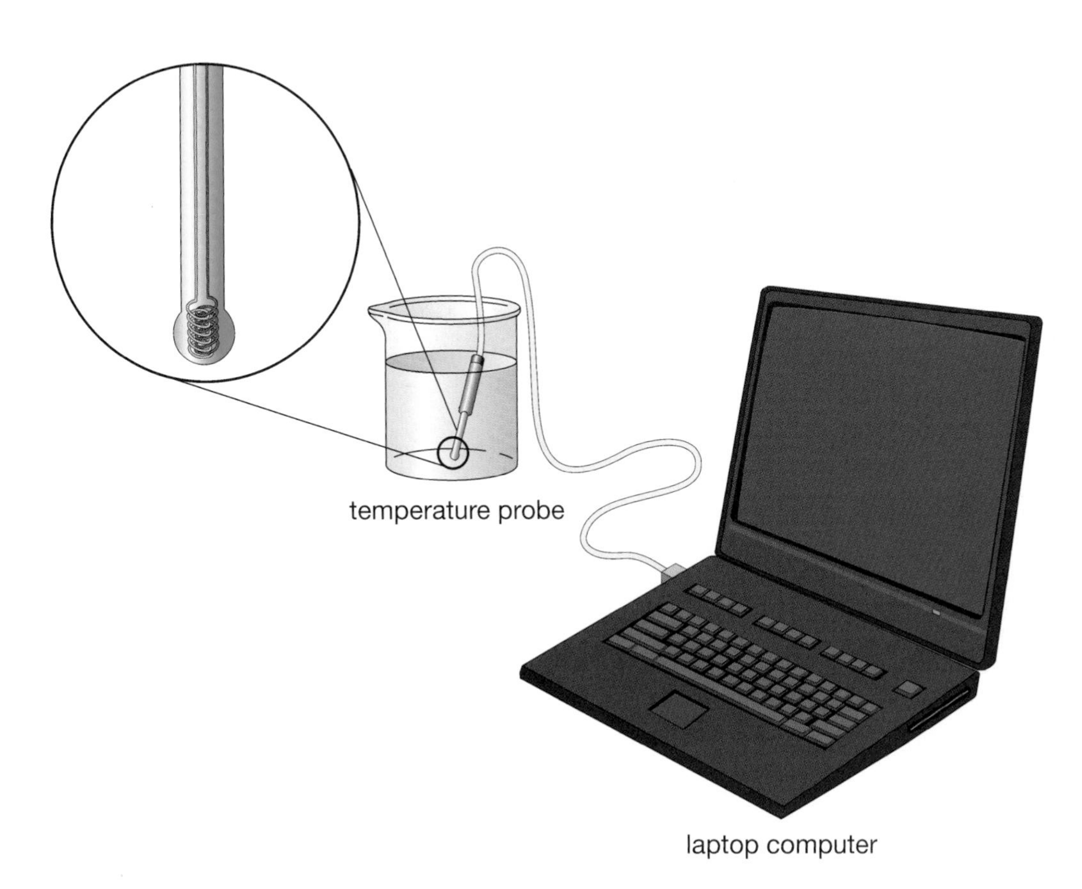

Figure 7.6 A thermocouple being used to measure the temperature of a liquid

The Bimetallic Strip

A bimetallic strip is made of two different metals joined firmly together. When the bimetallic strip is heated, one metal expands more than the other, causing the strip to bend. When the bimetallic strip cools, one metal contracts more than the other, causing the strip to straighten or even bend in the other direction. When the strip is formed into a coil, the uneven expansion or contraction of the metals causes the coil to tighten or loosen. In a thermometer, the moving end of the strip can be attached to a pointer that points to a temperature scale.

Word CONNECT

Recording thermometers are sometimes called "thermographs." The syllable "therm" comes from the Greek word for temperature or heat, and the syllable "graph" comes from the Greek word for writing. Thus a thermograph is a "temperature writer." List as many other "therm" and "graph" words as you can, along with their meanings.

Figure 7.7 The coil in this household thermostat is a bimetallic strip. When the temperature changes and the coil tightens or loosens, it tilts a glass capsule containing a drop of liquid mercury. When the mercury rolls to one end, it turns the furnace on. When it rolls to the other end, it turns the furnace off.

The Recording Thermometer

In one type of recording thermometer, a hollow probe is connected by an airtight tube to a flexible metal bellows. As the temperature changes, the air warms or cools, pushing the bellows apart or pulling them together. One end of the bellows is attached to a long, light metal lever which ends in a special pen. Tiny movements of the bellows cause much larger movements of the free end of the lever and the pen. The pen traces a rising and falling line on a strip of paper attached to a slowly turning drum. The drum usually makes one turn every seven days, so each strip of paper contains a record of temperature changes for an entire week. (You will find out about another instrument that works in a similar way in Unit 4.)

Figure 7.8 This recording thermometer uses a bimetallic strip to detect changes in temperature. The end of the coil is attached to the short end of a lever. The long end of the lever is attached to a pen that makes a permanent recording of the temperature on graph paper attached to a rotating drum.

The Infrared Thermogram

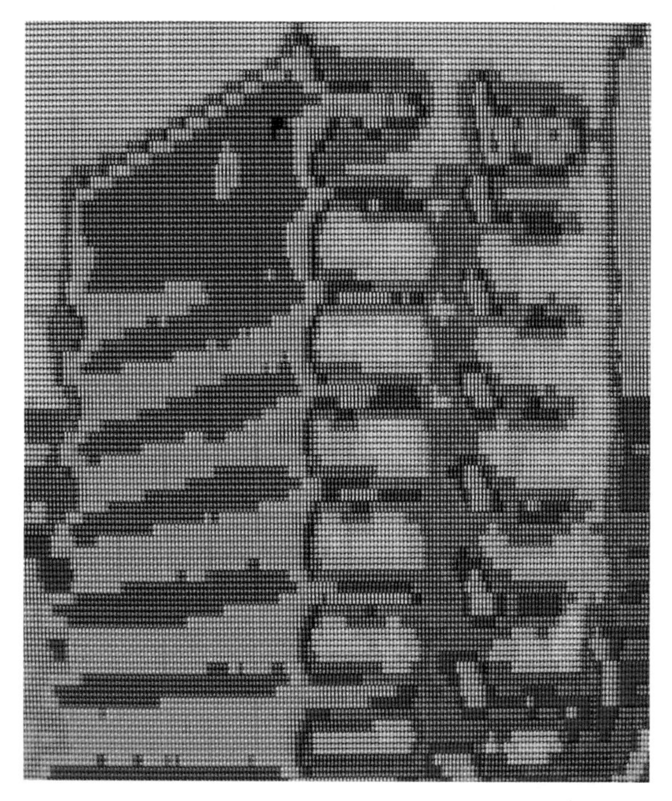

Figure 7.9 The photograph on the left shows an infrared image; the one on the right is a normal photograph of the same image. What colour indicates the highest temperatures in the infrared photograph? What colour shows the coolest temperatures?

Objects do not have to be glowing red hot to give off radiation. Anything that is warmer than absolute zero gives off infrared radiation (IR), a type of radiation similar to light, that your eyes cannot detect. Your skin can detect infrared radiation when you are near hot objects. Even if you are not actually touching the object, you can feel the warmth. Special camera films can photograph IR, and electronic sensors can detect it and display infrared images on television screens.

INTERNET CONNECT

www.school.mcgrawhill.ca/resources/

Warm objects, such as your body, give off more infrared radiation than cool objects. Thermograms of certain body parts can help physicians diagnose some medical problems. To explore more about this topic go to the web site above. Go to **Science Resources**, then to **SCIENCEPOWER 7**, to find out where to go next.

DidYouKnow?

Some kinds of crystals turn certain colours at different temperatures. You may have seen these crystals in strips used to take your temperature. When you place the strip on your forehead, the crystals that turn colour will show the temperature of your skin.

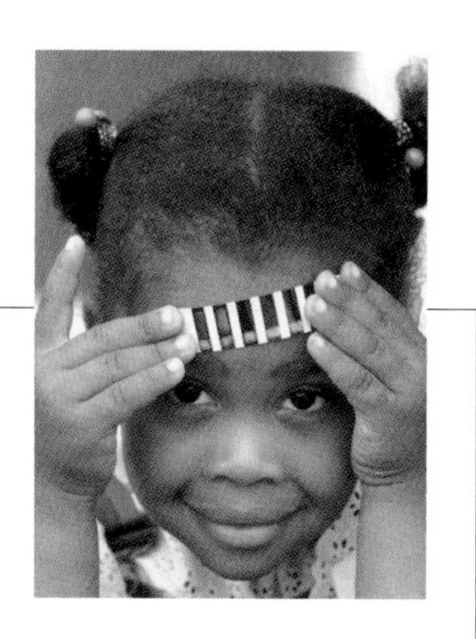

Check Your Understanding

1. Show that you understand the meaning of each term below by using it correctly in a sentence.
   (a) thermometer (b) temperature
   (c) temperature scale

2. Identify a material or an object that has a temperature of
   (a) 0°C (b) 100°C
   (c) 37°C (d) about 20°C

3. Suppose that you were present on the hottest day ever reported in Canada.
   (a) What would your body temperature have been?
   (b) If the air temperature had dropped by 5°C, would you have felt warm or cold?

4. **Apply** What might be the advantages and disadvantages of using a thermocouple instead of a regular lab thermometer?

5. **Apply** Many household appliances, such as the hair dryer below, are heated electrically. They usually contain a thermostat that switches electricity on and off to keep the appliance at a constant temperature. Think of at least three examples of this use for thermostats.

6. **Thinking Critically** Choose the most appropriate temperature-measuring instrument to use in each situation below. In each case, explain your choice.
   (a) controlling an electric frying pan
   (b) making long-term temperature records at a weather office
   (c) detecting small forest fires before they spread
   (d) monitoring temperatures inside a furnace
   (e) checking trains for overheating wheel bearings as they pass by a station
   (f) studying temperature changes inside a building over a 24 h period

7.2 Particle Theory, Temperature, and Thermal Energy

In Conduct an Investigation 7-B, you saw that different liquids respond quite differently when they are warmed or cooled. Early scientists tried to explain why this happened. How does a material change when it is warmed or cooled? How does warming or cooling affect the tiny particles of which everything is made? You have already observed one important clue when you warmed your hands by rubbing them together at the start of this chapter.

Find Out ACTIVITY

Detect a Connection

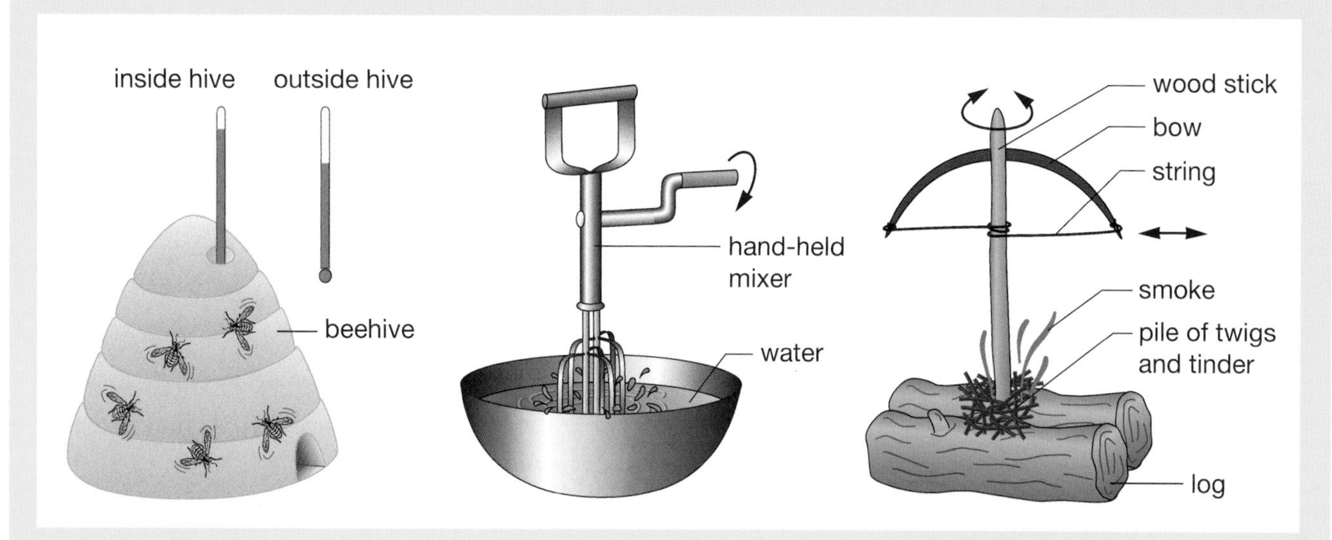

What to Do

Carefully examine each picture. Then answer the following questions.

1. One way that bees control the temperature in their hive is by beating their wings vigorously. Explain what happens to
   (a) the motion of the air particles in the hive
   (b) the air temperature in the hive
2. In a famous experiment, James Joule observed the temperature of water as a mechanical mixer stirred it vigorously. Explain what happened to
   (a) the motion of the water
   (b) the water temperature
3. To start a fire, early people used a fire drill to twirl a stick pressed against a piece of wood.
   (a) What happened to the temperature at the pointed end of the drill?
   (b) What do you think caused the particles of wood to change temperature?

What Did You Find Out?

1. What common feature caused the changes in temperature in each example you examined?
2. Identify at least two other situations that are similar to the three examples in this activity.

Temperature and the Particle Theory

Do you remember the key features of the particle theory of matter, from Unit 2? Modern scientists use two of these features to explain changes in temperature.

- All substances are made of tiny particles.
- The particles are always in motion — vibrating, rotating, and (in liquids and gases) moving from place to place.

As you probably noticed in the last activity, if the motion of the particles in a substance changes, the temperature of the substance changes, too. When a substance warms — when its temperature increases — its particles are moving faster. When a substance cools — when its temperature decreases — its particles are moving more slowly.

It is not easy to test this idea directly. The smallest particles of matter are too tiny to observe clearly. In any substance, some particles always seem to be moving faster than average. Other particles seem to be moving unusually slowly. The best that scientists can do is observe the average behaviour of large groups of particles. Such observations have led to this definition: temperature indicates the average speed of particle motion in a substance.

Did You Know?

Antoine Lavoisier (1743–1794) believed that an invisible substance called caloric fluid caused changes in temperature. Fires, for example, had a lot of caloric fluid, so they were hot. If caloric fluid moved from a fire to a cooking pot, the pot warmed up. For many years, scientists tried to detect and measure caloric fluid. No one could. Finally scientists stopped looking for it and abandoned Lavoisier's theory.

Today we know that warming and cooling do not add or take away any matter from a substance. It is just a sign of faster or slower motion of the particles making up the substance.

What is Energy?

As the particle theory of matter was developed, scientists wanted to find a useful way to describe the motion of the particles in a substance. For everyday objects, such as a moving bird or a soccer ball, we can describe position, speed, and direction of motion. These observations, however, are not possible for the very small particles that make up matter.

What *is* possible to find out is how much the tiny particles can change other things. **Energy** is a measurement of something's ability to cause changes or to make something else move. Study the illustrations in Figure 7.10 to find out how scientists describe energy in everyday situations.

Figure 7.10A Fully-charged batteries can power a stereo; dead batteries cannot. A charged battery stores more energy than a dead battery.

Figure 7.10B A stick of dynamite can blast a hole in solid rock; a pile of ashes cannot. The ashes store much less chemical energy than the dynamite.

Figure 7.10C Catching a heavy, fast-moving baseball stings more than catching a light, slow-moving ping-pong ball. The baseball has much more energy of motion than the ping-pong ball.

Word CONNECT

The term "thermal energy" has a precise scientific meaning, but it is not used very much in everyday language. Scientists sometimes use the word "heat," but they give it a specific meaning: thermal energy being transferred because of temperature differences. To avoid confusion, this textbook uses the scientific terms "thermal energy" and "energy transfer" whenever possible.

Can you think of other words that have slightly different meanings in science and everyday life?

Energy descriptions are particularly useful because scientists have learned how to calculate energy values in many different situations. Scientists can use these calculations to make detailed and successful predictions about what will happen when different objects interact. For example, the idea of energy is the key to explaining the behaviour of hot and cold substances.

Boiling-hot water can move a steam engine or turn an electrical generator. Cold water cannot cause these changes. Using the idea of energy, we could say that the cold water has less thermal energy than the hot water. **Thermal energy** determines a substance's ability to cause changes because thermal energy moves from an object at a high temperature to an object at a low temperature. In the next investigation, you will look for ways to control thermal energy transfer.

Cool Tools

Energy is measured in joules (J), in honour of James Joule (1818–1889), an amateur scientist who devoted his life to studying energy. To investigate the connection between energy and temperature changes, Joule built many ingenious devices. One was a set of paddle wheels that stirred water as they were turned by falling weights. The temperature of the water increased a small, but measurable, amount. If you have a sensitive computerized temperature probe, you could repeat Joule's experiment using an electric mixer or a blender to stir the water.

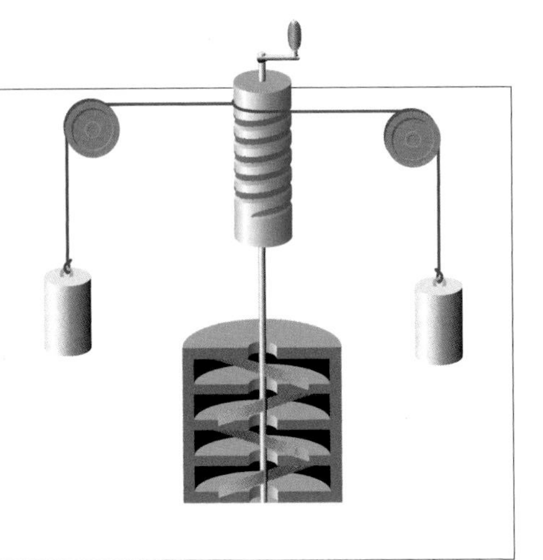

DESIGN & DO

INVESTIGATION 7-D

Build a Better Energy Barrier

Have you ever shivered under a too-thin blanket or sleeping bag on a cold night? Trapping thermal energy and staying warm is sometimes more than a matter of comfort. In a wilderness survival situation, it can make the difference between life and death. As you work at this design challenge, use ways that you already know to keep things warm. In the next two chapters, you will learn how to make an even better design.

Challenge

Design and construct a device that will keep hot water in a tin can warm for the longest possible time.

Materials

laboratory thermometer

tin can

drinking straw (large enough to allow the thermometer to pass through)

water

any household materials you can use to build your device

Safety Precautions

Use oven mitts or hot pads to handle the tin can of hot water.

Design Criteria

A. You can recycle any household materials to build your device. You may *not* use a commercial device, such as a Thermos™ flask.

B. Your device must fit on the top of your desk.

C. You must be able to open your device to fill the tin can with hot water.

D. The drinking straw must be used to provide an opening through which the thermometer can be inserted to measure the water temperature.

E. Your device may not contain a heater or hot materials of any kind, except for the can of hot water.

Plan and Construct

1. With your group, brainstorm possible materials and designs for your device. Make a list of all the suggestions as they are made.

2. From the list, choose materials that you can obtain easily. Decide on a design that you will be able to build.

3. Prepare a one-page summary, describing your design and materials. Include a neat, labelled sketch, showing how your device will be built. Submit your summary and sketch to your teacher for approval. Then keep it to display on testing day.

4. You will have 20 min in class to assemble your device before the test starts. If assembling your device will take longer than this, decide when and where you will do the assembly.

Skill POWER

For tips on the design process and technological drawing, turn to pages 470 and 496.

Evaluate

1. On the day of the test, obtain 200 mL of hot water from your teacher, put it in your device, and close your device. Make sure that the straw provides an opening into the hot water.
2. Insert the thermometer through the straw, and measure the initial temperature of the water. Then measure the temperature every 5 min for 20 min. Record your observations in a two-column data table.
3. Between temperature measurements, examine (but do not disturb) the devices your classmates built. Make a list of the features included in each device to keep the water from cooling too rapidly.
4. After 20 min, calculate the difference between the initial and final temperatures of the water in your device. The winning device is the one that allowed the smallest temperature change.
5. Write a short summary of how well your device worked and how it could be improved if you repeated this investigation.
6. What problems did your group overcome as you worked together on this investigation? Identify the most important change you could make to improve the way your group tackles its next design task.

Across Canada

Bicycles, coins, bridges, steel beams, copper wire — thermal energy plays a key role in processing metals for these and many other useful objects. Dr. J. Keith Brimacombe pioneered better ways of using thermal evergy in metal-treating processes such as flash smelting, steel-billet casting, and gas injection to recover copper and zinc oxide. Keith was a professor of materials process engineering at the University of British Columbia and president of the Canada Foundation for Innovation. He spent time in industrial plants, studying searingly hot blast furnaces and kilns. He and his students spent even more time doing laboratory experiments and using computers to analyze their observations mathematically.

Born in Windsor, Nova Scotia, Keith attended St. Francis Catholic High School in Lethbridge, Alberta. Of his teachers, he said, "Beyond kindling my interest in mathematics, physics, chemistry, and history, they showed me commitment and purpose . . . I ended up playing basketball and football, although neither particularly well. But most importantly, I experienced what sports are really all about — teamwork, perspiration, dedication, and a strong desire to win." Keith carried these lessons into his award-winning engineering and science career.

Dr. J. Keith Brimacombe

Sadly, that career was cut short when Keith died of a heart attack in 1997, at age 54. He had passed his sense of commitment and purpose on to his students, however, and several of them today continue the ground-breaking work he began. Keith was awarded the Canadian Institute of Mining, Metallurgy, and Petroleum's INCO medal in 1998, the year after his death. His daughters, Jane and Kathryn, accepted it on his behalf.

Pause & Reflect

The ideas in this section are tricky! Use them to write a short explanation in your Science Log that demonstrates your understanding of temperature, particle motion, and energy (especially thermal energy). Use your own words, and include examples or diagrams. With a partner, take turns reading your explanations. How are your ideas similar? How are they different?

Check Your Understanding

1. List the main points of the particle theory of matter that were presented in this section.
2. Why is it so hard to test the particle theory to see if it is correct?
3. Describe two situations in your life in which an object with a high temperature has caused changes in something.
4. Name an important discovery or idea contributed by each of these scientists.
   - (a) James Joule
   - (b) Anders Celsius
   - (c) Lord Kelvin
5. **Thinking Critically** Modern scientists do not use Lavoisier's "caloric fluid" theory (see the Did You Know? on page 200). If this theory is wrong, why do you suppose it is discussed in many science textbooks?

7.3 More About Energy

How are thermal energy and temperature related to each other? You began to look at this question earlier. To answer it completely, try to think in terms of the particles that make up hot and cold substances. The particle theory suggests that particles of matter are always in motion. Motion is a sign of **kinetic energy**: the energy of moving objects. Faster motion means more energy. The particle theory also proposes that in any substance, most particles have about the same speed but some particles are moving faster and some are moving more slowly. Combining all these ideas, scientists have developed two important concepts:

- The thermal energy of a substance is the *total* kinetic energy of all of its particles, added together
- The temperature of a substance is a measure of the *average* kinetic energy of its particles

These concepts are theoretical, but a comparison might help you to visualize them. Imagine going ten-pin bowling. You fire the bowling ball down toward the pins, which split up as the ball hits. The different pins go in many directions. No matter how hard you strike the pins, some may not move at all, even though others may fly off at high speed.

Word CONNECT

How would you use the words "energy" and "temperature" in an everyday context? How is this different from their scientific meanings? For each word, think of and write down a sentence which uses the word in an everyday sense. Now try to "translate" your sentences into scientific descriptions.

The total movement of all the pins just after the bowling ball strikes them is like the thermal energy of a substance. The harder you bowl, or the more "thermal" energy you put in, the more movement and thus kinetic energy there will be overall. In a real substance, however, it would be impossible to measure the speed of each individual particle — they are far too small, and there are far too many of them. Thus we cannot measure thermal energy directly using this definition, even though it is scientifically accurate.

What scientists can do is use thermometers to measure changes in temperature. From this and knowledge about the properties of the substance, they can calculate changes in thermal energy. It is even possible to calculate information about the average motion of particles of matter, even though they are too small to be seen! As in many other parts of science, mathematics is a very useful tool for obtaining information that is difficult to observe directly.

What Energy Is . . . and Is Not

"I just don't have enough energy to do my homework."
"I'm so hungry! I need a big meal to get enough energy for the soccer game."
"You look exhausted! Did cleaning your room use up all your energy?"

Energy is *not* a substance. It cannot be weighed. It does not take up space. Energy describes a quality or condition. Think about words that describe other qualities or conditions. You might describe the drums in a band as "loud," but that does not mean they are filled with extra "loudness." If the guitar is played softly, that does not mean its "loudness" is almost used up.

What is energy? Energy is a property or quality of an object or substance that gives it the ability to move, do work, or cause changes. Energy is the topic of one of the most important laws of nature. *Energy cannot be created or destroyed. It can only be transformed from one type to another or passed from one object to another.*

Is it still okay to say, "Wow, I'm feeling full of energy today"? Of course! Everyday language is fine for everyday life. Just remember to be more precise when you are giving a scientific description. For example, think back to the definition of energy on page 200.

The Big Picture

Do you want to make a mug of hot chocolate? A microwave oven can easily increase the thermal energy of the particles in the drink until it is steamy hot. How about having some cold juice on a summer day? Just add a few ice cubes. They reduce the thermal energy of the particles in the juice until it is cool and refreshing. Warming or cooling the small amount of matter in a mug is not hard.

Warming or cooling the water in a lake is a much bigger energy change. What about warming the entire world? That would be a really big event, important enough to make the news — and it has! Large-scale warming and cooling, involving enormous amounts of energy, are taking place all the time, with important consequences for all of us. The greenhouse effect is one important example that you learned about in Chapter 3.

Thermal Pollution

Wherever people use machines that burn fuels, they warm the atmosphere. Cars, trucks, trains, and aircraft give off hot exhaust gases that warm the air around them. Hot motors warm even more air. In fact, around 75 percent of the chemical energy from a car's burning fuel actually ends up warming the atmosphere instead of moving the vehicle.

The heating system in a building warms the furnace, the chimney, and the air in the chimney, as well as the air in the building. Some of this warm air escapes easily through open doors and cracks. Some of it warms the walls and floor of the building, which in turn warm the air outside. In the summer, an air conditioning system reduces the thermal energy of the inside air so that the air stays cool. But the thermal energy does not just vanish. It is transferred to the air outside and warms the air. So all year round, modern buildings cause **thermal pollution** — accidental warming of the environment.

Figure 7.11 Thermal energy from buildings contributes to thermal pollution.

Power stations and many large industries are a third source of thermal pollution. Their machinery is often cooled by water pumped from rivers or wells. If the warmed water is returned directly to rivers or lakes, it can cause great changes in the ecosystems. For example, fish may suffocate because the warm water does not hold enough dissolved oxygen. To reduce such problems, the warmed water is often released into artificial ponds where it can cool before being returned into the natural water system. The cooling ponds themselves can cause problems, however, by increasing the flow of water through soil around the plant or by changing the rate at which soil minerals are dissolved and deposited.

Figure 7.12 Cooling ponds allow warm water from factories and power stations to cool before being released into rivers and lakes.

Sometimes waste heat from one industry is used by another industry. For example, cooling water from a power plant can be used to warm a nearby greenhouse. This important method of energy conservation is called **cogeneration**. If you can think of an example of cogeneration in your community, you may wish to report on it in the next activity.

Just the Facts

Have you ever wondered how radio and television interviewers know so much about all of the topics they discuss? Part of the answer is that they have help. Research assistants find information about topics an interviewer will discuss "on air." They prepare "backgrounders": short summaries of important facts about a topic. The interviewer uses these backgrounders to plan questions or conduct an intelligent conversation with an expert.

What to Do

Use library or Internet resources to prepare a backgrounder on a phenomenon that involves large-scale warming or cooling. You may choose one of the topics in this section or any other topic that your teacher approves: for example, the weather phenomenon called El Niño, the reasons why the planet Venus is so hot, or the causes of the ice ages.

Your backgrounder is limited to one piece of looseleaf paper, so organize it carefully. On a separate page, list the information sources you consulted, in the form your teacher prefers.

Skill POWER

For more information on how to do research using print resources or the Internet, turn to page 497.

Check Your Understanding

1. Write a definition of thermal energy.
2. (a) What SI unit is used to measure thermal energy?
   (b) What SI unit is used to measure temperature?
3. (a) How does kinetic energy differ from other forms of energy?
   (b) Give three examples of objects with high kinetic energy. Use your own experience; do *not* repeat examples from this textbook.
4. How is thermal energy related to kinetic energy?
5. How is thermal energy different from temperature? Hint: You have studied three answers to this question so far.
6. **Thinking Critically** Think of a form of energy that you knew by name before you studied section 7.3.

CHAPTER at a glance

Now that you have completed this chapter, try to do the following. If you cannot, go back to the sections indicated.

Identify these parts of a laboratory thermometer: bulb, scale, bore. (7.1)

Which part of a laboratory thermometer is placed in the material being measured? (7.1)

Measure the temperature of an object with a laboratory thermometer. Read the scale carefully and correctly. (7.1)

Give a reasonable temperature (in degrees Celsius) for each of the following situations: freezing water, room temperature, normal human body temperature, boiling water. (7.1)

Describe three steps in calibrating a thermometer. (7.1)

Explain how to make a rechargeable battery have each of these forms of energy: electrical energy, kinetic energy, thermal energy. (7.2)

With a partner, create a short skit to show the behaviour of particles of matter in each of these situations: low temperature, warming up, small amount of thermal energy, a large amount of thermal energy. (7.2, 7.3)

Identify some sources of thermal pollution. (7.3)

Prepare Your Own Summary

Summarize this chapter by doing one of the following. Prepare a graphic organizer (such as a concept map), produce a poster, or write the summary to include the key chapter ideas. Here are a few ideas to use as a guide:

- What are three effects that warming or cooling can have on matter?
- Identify two methods for estimating temperature. For each method, give a situation where it is useful and a situation where it is not appropriate to use it.
- Name two temperature scales that are commonly used today. For each scale,
  - **(a)** what number represents the temperature of freezing water?
  - **(b)** what number represents the temperature of boiling water (at sea level)?
  - **(c)** where is this scale often used?
- Give a general definition of energy. Then describe the scientific idea of what energy is and what energy is not.
- What forms of energy were discussed in this chapter? Give a brief definition and an example of each form.
- Use the illustration below. What is the connection between
  - **(a)** temperature and particle motion?
  - **(b)** thermal energy and particle motion?
  - **(c)** temperature and thermal energy?

- What happens to the temperature of a substance when you
  - **(a)** add thermal energy?
  - **(b)** prevent gain or loss of thermal energy?
  - **(c)** remove thermal energy?

CHAPTER 7 Review

Key Terms

temperature
thermometer
scales
calibrate
Celsius scale
Kelvin scale
energy
thermal energy
kinetic energy
thermal pollution
cogeneration

Reviewing Key Terms

If you need to review, the section numbers show you where these terms were introduced.

1. Write a sentence that uses at least three of the key terms correctly. (7.1, 7.2, 7.3)
2. Write a short paragraph (three to five sentences) that uses at least six other key terms. The paragraph does not need to be serious, but you must use the terms correctly. (7.1, 7.2, 7.3)

Understanding Key Ideas

Section numbers are provided if you need to review.

3. **(a)** Why does a thermometer need a scale? (7.1)
   (b) Name the most common temperature scales used today. (7.1)
4. Identify the particular situation that matches a temperature of "zero degrees" (7.1)
   (a) on the Celsius temperature scale
   (b) on the Kelvin temperature scale
   (c) in your everyday experience
5. Do all substances warm or cool in the same way? Give evidence to support your answer. (7.1)
6. **(a)** What do thermometers measure? (7.1)
   (b) What do thermometers actually detect about the moving particles that make up a sample of matter? (7.2)
7. In Unit 2, the particle theory of matter was introduced on page 114. (7.2)
   (a) What features of the particle theory did you use in this chapter? List them in point form.
   (b) Give one piece of evidence for each feature of the particle theory in your list.
8. In your notebook, copy and complete the following table to explain the meaning of thermal energy. Give your table a title. (7.2)

	Substance with a large amount of thermal energy	Substance with a small amount of thermal energy
Average speed of particle motion		
Kinetic energy of particles		
Temperature		

9. In your notebook, copy and complete the following table to compare thermal energy and temperature. Give your table a title. (7.2)

	Thermal energy	Temperature
SI units of measurement		
What it tells about particles of matter		
Measuring device or method		

10. **(a)** What are you describing if you tell someone an object's energy? (7.3)
    (b) Identify three forms of energy. Give an example of a situation where each form would be measured. (7.3)
11. Name the types of objects that have
    (a) a large amount of kinetic energy (7.3)
    (b) a small amount of kinetic energy (7.3)

12. Identify at least two differences between matter and energy. (7.3)

13. Identify and describe two examples of thermal pollution. (7.3)

Developing Skills

14. Invent a simple story about students in a classroom that can be used as a model for the behaviour of particles of matter at different temperatures.

15. The diagram below describes what happens to a large ice cube as it falls into a saucepan of hot soup boiling on a stove. Copy the diagram onto a full sheet of paper. Add information to each section about particle motion, temperature, and thermal energy. Leave space for more information, and keep the diagram to use in the next chapter.

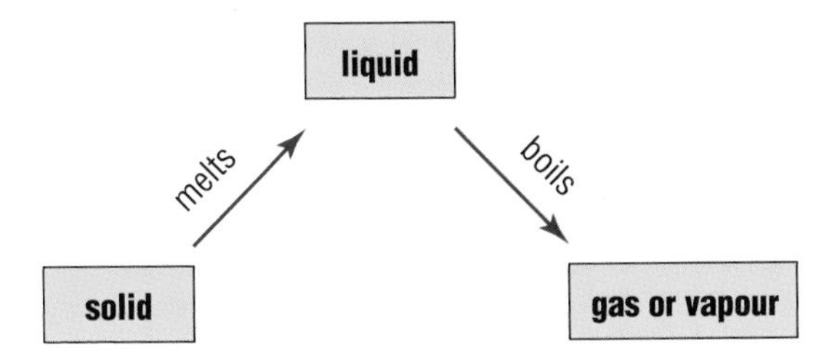

Problem Solving/Applying

16. Review your observations in Conduct an Investigation 7-B: Tracking Temperature Changes. Then predict which would warm up faster on the stove, a pot of water or an identical pot filled with the same amount of cooking oil.

17. What type of thermometer would be appropriate for measuring temperature in each situation below? Give a reason for your answer.
    (a) measuring very low temperatures in the Arctic
    (b) regulating the temperature in a hot oven
    (c) finding if the temperature of a flame is high enough to melt the metal parts above it

18. Imagine that you are a particle of water living with many friends in a pitcher in a refrigerator. Write a short story describing how you and your friends are behaving and how this behaviour changes when you are taken out of the refrigerator, heated in a microwave oven, and left to cool on the kitchen counter.

Critical Thinking

19. The particle theory of matter tries to explain everyday happenings by describing the behaviour of things that are too small to see clearly.
    (a) Why are scientists convinced that the particle theory is correct?
    (b) Which parts of the particle theory seem reasonable to you? Which parts do you find hard to accept?

20. Why is energy a useful characteristic to measure or calculate when you are preparing a scientific description of an object or a system?

Pause& Reflect

Go back to the beginning of this chapter on page 184 and check your original answers to the Getting Ready questions. How has your thinking changed? How would you answer those questions now that you have investigated the topics in this chapter?

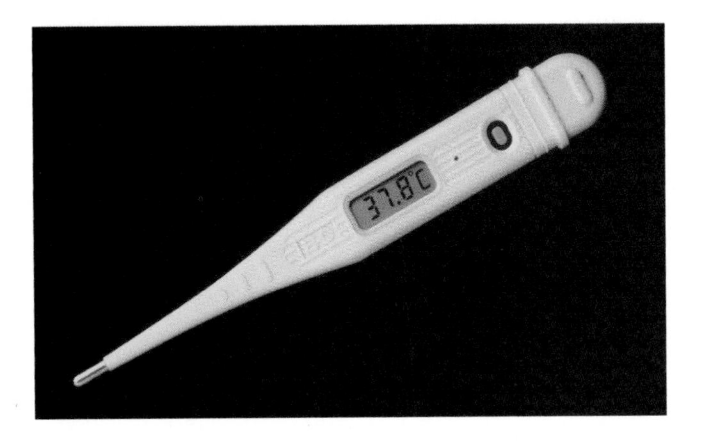

CHAPTER

8 Thermal Effects on

Getting Ready...

- Why does sand on a beach feel so hot while the water feels cool?
- Why do overhead power wires usually sag between the poles?
- What are particles of water doing when they freeze and form an ice cube?

Science Log

Brainstorm at least two possible answers to each question above. (It may help to trade ideas with a partner.) Then record your ideas in your Science Log. If you are sure that one answer is at least partly correct, explain how you know.

A bit of warming or cooling — a slight change in a substance's thermal energy — can certainly change the condition of your surroundings. Warm air and sunlight can destroy a winter wonderland; cooler air and a cloudy day can create one. These are common thermal effects. You depend on other common thermal effects whenever you cook food, use a hair dryer, or make (or even eat) ice cream. By investigating thermal effects in a planned way, scientists have learned how to produce and control them. Scientists have also used the particle theory of matter to explain why thermal effects occur.

In this chapter, you will make planned, scientific observations of common substances while you add or remove thermal energy. You will then try to infer how the particles of matter in your samples are behaving. Sometimes you may be able to predict the results, so your findings will confirm your everyday experiences. At other times, your observations may surprise you!

Matter

Spotlight On Key Ideas

In this chapter, you will discover

- what happens to solids, liquids, and gases as they warm and cool
- how the particle theory of matter explains what happens when matter changes state
- how evaporation of a liquid can be related to either heating or cooling

Spotlight On Key Skills

In this chapter, you will

- use the process of science inquiry to predict, observe, and analyze changes in temperature and state
- describe changes of state using scientific terms
- design your own experiment to investigate rates of cooling
- construct and interpret graphs that display temperature changes during an experiment

Starting Point ACTIVITY

Quick Cool-Down

Before giving you an injection, a nurse or doctor cleans the area of skin that the needle will puncture. Besides killing germs, the cleaning procedure affects temperature, thermal energy, and particle motion in your skin.

What You Need

ethyl alcohol at room temperature

water at room temperature

2 cotton balls

paper towel

CAUTION: If you have skin allergies, do this activity with water only.

What to Do 

1. Work with a partner. Use the cotton balls to sponge a small area of your partner's forearm with alcohol and a nearby area with water.
2. For about 2 min, have your partner report how the skin feels and what changes occur.
3. Switch roles with your partner, and repeat the activity.

What Did You Find Out?

1. Describe all the temperature changes of the sponged areas of skin you observed.
2. Describe any differences between the effects of the two liquids that you and your partner felt and saw.
3. Why do first-aid manuals recommend sponging the skin of a feverish person with a moist towel or washcloth?
4. Suggest what could be happening to the particles of water or alcohol on your skin to cause the effects you observed.

8.1 Expansion and Contraction

As materials warm up, the particle theory of matter says that their particles move faster and spread apart. Therefore substances should **expand** (increase in volume) as their temperature increases. They should shrink or **contract** (decrease in volume) as their temperature decreases.

Before checking these predictions, you need to have a clear idea of some important features of the materials you will be observing. Remember that a pure substance is a type of matter that is made of only one kind of particle. On Earth, a pure substance may exist as a solid, liquid, or gas. These are the three **states** or **phases** of matter. Examine the illustrations to review the key characteristics of each state.

DidYouKnow?

Not everything we observe has the properties of matter. Light does not take up space or have mass; neither does sound. When you turn on a lamp or stereo, and fill a room with light or sound, the room does not become heavier. Light and sound are forms of energy, and like all forms of energy, they do not have mass or volume. With this in mind, make a prediction about whether the mass of an object changes when it warms or cools (that is, when its thermal energy changes).

Figure 8.1 The three states of matter

In the solid state, materials keep their shape and size. **Solids** like ice cannot be compressed into a smaller space. In the liquid state, materials have a definite size (volume), but no fixed shape. **Liquids** like water settle to the bottom of their container and take its shape. Liquids cannot be compressed. **Gases** have no definite shape or size. They expand to fill all parts of their container and can easily be compressed into a smaller space. Many gases cannot be seen. For example, the space just above a kettle's spout is filled with invisible water vapour (steam). As the water vapour (water in the gaseous state) rises and cools, it forms a cloud of tiny drops of liquid water.

Pause& Reflect

How do you visualize the particles that make up a substance you are observing? In your Science Log, write some ideas to answer the questions at the right. Do not worry about being "correct." You can change your ideas as you work through this chapter.

(a) Are the ice, liquid water, and water vapour in the illustrations made of the same kind of particle, or are they different types of matter?

(b) How does the motion of particles in solids, liquids, and gases differ?

Expansion and Contraction of Solids

The lengths of solid bars of different materials can be measured at different temperatures using very precise equipment. Table 8.1 shows some of these measurements. You can see that the changes in a 100 cm long bar are very small. If the bar were twice as long, however, the changes would be twice as large. In a very long structure, such as a bridge or a train track, the small changes can add up and become very important.

Table 8.1 Expansion and Contraction of Solids

Material	Length at −100°C (cm)	Length at 0°C (cm)	Length at 100°C (cm)
lead	99.71	100.00	100.29
steel	99.89	100.00	100.11
aluminum	99.77	100.00	100.23
brass	99.81	100.00	100.19
copper	99.83	100.00	100.17
glass	99.91	100.00	100.09
Pyrex™	99.97	100.00	100.03

Career CONNECT

Room to Grow

Civil engineers have to think about thermal expansion and contraction when planning roadways, sidewalks, and bridges. Often they need to include expansion joints: gaps or sections of material that can compress easily. Without these expansion joints, the concrete or other building material could buckle upward or outward when it expands. As you have seen in your studies, different materials expand differently, some more than others. Builders and engineers need to know the properties of the materials they use so that they can plan for the correct amount of expansion.

Imagine that a builder or a civil engineer has been invited to speak to your class. Write five questions that you would like to ask this person about engineering as a profession. Make sure that your questions are clearly worded, and that they will help you find out what you want to know.

Find Out 

Stretch and Shrink

Are there similarities in how substances expand when heated? Are there similarities in how they behave when cooled? This activity will help you to identify any patterns.

What to Do

1. Examine Table 8.1, and use it to answer these questions.
   (a) What similarity do you see in how all the materials react as they warm?
   (b) In what way do the materials react differently as they warm?
   (c) Which material expands the most as it warms?
   (d) Which material expands the least as it warms?
2. Copy the list of materials in Table 8.1, but arrange them in order, starting with the material that expands the most and ending with the one that expands the least.
3. Does your list also correctly rank the materials as they cool and contract? That is, does the material that expands the most at a high temperature also contract the most at a low temperature?

What Did You Find Out?

1. If a material expands and contracts very little, is the attraction between its particles probably very strong or very weak?
2. Which material in Table 8.1 probably has the strongest attraction between its particles? Which material probably has the weakest attraction between its particles?

CONDUCT AN

INVESTIGATION 8-A

Expanding Solids

You can observe directly how solid materials contract and expand, if you look closely. Usually the changes are small and easy to overlook. The two special situations you will examine in this investigation make it easier to see what happens when solids are warmed and cooled.

Problem

What evidence can you observe of solid materials expanding as they are warmed, and contracting as they are cooled?

Safety Precautions

In Part 1, you will be working with an open flame and hot objects. Be careful!

Apparatus

long copper or iron wire
small hooked mass (200 g or 500 g)
metre stick
ball-and-ring apparatus
laboratory burner

Materials

candles
matches
cold water

Part 1

The Sagging Wire

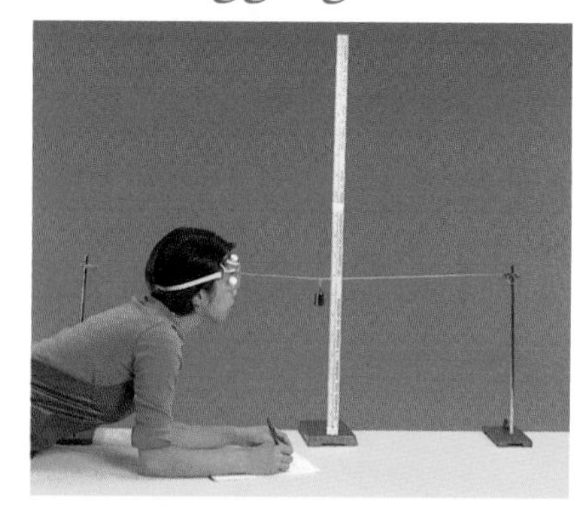

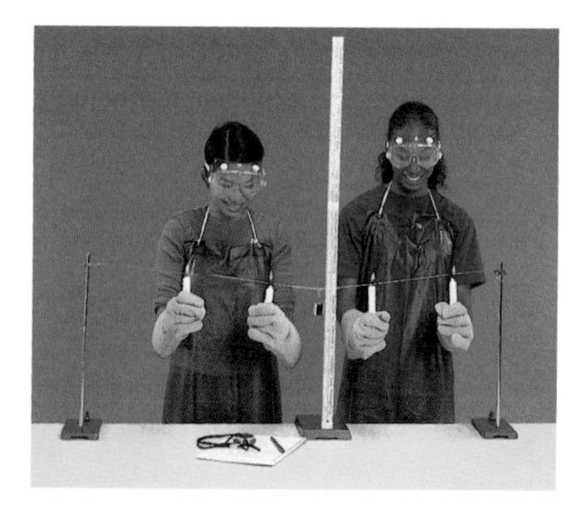

Procedure

1. Make a table for recording your data.
2. Stretch the wire tightly between two firm supports. Place the small mass in the middle of the wire. Put the metre stick behind the mass, and record its height.
3. Light the candles, and use them to warm the entire length of the wire for several minutes. As you do, observe and carefully record the height of the mass after each 30 s of heating.
4. Stop warming the wire. Observe and record what happens to the height of the mass during the next 2 or 3 min.

Analyze

1. (a) If the wire sags, the mass moves down. Does this mean that the wire is getting longer or shorter?
   (b) What is happening to the length of the wire if the mass moves up?
2. Do your observations in this investigation support the prediction in the problem?
3. If the wire sags, are its particles getting farther apart or closer together? Explain why they would do this. (Hint: Think about their motion.)

Part 2
Your teacher will do the heating in Part 2 as a demonstration.

The Ball and Ring

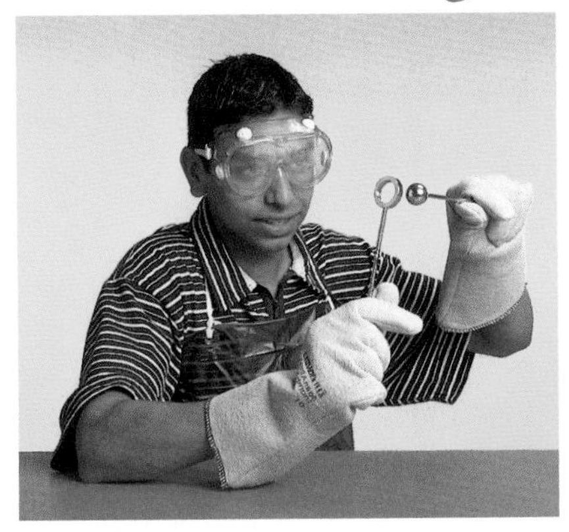

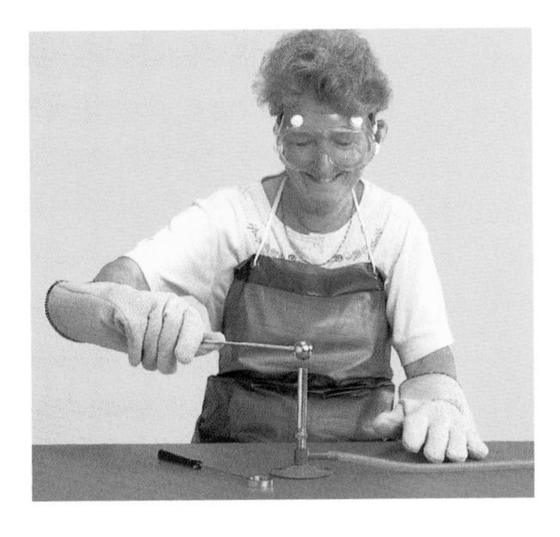

Procedure

1. Observe whether the brass ball fits through the brass ring when both the ball and the ring are at room temperature.
2. Does the ball fit through the ring when your teacher warms only the ring in a hot flame for 30 s?
3. What happens when your teacher warms both the ring and the ball?
4. As a class, brainstorm possible ways to make the ball fit through the ring. You or your teacher will test the ideas until one method works. With the ball through the ring, cool both the ball and ring. Can you or your teacher pull the ball back through the ring? If not, find a way to separate them by warming or cooling.

Analyze

1. How did the demonstration give evidence that solids can expand? Explain what you did to cause the expansion and which part of the apparatus (the ball, the ring, or both) expanded.
2. How did the demonstration give evidence that solids can contract? Explain what you did to cause the contraction and which part of the apparatus (the ball, the ring, or both) contracted.
3. Explain precisely what you think happens to the particles that make up the apparatus when the ball and ring are heated. What do you think happens to the particles when the apparatus is cooled?

Conclude and Apply

4. Predict how the position of the electric transmission lines in the photograph below, taken in summer, would change as the temperature dropped in winter. Why would it be a bad idea to stretch the transmission lines more tightly between the towers so they would sag less in the summer?

Expansion and Contraction in Gases

Because most common gases are colourless, they are difficult to observe. As well, gases have no fixed shape or size. (Remember that they always take the shape and size of their container.) If you put gases in a flexible container such as a balloon, however, you can see that they expand and contract much more than solids when the temperature changes. Warming a sample of helium from 0°C to 100°C, for example, increases its volume by about one third. Unlike the particles in solids, the particles in gases are far apart and moving fast and freely. The forces holding gas particles together are quite weak, so it is easy for the particles to move farther apart.

Figure 8.2 The particle theory predicts that warming air will cause its particles to move faster and spread farther apart. When the air in the flask is warmed, the air expands and fills the ballon.

Pause& Reflect

You have observed gases expanding, in your work in this unit and in everyday life. Think about the situations below, and answer the questions in your Science Log.

(a) Look back at your observations in the Starting Point Activity at the beginning of Chapter 7. Which part of the activity showed that gases expand when heated? Explain which gas was expanding and how you know it was expanding.

(b) A spray can, even when it is almost empty, contains compressed gases. Why does the safety warning on the label tell you not to dispose of the can by putting it in a fire?

(c) The tires on a car are filled with compressed air. In the winter, when the air temperature drops very low, the tires become slightly flat, even when they are not leaking. Why?

Bulging Balloons

In science, even ideas that seem like common sense are checked to see if they agree with observations and the rules for logical thinking. Can you find evidence to support the following statement, which you read earlier in this section?

If you put gases in a flexible container, such as a balloon, you can see that they expand and contract.

What You Need

2 identical balloons

refrigerator or freezer

hair dryer, electric heater, or toaster

What to Do

1. Blow up the balloons several times to stretch them. Then blow up both of them to the same size, and tie them so that no air can get in or out.
2. Put one balloon in the refrigerator or freezer to cool it. Leave the other balloon at room temperature. After an hour, compare the size of the two balloons.
3. Warm the cold balloon by blowing warm air from a hair dryer over it, or by holding it in warm air from a heater or above a toaster. Observe what happens as the air in the balloon warms. Continue until the balloon feels much warmer than room temperature.

What Did You Find Out?

1. Describe what you observed using the words "expand" and "contract."
2. How well do your observations support the statement you were testing: completely, partially, or not at all?
3. Describe any differences between what you expected to happen and what did happen.
4. In this activity, one balloon is called the *control* and the other is called the *test*. Which is which? Why?
5. At which point in this activity were air particles in one balloon farthest apart? When were they closest together?

Expansion and Contraction in Liquids

Imagine watching a laboratory thermometer as its temperature changes. As the thermometer liquid moves up the bore, it takes up more space. In other words, the liquid expands as it warms. As the thermometer cools, the liquid contracts, so it moves back down the bore. The liquid must be contracting as it cools. Do all liquids expand and contract in this way? Do some liquids change volume more than others as they warm and cool? Follow the next activity carefully to find out.

Find Out **ACTIVITY**

Race for the Top

You have already observed a liquid in a thermometer expanding and contracting. Do all liquids behave the same way? Write a hypothesis, and explain why you think as you do. Check your hypothesis by observing the behaviour of liquids in this activity. [Your teacher may choose to demonstrate some or all of the steps for you.]

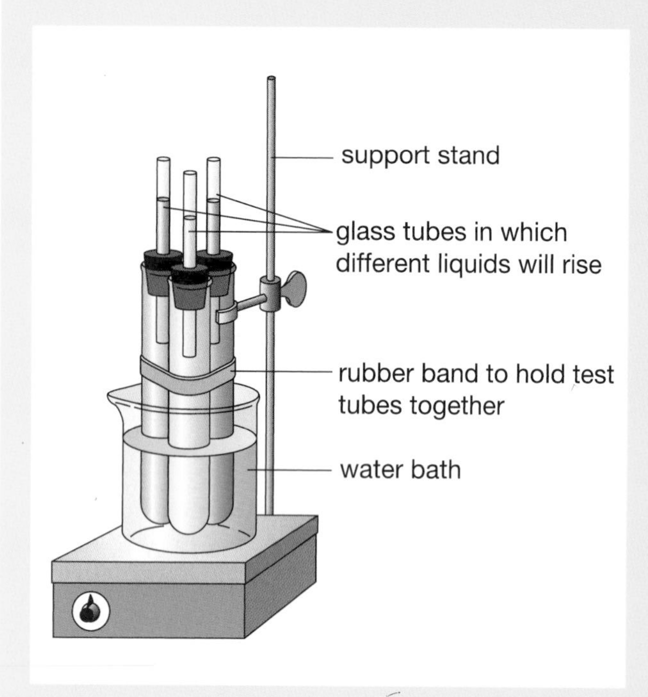

What You Need

3 liquids (coloured water, ethyl alcohol, and cooking oil)

3 large test tubes

3 one-hole rubber stoppers, with 50 cm pieces of glass tubing inserted

laboratory stand and clamps

rubber bands

markers

2 large tin cans or 500 mL beakers

very hot water

ice-cold water

What to Do

1. Completely fill one test tube with coloured water, the second with ethyl alcohol, and the third test tube with cooking oil. Insert a stopper in each test tube so there are no air bubbles and the liquid rises a few centimetres up the glass tubing. Hold the test tubes together with the rubber band so the liquids are at the same level in the glass tubing, and arrange the apparatus as shown in the diagram.
2. Use the markers to mark the starting height of each liquid on the glass tubing.
3. Pour the hot water into the beaker around the test tubes. Watch the height of the liquids closely as the liquids warm.
4. Before the liquids overflow the glass tubes, lift the apparatus out of the hot water and put it into the ice-cold water. Keep watching the height of the liquids as they cool.

What Did You Find Out?

1. Did all the liquids expand by the same amount as they warmed? If not, answer the following questions.
   (a) Which liquid expanded more?
   (b) Did the liquid that expanded more as it warmed also contract more as it cooled?
2. At the end of the activity, did the liquids return to their original heights in the tubes? Did you expect them to? Explain.
3. Based on your observations, do all liquids expand and contract the same amount when they are heated and cooled?
4. Explain which of the liquids you tested would be most suitable for making a thermometer that could be used to
   (a) show small changes in temperature very clearly;
   (b) measure large changes in temperature, without the thermometer being too large.

Astrophysicists (scientists who study the way stars work) have found evidence that most stars, late in their lifetimes, go through a phase where they cool down and expand at the same time. A star that has expanded in this way is called a *red giant*. It usually becomes many, many times larger than its original diameter, and it may swallow up planets in orbit around it. Why do stars go through this strange behaviour? Is our own Sun expected to become a red giant, and if so, when? Do some research in a library or on the Internet to find answers to these questions.

Check Your Understanding

1. Name the three states of matter. Give examples of three substances that are each in a different state at room temperature.
2. From your observations in this section, write a general description of what happens to solids, liquids, and gases as they are
   (a) warmed
   (b) cooled
3. Which state of matter shows the largest change in volume when warmed or cooled? Which state shows the smallest change?
4. The graphs below show the volume of mercury in a thermometer.
   (a) Which graph could be called a warming curve? Explain why.
   (b) Which graph could be called a cooling curve? Explain why.
   (c) Which graph shows what happens when a thermometer is placed in hot soup?
   (d) Which graph shows what happens when a thermometer is placed in ice cream?

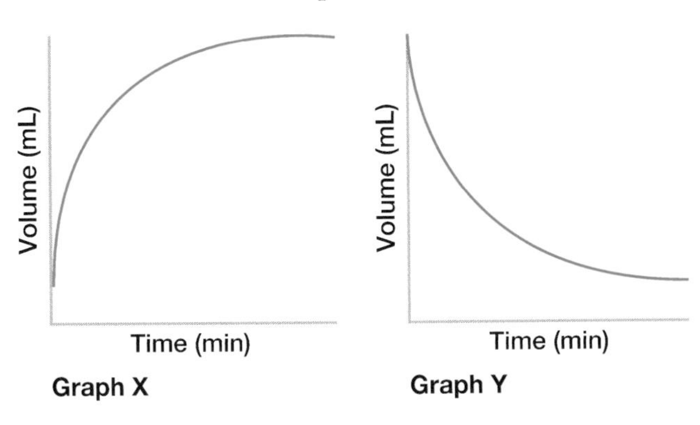

5. **Apply** Bridges are made from materials that contract and expand as the temperature changes, so they cannot be fastened firmly to the bank of a river or lake. The photographs below show an expansion joint at the end of a bridge in winter and in summer.
   (a) Which season is shown in each picture? Explain how you know.
   (b) Why do you suppose concrete roadways and sidewalks are laid in sections with grooves between them?

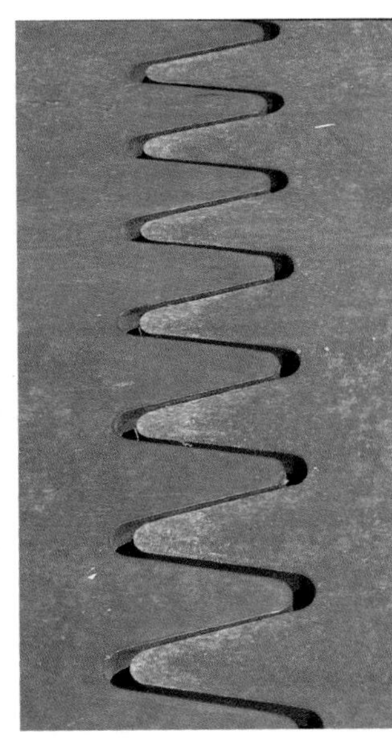

8.2 Temperature Changes and Specific Heat Capacity

Figure 8.3 The same hot Sun beats down on both sand and water at the beach. The sand warms up quickly. The temperature of the water changes much more slowly.

You have seen that different materials expand by different amounts as they warm up. Another difference in the way that materials respond when they are heated, is the amount that their temperature rises when a certain amount of thermal energy is added. Some materials, such as sand, warm and cool quickly. Under identical conditions, other materials, such as water, warm and cool slowly.

The **heat capacity** of an object is a number that tells you how much thermal energy must be added to the object to make its temperature increase by one degree Celsius. If you remove the same amount of thermal energy from the object, its temperature will decrease by one degree Celsius. If an object has a large heat capacity, that means that you must add a large amount of heat to make its temperature increase just a little. For example, it takes more than four times as much added thermal energy to raise the temperature of water one degree Celsius than it takes to raise the temperature of the same amount of copper by one degree Celsius. The heat capacity of water is more than four times as large as the heat capacity of copper.

Of course, temperature changes also depend on the amount of material in an object. A small glass of water warms up much more easily than a large saucepan or a whole lake. To make fair comparisons between different materials, you must warm or cool samples that have the same mass. The **specific heat capacity** of a material is the amount of energy you must add to a standard amount of the material (one gram or one kilogram) to increase its temperature by one degree Celsius. For example, the specific heat capacity of copper is 0.39 J/g °C. This means that to warm up one gram of copper by one degree Celsius, you must add 0.39 J of thermal energy. To warm it by two degrees, you would need to add 2×0.39 J or 0.78 J of energy.

INVESTIGATION 8-B

Hot Stuff!

Think About It

Scientists have measured the specific heat capacities of many common materials. Reference books contain tables of their results. People who work with heating and cooling systems use these tables to predict how their designs will work. In this investigation, find out if you, too, can understand and apply information about specific heat capacities.

What to Do 

1 The diagrams below show the results of an experiment to compare the specific heat capacities of four different materials. Examine the diagrams closely to find answers to the following questions.

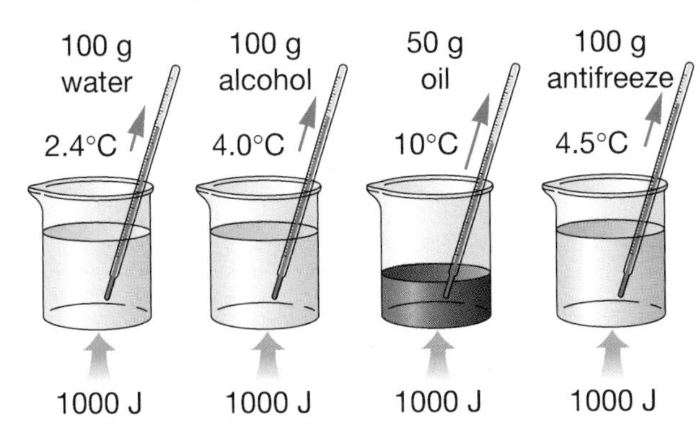

(a) This experiment is not a fair test of all four materials. Which sample cannot be compared with the others? Explain why. Do not use this sample to answer the rest of the questions.

(b) Why can the temperature changes in the remaining three materials be compared fairly?

(c) Rank the three materials according to temperature change. List the material with the greatest temperature change first.

(d) Which material has the largest specific heat capacity?

(e) Which material has the smallest specific heat capacity?

2 The table below shows the temperature increase in several materials when 1.0 kJ of energy was added to 1.0 kg of the material. Use the table to answer the following questions.

Temperature Changes in Materials

Material	Temperature increase (°C)
aluminum	1.1
concrete	0.3
copper	2.6
mercury	7.0
steel	2.2
water	0.2
ice	0.5

(a) Notice that the table describes identical 1.0 kg samples of all of the materials in the table, and that the thermal energy of each sample is increased by 1.0 kJ. Which material shows the greatest temperature increase? Which shows the smallest temperature increase?

(b) Which material has the largest specific heat capacity? Explain your reasoning.

3 List the materials in order of their specific heat capacities, from highest to the lowest.

Analyze

Look back at the beach scene shown in Figure 8.3. What can you infer about the specific heat capacity of sand to explain why the sand and the water are at such different temperatures after warming in the Sun all morning?

CONDUCT AN

INVESTIGATION 8-C

Cool It! 

A drop in temperature means that thermal energy has been removed from the object. Particles are moving more slowly than before. Would this happen more easily in materials with a low specific heat capacity or in materials with a high specific heat capacity? In other words, do materials that warm up easily also cool down easily?

Problem

Which material, sand or water, will cool more quickly under identical conditions?

Safety Precautions

Handle containers of boiling water with care. If boiling water touches your skin, hold the burned skin under a stream of cold water for several minutes.

Apparatus and Materials

Have your teacher check your list of apparatus and materials to make sure that all the equipment you need is available.

After reading the procedure, and before you begin this investigation, make a complete list of the apparatus and materials you will need.

Procedure

1. Examine the table below of specific heat capacities. How does the specific heat capacity of water compare with the specific heat capacity of most other materials? Form a hypothesis about which material, sand or water, will cool most easily.

Specific Heat Capacities

Material	Specific heat capacity (J/g°C)
aluminum	0.92
concrete	3.00
copper	0.39
mercury	0.14
steel	0.44
water	4.19
ice	2.10

2. After thinking about the following points, write a step-by-step procedure. Have your teacher approve your procedure. Obtain the necessary equipment and carry out the investigation.
   - **(a)** In what ways will your samples of sand and water have to be the same in order for you to conduct a fair test of the way they cool?
   - **(b)** One way to warm samples equally is to put them in a container of hot or boiling water (a "hot water bath").
   - **(c)** To make sure that all parts of the samples (especially the sand) warm and cool equally, you will need to stir the samples.
   - **(d)** Remember to follow safety precautions. Make a list of apparatus and materials as you write your procedure.
   - **(e)** Plan what you will need to observe, and make a data table before starting the investigation. Give your table a title.

Computer CONNECT

Old computers can be recycled and turned into useful lab instruments by equipping them with electronic sensors. A computer with temperature probes, for example, can automatically record and graph temperatures over a long time period. If your school has a computer with temperature probes, you could use it for this experiment or for similar experiments comparing the warming and cooling of other materials. If your school does not have such a computer, you could still make observations by hand and enter them into a spreadsheet program. Then you could create a graph of your results from the spreadsheet.

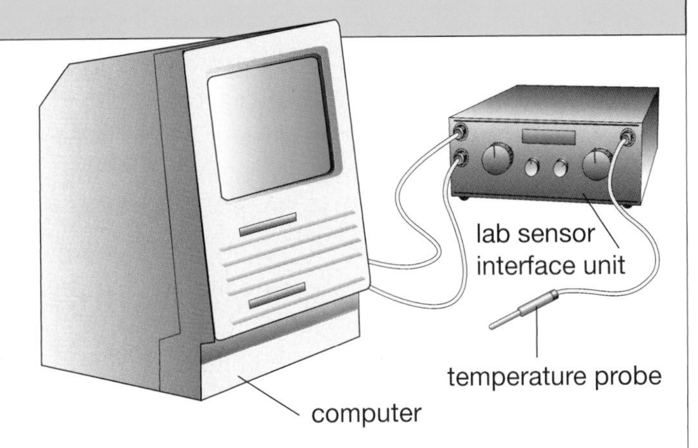

Analyze

1. Make one graph to show the temperature change in the two samples during the investigation.

2. Compare your results with those of your classmates. Did all the groups have similar results? If not, what could be causing the differences?

Skill POWER

To choose the best type of graph and to review how to draw it, turn to page 486.

Conclude and Apply

3. Write a short conclusion, summarizing the evidence you found to support or reject your hypothesis.

4. Use your results to write a conclusion to the beach scene in Figure 8.3, at the beginning of this section. In your conclusion, describe conditions at the beach an hour after the Sun is completely hidden by a large cloud.

5. Describe one thing you could do to help your group work more efficiently in the next investigation or activity.

Temperature and Thermal Energy

Have you developed a clear understanding of the difference between temperature and thermal energy? Thermal energy is the sum of the kinetic energies, or energy of motion, of all of the particles in an object. Temperature depends not only on the amount of thermal energy of an object, but also on the material that makes up the object. The extent to which the temperature will rise when energy is added to an object depends on the specific heat capacity of the material making up the object.

You have probably noticed that water has an unusually large specific heat capacity compared to most other materials. As a result, you can add large amounts of energy to water without causing a large temperature change. Since water covers more than 70 percent of Earth and since living organisms are made up of mostly water, this quality of water has a major effect on life on Earth.

Nevertheless, if you add enough thermal energy to water, or any other material, the temperature will rise higher and higher. What happens when you continue to add thermal energy to water after the water has reached a high temperature, for example, 100°C? It will boil. When liquid water boils, it becomes a gas. This type of change is called a phase change, or change of state, and is the topic of the next section.

Check Your Understanding

1. (a) Write a sentence using the word "capacity" in its everyday sense.

   (b) Write a sentence using the term "specific heat capacity." Your sentence must show that you understand its meaning.

2. Look closely at the descriptions of heat capacity and specific heat capacity at the beginning of this section.

   (a) What is the difference between these terms?

   (b) Which characteristic (heat capacity or specific heat capacity) is more useful for comparing different materials? Explain why.

3. **Apply** The specific heat capacity tells you the amount of thermal energy needed to warm 1 g of material by 1°C. Suppose that you want to heat a mug of hot chocolate (about 125 g) from room temperature (20°C) to 44°C. This is a temperature change of 24°C. Use the following procedure to find the energy needed.

   (a) Hot chocolate is mostly water. Find the specific heat capacity of water in the table on page 224.

   (b) Multiply the specific heat capacity by 24°C to find how much energy is needed to warm 1 g of water by 24°C.

   (c) Multiply your answer in part (b) by 125 g to find the amount of energy needed to warm all 125 g of hot chocolate to the proper temperature.

4. **Apply** Use the procedure in question 5 to find

   (a) the amount of energy needed to warm 250 g of concrete by 15°C

   (b) the amount of energy needed to warm a 1 kg brass mass from 25°C to 55°C

5. **Thinking Critically** Find the specific heat capacities of aluminum and steel in the table on page 224. Suppose that an aluminum frying pan and a steel frying pan of equal mass are heated identically on a stove.

   (a) Which pan will reach cooking temperature first?

   (b) Which pan will be more likely to burn food if it is heated a little too long?

   (c) After the stove is turned off, which pan will cool enough to wash first?

8.3 Changes of State

Figure 8.4 At the top of the candle, solid wax melts into a liquid, which flows up the wick. There the liquid wax vaporizes and burns.

In a candle, the same substance — wax — changes among all three states of matter: solid, liquid, and gaseous states. You can observe the same phase changes or changes of state with another common substance — water. Everyday changes in temperature cause water to **melt** (turn from solid ice into liquid water) or **freeze** (turn from liquid water into ice). Temperature changes can also cause water to **evaporate** (turn into invisible water vapour, the gaseous form of water) and **condense** (turn back from a gas into liquid water).

Most other substances are not so easy to study. Hydrogen, for example, is a gas, even at the coldest winter temperatures. If you want to make liquid hydrogen, you need to cool hydrogen gas to –253°C! To make solid hydrogen, you need even lower temperatures, as well as extremely high pressures.

Any pure substance can exist in all three states of matter. You can cause any substance to change state if you warm or cool and, possibly change the pressure of the substance enough. Changes in temperature, however, are just a sign of changes in particle motion, which means changes in thermal energy. In this section, you will explore links among these three ideas: state of matter, temperature, and thermal energy. You will experiment with water, but your conclusions will apply to other substances, as well.

DidYouKnow?

Scientists have found evidence to show that most of the giant planet Jupiter, just below its cloudy surface, is made up of a strange substance: liquid metallic hydrogen. At pressures of more than 4 000 000 times that of Earth's atmosphere, hydrogen not only flows like a liquid but also conducts electricity like a metal.

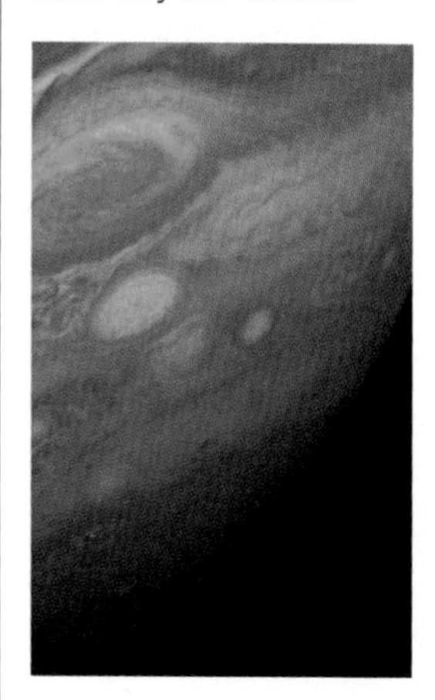

CONDUCT AN
INVESTIGATION 8-D

The Plateau Problem

Puddles of liquid water on the road can freeze on a wintry night, but this does not happen instantly. Water boiling in a kettle is turning into water vapour, but this takes time. What happens while the water is changing state?

Problem

What happens to the temperature of water while it changes state?

Safety Precautions

- Use oven mitts, hot pads, or tongs to handle the beaker of boiling water.
- Unplug the hot plate at the end of the investigation, and let it cool before putting it away.

Apparatus

2 laboratory thermometers
stirring rod
hot plate
kettle
2 beakers (250 mL)
clock or watch

Materials

crushed ice
ice-cold water
hot water (almost boiling)

Procedure

1. Think about familiar situations in which ice is melting or liquid water is boiling. Then form two hypotheses by completing the following two statements. Add reasons for your hypotheses.
   (a) While water melts from solid ice to liquid water, the temperature will (drop/stay the same/increase), because . . .
   (b) While water boils from liquid to gas, the temperature will (drop/stay the same/increase), because . . .
2. Make a data table like the one shown here. You will need space for at least five observations (more if time permits).

3. Fill one beaker with hot water from the kettle, and put it on the hot plate to boil.

4. In the other beaker, make a slush of crushed ice and a little cold water.

Time (min)	Temperature of melting ice (°C)	Temperature of boiling water (°C)

Skill POWER

For some tips on the correct way to use a thermometer, turn to page 485.

5 With a stirring rod, stir the contents of each beaker for several seconds, and then measure and record the temperature. Lift the thermometer off the bottom of the beaker to ensure that you are measuring the temperature of the contents not the container.

6 Repeat the temperature measurements every 3 min. For a fair test, make sure that you stir and measure exactly the same way each time. Record each result.

7 Stop heating the boiling water *before* it all boils away. Unplug the hot plate, and carefully set aside the hot beaker to cool.

Skill POWER

For help in drawing your line graphs, turn to page 486.

Analyze

1. In this activity, you measured time and temperature.
   (a) What was your dependent variable? (Which value was unknown until after you made an observation?)
   (b) What was your independent variable? (What value did you select before making an observation?)
2. Draw two line graphs to show your temperature time observations: one for the melting ice and one for the boiling water. Instead of joining the points dot-to-dot, draw a smooth line or curve that passes through or between the points (a best-fit line).
3. On your hot-water graph, mark the part where
   (a) the water was hot but not yet boiling
   (b) the hot water was boiling vigorously (called a "full rolling boil" in cooking)
4. Label any plateaus (flat, horizontal segments) on your graphs.
5. Compare the temperature of your melting slush with the "official" temperature you learned in Chapter 7.
   (a) If the two temperatures are almost the same, any small difference might be caused by errors in your equipment or measurements. Suggest at least two specific errors of this sort that might occur.
   (b) If the two temperatures are quite different, the conditions in your laboratory or your sample may be responsible. Suggest at least two specific conditions that might cause this type of error.
6. Imagine that you combined both parts of this investigation. Sketch a third graph that shows what would probably happen if you heated one sample from ice to water and then to water vapour.
7. On the temperature scale of your third graph, mark the melting point and the boiling point of your samples, according to your observations.
8. Combine all the results from your class to find the average melting point and the average boiling point for water. Compare these values to the "official" values. Are they closer than your individual group values? If they are closer, explain why.

Conclude and Apply

9. From your observations, write a clear answer to the problem at the beginning of this investigation.
10. How well do your observations support your hypotheses in step 1?
11. **(a)** Identify any problems you had with apparatus, procedure, or the way you organized and worked together in your group.
    (b) Describe one improvement your group could make the next time you work together.

INVESTIGATION 8-E

Learn the Lingo

Think About It

There are six possible changes of state, as shown by the arrows on the diagram. Several of the changes have common names, which you probably know. Some of the changes are identified by technical terms when described scientifically. It will be easier for you to describe and read about changes of state if you learn and practise using both their common and technical names.

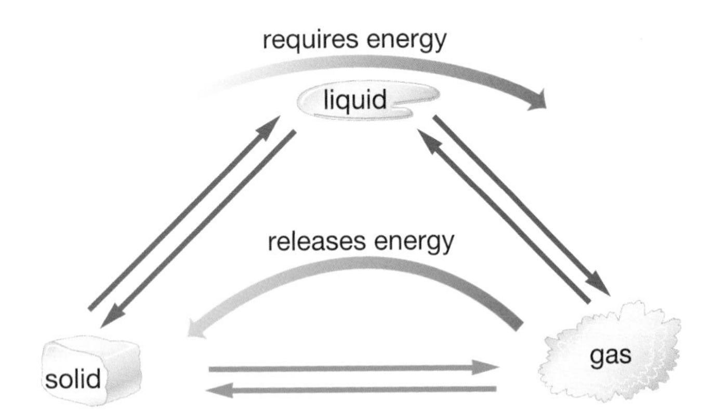

What to Do

1. Copy the diagram above onto a clean sheet of paper.

2. Above the correct arrow in the diagram, write each technical term given in the table below. Also write (in brackets) the common term for the change, if there is one. Notice that one technical term is used to describe two different changes.

Changes of State

Term	Change
freezing (solidification)	liquid to solid
condensation	gas to liquid
sublimation	gas directly to solid
evaporation	liquid to gas
fusion	solid to liquid
sublimation	solid directly to gas

3. On the back of the page with your diagram, copy the technical names for the changes of state, from step 2, in alphabetical order. Without looking back, write a definition for each term beside it. When you are done, check your work using your diagram.

4. Read statements (a) through (d) below. Then describe each statement using a technical term, as in the example.

   Wet clothes dry in the sunshine.
   Description: evaporation of water

   (a) Melted wax in a candle hardens when the candle is blown out.

   (b) A warm wind makes snow on the ground disappear, but no puddles of water form.

   (c) In the winter, invisible moisture in the air sometimes forms frost on car windshields.

   (d) On a cold day, you can "see your breath."

Analyze

1. Classify the six changes of state according to the change of thermal energy they involve. Arrange your answers in a table, with the headings "Receiving thermal energy" and "Releasing thermal energy."

2. When you add energy to an object that is not changing state, the kinetic energy of its particles increases. When the average kinetic energy of particles increases, the temperature rises. In the last investigation, you discovered that, during a change of state, adding energy to ice water or boiling water did not cause an increase in temperature. Therefore, the average kinetic energy of the particles could not be changing. Formulate a hypothesis to explain what happens to the energy that you add to ice water or to boiling water, if it does not become kinetic energy of the particles.

Melting and Boiling Points

The melting and boiling points of a substance are vital pieces of information, and not only for scientists. You have already seen that water has an unusually high heat capacity. Another unusual feature of water as a substance, and one that is even more important for Earth's climate, is the temperature range at which water is a liquid. A glance at Table 8.2 will show you that most common substances are either gases or solids at everyday temperatures on Earth.

Table 8.2 Melting and Boiling Points of Pure Substances

Substance	Melting point (°C)	Boiling point (°C)
oxygen	−218	−183
mercury	−39	357
water	0	100
tin	232	2602
lead	328	1740
aluminum	660	2519
table salt (sodium chloride)	801	1413
silver	962	2162
gold	1064	2856
iron	1535	2861

Word CONNECT

Have you noticed that the same temperature can have two different names? For example, as you warm candle wax, it melts at about 71°C. You can call 71°C the melting point of the wax. When the melted wax cools again, it solidifies at 71°C. Therefore this temperature can also be called the freezing point of the wax. The same thing is true for boiling and condensation points. The boiling point is the same temperature as the condensation point.

If liquid water did not change state at 0°C and 100°C, life on Earth would be very different, perhaps impossible. Each statement below gives a new freezing or boiling point for water, along with one possible effect it would have. Copy the statement, and add at least two other possible effects. Be creative, but be ready to explain why your effects are scientifically correct.

(a) If water froze at −70°C . . . there would be no outdoor skating rinks in Canada.

(b) If water froze at 45°C . . . ice cream would not melt on a hot day.

(c) If water boiled at 15°C . . . your blood would be boiling right now!

Find Out **ACTIVITY**

State the State

Different pure substances have different melting and boiling points. Table 8.2 on page 231 gives some examples. Scientists measured and used these temperatures to identify substances long before they could explain the differences. You can probably guess that modern explanations involve the strength of the forces that hold the tiniest particles of a substance together.

What to Do

Use the information in Table 8.2 to answer each of the following questions.

1. In what state (solid, liquid, or gas) would each substance be at the given temperature?
   (a) oxygen at –50°C
   (b) aluminum at 800°C
   (c) table salt at 800°C
   (d) gold at 3000°C
   (e) iron at 2000°C

2. What change of state would each substance go through during the given temperature change?
   (a) mercury cooling from –10°C to –45°C
   (b) silver cooling from 1000°C to 950°C
   (c) tin warming from 2200°C to 2300°C
   (d) mercury warming from 300°C to 350°C
   (e) iron cooling from 1600°C to 1500°C

3. Sketch a heating curve (time-temperature graph) to show what happens to lead as it is warmed from 0°C to 2000°C. On the curve, mark the sections where the lead is
   (a) a solid
   (b) melting
   (c) a liquid
   (d) boiling
   (e) a gas

4. Sketch a cooling curve for mercury from 400°C to –100°C. On the curve, mark the same changes of state that you did for lead in question 3.

What Did You Find Out?

Compare the melting points of aluminum and tin. Which metal do you suppose has stronger forces holding its particles together? Explain your reasoning.

You are familiar with three states of matter. There is a fourth state of matter, called *plasma*. To change a material into a plasma, extremely high temperatures are required, like those inside the Sun — millions of degrees Celsius! In a plasma, individual particles that make up the material start to break apart into tinier pieces called electrons and ions. Plasmas can be produced on Earth, but only under extreme conditions. Matter on Earth exists as a solid, liquid, or gas, almost all of the time.

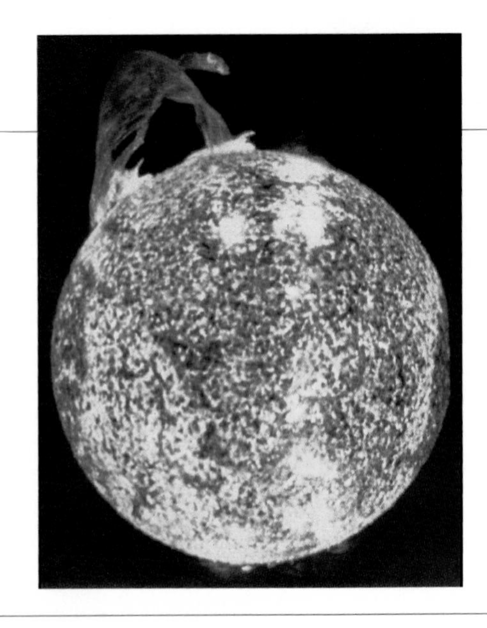

Potters need to check the very high temperatures inside the kilns that bake and harden their pottery. To do this, they use small ceramic pyramids called "pyrometric cones" like the ones shown in the photograph. Sets of four cones are placed in the kiln along with the pottery being fired. Two of the cones soften and bend over as the kiln heats up. The third cone bends at the desired temperature. If the fourth cone bends, the kiln has overheated and the pottery may be damaged. Potters refer to cones by code numbers. For example, a number 022 cone bends at 585°C, a number 1 cone bends at 1125°C, and a number 26 cone bends at 1595°C.

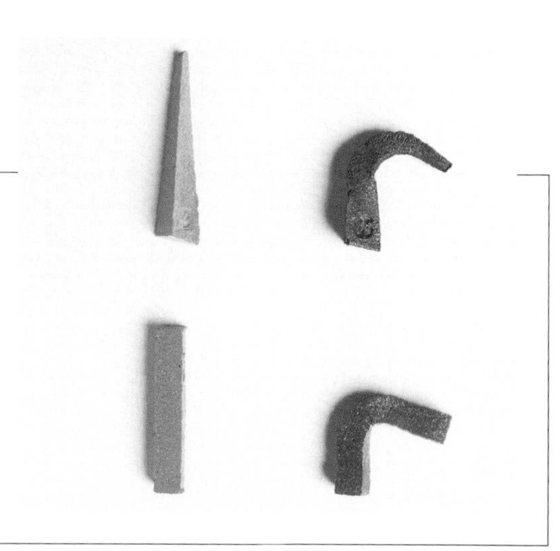

Check Your Understanding

1. In your notebook, indicate whether each statement below is true or false. If the statement is false, make it true by changing the *italicized* word(s).
   - **(a)** All pure substances have *the same* melting and boiling points.
   - **(b)** A plateau is a *flat* region on a graph.
   - **(c)** While a substance is changing state, its temperature is *changing*.
   - **(d)** Melting point and freezing point are *the same* temperature.
2. Use Table 8.2 on page 231 to find a temperature at which
   - **(a)** oxygen is a liquid
   - **(b)** table salt is a gas
   - **(c)** tin is a liquid
3. List the six changes of state, and give a name for each one. Try to complete your list from memory.

8.4 Particle Theory and Phase Changes

Evaporation

Now that you have studied temperature, energy, and changes of state, look back at the Starting Point Activity at the beginning of this chapter. Remember how cold a drop of alcohol felt on your skin? The particle theory can explain this!

In a drop of liquid, particles are moving at many different speeds. At the surface of the drop, some of the faster-moving particles are able to escape into the air. Slower-moving particles stay in the liquid state. Slower motion means lower average kinetic energy, however, and this means lower temperature. As high-energy particles evaporate, the remaining liquid cools. The cool liquid then cools the surface on which it is resting. Scientists call this phenomenon **evaporative cooling**.

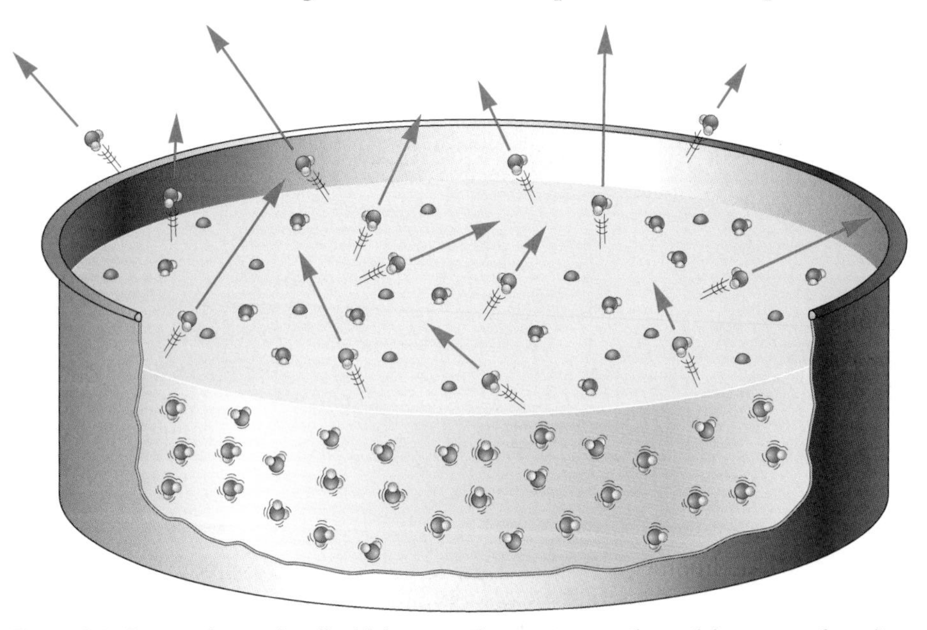

Figure 8.5 Evaporation cools a liquid, because the most energetic particles escape from its surface.

Evaporative cooling is common and can be very useful. Can you think of other examples besides the ones below?

- Joggers feel cold as their clothes dry out after getting soaked in a rainstorm.
- A home-owner sprays the roof of a house with water to cool the house on a hot summer day.
- A first-aid worker puts a wet cloth on the forehead of a person with a high fever.
- One type of air conditioner blows hot air into pipes that have a thin layer of water trickling over them. The air cools as it goes through the cold pipes.

Figure 8.6

How Low Can It Go?

At the beginning of this chapter, you saw that alcohol evaporates more rapidly than water. If you compare the temperature change when the two liquids evaporate, the results may surprise you.

What You Need

lab thermometer or computer temperature sensor

electric fan

2 strips of cloth or paper towel

room-temperature water

room-temperature alcohol

Safety Precautions

Pure alcohols are harmful to the body. Do not taste these chemicals, and do not breathe their vapours. Do not use alcohols near open flames, as the alcohol vapours may catch fire or explode.

What to Do

1. Measure and record the temperature of the liquid water and alcohol.
2. Wrap the cloth strip around the thermometer bulb, and soak it in the water. Hold the thermometer near the fan to speed up evaporation from the wet cloth. Record the temperature every 30 s until it stops dropping.
3. Repeat step 2, using a second cloth strip and alcohol.
4. To compare your observations for the two liquids, draw a graph with temperature on the vertical axis and time on the horizontal axis. Plot both sets of data on the same graph.

What Did You Find Out?

1. In which liquid were the particles evaporating faster? How do you know?
2. Which of the two cloths would take longer to dry completely? What would happen to temperature of the cloth after all of the liquid had evaporated?
3. In which liquid do you suppose the particles have a stronger attraction for one another? Why do you think so?

Extension

4. A fundamental law of science states that an energy loss in one place is always balanced by an energy gain somewhere else. A refrigerator, for example, removes the thermal energy from everything inside it — it cools things. So somewhere outside every refrigerator, the energy of other particles must be increasing. What is happening to the temperature of these particles? Try touching the sides of a refrigerator or freezer and feel the air near the bottom. Then record your observations and ideas.

Potential Energy

What, exactly, is happening to particles of a substance during a phase change? For example, what happens to water particles in ice crystals as the ice melts? According to the theory you studied in Chapter 7, the average speed of the particles cannot be changing because temperature stays constant during a phase change. If the speed of the particles changed, the temperature would have to change, too.

What does change, according to modern particle theory, is the *arrangement* of the particles. Study Figure 8.7 to visualize how this happens. Particles become *less* organized as their energy increases, so the substance changes from a solid to a liquid, and then to a gas. Particles become *more* organized as their energy drops, so a gas will change to a liquid and then to a solid.

Pause& Reflect

When you take a shower, beads of water may form on the bathroom mirror and other cold surfaces far from the shower.

(a) Where did the water in the beads come from?

(b) In what state was the water that formed the beads as it travelled to the mirror?

(c) What change of state is occurring when the beads of water form?

Energy associated with particle position and arrangement is called **potential energy**. During a phase change, the potential energy of a substance increases or decreases as its particles change their arrangement. The average kinetic energy of the particles, however, does *not* change. You can observe the results of a potential energy change as an ice cube melts. Since thermometers cannot detect changes in potential energy, however, the temperature of the substance stays constant. The energy change is hidden from the thermometer, so it is called "hidden heat" or "latent heat."

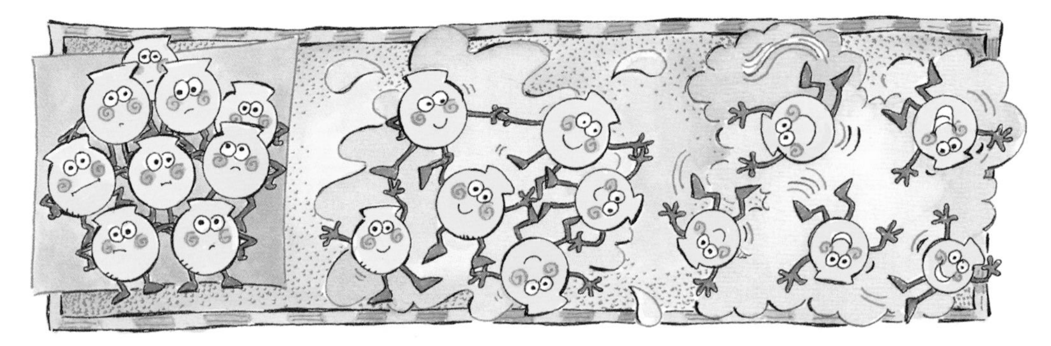

Figure 8.7A Particles in many solids have a regular arrangement. They move by vibrating in the same spot.

Figure 8.7B Particles in a liquid move freely, but they are still held loosely together. They can vibrate and rotate, but they can only move a short distance before colliding with a nearby particle.

Figure 8.7C Particles in gases move independently and are separated by large spaces. They can vibrate, rotate, and travel longer distances between collisions.

Check Your Understanding

1. Name a change of state in which particles become
   (a) more organized
   (b) less organized
   (c) able to move more freely
   (d) held more tightly together
2. Define potential energy, and give two examples of different types of potential energy.
3. Imagine that you can see the moving particles in a drop of liquid on your skin. Describe
   (a) the speed of the particles that are able to escape from the surface of the drop
   (b) the speed of the particles that are left behind in the drop
   (c) the temperature change of the drop as particles continue to escape
   (d) the change of state that is occurring
4. **Apply** Anyone who falls into a lake fully clothed may develop hypothermia (dangerously low body temperature) after being rescued. No matter whether the water or the weather is warm or cold, first-aid experts say that the victim's wet clothing should be removed immediately. Use your knowledge of energy and change of state to explain why.

CHAPTER at a glance

Now that you have completed this chapter, try to do the following. If you cannot, go back to the sections indicated.

Describe one observation to show that
(a) solids expand when they warm
(b) liquids usually contract when they cool (8.1)

Explain how
(a) sagging electrical transmission lines show that solids contract as they cool
(b) a balloon can show that gases expand as they warm (8.1)

The specific heat capacity of iron is 0.450 J/g °C. The specific heat capacity of brass is 0.380 J/g °C. If you have a 100 g sample of each metal, identify which will
(a) warm up faster, if both samples are heated equally
(b) absorb more energy when it warms by 10°C
(c) transfer more energy to the surroundings as it cools by 10°C (8.2)

One student says that water boiling on a stove gets hotter and hotter the longer it boils. Another student says that the water stays the same temperature as long as it boils. Who is correct? Explain how energy from the stove affects the particles of water as they boil. (8.3, 8.4)

The graph shows a warming curve for an unknown substance. In section A, the substance is a solid. In section C, it is a liquid and in section E, it is a gas. Identify by letter the section(s) of the curve in which the substance is
(a) boiling
(b) melting
(c) increasing in kinetic energy
(d) increasing in potential energy (8.3, 8.4)

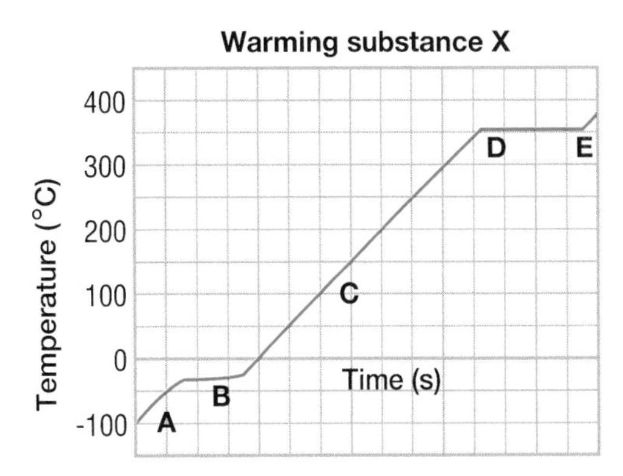

Use the information in Table 8.2 on page 231 and the graph above to identify the unknown substance. (8.3)

Prepare Your Own Summary

Summarize this chapter by doing one of the following. Use a graphic organizer (such as a concept map), produce a poster, or write the summary to include the key chapter ideas. Here are a few ideas to use as a guide:

- Under what conditions do most materials expand? Under what conditions do most materials contract?
- What three factors affect how much an object expands or contracts as its temperature changes?
- What does specific heat capacity measure? On what does it depend? In what units is it expressed?
- How can you use a substance's specific heat capacity to predict the following: how easily the substance will warm or cool, and how much energy it gains or loses as it warms or cools.
- What information is shown on a warming or cooling curve? Identify as many facts as possible that you can determine by examining the curve, and explain how to find each fact. (A sketch may help.)
- What differences are there in the motion and potential energy of a material in its solid, liquid, and gaseous states? (A chart may be useful to summarize your answer.)

CHAPTER

8 Review

Key Terms

expand
contract
state or phase
solid
liquid
gas
heat capacity
specific heat capacity
melting
freezing
condensation
plateau
solidification
sublimation
boiling
evaporation
fusion
evaporative cooling
potential energy

Reviewing Key Terms

If you need to review, the section numbers show you where these terms were introduced.

1. Explain one difference between
   (a) boiling and evaporation (8.3)
   (b) freezing point and melting point (8.3)
   (c) heat capacity and specific heat capacity (8.2)

2. Complete the crossword puzzle below using new terms you learned in this chapter.

Across

1. cooling caused by a wet blanket (8.4)
2. increase in size (8.1)
5. never slanting, always flat (8.3)
7. phase change involving no liquid (8.3)
8. state of greatest potential energy (8.1)
9. energy change when the temperature of an object changes by 1°C (8.2)
10. cannot be compressed (8.1)

Down

1. a drying puddle (8.3)
3. state not found on Earth (8.3)
4. happens when ice is heated (8.3)
6. opposite of evaporation (8.3)

Understanding Key Ideas

Section numbers are provided if you need to review.

3. The two graphs below show the volume of liquid in a laboratory thermometer during a temperature measurement. (8.1)
   (a) Which graph is a warming curve, and which is a cooling curve? Explain how you know.
   (b) Which graph shows what happens when the thermometer is placed in water in a pot on a stove and the burner is turned on? Which shows what happens when the thermometer is placed in a jug of water that was just put in a refrigerator?

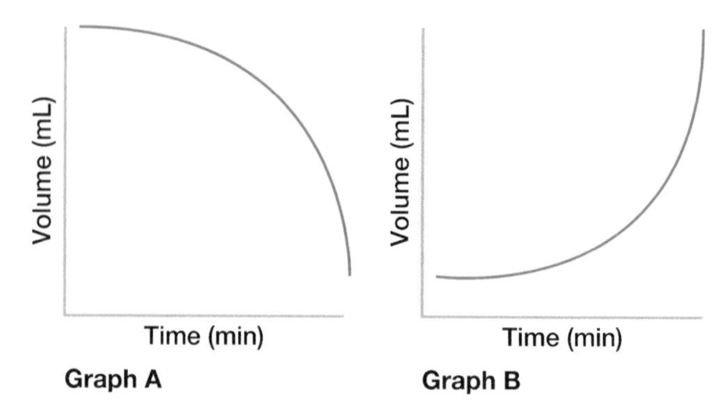

4. Carbon dioxide gas sublimes at –78.5°C. (8.3)
   (a) If you cool carbon dioxide below –78.5°C, in what state will it be?
   (b) Could you produce liquid carbon dioxide by cooling the gas or by warming the solid?
   (c) Solid carbon dioxide is called "dry ice." Why is this name appropriate?
5. 1000 J of thermal energy will warm 1.0 kg of mercury by 7.0°C or 1.0 kg of copper by 2.6°C. (8.2)
   (a) Which substance has greater specific heat capacity, mercury or copper?
   (b) Suppose that you have equal masses of mercury and copper, both at 55°C. Which material will cool to room temperature faster?

Developing Skills

6. Imagine that the students in your class represent particles of matter. The school gym represents a container. By their position and movement, the students will demonstrate the "gymnasium model of matter." Write directions for the students to model each change of state.
7. Imagine hitching a ride on a particle of candle wax as it melts and then evaporates in the candle flame. Describe what the ride would be like. Include as many details as you can about the changes in position and motion of the particle.

Problem Solving/Applying

8. One way to preserve fruit is to "can" it. Fruit and syrup are placed in a clean glass jar and then heated in boiling water to kill harmful bacteria. An airtight lid is fastened onto the jar, which is then left to cool.
   (a) Should the jar be filled completely to the brim before it is heated? Explain why.
   (b) After the jar cools, there is always some empty "head space" at the top. Does this mean that the contents leaked out during the canning process? Explain your answer.
9. Fruit farmers sometimes spray water over their orchards on cold, frosty evenings. As the water freezes, it protects the fruit from frost.
   (a) When the water freezes, does the energy of the water particles increase or decrease?
   (b) Is this a change in kinetic energy, potential energy, or both? How do you know?
   (c) Explain how a change in the energy of the water particles can protect the fruit that the water is covering.
10. A steel bridge is 1000 m long at 20°C. Each metre of steel expands or contracts by 0.11 cm when the temperature changes by 1°C.
    (a) How long will the bridge be in summer at 30°C? (Hint: Convert the length of the bridge to centimetres first.)
    (b) How long will the bridge be in winter at 0°C?

Critical Thinking

11. To loosen a very tight lid on a glass jar, people often hold the jar and lid under hot running water. After a minute, the lid twists off easily.
    (a) What must be different about the behaviour of the metal and the glass to explain why the lid loosens?
    (b) What would probably happen if a jar with a loose-fitting lid were placed in a freezer? Explain why.
12. A large body of cool water can easily have higher total thermal energy than a small amount of hot water, such as a hot water bottle. Explain how this is possible.

Pause& Reflect

Go back to the beginning of this chapter on page 212, and check your original answers to the Getting Ready questions. How has your thinking changed? How would you answer those questions now that you have investigated the topics in this chapter?

CHAPTER

9 Energy Transfer

Getting Ready...

- What provides the energy to power violent thunderstorms, huge hurricanes, and destructive tornadoes?
- How is energy from the Sun transferred through space to Earth?
- What makes a furnace, a fireplace, or a car "efficient"?

Science Log

When most children are five or six years old, they are full of questions. In your Science Log, try to explain to an imaginary six-year-old, your ideas about one of the Getting Ready questions. Make your explanation simple and clear so that the child will understand, and remember to include good examples.

Whenever changes occur in the physical world, energy is involved. For example, experimental aircraft like these need energy to resist the force of gravity and stay aloft. Designers hope that solar cells will be able to harness enough energy from sunlight during the day to run the larger plane's electric motors and also charge its batteries. If these solar cells are effective, the plane will use solar power during the day and batteries at night and transmit weather data and communications signals for many days.

Super-efficient motors in the smaller plane need the chemical energy from less than 8 L of fuel to carry the aircraft right across the Atlantic Ocean! Aircraft like this could improve weather forecasting by gathering wind and temperature data over the oceans at very low cost.

We can think of these aircraft as energy transfer systems. A **system** is a set of things in the natural or human-made world that are organized and interact (affect each other) so much that they can be described as a single unit. Energy transfer systems support life and accomplish useful tasks. Most systems also waste energy, however, and sometimes they change the environment in unexpected ways. In this chapter, you will apply your knowledge of temperature changes and thermal energy to investigate and analyze systems in which energy is transferred and transformed.

Systems

Spotlight On Key Ideas

In this chapter, you will discover

- three methods of transferring energy
- how energy is transformed into useful forms from its original sources
- the role of energy in Earth's weather systems

Spotlight On Key Skills

In this chapter, you will

- analyze characteristics that influence how materials absorb and transmit energy
- analyze feedback systems that control temperature in living organisms and manufactured devices
- examine some natural and manufactured thermal energy systems
- design and construct an energy transfer system

Starting Point ACTIVITY

Cold Hands, Warm Heart

Have you ever used a hand-warmer? You can find out how this kind of energy transfer system works by making your own hand-warmer from everyday materials.

What You Need

25 g iron powder
measuring spoon
small plastic bag
sodium chloride (table salt)
vermiculite
water
twist tie

Safety Precautions

Handle iron powder with care. It is an irritant to eyes and skin.

What to Do

1. Place about 25 g of iron powder in the small plastic bag. Add about 1 g of sodium chloride and about 15 mL of vermiculite.
2. Close the bag, and shake it gently to mix the chemicals.
3. To activate the hand-warmer, add 5 mL of water to the bag and seal it again, making sure that a lot of air is trapped inside. Squeeze and shake the bag to moisten and mix all of the contents with water and air. Wait 1 or 2 min.

What Did You Find Out?

1. After a couple of minutes, what do you notice about the temperature of the hand-warmer?
2. How long did the hand-warmer stay noticeably warm?
3. What do you think caused thermal energy to be released by the hand-warmer?

9.1 Methods of Thermal Energy Transfer

Imagine holding your hand near a light bulb or in front of a hot fire. You can feel the warmth. Your skin warms up because it receives thermal energy from the bulb. The light bulb is an **energy source**: an object or material that can transfer its energy to other objects. In this section, you will study three ways in which energy can be transferred: radiation, conduction, and convection.

Radiation Transfers Energy

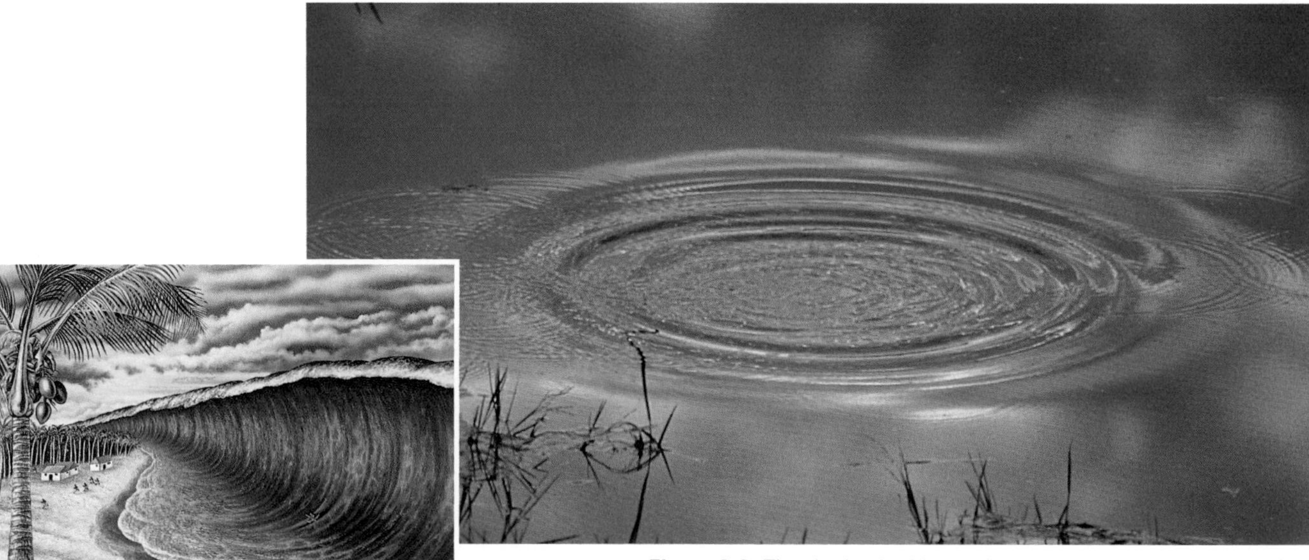

Figure 9.2 The ripples in this pond are evidence of energy transfer.

Figure 9.1 A tsunami carries enormous amounts of energy from its source, an underwater earthquake, across thousands of kilometres of ocean. When the wave hits land, the energy can devastate buildings and the natural environment, as well as sometimes costing thousands of lives.

If you toss a stone into a pond, energy is transferred from the stone to the water. A fast-moving stone has a great deal of kinetic energy. When the stone hits still water, it slows down — it loses energy. Some of the stone's energy is transferred to the water, so the water gains kinetic energy. As ripples spread across the water, the kinetic energy of the stone (the energy source) is transferred to, and across, the water. Ripples, like all waves in water, air, or even empty space, transfer energy from one place to another.

Of course, water waves can transfer energy through water. **Radiation** is the transfer of energy in a special form of wave that can travel through many materials or empty space. Energy that is transferred in this way is called **radiant energy** and it is carried by **electromagnetic radiation (EMR)**. There are many different forms of EMR, including radio waves, microwaves, visible light, and X rays. If the energy source is a warm object, such as the Sun or even your body, some of its thermal energy is transferred as a type of EMR called **infrared radiation (IR)** or "heat radiation." All of the different forms of radiant energy share several characteristics:

- They behave like waves.
- They can be absorbed and reflected by objects.
- They travel across empty space at the same very high speed: 300 000 km/s.

Find Out **ACTIVITY**

Absorb That Energy

If an object absorbs radiant energy, what happens to its temperature?

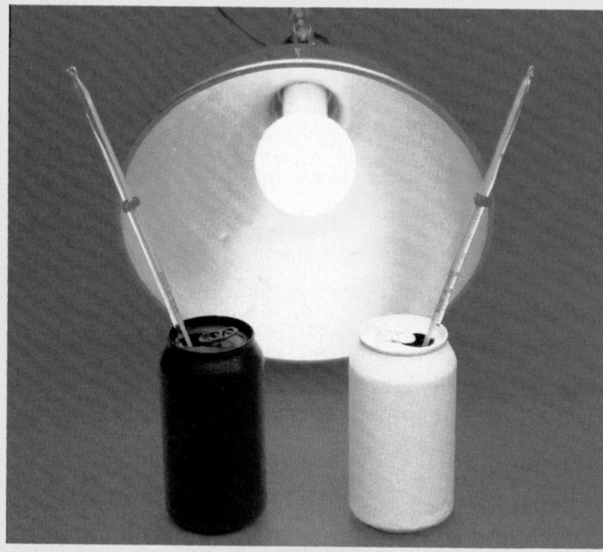

What You Need  

2 empty pop cans
2 thermometers
light (at least 100 W)
ruler
dark- and light-coloured cloth, or black and white paint
aluminum foil
200 mL cooking oil
tape or rubber bands

What to Do

1. Think of summer sunlight beating down on different materials. Use your own experience to write a prediction about which type of surface absorbs the radiant energy best:
   (a) dark or light
   (b) shiny or dull
2. Use an appropriate choice of materials to cover the pop cans so that you can test one of your predictions.
3. Pour 100 mL of cooking oil into each can. Place the cans an equal, short distance from the light. (Try 10 cm.)
4. For each can, record the initial temperature of the oil and the temperature of the oil every 5 min for 15 min.
5. Calculate the temperature change of the oil in each can by subtracting the intial temperature from the final temperature.

What Did You Find Out?

1. Compare the temperature change of the oil in the two cans. Do your observations support your prediction?
2. If several groups tested the same prediction, how well did their results agree?
3. What other factors, besides the one that you tested, may be affecting the temperature change in the oil?
4. According to scientific theory, the same materials that absorb radiant energy well should also radiate energy well. Suppose that you have pairs of similar objects with different surfaces, as listed below. You heat them to the same high temperature. Which type of surface radiates energy better and thus cools down more quickly?
   (a) a light coloured surface or a dark surface
   (b) a dull surface or a shiny surface

You know, from your own predictions or from the activity above, that some materials absorb radiant energy well, and some materials reflect well. Do the same materials make good reflectors and bad absorbers of radiant energy? In the next investigation, you can think about this and similar questions, and infer some everyday-life consequences.

INVESTIGATION 9-A

Comparing Surfaces

Think About It

From your own experience, can you think of examples of the following scientific observations?

- Dark-coloured surfaces absorb and radiate energy better than light-coloured ones.
- Dull surfaces absorb and radiate energy better than shiny ones.
- Shiny surfaces reflect radiant energy better than dull ones.
- Light-coloured surfaces reflect radiant energy better than dark-coloured ones.

What to Do

1. In your notebook, make a table like the one below. Give your table a title. Complete the table by writing "better" or "worse" to describe the behaviour of each surface compared to its opposite.

Surface	Ability to absorb	Ability to radiate	Ability to reflect
light coloured			
dark coloured			
shiny texture			
dull texture			

2. Identify the combination of colour and texture that would be

   (a) the best reflector **(b)** the worst reflector

   (c) the best absorber **(d)** the worst radiator

People radiate energy. Have you ever been cool and comfortable at the start of a concert or school assembly and then, after an hour or so, found yourself getting unbearably hot? Thermal energy from the crowd of warm bodies was probably to blame. Each person acted like a miniature furnace, warming nearby air and furniture. Without air conditioning to transfer the thermal energy elsewhere, a crowded room can quickly become uncomfortable.

Analyze

1. Use your answers in steps 1 and 2 to explain why
   (a) black car seats get very hot in the summer sunlight
   (b) dairy trucks have shiny white or silver-coloured refrigerator tanks to hold the milk
   (c) a concrete sidewalk stays cooler than a dark asphalt road in the summer sunlight
   (d) tennis players wear white or light-coloured clothing
2. The Russian government once experimented with a method to speed up the melting of snow on northern farmland, so that crops could be planted earlier in the spring. Black coal dust was dropped on the snow from low-flying aircraft.
   (a) Explain why the coal dust was expected to speed up snow melt.
   (b) The snow did melt sooner, but the method was never actually put into use. Think of some reasons why it would be impractical.

Conducting Energy Through Solids

In solids, where particles are close together, thermal energy can be transferred directly from one particle to the next. **Thermal conduction** is the process of transferring thermal energy through direct collisions between particles. Study Figure 9.3 to see how conduction transfers energy.

Figure 9.3A Particles near the heat source absorb energy from it and begin moving more rapidly.

Figure 9.3B The fast-moving particles bump into neighbouring particles, increasing their energy and motion.

Figure 9.3C In this way, thermal energy is transferred throughout the material.

Most metals, especially gold and copper, are excellent heat conductors. A hot stove burner touching one part of a copper saucepan, for example, soon heats the entire pan. Other solids, such as glass and wood, are much less efficient at transferring thermal energy by conduction. Poor conductors are called **heat insulators**. Foam plastic cups and containers, the puffy inner layers of winter clothing, and the insulation in walls are all poor conductors. When insulators are wrapped around an object, they slow down the transfer of thermal energy to or from the surroundings. The object stays warm or cold longer.

Find Out **ACTIVITY**

The Super Stirrer

Can you predict how well a substance will conduct heat, so you can choose the most suitable material for a particular use? Try it. Find the material that will make the best stir stick.

What You Need

equal-length pieces of plastic from a pen, pieces of copper wire, long iron nails, wooden craft sticks, or wooden pencils

plastic cup of very hot water

Safety Precautions   

Handle the hot water with care.

What to Do

1. Predict which of your sample stir sticks will be the
   (a) best conductor
   (b) worst conductor
   (c) best insulator
   (d) worst insulator
2. Place one end of each sample in the hot water. Wait 1 min.
3. Touch the inside of your wrist to the top of each sample to identify the warmest one (the best conductor). Remove it from the cup and record which materials made the best conducter.
4. Wait another minute. Then repeat step 3 to find the second-best conductor. Continue to repeat step 3 until you have ranked all of the samples in order, from the best to the worst conductor.

What Did You Find Out?

1. Explain which of your samples would be the best for making
   (a) a stir stick
   (b) the bottom of a frying pan
   (c) the handle of a frying pan
   (d) a container for delivering hot pizza
2. How might the particles in your best insulator differ from the particles in a conducting material?

At Home **ACTIVITY**

Keeping in the Warmth

Fibreglass building insulation

Insulators that are used in building construction are rated by their RSI value. This value describes the resistance of a 1 cm thickness of a material to heat conduction. Materials with higher values are better insulators. Some typical values are given in the table below.

Material	RSI per cm
blue plastic foam panels	0.35
white plastic foam panels	0.29
fibreglass	0.24
vermiculite	0.16
plywood	0.087
glass	0.017

Extra thickness increases the RSI value. For example, a 3 cm thickness of fibreglass would have an RSI value of 3 x 0.24 = 0.72. Only 2 cm of blue plastic foam would provide about the same resistance to heat conduction (2 x 0.35 = 0.70).

What to Do 

Try to find out what type of insulation is used in your home and how thick it is. Then calculate its total RSI value. If the material is not listed in the table, check with a building materials store, in the library, or on the Internet to find its RSI value.

Convection, Energy on the Move

Thermal energy can be transferred in a third way by **fluids**: materials that can be poured or that flow from place to place. A hot, fluid may force its way up through a colder fluid. In **convection**, the warm fluid, itself, moves from place to place, carrying the thermal energy with it. The moving fluid is called a **convection current**. Study Figure 9.5 to identify the different parts of a convection current. Then read on to learn the details of how a convection current operates.

Figure 9.4 Smoke trails in this apparatus show how air moves in a convection current.

Why do fluids, at different temperatures, rise, sink, and create convection currents? Remember that materials expand as they warm up. Their particles move farther apart. Each section of the warmed material is left with fewer particles than when it was cold, so each section is a bit lighter than it used to be. In other words, the warmed material becomes less dense. Colder, denser fluid sinks down and pushes nearby warmer fluid upward. Then this cold fluid, too, is warmed and pushed upward.

As warm fluid rises and moves away from the heat source, it cools. It contracts as its particles move closer together. It becomes denser and sinks back down toward the heat source, where it is warmed and forced upward. As the whole process repeats, a continuous movement — a convection current — forms.

Convection currents occur throughout Earth's oceans and atmosphere, distributing solar energy around the planet. As you will see in Unit 4, scientists believe that convection currents in molten rocky material within Earth carry entire continents slowly from one location to another. Much smaller-scale convection currents are important in the operation of home heating systems, air conditioners, and even ovens in stoves.

heat source

warm air

cool air

A Warmed air particles expand.

B Less dense, warmer air rises.

C The rising air cools and contracts.

D The cool, denser air sinks.

E The cool air moves in to replace the rising warm air.

Figure 9.5 All convection currents display the features shown here.

Career CONNECT

Setting the Standard

Steve Reid knows that convection currents cannot warm a house unless the heat source is working. Steve is a gas appliance technician. His job is to install and repair any type of furnace, boiler, stove, dryer, or other appliance that uses natural gas as its fuel.

A mistake could turn a safe appliance into a hazardous one, so Steve's work must be done with great care. To train for his work, Steve completed a college course in heating, refrigeration, and air conditioning. Then he passed a government test in order to receive a gas fitter's licence. Without this licence, Steve could not legally work on gas appliances.

To ensure everyone's safety, the government has set standards that must be met by people who want to work in occupations that are potentially dangerous to the worker or the public. In some other occupations, an organization of people who already work in the occupation create a test that others must pass to become a registered or certified member. This process ensures that everyone who works in the occupation is well qualified.

With a partner, brainstorm at least three occupations that may require people to be licensed, registered, or certified. Start by thinking about occupations that could involve danger to the worker or the public. You and your partner could each choose one of the occupations, and, after checking with your teacher, contact someone in that field of work for comments on how licensing is done and its importance to the occupation.

Displaced Drops

You can create a small-scale model of parts of a convection current. Observe what happens in this activity, and compare it with the explanation you have just read.

What You Need

dropper
250 mL beaker of room-temperature water
100 mL beaker of coloured, ice-cold water
100 mL beaker of coloured, very hot water

CAUTION: Handle the hot water with care.

What to Do

1. Make sure that the beaker of room-temperature water is completely still.
2. Fill the dropper with ice-cold water, and hold it just above the surface of the room-temperature water. Gently squeeze out one drop of cold water. Watch to see if it can force its way to the bottom of the beaker.
3. Repeat step 2 several times. Then make a careful diagram showing what usually happens to the drop of cold water.
4. Repeat steps 2 and 3 using very hot water in the dropper.

What Did You Find Out?

1. Did the drops of very hot water appear to be more or less dense than the room-temperature water around them? How do you know?
2. How did the density of the drops of ice-cold water compare with the density of the room-temperature water? How do you know?
3. In what way do your observations agree or disagree with the text description of what happens to warm and cold fluid in a convection current?
4. Why did the drops of hot or cold water not move in a complete convection current?

Pause& Reflect

In your Science Log, make an outline or a concept map that summarizes the important features of the three methods of energy transfer discussed in this section. For each method, include

- its name
- the types of materials in which it occurs
- a short description of how it occurs
- an important example of it
- one other important fact about it

Check Your Understanding

1. Which type of heat transfer does each situation below model?
   - (a) At a sports event, the spectators do "the wave." Starting at one end of the stadium, they stand up and then sit down as the next person stands. A ripple of motion spreads across the crowd.
   - (b) Several dominoes are standing close together on edge. You give the end domino a small push, increasing its kinetic energy. It falls onto the next one, which falls onto its neighbour, and so on. Soon all the dominoes have fallen.
   - (c) On a ski hill, people line up at the bottom of the hill, take the lift to the top, and then ski down again to join the crowd waiting for the lift.
2. The oven in most stoves has an electric or gas heat source near the bottom. Air in the oven moves freely around the heat source and the food.
   - (a) Explain how all three methods of energy transfer might occur in an oven?
   - (b) Describe the natural movement of the air in the oven.
   - (c) If the food is inside a cooking pot, the oven can only directly warm the pot. How is thermal energy transferred from the pot to the food to cook it?
3. An outdoor basketball court is paved with black asphalt. The roof of a nearby refreshment stand is white. It is a sunny day.
   - (a) Which method of heat transfer will warm both the court and the roof of the stand?
   - (b) Which surface will become hotter, the court or the roof? Explain why.
4. A mixture of fruit and sugar can easily reach a temperature of 105°C when it is being made into jam.
   - (a) Which type of spoon could be used to stir the mixture without becoming soft or burning your fingers: wooden, metal, or plastic?
   - (b) Which method of heat transfer must be prevented in the spoon?
5. **Thinking Critically** Look back at the models of heat transfer in question 1. Describe one way in which each model is different from the method of energy transfer that it represents.

9.2 Analyzing Energy Transfer Systems

Figure 9.6 What are the parts of this energy transfer system?

A volleyball rockets across the net. This fast movement means a large amount of kinetic energy. The source of the energy was Carrie's hard-swinging fist. The impact of her fist on the ball transferred energy to the ball.

Most of the kinetic energy of Carrie's fist ended up as the kinetic energy of the ball — most, but not all. The other players heard the smack of Carrie's spike and the thud of the ball hitting the floor. If you asked Carrie, she would report that her fist stung and felt warmer after her shot. The part of the ball that she hit also warmed up a bit. When the ball hit the floor, the ball warmed up a bit more. So did the spot on the floor where the ball landed. A few moments later, the floor nearby and even the air above it was a tiny bit warmer.

Can you explain what is going on in this energy transfer system? Examine Figure 9.7 without reading the information below it. Try to explain what is happening at each letter, in terms of energy, particle movement, and temperature. Then check to see if your explanation included everything in the caption.

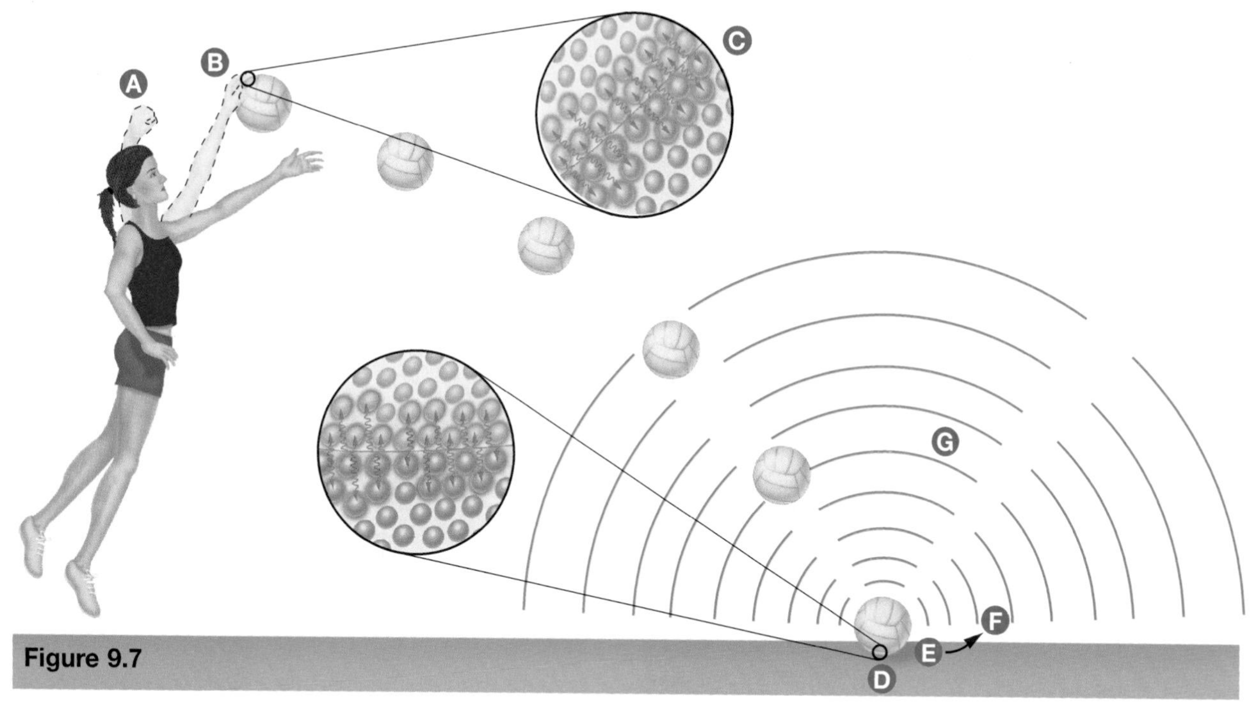

Figure 9.7

A The moving fist (energy source) has a large amount of kinetic energy.

B Most of the energy is transferred to the entire volleyball, which moves away rapidly.

C Some energy is transferred to individual particles in the skin and the volleyball, which vibrate more rapidly, producing a temperature increase.

D Energy is transferred from the ball to the floor. Particles that make up the floor vibrate more rapidly, producing a temperature increase.

E Energy is transferred by conduction to nearby particles in the floor.

F Convection currents transfer energy through the air.

G Energy is also transferred to air particles by compressing them and starting a sound wave. The sound wave distributes this energy throughout.

Features of Energy Transfer Systems

All energy transfer systems have similar features. Hair dryers, bicycle brakes, weather systems, and ocean currents — they all resemble Carrie's volleyball spike in these five ways: energy source, direction of energy transfer, transformations, control systems, and waste heat.

- *Energy Source* Some part of the system acts as an energy source, supplying energy to the rest of the system. Some systems have *mechanical* energy sources, such as Carrie's fist or a tightly wound spring in a toy. As the source of mechanical energy moves, the energy is transferred to other parts of the system. Cars, trains, and even humans and animals depend on *chemical* energy sources, such as gasoline, diesel fuel, or food. As the chemicals react with each other, the chemical energy is transformed into other forms of energy. Stars, atomic bombs, and nuclear power plants use *nuclear* energy sources: substances whose smallest particles can fuse together or break apart, releasing large amounts of energy. Radios, power tools, and plug-in appliances use *electrical* energy sources: batteries or generators in a power station.
- *Direction of Energy Transfer* Energy is always transferred *away* from concentrated sources. Changes in non-living systems always spread energy around more evenly. For example, thermal energy is always transferred from hot objects (heat sources) into colder objects. As this happens, the hot object cools down, the cold object warms up, and the whole system ends up closer to the same temperature.

Word CONNECT

Recall from Chapter 7 that in science and engineering, the word "heat" has a very precise meaning: thermal energy transferred as a result of temperature differences. Heat is not a substance or a thing. It is the name of a particular kind of energy transfer. What do you suppose the scientific meaning of "cold" might be? Use a good dictionary to check your idea.

- *Transformations* In ordinary life, energy never just appears or vanishes. As the energy of one part of a system changes, there is an equal and opposite change elsewhere. The total energy of the system stays the same — it is conserved. Energy does not necessarily keep the same form as it is transferred from place to place, however. Energy can be transformed. When Carrie hit the volleyball, the original kinetic energy of her fist was only partly transferred to the ball. Some energy was transformed into thermal energy of the ball and Carrie's fist, as well as mechanical energy of the air in the sound wave.

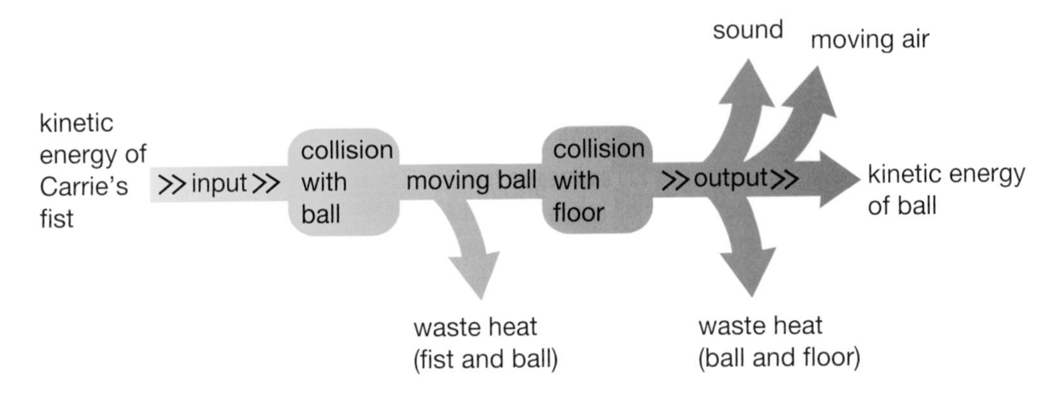

Figure 9.8 These are the energy transformations that take place when a volleyball is spiked.

- *Control Systems* Many natural and manufactured energy transfer systems are able to stay in a particular state despite changes in their environment. For example, your body uses some of the energy released from your food to stay at a healthy temperature. (Do you remember what this temperature is?) If too little energy is released, your temperature starts to drop. This signals your body to release more thermal energy until you warm up. Your body does so by starting to shiver. The resulting kinetic energy of your muscles is transformed into thermal energy.

  If you get too warm, less thermal energy is released until you cool down. Extra blood may flow to your skin so that excess thermal energy can be transferred to the surrounding air. If you perspire, evaporation of the perspiration cools you even more. Your body has used **feedback** — to control its energy transfer system and stay at an efficient operating temperature. Detectors sense the temperature of your blood and skin and "feed" the information "back" to your brain. Then your brain tells your body how to correct its temperature.

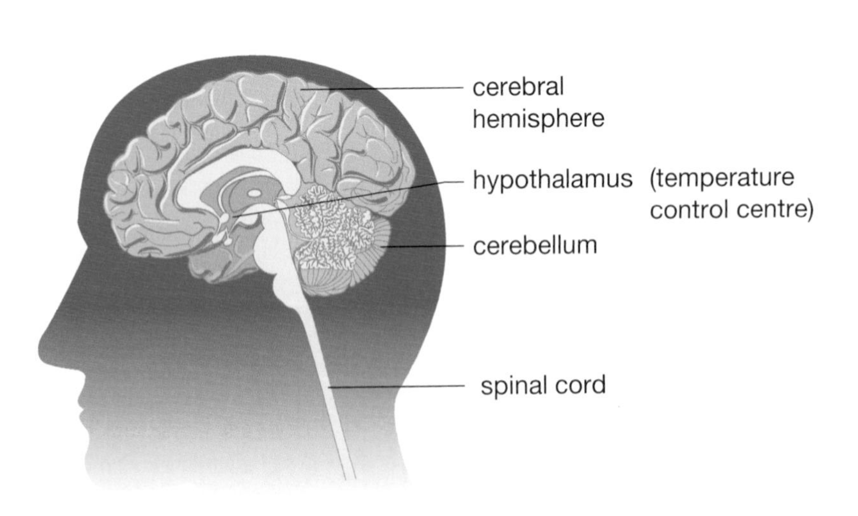

Figure 9.9 Your brain receives information about your body temperature in two main ways. It samples the temperature of your blood, and it receives temperature signals from your skin and other body parts.

• *Waste Heat* Almost all energy systems transfer at least a little thermal energy into the surroundings. When Carrie hit the volleyball, she did not intend to warm it up, but she did. This thermal energy was not available to increase the speed of the ball. A home furnace transfers thermal energy from burning fuel to air in the house. It is not designed to warm the chimney and the cold air outside the house, but it does. As a result, less thermal energy is available to keep the house warm. In a complex system, such as the generating station in Figure 9.10, there can be many sources of waste heat. All of them reduce the amount of energy available for whatever work the system is designed to do. Reducing waste heat — making energy transfer systems more efficient — is an important step in energy conservation.

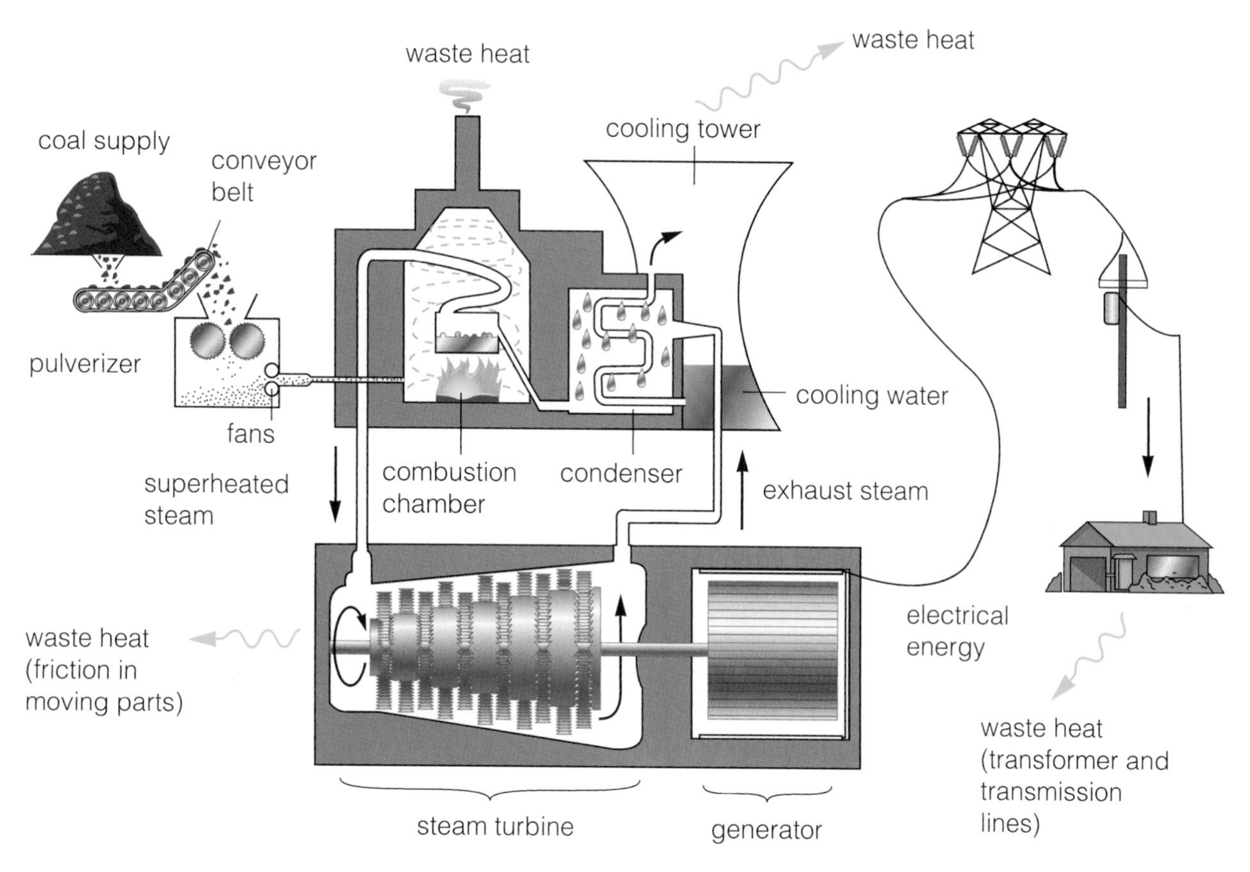

Figure 9.10 Only part of the energy released in a generating station actually ends up as useful electrical energy in your home.

Pause& Reflect

In your Science Log, draw a concept map or a tree diagram to summarize the five features of all energy transfer systems, as discussed in this section. Add important vocabulary, details, or examples for each main idea. Colour-code your diagram so that you can quickly pick out each type of information. For example, use black ink for main ideas and blue ink for vocabulary words.

INVESTIGATION 9-B

Tracking the Transfers

Think About It

Tracing energy transfers all the way through mechanisms or natural phenomena is serious scientific detective work. Remember James Joule? In the nineteenth century, he traced energy transfers through an entire cannon factory, from the horses that turned the machinery to the waste heat that was produced when holes were drilled in metal gun barrels.

In this investigation, you can practise the same kind of thinking. As you work, pay special attention to the **input**, the energy transformations, and the **output**. The input includes all of the materials and forms going in to the system. The output includes the products, both materials and forms of energy.

What to Do

1. Read the descriptions of the two systems below. Each letter identifies an important energy transfer or transformation.

2. Identify the energy input and useful energy output of each system.

3. In your notebook, list the letters in each description. Identify the energy transfer or transformation that each letter represents.

   Example: (A) chemical energy of candle wax to radiant energy of the flame

4. Draw a diagram of the energy flow through each system. Use Figure 9.10 on page 253 as a pattern. Be sure that your diagram shows waste energy being removed wherever this occurs.

5. Answer the Analyze questions for each system. The last question for each system will require some additional research. Summarize your findings as your teacher directs.

Skill POWER

To find out how to use reference materials or the Internet to do research in science, turn to page 497.

System 1

Hot Water Heating System

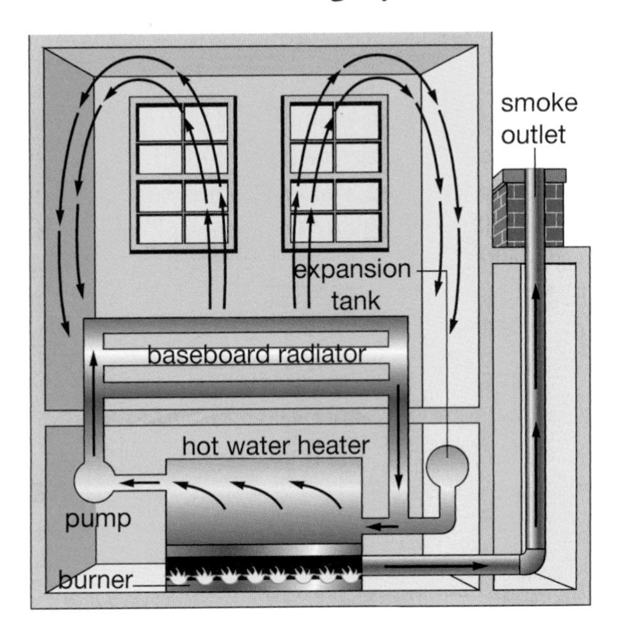

Oil, natural gas, or another fuel burns (A) in a furnace, heating water in the boiler (B). An electric motor turns the pump (C), which circulates the hot water through the heating system pipes (D). In the radiators, the hot water warms flat metal fins (E), which, in turn, warm nearby air (F). The hot air circulates throughout the room by convection (G). A thermostat controls the water flow through the radiator, turning it off when the room is warm enough.

Analyze

1. Name two energy sources for the heating system.

2. What provides feedback to control the heating system?

Extend Your Knowledge

3. Find out exactly how a thermostat controls a heating system. (Remember that thermostats are found in many systems and devices, not just in home heating systems.)

System 2

The Water Cycle

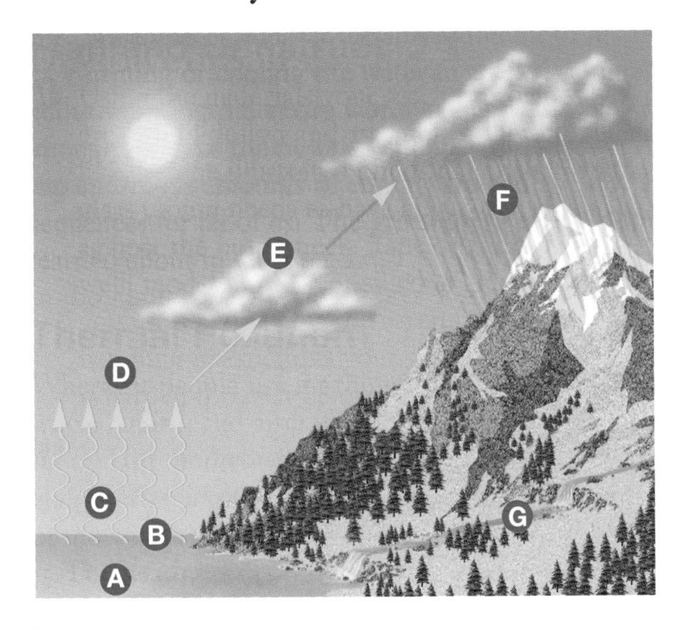

Water in lakes and oceans absorbs sunlight and becomes warmer (A). In turn, the warm water heats some of the air above it (B). Some of the warmed water evaporates into water vapour (C). The warm air rises in a convection current, carrying the water vapour with it (D). The convection current moves water vapour through the cold upper atmosphere, where it condenses again, releasing thermal energy that warms the air (E). The condensed drops of liquid water fall to the ground as rain or snow (F). Eventually most of this water runs back into rivers, lakes, and oceans (G) and the cycle continues.

Analyze

1. Sunlight is much more intense in tropical parts of Earth than it is farther north and south. Would you expect the water cycle to distribute energy from the tropics toward the north and south poles, or from the poles toward the tropics?
2. If the water cycle suddenly stopped, what would happen to the temperature in the tropics? What would happen to the temperature at the poles?

Extend Your Knowledge

3. Near the shore of a lake, solar heating causes another phenomenon. Do some research to find out what "land breezes" and "sea breezes" are, when each occurs, and why they occur.

Many older schools were originally lit with light bulbs like the ones you have at home. To conserve energy, the bulbs are now being replaced with fluorescent lights, which use much less electricity. After this is done, however, the furnace in the school often operates more frequently, because the new lights do not give off as much heat as the old bulbs.

(a) Identify at least two energy transformations mentioned.

(b) Would it be correct to say that the original light bulbs produced "waste" heat? Give reasons for your answer.

INVESTIGATION 9-C

Making a Transfer

You know more about energy than most people who have ever lived! Almost all of the ideas you have been studying were developed in the last 200 years. Can you apply all this new knowledge in a practical way?

Challenge

Design and build a simple but efficient device to harness and transform energy: a candle-powered water heater.

Safety Precautions

- All nonflammable materials must be approved by your teacher.
- During and after heating, handle the apparatus with care. It may be hot enough to burn you.
- Candle flame soot is hard to wash off clothing. Wear an apron, and wash your hands immediately with soap and hot water if you get soot on them.
- Have water or a fire extinguisher nearby.

Materials

thermometer
birthday candle
100 mL room-temperature water
nonflammable containers, fasteners, and insulation
matches

Design Criteria

A. Your goal is to raise the temperature of the water as much as possible.

B. The water may be heated directly with the candle or indirectly using the candle to heat something else, which will then heat the water.

C. Your energy transfer device must be nonflammable and movable so that it can be safely placed over the candle after the candle is lit.

D. The candle will be allowed to burn for only 3 min during your demonstration.

Plan and Construct

1. Brainstorm ideas about how to build the most efficient heater. Think about
   - energy transfer by convection, conduction, and radiation
   - prevention of heat loss to the surroundings
   - specific heat capacity (you will need to make sure that the heater itself does not absorb too much energy)
   - possible materials to use (remember that paints, plastics, glues, and tape are flammable, so they do not meet the design criteria)

2. Choose the most practical ideas, and write a design proposal for your teacher that includes:
   - a list of materials
   - a labelled sketch of your device
   - a task list and a time line to show how each group member will contribute to the project

3. Assemble the materials, and build your device. You may test and modify it, but the candle can be burned for only 1 min during a test. Keep a written record of any design changes you make.

4. For the demonstration, be ready to give a brief explanation of the design features of your device. Then show how it works!

Evaluate

1. What knowledge from this unit did you use when designing your device?

2. How could your device be improved if you had more time to work on it?

3. What extra resources would have helped you to do a better job on this project?

4. What could your group do differently in the next design project to be more efficient?

In the 1970s, the price of fuels rose suddenly. People and governments became very concerned about conserving energy and using fuels efficiently. The auto industry started making vehicles that travelled much farther on each litre of gasoline. Architects designed buildings with more insulation and better seals around windows and doors. Today there is less emphasis on energy conservation. Few headlines or television newscasts focus on this topic. Do you think people should still be concerned about energy conservation? Why? Find out about the United Nations Earth Summits, held every five years.

Check Your Understanding

1. Define the term "energy source," and list four common types of energy sources. Give an example of each type.

2. List five features of all energy transfer systems.

3. **Apply** 100 mL of hot water (50°C) is mixed with 100 mL of cold water (10°C).
   (a) Predict the temperature of the mixture after it is well stirred.
   (b) In which direction was energy transferred?

4. **Apply** In the winter, pioneer families spent much of their waking time in the kitchen because it was the warmest room in the house. A wood box beside the large cast-iron stove held the fuel for the stove. The stove pipe passed through a hole in the ceiling and went up through the upstairs hallway and out through the roof.
   (a) What was the source of heat for the house?
   (b) How was thermal energy released from the source?
   (c) Explain how thermal energy was transferred to
      - the iron stove
      - the kitchen
      - the bedrooms
      - the rest of the rooms in the house
   (d) How did this heating system release waste heat to the surroundings?
   (e) How is this heating system an example of thermal energy spreading out?

9.3 Thermal Energy in a Cold Climate

Figure 9.11 Only 60 years ago, most Canadian buildings were heated by coal-burning furnaces, and trains were pulled by coal-burning engines. Mining and transporting coal were major industries. These boys had to work long hours sorting freshly mined coal.

Living in a cold climate is a challenge! Animals migrate, hibernate (sleep through winter), or depend on adaptations such as fur to insulate themselves from the life-threatening winter weather. People depend on technology. For example, open fires gave early humans a source of extra thermal energy. Heat from a fire is transported mainly by radiation, however. Around a campfire, your face gets warm but your back stays cold. An open fire is inefficient and difficult to control. Fortunately people developed other technologies to help keep warm.

Home heating systems, as we know them today, are a fairly recent development. Clean fuels, furnaces, thermostats that run automatically, and houses that are warm all over — these things have been developed in the last two centuries, mostly in the last 50 years.

Heating Homes

Most heating systems used in Canada today have the same basic parts.

- an energy source: usually a burning fuel or electricity
- an energy conversion system: a device to release thermal energy and a feedback system to control the temperature
- a distribution system: a method to transfer thermal energy to all parts of the building, using a combination of convection, conduction, and radiation

What to Do

1. Identify the basic parts of three different heating systems: one that was used in Canada in the past, one that is used now, and one that could be used in the future.
2. Use a variety of information sources to find out about these heating systems. For example, find out how Canadian families kept their houses warm in the past, and how your home and school are heated now. Do library or Internet research to learn about an alternative energy source or heating technology that might be used in the future, when our existing fuels become scarce and expensive.
3. For each heating system, find out at least one improvement that could make the basic system more efficient or less expensive.
4. Summarize your findings using a chart or another method approved by your teacher.

Skill POWER

To learn how to use the Internet or library resources for science research, turn to page 497.

Heating Living Creatures

Birds and mammals, including humans, are sometimes called "warm-blooded" because their bodies are usually at a constant temperature of 37°C (or higher for some birds). The source of the thermal energy in your body is the food you eat. Every cell in your body uses the food energy to carry out the cell's special function, and each one of these functions releases thermal energy as a by-product. A complicated control system, directed by your brain, keeps enough of that thermal energy inside your body to maintain the correct temperature. Sometimes you have to help out that control system, by wearing insulated clothing to keep thermal energy in your body when you are exposed to cold weather.

Fish, reptiles, and amphibians, such as frogs, are sometimes called "cold-blooded." This is not a good term though, because their bodies are sometimes quite warm. However, these animals cannot produce enough energy inside their bodies to maintain a constant body temperature. They have to seek out an environment with a warm temperature, use radiant energy form the Sun to warm their bodies, or simply tolerate a low body temperature. Some lizards can bask in the Sun and absorb enough radiant energy to raise their body temperature several degrees above the air temperature. However, when they must tolerate a lower body temperature, these animals become less active. Snakes, for example, become sluggish when air temperatures drop much below 20°C.

Figure 9.12 Insulated protective clothing can reduce heat transfer enough for people to survive and even work in arctic conditions.

DidYouKnow?

Thermogenic plants can raise their own temperature, sometimes by an astonishing amount. Skunk cabbage, arum lily, and the metre-tall voodoo lily are all thermogenic. On a day when the air temperature is 20°C, a voodoo lily can heat itself to 35°C, only 2°C less than human body temperature! Even a tiny crocus plant, when it is growing rapidly, can release enough thermal energy to melt an opening through the snow.

Figure 9.13 Clustering together in cold weather is one way to stay warm. Individuals near the inside of the mass are insulated and warmed by those on the outside.

Creatures that live in northern climates have developed several methods to cope with severe winter temperatures. Special body features like thick fur or fat can insulate and reduce heat loss. Many species of birds, and some insects, migrate to warmer climates to avoid winter entirely. Other creatures reduce their energy requirements and energy losses by hibernating through the winter. Some animals even use snow, which is a good thermal insulator, to keep warm. By living in tunnels under the snow, small animals such as mice, can avoid extreme temperatures.

INTERNET CONNECT

www.school.mcgrawhill.ca/resources/

Scientists track the seasonal migrations of birds, fish, mammals, and even insects. For example, monarch butterflies make spring and fall journeys between Mexico and the United States or Canada. Find out more by going to the web site above. Go to the **Science Resources**, then to **SCIENCEPOWER 7** to know where to go next.

Insulation

To survive winter, you need to trap thermal energy in the animal, the person, or the room that needs to stay warm. Fur, clothing, and building insulation all have the same function — they prevent energy transfer. This means preventing conduction, convection, and radiation. In addition, the movement of heated air, and especially water vapour, often needs to be controlled so that it does not carry its "hidden heat" away. Each method of heat transfer is controlled in a different way.

Figure 9.14 Polar bear fur is excellent insulation — and more. Each hair in the fur is actually transparent. Radiant energy from the Sun enters the hair, bounces down to the bear's black skin, and is absorbed, helping to warm the bear. Light that is reflected from the mass of hairs makes the animal appear white.

- *Radiation* Shiny silver-coloured metals or coatings reflect radiant energy. Glass thermos bottles, lightweight survival blankets, and some sleeping bags use this principle to reduce loss of heat by radiation. Some types of building insulation have a thin covering of metallic foil that does the same job. Even window glass can be given a reflective metal coating, just like mirror sunglasses.
- *Conduction* Air is a very poor heat conductor, so many types of insulation are filled with air spaces or bubbles. Animal fur, feathers, clothing, and house insulation all contain air spaces that reduce heat loss. The solid part of the insulation (usually plastic, glass fibres, or wood) is also a poor conductor, so it helps to reduce energy transfer by conduction. Metal window frames and door frames, which are good conductors, are often made like a sandwich. The inside and outside metal sections are separated by a solid plastic or wood "thermal break." Since the thermal break is a poor conductor, it reduces heat transfer through the metal.
- *Convection* Air itself may be a good insulator, but large spaces filled with air are not. Examine Figure 9.16, which shows heat loss in the space between the inside and outside walls of a house frame. One way to stop such heat loss by convection is to prevent motion of the air by breaking up air spaces into many little pockets. That is the main reason why exterior house walls are often filled with vermiculite, cellulose fibre, or fibreglass insulation. The insulation stops the formation of large convection currents by creating many small air spaces.

Figure 9.15 The outside of this building is mostly glass, with a reflective coating to control the transfer of radiant energy.

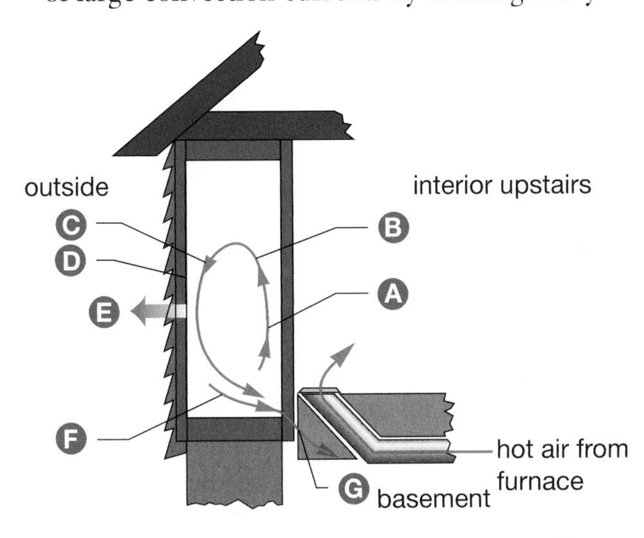

Air next to the inside wall is warmed by conduction (A). The warmed air starts to move in a convection current (B), and ends up next to the cold outside wall (C). Much of the air's thermal energy is transferred to the outer wall (D) and then dissipated in the cold outside air (E). Then the cooled air moves back against the inside wall (F), or even through it into the house (G), where the furnace warms it up.

Figure 9.16 Convection currents in a house wall contribute to heat loss.

- *Air Leakage* If warm air escapes through an opening in a building or clothing, colder air will replace it. Sealing the cracks in a building's frame is an important part of creating a well-insulated structure. Elastic cuffs and ties on clothing help to create a tighter seal against the wearer's body, reducing heat loss.
- *Controlling Water Vapour* Molecules of water vapour are small enough to squeeze through most building materials. In the winter, as water vapour travels through outside walls, it cools and condenses. This change of state transfers thermal energy away from the inside of the building. It also moistens the material in the walls, making it a better heat conductor and increasing heat loss even more. What is worse, the wet material may rot. To prevent this, houses in cold regions have a moisture-proof "vapour barrier," usually made of thin, flexible plastic film, sealed onto the inner surface of their exterior walls.

Find Out ACTIVITY

Show and Tell 

Choose an everyday object or device that uses some type of insulation or sealing to control energy transfer. The object may be designed to keep something from losing thermal energy so that it stays warm, or from gaining thermal energy so that it stays cool. Analyze the materials and the methods that the object uses to do its job. In a poster or an oral presentation, explain how the object or device controls thermal radiation, conduction, convection, movement of air and water vapour, and any other phenomena that could cause heat loss or gain.

Clothing designers face just the opposite problem. Clothing must be able to "breathe": that is, to let the water vapour given off from skin escape. Otherwise, the water vapour will condense in the clothing, making it damp and uncomfortable. The clothing will also be cold because air spaces that used to insulate the wearer are now filled with water, which is a better heat conductor. Making a fabric that lets water vapour escape but prevents cold winds from entering is a real challenge!

Figure 9.17 A vapour barrier is stapled and glued to the inside surface of the wall of this house before the finishing drywall is fastened in place and painted.

INVESTIGATION 9-D

Earth's Weather Systems

Think About It

Why do some parts of the world enjoy long periods of tropical weather while others have long, cold winters? The same source of thermal energy — the Sun — shines on the whole world. Why, then, is Earth warmed so unevenly?

What to Do

Some important ideas about Earth's weather systems are shown in the diagrams in this investigation. Examine them follow the red arrows, and then answer the Analyze questions on page 265.

Radiant energy passes right through Earth's atmosphere without warming it very much.

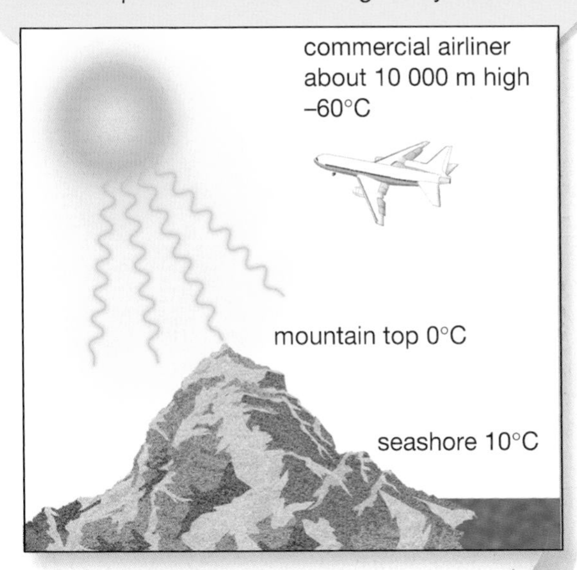

Materials on Earth's surface absorb the incoming radiant energy and warm up. The air just above them is warmed, in turn, by conduction and by absorption of the radiant energy given off by the warmed surface. For example, you can feel hot air above a paved parking lot on a sunny day. This is called re-radiated energy.

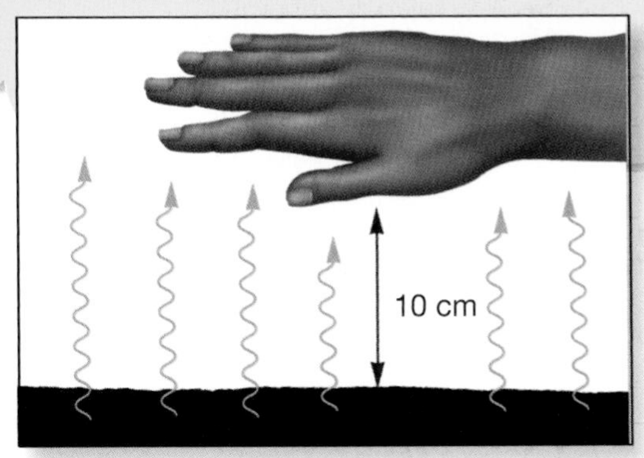

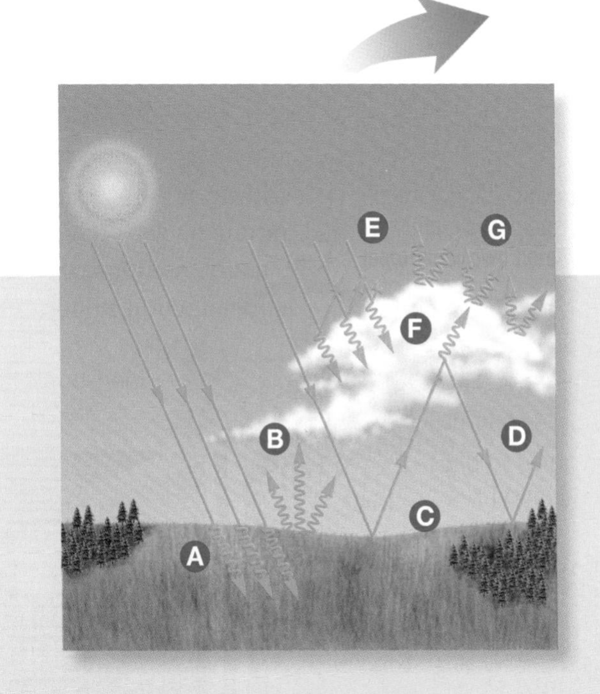

A Solid surfaces absorb thermal radiation and warm up.
B Warm surfaces radiate thermal energy that warms air nearby.
C Air nearest the ground gets warmest.
D Air farther above the ground stays cooler.
E Some solar radiation is reflected off clouds back into space.
F Some thermal energy is absorbed, warming water in clouds.
G Clouds radiate some thermal energy, warming air.

The atmosphere is warmed by re-radiated energy from Earth's surface. Some of this energy is trapped near the surface by clouds, some is absorbed by materials in the atmosphere (greenhouse gases), and some escapes back into space.

CONTINUED ▶

Atmospheric warming is unequal. Radiant energy is more concentrated at the equator, which it strikes directly, than near Earth's poles, which it strikes at an angle. In some places, light-coloured sand, snow, and clouds reflect solar energy. Ocean water, forests, bare ground, and rocky mountains all have different colours, textures, and specific heat capacities, so they warm up differently.

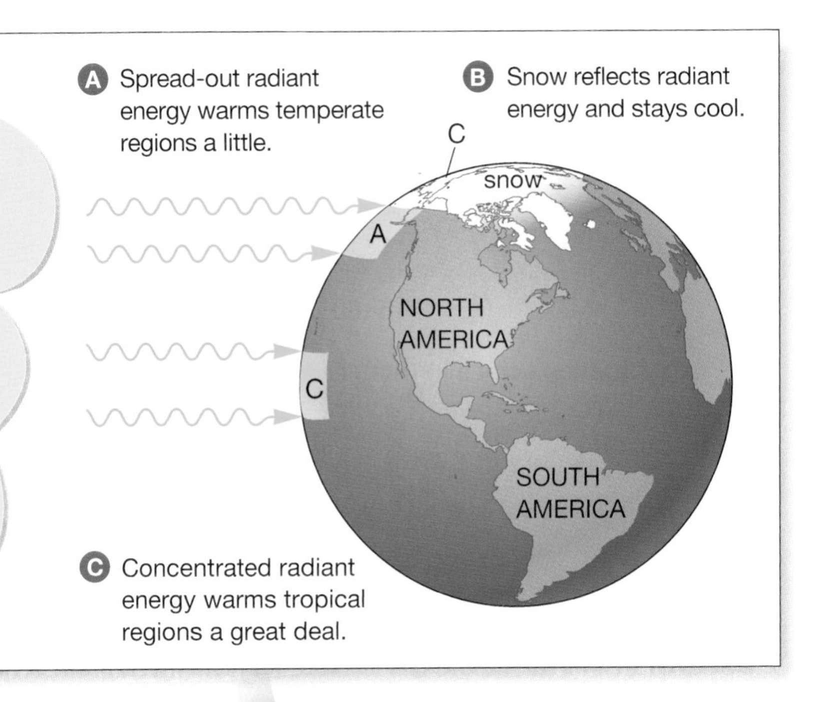

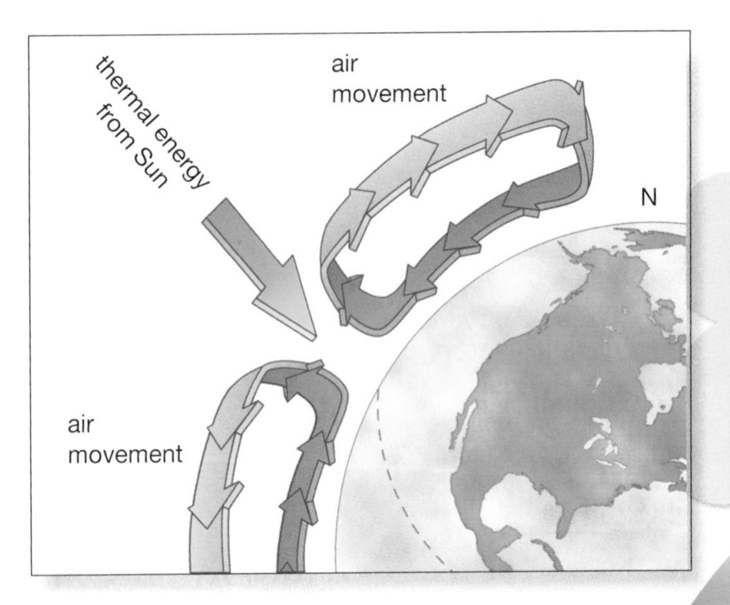

Unequal warming of the atmosphere leads to small and large convection currents. Small convection currents, such as sea breezes and land breezes, are caused by small local differences in heating. Thunderstorms are much larger convection currents. Largest of all are Earth's great wind systems, such as the trade winds, which start with rising moist air above the tropical oceans.

The water cycle plays an important role in Earth's energy systems. Moist air moving in huge systems of convection currents transfers solar energy (from the Sun) from the tropics to other parts of Earth. Water vapour from tropical oceans may be carried thousands of kilometres north or south before it condenses, releasing thermal energy that warms the air and ground nearer Earth's poles. This release of energy can also cause violent storms, such as hurricanes and tornadoes, when it occurs nearer the equator.

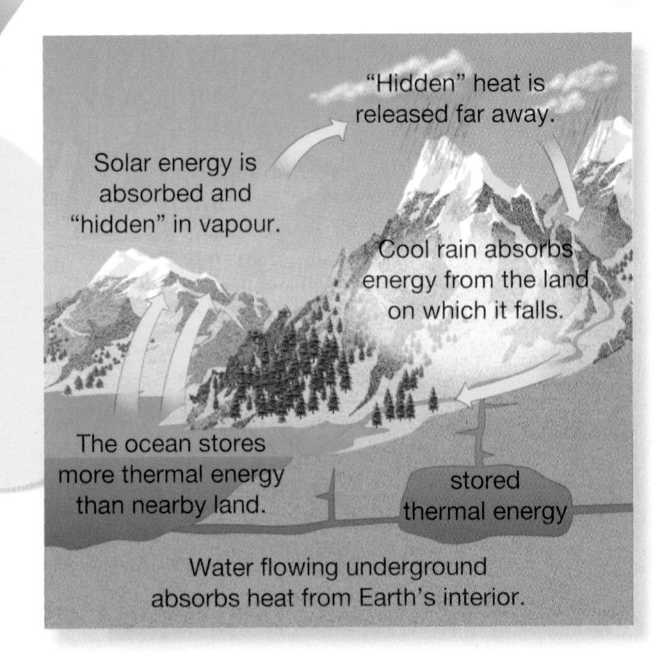

More than half of Earth's surface is covered by water. Unequal heating causes convection currents in the water, just as in air. Ocean currents transfer large amounts of thermal energy. Currents of warm water from the tropics warm some locations, while currents of cold water from the Arctic and Antarctic reduce temperatures in other areas.

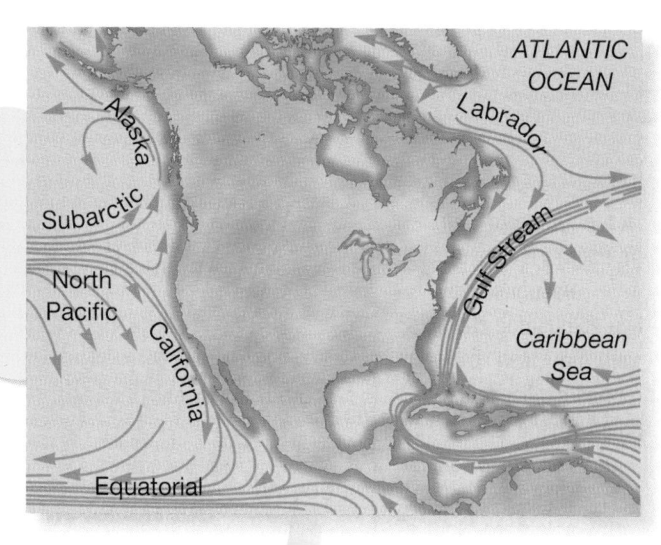

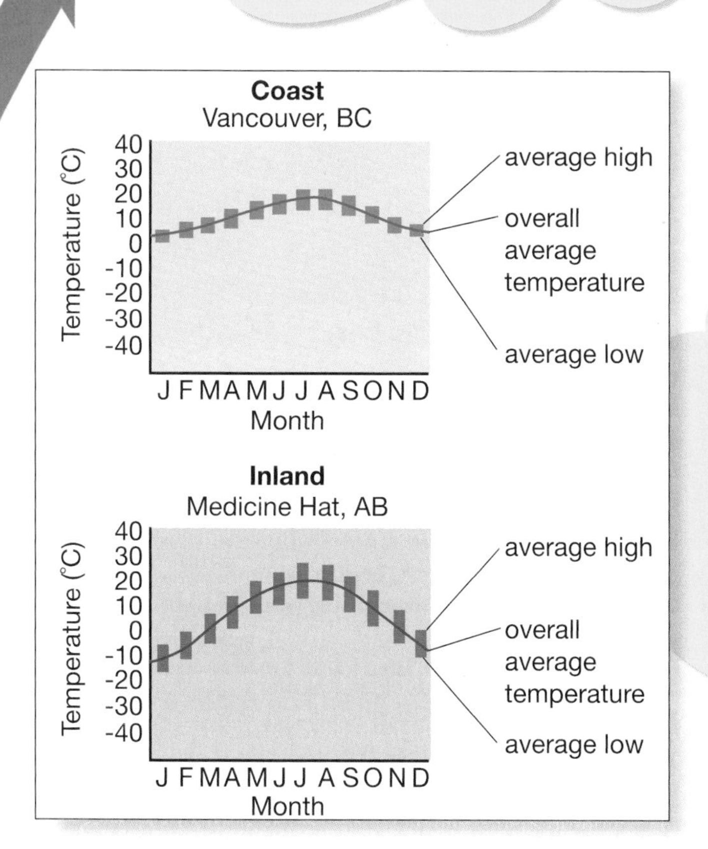

Liquid water in lakes and oceans has another important effect on climate. Because water has such a large specific heat capacity, its temperature changes relatively little as it absorbs or releases thermal energy. The temperature of the air near the water does not change very much either. As a result, coastal regions generally have more moderate climates than similar regions farther inland.

Analyze

1. Explain how thermal energy is transferred
   (a) from the Sun to Earth
   (b) from Earth's surface to the atmosphere
   (c) from one part of the atmosphere to another
   (d) from one part of the ocean to another
2. (a) Which has greater specific heat capacity, 1 kg of crushed rock or 1 kg of water?
   (b) Which will feel warmer after sitting in the Sun for an hour, the rock or the water?

Extension

3. A young child might think that the top of mountains must be warmer than the valleys below because mountain tops are closer to the Sun. How could you show the child that this idea is incorrect?
4. Describe three ways in which water affects Earth's weather systems. What property of water is important in transferring energy?
5. Find out about the climate of the planet Venus, and write a short paragraph comparing it with Earth's climate. How would Earth's weather change if a thick layer of clouds developed over the entire planet?

Did You Know?

Can you imagine plants and bacteria taking care of the world? According to British scientist James Lovelock, Earth would quickly become uninhabitable without the influence of living organisms. In 1968 Lovelock and an American colleague, Lynn Margulis, proposed that our planet's temperature, atmospheric gases, and other key factors remain stable because of the interaction of living and non-living systems. They called their idea the Gaia (GUY-uh) hypothesis, after the Greek goddess of Earth.

According to the Gaia hypothesis, we can picture Earth as a single living "super-organism." Like all living creatures, Gaia (Earth and all the life forms on it) has many complex feedback system. For example, for millions of years the level of oxygen in the atmosphere seems to have stayed almost constant. If it rose much higher, uncontrollable fires would disrupt or destroy everything on the planet. Gaia has many ways of preventing this disaster. For example, micro-organisms produce methane gas, which combines with oxygen in the air to form carbon dioxide. This helps to keep the amount of oxygen at a safe level and replaces the carbon dioxide which plants use up.

When Lovelock and Margulis first proposed their ideas, many scientists did not take them seriously. Does the Gaia hypothesis seem likely to you? What does it suggest about the way we should treat and care for Earth?

Check Your Understanding

1. Name the three basic parts of a home heating system. Using the heating equipment in your own home (or school) as an example, identify the specific devices used for each part of the system.
2. Explain how warm-blooded and cold-blooded creatures differ in terms of
   (a) body temperature
   (b) response to cold temperatures
3. Draw a concept map showing how insulation stops each of the three forms of heat loss. Include examples of both natural and manufactured insulation.
4. **Apply** Emergency survival blankets often have one shiny, foil-coated side and one dark, plastic side. If a survival blanket is wrapped around you, explain which side should face out
   (a) to keep you warm on a cold night
   (b) to keep you cool at the edge of a forest fire
5. **Thinking Critically** Building designers have to decide where to put heating devices (for example, radiators and hot air outlets) and thermostats. There are two possible locations: on outside walls and on inside walls. Think of buildings you know and the information you have studied in this unit. Then suggest
   (a) where each device usually seems to be located
   (b) why this location is appropriate
   (c) why hot air outlets in a basement ceiling do not keep the basement at a comfortable temperature
   (d) why thermostats are not placed near the floor or ceiling

CHAPTER at a glance

Now that you have completed this chapter, try to do the following. If you cannot, go back to the sections indicated.

Describe the method of energy transfer through each material below. (9.1)
(a) a pot of soup warming on the stove
(b) the space between you and a bonfire
(c) a hot dog being cooked over a campfire

Predict which surface would you expect to warm faster in bright sunlight, a shiny black car or dull black pavement. Explain why. (9.1)

Predict which object would cool faster when taken out of a hot oven, a shiny black ceramic casserole dish or a shiny white casserole dish. Explain why. (9.1)

For each use below, choose either a good thermal conductor or a good insulator. Give an example of a suitable material for each use. (9.1)
(a) the handle of a marshmallow toasting fork
(b) the shell of a picnic cooler
(c) a cake pan

A hair dryer is an energy transformation system. Identify each part of this system. (9.2)
(a) the input energy
(b) the output energy
(c) the converter or energy-transforming device
(d) the feedback device

Describe how a hair dryer illustrates each principle below. (9.2)
(a) the spreading of energy
(b) energy transformation
(c) energy conservation
(d) waste heat

State the feature of energy transfer that is illustrated in each weather example below. (9.3)
(a) In the spring, a dark plowed field is warmer than a snow-covered field nearby.
(b) "Heat waves" are sometimes visible above a dark road, but they are seldom seen above a light-coloured sidewalk.
(c) Winter temperatures in Vancouver, on Canada's west coast, are milder than in Halifax, on the east coast, even though Halifax is much farther south than Vancouver.

Prepare Your Own Summary

Summarize this chapter by doing one of the following. Use a graphic organizer (such as a concept map), produce a poster, or write the summary to include the key chapter ideas. Here are a few ideas to use as a guide:

- What are three methods of energy transfer? For each method, give a key feature that allows you to identify it, as well as the types of materials (solid, liquid, gas, vacuum) that it operates in.
- What are the three main parts of an energy transfer system? Illustrate them with a diagram.
- What are the four most important features of all energy transfer systems?
- What five factors are important when designing and constructing well-insulated buildings and clothing?
- Under what conditions is air a good insulator? Under what conditions is air a poor insulator?
- Answer the following questions for Earth's weather system: What is the source of energy? How is the surface warmed? How is the atmosphere warmed? How do convection currents in the atmosphere and oceans affect the weather? How do changes of state of water affect the weather and climate? How do lakes and oceans affect climate?

CHAPTER

9 Review

Key Terms

system
energy source
radiation
radiant energy
electromagnetic radiation (EMR)
infrared radiation (IR)
thermal conduction
heat insulators
fluids
convection
convection current
feedback
input
output

Reviewing Key Terms

If you need to review, the section numbers show you where these terms were introduced.

1. For each group of words, write a sentence that uses all the words correctly.
   (a) input, output, transform, system (9.2)
   (b) energy source, fluid, convection current (9.3)

2. From the list of key terms, choose a term with the opposite meaning of each term below.
   (a) input (9.2)
   (b) conductor (9.3)
   (c) solid (9.1)

Understanding Key Ideas

Section numbers are provided if you need to review.

3. A block of ice is placed on the pavement in a parking lot on a sunny day. An hour later, there is a puddle of liquid water — no solid ice is left. Two hours later, there is only a damp spot on the pavement. (9.2)
   (a) List all of the changes of state involved in this event.
   (b) Identify the source of energy for the changes.
   (c) Name the energy transformations that occur.
   (d) Explain how energy was distributed during the event described.

4. You pour 100 mL of hot water at 50°C into 200 mL of cold water at 10°C. (9.1)
   (a) In which direction is thermal energy transferred?
   (b) Which methods of energy transfer occur?
   (c) What happens to the temperature of the hot water and the temperature of the cold water?
   (d) If you leave the water mixture in a glass beaker for 24 h in a room with a temperature of 20°C, what will happen? What general feature of energy transfer does this illustrate?

5. A new "super insulation" is made from plastic foam with a shiny, metal foil covering (see diagram below). All the air is pumped out of the holes in the plastic foam. The foil prevents air from entering the foam again. Experts estimate that widespread use of this new material could save as much as one billion dollars' worth of fuel each year. (9.3)
   (a) Explain how the super insulation prevents heat transfer by
   - convection
   - conduction
   - radiation

   (b) Suggest one problem that might occur with the new insulation.

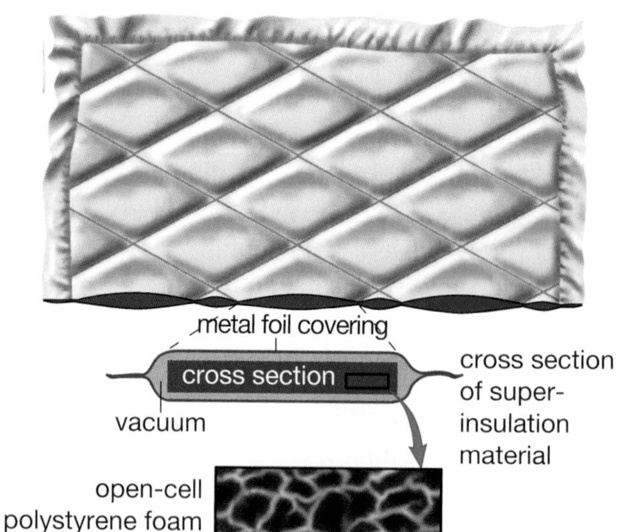

Developing Skills

6. Recognizing systems is an important step in understanding the world around you. In this chapter, you analyzed heating systems in your body and in buildings. Choose a different system in your body, and explain
   (a) how it fits the definition of a system
   (b) what its main parts are and what each part does
   (c) how energy is involved in the system's operation

7. To *analyze* a system, you break it down into smaller, simpler parts and try to understand how each part operates and affects the others. Sometimes comparing a complicated system (such as Earth's weather system) with a simpler one (such as a home heating system) is helpful. Make a table like the one below to analyze these systems. Give your table a title.

	Home heating	Earth's weather
Energy source		
Method of increasing thermal energy		
Method of distributing thermal energy		
Control system		

8. Draw a diagram that traces the path of energy through one of these systems: a bicycle, the hot water system that supplies your shower, a pet. Start with the source of almost all energy on Earth — the Sun. Remember to include waste heat. Label different energy transformations and transfers using the vocabulary you have learned in this chapter.

Problem Solving/Applying

9. Identify the signals that your body's temperature control system would probably receive and send out while you spent an afternoon
   (a) playing beach volleyball
   (b) ice fishing

10. A solar heating system for a home is designed with dark-coloured roof panels containing liquid-filled pipes to absorb solar energy. The warm liquid transfers energy to the interior of the house.
    (a) What liquid would be best for the system: water, alcohol, or a mixture of water and antifreeze? Explain your choice.
    (b) Extra thermal energy can be used to warm an underground water tank, an underground bin of gravel, or a big brick wall in the middle of the house. Which energy storage method do you think would be best? Why?

Critical Thinking

11. **(a)** How does bowling model energy transfer by conduction?
    (b) How does swimming model energy transfer by radiation?
    (c) How does dancing model energy transfer by convection?

12. Birch bark and paper both burn well. Yet water in a paper bag or a birch bark container can be warmed and even boiled over an open flame without catching fire. How is this possible?

Pause& Reflect

Go back to the beginning of this chapter on page 240, and check your original answers to the Getting Ready questions. How has your thinking changed? How would you answer those questions now that you have investigated the topics in this chapter?

UNIT 3 Ask an Expert

Planning a camping trip? Talk to someone like Mario Patry. Mario, who works at a camping equipment store in Ottawa, Ontario, knows a lot about outdoor equipment. He can help you figure out which of the store's 30 different sleeping bags you will need to keep you warm.

Q How do you help a customer choose a sleeping bag?

A The first question I ask is where and when they intend to use the bag. If you are backpacking in southern Canada in July, you will want a very different bag than if you are backpacking in Yellowknife in April.

Q Can a sleeping bag really keep you warm in Yellowknife in April?

A The right kind of sleeping bag can. Not all of the bags we sell would be warm enough in that situation, especially those designed for summer camping. Our warmest bag, though, has a temperature rating of –40°C. It can keep you warm on very cold nights, if you use it in a tent for protection from the wind and on top of a sleeping pad for insulation from the cold ground.

Q What makes these bags so much warmer than the summer ones?

A Many factors affect how well a sleeping bag can keep the sleeper warm. Probably the biggest factor is what kind of insulation it uses. By that I mean what the bag is filled with. There are two main types of sleeping bags: those that are filled with down (the fluffy layer under the feathers of water birds) and those that have synthetic fills (fibres made by machine).

Q Which type of fill keeps you warmer?

A If you compare a down sleeping bag with a synthetic sleeping bag of the same thickness, the down bag is warmer.

Q Why is that?

A Down puffs up very high — we call this effect "loft" — because of the many, many tiny air pockets in-between the bits of down. These air pockets are excellent insulators. They warm up with heat from your body and hold on to this warmth instead of letting it seep out of the sleeping bag. If you unroll your sleeping bag when you first set up camp, it has a chance to puff up with as many air pockets as possible before you sleep in it.

Q What about synthetic bags?

A Synthetic fill doesn't have as much puffiness, or loft, as down because synthetics have fewer air pockets. Fewer air pockets mean less trapped heat.

Q Why would a customer choose a synthetic bag, then?

A For any of several good reasons. Down is very expensive, sometimes twice the price of a comparable synthetic bag. Also, synthetic is a better choice if there's a chance your sleeping bag will get wet. A down bag takes much longer to dry out than a synthetic bag. While it's wet, the feathers are stuck together. That means fewer air pockets and not as much warmth. And, of course, people who are allergic to down need to buy synthetic bags.

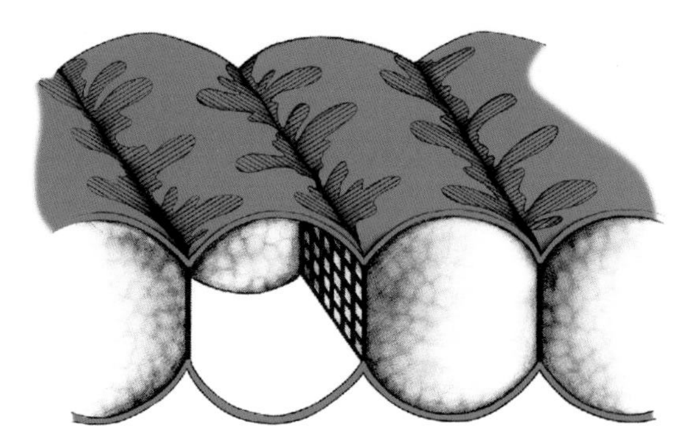

Cross section of synthetic sleeping bag insulation

Q **How did you learn so much about sleeping bags?**

A A lot of what I know about equipment I've learned from reading backpacking books and magazines and from trying out the equipment myself. I've been cycling, skiing, and winter camping for years. When I began working here, I completed two weeks of training to learn about the specific products the store offers.

EXPLORING Further

Who Is Warmer?

Imagine this situation. Sydney and Yasmine arrive with their school group at a camp site cold and damp from a day of canoeing. Luckily their equipment is dry. Yasmine unpacks her synthetic sleeping bag, pad, and one-person tent, and quickly sets up camp. Sydney ignores her down sleeping bag and gets the stew cooking. As the Sun sets, they feel the temperature dropping. After supper is cleared away, Sydney unrolls her sleeping bag on the ground. Then both campers climb into their sleeping bags and go to sleep.

If Yasmine's and Sydney's sleeping bags are both rated to −10°C, who do you think will be warmer during the night? Why?

An *Issue* to Analyze

A DEBATE

Why Are Prairie City's Summers Getting Warmer?

Think About It

There is no question about it: summers in Prairie City are warmer than they were 40 years ago. Weather records show that average summer temperatures in the last decade have been 2°C to 3°C higher than in the 1950s. Some people assume that this is because of global warming. Those who go camping or have cottages know that summer temperatures away from the city have not increased, so the higher temperatures are not "global." There must be another reason for the change.

How has Prairie City changed in the past 40 years? First, the city is much larger, with many more buildings and roads. Materials such as pavement and brick absorb heat more readily than natural ground cover, and they release this heat throughout the day and night.

A second major change is the number of air conditioners. In the 1950s, only the largest stores and newest office buildings were air conditioned. Today nearly every public building and private home has an air-conditioning system. These air conditioners run 24 h a day all summer, pumping warm air into the surrounding neighbourhood. The result is higher outdoor temperatures, which cause people to run their air conditioners even more. This has become a "vicious cycle," and an influential group of citizens thinks the time has come to do something about it.

Resolution

Be it resolved that all air conditioners in Prairie City should be registered with the city government and licensed at a fee of $2000 a year.

What to Do

1. Read the points in favour of and against the resolution on the opposite page. Begin to think about other points that could be made.
2. Four students will debate the topic. Two students will speak in support of the resolution and two will speak against it. Note: No matter what side you actually believe in, you must try your best to convince the jury, or debate audience, of the point your side is defending.
3. To aid the two teams, two other students will work with each team to gather the background information needed to put forward a strong case for the point that side is defending.
4. The rest of the class will act as the jury in hearing the debate. In preparation for the debate, they should do their own research in order to understand the science and technology behind the issues raised.
5. Your teacher will provide you with *Debating Procedures* to follow.

In Favour

- This action will discourage use of air conditioners. The average summer temperatures will be brought down to 1950s levels. The streets will be cooler.
- Registering and licensing air conditioners will reduce the summertime use of electricity. This will be better for the environment, whether the electricity is produced by a nuclear power plant or a hydroelectric generator.
- The registration fee will raise money so that city taxes can be lowered.
- People can be educated in natural ways to cool off through the use of outdoor blinds and appropriate choices of exterior colours. (For example, brick absorbs far less heat if it is painted white.)

Against

- Poor people, especially the elderly and chronically ill, will suffer most. Every summer elderly people without air conditioning die of heat related disorders. Only those with enough money will be able to stay healthy.
- This action will create a huge environmental problem as air conditioners which cannot be used are removed and dumped where they are likely to release potentially harmful refrigerants into the air.
- People who rent their accommodation will have no control over their landlord's decision. Either their rents will rise or their comfort level will fall.
- The city should not be regulating people's lives in this way.

Analyze

1. Which side won the debate?
2. (a) What accounts for the winning side's success: logical arguments or emotional appeal?
   (b) What should determine who wins a debate? Give reasons for your answer.
3. Which neighbourhoods are likely to be warmer at night without air conditioners: a zone of high rise apartment buildings or a zone of single family dwellings on separate lots?

Extensions

4. Should scientific research be the most important factor in making public decisions?
5. List any problems with the licensing scheme that you discussed during your debate. Try to think of an alternative to the scheme that might solve some of these problems.
6. Do some research, at a library or on the Internet, into cities which have recorded summer temperature increase; for example, Phoenix, Arizona, or Perth, Australia.

An air conditioner removes thermal energy from a room or a house. Where does the thermal energy go?

UNIT 3 PROJECT

Cooking with Sunlight

Not all cookers and ovens need a supply of gas or electricity. In many parts of the world, gas and electrical power are not easily available, so a cooker that runs on solar power can be very useful. Heating water in a solar cooker to 65°C or more for half an hour will kill disease-causing micro-organisms. In bright sunlight, some solar cookers can reach temperatures of 90 to 135°C. This is hot enough to prepare foods like stews of rice or beans, which need to simmer for several hours in hot water.

Figure A

Figure B

Panel cookers (see Figure A) use flat, shiny surfaces to bounce sunlight onto a dark cooking pot. To trap solar energy, the cooking pot is placed in a plastic oven bag or under a glass bowl. Panel cookers are cheap, easy to make, and easy to use. They heat food gently, and you do not need to stir the food or turn the cooker frequently. Parabolic cookers (Figure B) use curved reflectors to focus sunlight on a small spot. A dark-coloured cooking pot placed at this spot absorbs radiant energy and warms up. The curved parts are complicated to make and hard to join together. The cooker has to point directly at the Sun, so it must be turned every few minutes as the Sun's position changes. The cooking pot can get hot enough to burn food, so the food has to be stirred frequently. Parabolic cookers can heat food more strongly, however.

Challenge

Apply what you have learned in this unit to build and test a practical solar cooker.

Apparatus

lab thermometer

glass bowl or similar container

Materials

a variety of materials with different reflective or insulating properties

structural materials, such as card, plywood panels, and construction tape

Design Criteria

A. Your cooker must be able to heat a sample of water to at least 65°C.

B. Your cooker should be easy to assemble, and you should include clear instructions and a good sketch with your design proposal.

Plan and Construct

1. Use the photographs and information above, as well as library or Internet research, to find out how solar cookers can be built and used. Keep a written record of what you find out. Be sure to keep track of your sources in a format approved

by your teacher, and plan which topics each group member will be responsible for.

INTERNET CONNECT

www.school.mcgrawhill.ca/resources/

Find out more about solar cooking by going to the web site above. Go to **Science Resources**, then to **SCIENCEPOWER 7** to know where to go next. Use the information you find in this project.

2. Based on your research, decide on the best kinds of materials for your solar cooker. What could you use for the outer box? How could you make the material more efficient? What would be the best material for the window? What could you use for insulation? Your teacher may be able to supply you with suggestions, as well.

3. Plan your design. Use your chapter summaries for this unit to find features of energy transfer devices that apply to solar cookers. Write a list of specific questions that you need to answer as you design your cooker. For example:
   - What type of cooker will you build?
   - What type of material would make the best cooking pot? Should it have high or low specific heat capacity?
   - Where in your cooker is radiant energy being transferred by reflection, convection, and conduction? What colours and surfaces will make each process most efficient?
   - Could insulation improve your design? Where should it be placed, and what should it be made of?
   - How large should your cooker be? (For example, a pizza-box-sized cooker will get almost as hot as a larger model, but it will not hold very much food.)

   With your group, answer these questions and any others you think of. Record your decisions.

4. Prepare a proposal with a sketch of your cooker design and a list of materials you used (or label your sketch).

5. Assemble the materials you will need, and build your cooker. Keep a record of the construction problems you overcome.

6. Now test your cooker by heating the water sample.

Evaluate

1. Use your background information, your proposal, and your written records to prepare a summary of your project. Use any format approved by your teacher: for example, a written or oral report, a display, or a videotaped or computer presentation. One part of your summary must be a diagram that shows how your cooker functions as an energy transfer system. A second part must be a sketch of your cooker showing all the features you built into its design. Finally, find an answer to at least one question that you wondered about during your project.

2. Are there ways you would improve your solar cooker if you could design and build again? If so, what are some improvements you would make?

3. Imagine that you have a chance to work with low-income people in a desert region to build practical solar cookers for everyday use. What have you learned from your solar cooker project that would be useful in this new project?

Extension

Design Your Own Think of a research question to investigate using your completed cooker. Design an experiment to find answers to your question. Have your teacher approve your experimental procedure.

UNIT 4

Earth's Crust

What forces unleash the monstrous power of volcanoes and earthquakes? What processes form Earth's mountains, its rocks and boulders, and the minerals and fuels deep within it, which we mine and use? What processes form the fertile soils that produce our food?

The answers to these questions lie in Earth's crust — the thin, ever-changing, outermost layer of our planet. Throughout our history, scientists have been trying to understand the forces that shape and change Earth's crust. In this unit, you will see how theories about Earth's crust were developed and then discarded, as scientists made new observations and saw new connections between their past observations. You will see how recent technology, which allows scientists to see the sea floor and to "X-ray" Earth's interior, has led to the current theory of a shifting crust. You will also have a chance to think about how our demands for food, minerals, and fuel harm Earth and to consider how we might take better care of Earth's resources.

Unit Contents

CHAPTER

10 Minerals, Rocks, and

Getting Ready...

- What makes a rock a rock?
- Why are some gemstones valuable while others are worthless?
- Why do plants grow in some places and not in others?

Use what you already know to answer the questions above in your Science Log. Look for more information to help you add to your answers as you read the chapter.

For tips on how to make and use a Science Log, turn to page 476.

Where Earth meets sea and sky, rocky coasts form one of the most majestic landscapes in the world, Peggy's Cove in Nova Scotia. Here waves crash endlessly against the boulders and spray reaches up to touch the lighthouse. Nearby, lush fields unfold in greens and golds.

Is there any connection between the barren rocks and the fertile farmland? If there is a process that changes the rocks to farmland, does the process end there, or does it continue? What valuable materials are buried in the ground? How do they get there, and why are they important to us?

Canada is rich in beauty and scenery, and in natural underground resources. It is a leading exporter of forestry, mining, and agricultural products, which depend on the rocks, minerals, and soils of Earth's crust. The gifts of Earth give Canadians strong economic support, but they will not last forever. It is important that we, as Canadians, manage these gifts wisely so that they will be available for us and for future generations to use and enjoy.

In this chapter, you will learn about the minerals, rocks, and soils on or near Earth's surface and find out about some of the processes that form them.

Soils

Starting Point ACTIVITY

Hard as a Rock

Most people know the saying, "as hard as a rock." What, besides hardness, makes a rock a rock? What materials make up rocks? How can you tell similar rocks apart?

What You Need

several unknown items

What to Do

1. With a partner, examine the items provided by your teacher. Decide how you might classify the items.
2. What does an item's hardness tell you? Should all of the "hard" items be classified as rocks? Are all rocks hard? What other observations can you make about each item? Why might you not classify all of the items as rocks? Explain your reasons for classifying each item as you did. Record your ideas.
3. Discuss with your classmates what makes a rock a rock. Record your ideas.
4. Start listing some uses you can think of for rocks. Think about why it might be important to know something about the properties of a particular rock.

Spotlight On Key Ideas

In this chapter, you will discover

- how minerals, rocks, and soils form and continue to change in a cycle
- how soil types and crops vary across Canada
- how human activity affects our environment positively and negatively

Spotlight On Key Skills

In this chapter, you will

- conduct tests on unidentified objects to identify minerals
- design your own experiment to find out about processes that shape Earth's crust
- work co-operatively to solve problems of resource and land management

10.1 Minerals

You have just been examining a number of items to decide which ones could be classified as rocks. What is rock? **Rock** is made up of one or more pure, naturally occurring, non-living solid materials called **minerals**.

Figure 10.1 Granite is a rock that is made up of an assortment of minerals. It is often polished and used in buildings and at the base of statues. Granite contains the minerals feldspar (sparkling grains), quartz (a glassy crystal), mica (greenish-grey flakes), and hornblende (dark flecks).

DidYouKnow?

Although glass is made up of the same elements as quartz, it is not a solid, so it is not a mineral. In fact, glass is like a very thick liquid that flows extremely slowly. Window panes in very old buildings are thicker at the bottom, because gravity has pulled the liquid downward. It takes hundreds of years for glass to flow only a few millimetres.

Most minerals are quite rare. Only a few, such as quartz and mica, are common, found throughout Earth's **crust** (the thin outermost layer of Earth). A mineral can be an **element** (a pure substance) or a compound (two or more substances). Quartz, for example, consists of the elements silicon and oxygen. No other mineral has these elements in the same arrangement and proportion. Sulfur, copper, gold, and diamond, on the other hand, are made up of a single element.

If you were a prospector digging for gold, how would you know if you had found it? "No problem," you say? We all know what gold looks like — or do we? Another mineral, pyrite, which is more common than gold, is almost identical to gold in appearance. The value of gold is high, while pyrite is almost worthless. How could you find out if that shiny, yellow metal you found is really gold? There are important clues in the properties of minerals.

INTERNET CONNECT

www.school.mcgrawhill.ca/resources/

Find out more about glass by going to the web site above. Go to the **Science Resources**, then to **SCIENCEPOWER 7** to know where to go next. Write a brief report based on your findings.

The Mohs Hardness Scale

You are probably aware that you can scratch a piece of chalk with your thumbnail, but you cannot scratch most other rock samples in the same way.

How can a substance's "scratchability" be used for mineral identification? This is a question that a German scientist, Friedrich Mohs, asked himself in 1812. He developed a scale of ten minerals with a "hardness" value of 1 to 10 (see Table 10.1).

How do you use the scale? Suppose that you have an unknown mineral that looks like talc or corundum. Scratch it with your fingernail. If it is talc, it will scratch easily, because your fingernail has a hardness value of 2.5 on the scale — much harder than talc, which has a hardness value of 1. If it does not scratch easily, it cannot be talc and must be corundum instead. The hardness of corundum is more difficult to test because corundum is so hard — it is harder than most other objects and minerals. What could you use to test the hardness of corundum?

Diamond is the hardest mineral. One of its uses is shown in Figure 10.2.

Table 10.1

The Mohs Hardness Scale		
Mineral	**Mineral hardness**	**Hardness of common objects**
talc	1 softest	soft pencil point (1.5)
gypsum	2	fingernail (2.5)
calcite	3	piece of copper (3.5)
fluorite	4	iron nail (4.5)
apatite	5	glass (5.5)
feldspar	6	steel file (6.5)
quartz	7	streak plate (7)
topaz	8	sandpaper (7.5)
corundum	9	emery paper (9.0)
diamond	10 hardest	

The Mohs Hardness Scale is a useful tool for mineral identification. Given that there are over 3000 minerals, however, other properties are also needed to identify them. Minerals sometimes occur as crystals. A **crystal** occurs naturally and has straight edges, flat sides, and regular angles. All of the minerals in Earth's crust can be grouped according to the six different crystal shapes shown in Table 10.2.

Table 10.2 The Six Major Crystal Systems

Examples	Systems
halite	cubic
wulfenite	tetragonal
corundum	hexagonal
topaz	orthorhombic
gypsum	monoclinic
albite	triclinic

Figure 10.2 Although the glass-cutter wheel looks like ordinary metal, its edge is actually embedded with a hard mineral, such as diamond.

Other Clues to Mineral Identification

Lustre

You have just seen that a mineral's crystal structure provides an important clue to its identity.

Some minerals, such as gold and other metals, appear shiny — another clue to their identity. Others, such as talc, can appear dull. Most dull minerals are non-metals. The "shininess," or **lustre**, of a mineral depends on how light is reflected from its surface. The surface of a mineral can reflect light in many different ways.

INTERNET CONNECT

www.school.mcgrawhill.ca/resources/

Find out which minerals are used in all Canadian coins by going to the web site above. Go to the **Science Resources**, then to **SCIENCEPOWER 7** to know where to go next. Make a pie chart for each coin to show the percentages of different minerals it contains.

Colour

Next to lustre, colour is one of the most attractive properties of minerals. The colour of a mineral can also be a clue to its identity. As in the case of gold and pyrite, however, colour alone cannot identify a mineral. In addition, not all minerals are the same colour all the time. For example, the mineral corundum (made of aluminum and oxygen) is white when pure. However, when it contains iron and/or titanium, it is blue (and is called a sapphire). When it contains chromium, it is red (and is called a ruby). See Figure 10.3.

Skill POWER

To learn how to make pie charts, turn to page 486.

DidYouKnow?

Graphite is a mineral that is used in pencils. Pencil marks are merely graphite streaks that are soft enough to be left on a piece of paper.

Figure 10.3 Oriental emerald (green), amethyst (purple), and topaz (yellow) are other coloured varieties of corundum due to impurities.

Streak

When a mineral is rubbed across a piece of unglazed porcelain tile, as in Figure 10.4, it leaves a streak. A **streak** is the colour of the powdered form of the mineral. Look-alikes, such as gold and pyrite, can be distinguished using a streak test. Gold leaves a yellow streak, while pyrite has a greenish-black or brown-black streak.

Minerals with a greater hardness than the porcelain tile (hardness value of 7) will not leave a streak. Such minerals, especially the black ones, can be crushed into powder. A surprising number of black minerals have lighter-coloured powders.

Figure 10.4 Colour is not always useful for mineral identification. Hematite, for example, can be dark red, grey, or silvery in colour. Its streak, however, is always dark red-brown.

Cleavage and Fracture

Hardness, lustre, colour, and streak are mineral properties that help geologists to identify many minerals fairly easily. Other properties provide further clues to their identity. Cleavage and fracture are two very useful ones.

The way a mineral breaks apart can be a clue to its identity. If it breaks along smooth, flat surfaces, or planes, it is said to have **cleavage**. Mica is an example of a mineral with cleavage (see Figure 10.5A). Separating the layers of mica is like separating the pages in a book.

Not all minerals have cleavage. Minerals that break with rough or jagged edges have **fracture** (see Figure 10.5B). Obsidian is an example of a mineral with fracture. While separating the layers of mica is like separating the pages in a book, breaking apart quartz is like ripping the book in half.

Figure 10.5A Mica is a mineral with a single cleavage direction that allows it to be pulled apart into sheets.

Figure 10.5B Obsidian is a mineral that fractures when broken apart.

CONDUCT AN

INVESTIGATION 10-A

A Geologist's Mystery

You are a geologist. You have just received a parcel from your company's field team in northern Alberta. The attached note reads, "New cave discovered. Enclosed are samples of minerals found there. Please identify."

How can the Mohs Hardness Scale and the classification of crystal structures help you to solve the mystery?

Problem

How can you identify different minerals? Which mineral properties would you examine, and in what order?

Safety Precautions

Be careful when handling materials with sharp points or edges.

Apparatus

numbered mineral samples
hand lens
iron nail
copper penny or piece of copper
utility knife
steel file
sandpaper
emery paper
streak plate
Tables 10.1 and 10.2 (page 281)
mineral guidebook

Procedure

1. Make a table like the one below.

2. Record the number of the first mineral sample in the first column of your table.

(a) Record the mineral's colour. You may use the hand lens to take a closer look. If you see any distinguishing crystal shapes, as in Table 10.2, record your observations under "Crystal Shape."

Characteristics of Some Common Minerals							
Mineral number	Colour	Crystal shape	Lustre	Streak	Hardness	Other	Mineral name

(b) Describe the lustre of the sample.

(c) Look at the sample, and make a hypothesis about what colour its streak might be. Then scrape the mineral once across the streak plate, and brush off the excess powder. Record the colour of its streak. If the mineral is too hard to leave a streak, leave this space blank.

(d) Look at the sample, and make a hypothesis about how hard it might be. Scratch the sample with your fingernail. If your fingernail does not leave a scratch, continue with the penny, and on up the scale, until something leaves a scratch. (If emery paper does not leave a scratch, assume the hardness is 10.)

(e) Record any other distinguishing features you observe under "Other."

3. Repeat all of Step 2 for the remaining mineral samples.

4. Identify the samples by comparing their properties to those in Table 10.2. Also use the Mohs Hardness Scale and other information from this section. Fill in any empty spaces on the chart with information from either of these sources.

5. Wash your hands thoroughly after completing this investigation, and clean your work surfaces.

Analyze

1. Before testing, which minerals looked the same?
2. (a) Which mineral was the softest? Which was the hardest?
   (b) Your hypotheses were based on appearance only. How correct were you?
3. (a) Which minerals were the same colour as their streak or powder?
   (b) Which streaks surprised you?
4. Were you able to use any other features or properties to help you identify some samples? Which features or properties?

Conclude and Apply

5. Were you able to identify all of the mineral samples? If not, can you suggest some other tests for further investigation?
6. (a) Which property was the most useful for identifying a mineral? Why?
   (b) Which property or properties were not very useful for identifying a mineral? Why?
7. How much does hardness seem to affect the similarity of a mineral's colour to the colour of its streak?
8. **Design Your Own** Write a procedure outlining the order for testing mineral properties in further investigations.

Extend Your Skills

9. Obtain a Mohs scale set of minerals. Use these minerals to test the hardness of your test samples more accurately.

Computer CONNECT

If you have access to a computer, set up a spread sheet to organize the data in Investigation 10-A.

Uses of Minerals

DidYouKnow?

Titanium is a silvery grey metal that resembles polished steel. It is the ninth most abundant element on Earth (0.63 percent of Earth's crust), but it is far more abundant on the Moon (almost 12 percent!). Although it was first discovered in 1791 by William Gregor, titanium was not obtained in its pure metal form until 1910. It could not be used on a wide scale until 1946, when scientists discovered an economical (affordable) purification process.

Why are minerals so important? As you have seen so far, minerals have certain properties that make them valuable and useful.

For example, gems are highly prized minerals because they are rare and beautiful. Many gems, which are brighter and more colourful than common samples of the same mineral, are cut and polished for use in jewellery (see Figure 10.6).

Diamond, a very valuable gem, also has practical uses. Because diamonds are the hardest of all minerals (no naturally occurring minerals can scratch a diamond except another diamond), they are used on drill bits and other instruments to cut through hard substances, such as steel and rock. Scientists are able to apply tiny rows of diamonds on the edges of surgical scalpels, razor blades, dental drills, diamond-tipped phonograph needles, and diamond-coated computer parts.

Metals are among the most useful minerals (see Figure 10.7). We use them in cars, appliances, and many household utensils. Metals are single elements, such as gold, silver, copper, aluminum, nickel, and iron. Metal deposits occur naturally in rocks, where they are mixed with other mineral deposits. Rocks that are rich in metals and metal oxides are called **metallic ores**.

Figure 10.6 Polished and unpolished amethyst

Figure 10.7 Titanium is as strong as steel and 45 percent lighter, making it especially suitable for use in aircraft and spacecraft.

DidYouKnow?

The difference between a gem and the common form of the same mineral can be slight. Amethyst is a gem form of quartz. It contains traces of manganese in its structure. Iron is what gives amethyst its desirable purple colour.

More Uses of Minerals

Iron, zinc, copper — you might think of these in terms of appliances and utensils, but what do they have to do with your body? They are important for your health. In order to survive, your body needs over 20 different kinds of elements found in minerals. For example, iron, from such minerals as magnetite and pyrite, helps blood carry oxygen. Calcium, found in calcite and dolomite, helps to make bones and teeth strong. Sodium, found in halite, helps to regulate water in the body's cells. Some of the vegetables and grains that we eat are rich in minerals.

Figure 10.8 This soybean crop is rich in calcium. Calcium in the soil dissolves in water and passes into the roots of the plants.

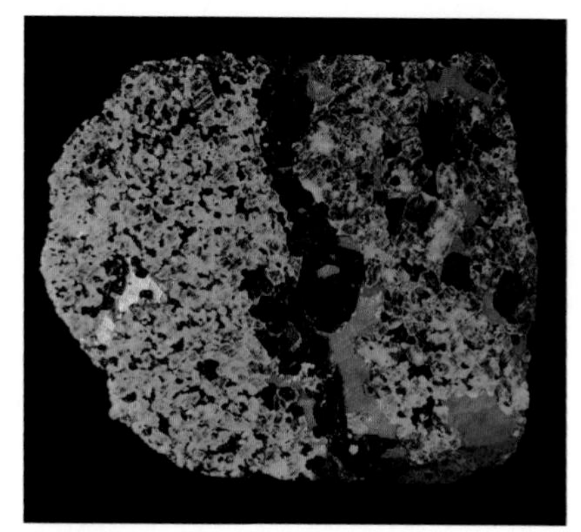

Figure 10.9 Some minerals can fluoresce (glow) under ultra-violet light or X-ray beams, and some can continue to glow (phosphoresce) for a short while after the light is removed.

Pause& Reflect

Imagine that you are on a camping trip and you find a cave that looks as though it has recently opened up because it has only a crevice for an opening. Small enough to squeeze inside, you are immediately bedazzled by a wall of shimmering gemstones. In your Science Log, make up a myth to explain something about this special wall. Include a description of the gems in the cave, using the terms "lustre," "hardness," "colour," and "streak."

Like you, crops need minerals to survive and stay healthy. Farmers must find ways to return minerals that crops have used up to the soil. A common method is to apply fertilizers. Organic fertilizers are made from plant and animal remains and waste. Inorganic fertilizers are made from essential elements and minerals.

Another method of preserving or returning minerals to the soil is crop rotation. Farmers plant a different crop in each field each year. This ensures that the minerals used by one crop are replaced the following year by a different crop. For example, soybeans will replace nitrogen that a previous corn crop has used up (see Figure 10.8).

Minerals have many other practical uses. When quartz is stimulated with an electric charge from a battery, it vibrates more than 30 000 times each second. Because the vibrations are so regular, they keep time very accurately. Today tiny pieces of quartz are used in most watches and timepieces.

Minerals that fluoresce have been used in the hands and numbers on clock and watch faces (see Figure 10.9).

Find Out **ACTIVITY**

What a Gem!

You probably know your own birthstone. Here is an opportunity to find out about others.

What to Do

Most birthstones are minerals. Work with a group of students who have the same birthstone as you. Research your birthstone, and use what you learn to make a greeting card for another person with the same birthstone.

Find Out ACTIVITY

Dig for Treasure

Where in Canada are minerals found?

What You Need

mineral map of Canada
research materials

What to Do

1. Your teacher will give you a map that shows where minerals are found all across Canada.
2. In a group, research and list the minerals found in Canada.
3. Choose one mineral, and research its characteristics (as you did in Investigation 10-A): where it is found, how it is mined, and what it is used for.

What Did You Find Out?

How might you explain to an industrial representative why Canada is a good place to develop an industry that needs a supply of this mineral?

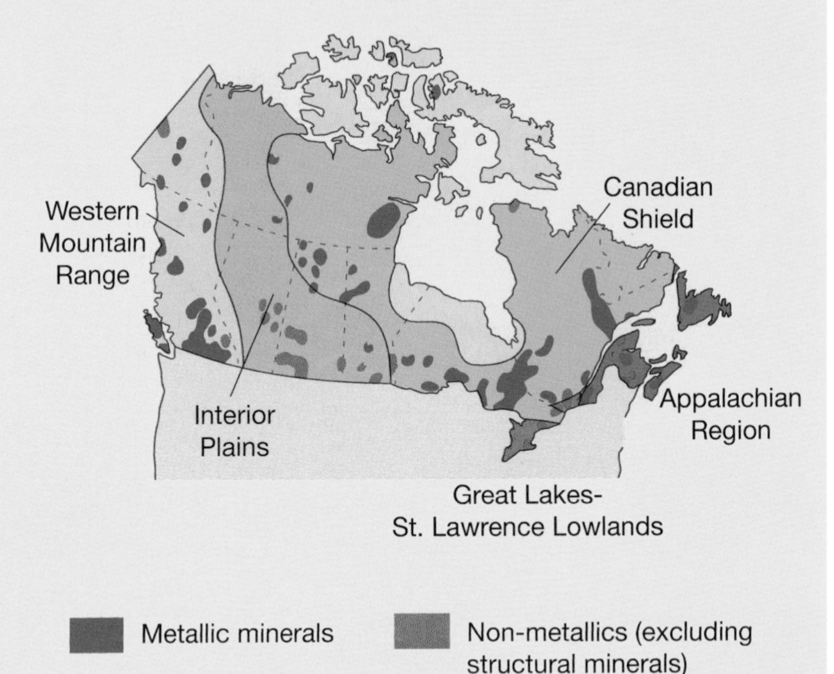

DidYouKnow?

Many of the world's major ore fields have been discovered by chance. The great nickel deposits of Sudbury, Ontario, were found accidentally in the 1880s by workers on the Canadian Pacific Railway. Before the development of modern exploration techniques, minerals were located by prospecting, as in "gold-panning," or searching indirectly for tell-tale signs, such as the "iron hats" of rusty, spongy rock that sometimes lie on top of sulfide ore deposits.

Check Your Understanding

1. Define the following: rock, mineral, element.
2. List the properties that are used to identify minerals.
3. Name three minerals that have practical uses. Describe how they are used.
4. **Apply** Suppose that you find a white, non-metallic mineral that is harder than calcite. You identify the sample as quartz. What are your observations? What is your inference?
5. **Thinking Critically** Many gemstones are polished so much that you can no longer detect a crystal shape. What could you do to a gemstone to determine its crystal shape?

10.2 Rocks and the Rock Cycle

Rock Families

As you have seen, rocks are made of minerals. In this section, you will examine how rocks and minerals form.

Scientists have grouped rocks into three major families, or types, based on how they form. The three families are igneous, sedimentary, and metamorphic rocks. Each can usually be identified by its appearance.

Figure 10.10 The "Balancing Rock" on Long Island, near Tiverton in Nova Scotia, is a spectacular basaltic sea stack. How might it have formed?

Word CONNECT

The word "igneous" comes from the Latin word, *ignis*, meaning "fire." Write what you think the word "ignite" means. Then check a dictionary to see how close you were.

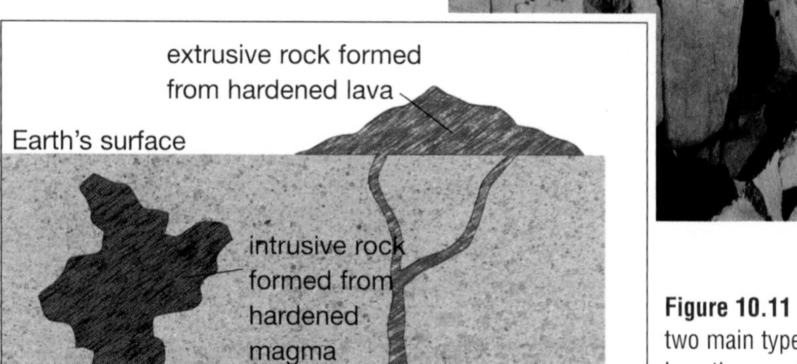

Figure 10.11 This diagram shows two main types of igneous rock and how they are formed.

Igneous Rock

Igneous rock forms when hot magma and when hot lava cool and solidify. **Magma** is melted rock found below Earth's crust, where temperatures and pressures are high. Any rock that is heated at great depths can melt into magma. Under high pressure, the magma can push away or dissolve the surrounding rock, making room for itself. Sometimes fingers of hot magma push up to the surface through cracks in Earth's crust.

Geologists classify igneous rock based on whether it was formed above or below Earth's surface. Magma can cool and harden below the surface. The resulting rock is called **intrusive rock**. Granite is an example of an igneous rock that formed very deep and very slowly in Earth's crust.

When magma breaks through Earth's surface, in the form of a volcanic eruption, it is called **lava**. Rock that forms when lava cools on Earth's surface is called **extrusive rock** (see Figure 10.11).

Magma can contain solidified crystals that can be seen in solidified rock. The appearance of the crystals in igneous rock samples can differ depending on how fast the rocks cooled. Write a hypothesis about speed of cooling and size of crystals, and test it in the next investigation.

Pause& Reflect

In your Science Log, hypothesize about how the "Balancing Rock" in Figure 10.10 was formed. Basalt is an example of igneous rock formed on Earth's surface. Does your hypothesis agree with the origins of this type of rock? Propose a new explanation of how this rock was formed.

Cool Crystals, Hot Gems!

Can you turn small crystals into larger, dazzling gemstones? You might, if you could re-create the formation of igneous rocks and, at the same time, control the conditions for crystal growth.

You can use a liquid solution to represent the important conditions and relationships involved in igneous rock formation. Crystal size and rate of cooling are the same as they are in melted rock.

Problem

How does the rate of cooling affect crystal size?

Safety Precautions

- Heat is used in this activity. Handle heated items with great care.
- Epsom salts are not to be eaten. Wash hands thoroughly after handling them.
- Be careful pouring hot liquids.

Apparatus

measuring cup or graduated cylinder
measuring spoons
small pot
heat source
stirring spoon
two small beakers
two bowls
hot water
labels
hand lens (optional)

Materials

Epsom salts
tap water
ice (crushed or broken)

Skill POWER

To review safety symbols, turn to page 492.

Procedure

1. Make sure that all containers and spoons are very clean.
2. Measure 100 mL of tap water using the measuring cup or graduated cylinder, and put the water in the pot.
3. Add 90 mL (6 tablespoons) of Epsom salts to the water in the pot. Stir with the spoon over low heat. Do NOT boil.

Skill POWER

To review working in groups, turn to page 478.

4. When most of the Epsom salts are dissolved (not all crystals may dissolve), remove the pot from the heat. Carefully pour off equal amounts of the solution into each beaker (about 50 mL, or $\frac{1}{8}$ cup, in each). Make sure that none of the undissolved crystals get into the beakers.
   - **(a)** Place one beaker in a bowl of ice and the other in a bowl of hot water. Label each one.
   - **(b)** Leave the beaker for 24 h to allow crystals to form.

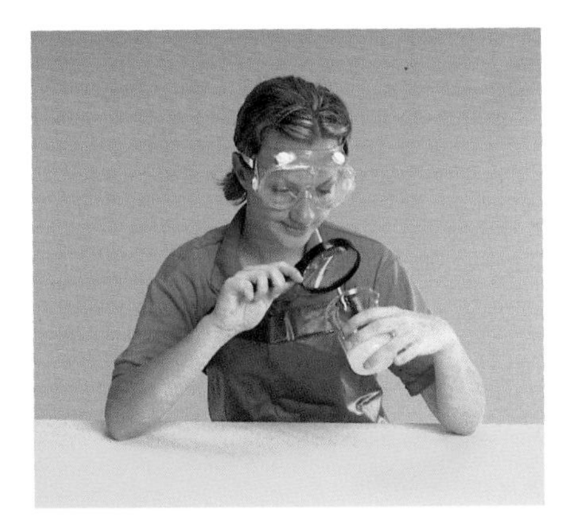

5. Tilt each beaker slightly, and, using a hand lens, examine the crystals in each glass. Make a drawing of what you see.

6. Wash your hands thoroughly after completing this investigation.

Skill POWER

To learn how to do scientific and technological drawing, turn to page 495.

Analyze

1. Which beaker formed larger crystals?
2. Which beaker formed smaller crystals?
3. Did you observe which beaker took longer to form crystals? If so, which one did?
4. Use the hand lens to help you identify the crystal shape according to the shapes in Table 10.2.
5. To be able to compare crystal size based on temperature of cooling only, all of the other conditions had to be the same, or controlled, for each beaker. List all of the conditions, or variables, that were controlled in this investigation.

Conclude and Apply

6. How did the rate of cooling affect the size of the crystals?
7. Which sample of crystals could represent extrusive rock? Why?
8. What is more likely to happen in intrusive rock?
9. Where do you conclude the larger gems might be found, on the surface of Earth or deep in the ground?

Extend Your Knowledge

10. Repeat the experiment, but divide your crystal solution evenly into *three* beakers. As before, place one beaker in ice and one in hot water. Leave the other beaker at room temperature. After 24 h, observe all three. Are the room temperature crystals different from the other two? If so, explain (and draw) how they are different. If the ice-water-cooled crystals represent crystals found in extrusive rock, and the hot-water-cooled crystals represent crystals found in intrusive rock, what natural conditions might the room-temperature crystals represent? If the crystals do not appear different, which of the other crystals, the hot or cool ones, do they more closely resemble?

Figure 10.12 In this gorge, layer upon layer of rock is visible. Rocks that break away from this rock face will also have layers. These layers provide clear evidence of how sedimentary rock forms.

Sedimentary Rock

As its name indicates, **sedimentary rock** is made from **sediment** — loose material, such as bits of rock, minerals, and plant and animal remains. These sediments become closely packed and cemented together (see Figure 10.12). Sedimentary rock makes up about 75 percent of all the rock we can see on Earth's surface. (Igneous rock occurs mostly below Earth's surface.) How does sedimentary rock form? Why do the fragments separate into layers? See if you can find some answers in the next activity.

Find Out **ACTIVITY**

Sedimentary Shake-Up

Sediments of different rocks can get churned up and mixed together in a rushing river or in larger lakes and oceans by the action of waves. How do they settle out into organized layers of rock?

Caution

Avoid inhaling dust from the clay and gravel.

What You Need

dampened clay powder
sand
fine gravel
coarse gravel
a large, clear glass jar (2 L or larger) with a screw-cap lid
water

What to Do

1. Half fill the jar with water.
2. Make a 1 L mixture of equal parts of all the sediments. Pour the mixture of sediments into the jar.
   (a) Cap the jar tightly, and shake it until all the sediments are moving about. Make a hypothesis about how the sediments will settle.
3. Set the jar down, and observe the sediments settling. Record your observations.
4. Let the jar sit undisturbed for half an hour to an hour. Re-examine the jar, and record your observations. Include a drawing of what you observe.

What Did You Find Out?

1. Which sediments started to settle immediately?
2. Which sediments took the longest to settle out of the water?
3. Based on your observations, can you label the layers in your sediment drawing?
4. Does the water look the same as when you first poured it into the jar? Why or why not?
5. What was it about the bottom sediments that caused them to settle first?

Formation of Sedimentary Rock

As you saw in the activity, rock sediment eventually settles, and, over the years, it starts to pile up. Most often this happens in lakes and oceans, where the larger, heavier fragments settle first and end up near the bottom. Sometimes wind, ice, or gravity move sediment to a place where it settles. As sediment slowly settles on top of other sediment, the rock takes on a layered appearance. These layers are called **beds**.

How does settled sediment become rock? Each layer of sediment is squeezed together by the weight of other sediment and the water on top of it. This process is called **compaction**. In some rocks, minerals dissolve as the water soaks into the rock, forming a natural cement that sticks the larger pieces of sediment together (see Figure 10.13).

Figure 10.13 Cement is used to bond or glue rocks and bricks together in building construction. "Roman cement" was a mixture made from equal portions of lime (calcium oxide) and volcanic ash or crushed tiles, mixed with water. Many buildings, like the famous Colosseum in Rome, were made from huge rocks glued together thousands of years ago with long-lasting "Roman cement."

Limestone is one of the most common and useful sedimentary rocks. It is also unique because many types are made up of large amounts of sediment from the remains of once-living things, called **fossils**. For this reason, limestone is in a separate class called **organic sedimentary rock**. Ocean animals, such as mussels and snails, make their shells mainly from the mineral calcite. When the animals die, their shells accumulate on the ocean floor, where most sedimentary rock is born. The appearance of a sedimentary rock can reveal what type of sediment formed it (see Figure 10.14).

Pause& Reflect

You have seen how crystals can be formed in igneous rocks. In which igneous rock would you expect to find the largest crystals? Why? Record your ideas in your Science Log.

DidYouKnow?

Evidence indicates that Earth, the Moon, and meteorites all formed at the same time, when the solar system formed. Although collisions in space and other processes might have altered the composition of some meteorites, the majority show strong chemical similarities to Earth rock. So, even if some rocks come from outer space, all rocks are related.

A shale

Figure 10.14A: Shale, or mudstone, is sedimentary rock formed from fine grains of clay or mud.

B sandstone

Figure 10.14B: A harder, rougher rock called sandstone is formed from larger granules of sand, mostly made of quartz.

C conglomerate

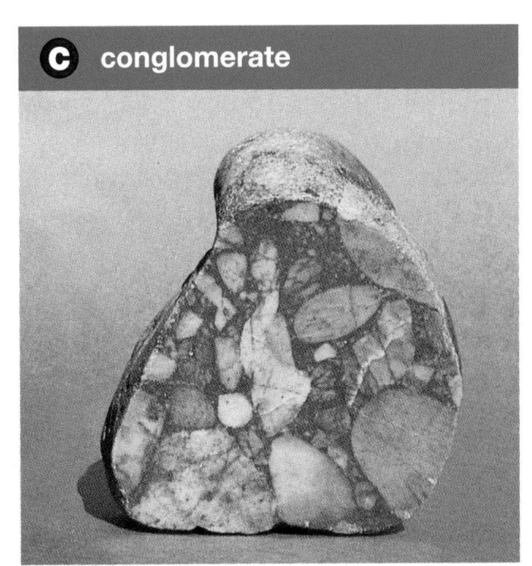

Figure 10.14C: Rounded pebbles and small stones, cemented together, form a type of rock called conglomerate.

A. sedimentary shale

B. slate

C. sedimentary limestone

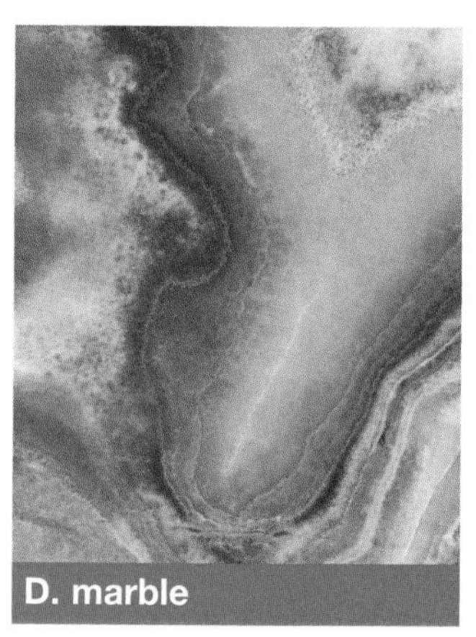
D. marble

Figure 10.15 Shale can become slate, and limestone can become marble. What causes these changes.

Metamorphic Rock

Once a rock is made, can it change its form? The answer is, yes, it can. Geologists have found rocks that resemble certain igneous and sedimentary rocks but differ from them in significant ways.

Look at the rocks in Figure 10.15. Slate and marble are just two examples of the third family of rock, called **metamorphic** (meaning "change form") **rock**. This rock may be formed below Earth's surface when extremely high pressure and heat cause the original rock, or **parent rock**, to change form. The type of rock formed depends on the amount of pressure applied. Shale, for example, can undergo several changes as pressure and temperature increase over time. This change results in the formation of slate → schist → gneiss from shale.

Metamorphic rock can change so completely that it no longer looks like the parent rock, but there are enough common characteristics that geologists know the two are related. For example, limestone and marble look different, but both have a hardness value of 3, and both are made of the mineral calcite.

Figure 10.16 shows how granite can be changed to form another rock, gneiss. Igneous granite can lie in large bodies, deep below Earth's surface. Diagram B shows what can happen to the mineral grains of granite as the pressure of heavy, overlying rock squeezes them closer together. Gneiss, the altered rock in this process, is an example of a metamorphic rock.

DidYouKnow?

Much of the central and eastern part of Canada consists of the Canadian Shield, a huge area of ancient igneous and metamorphic rock. Throughout Canada, enough limestone exists for it to be quarried and sold. Limestone is easy to cut and shape, so it is useful for constructing and decorating buildings. Crushed limestone is used in road construction, and it is also used in the pulp and paper industry. Look on the Internet to find out where limestone quarries exist in your province.

A. granite (igneous)

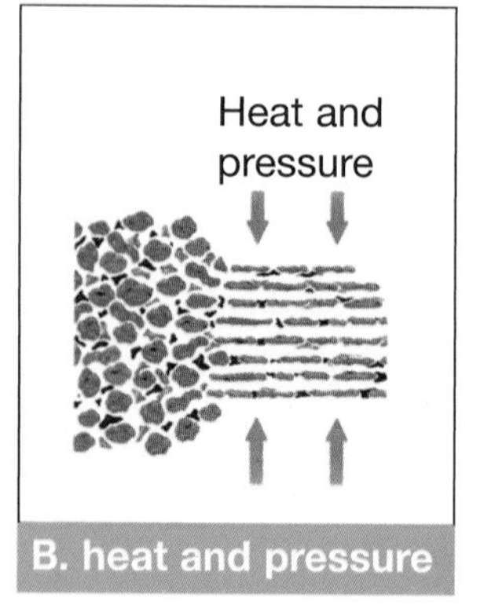

B. heat and pressure

C. gneiss (metamorphic)

Figure 10.16 The change from igneous rock to metamorphic rock is a long, slow process caused by heat and pressure over time.

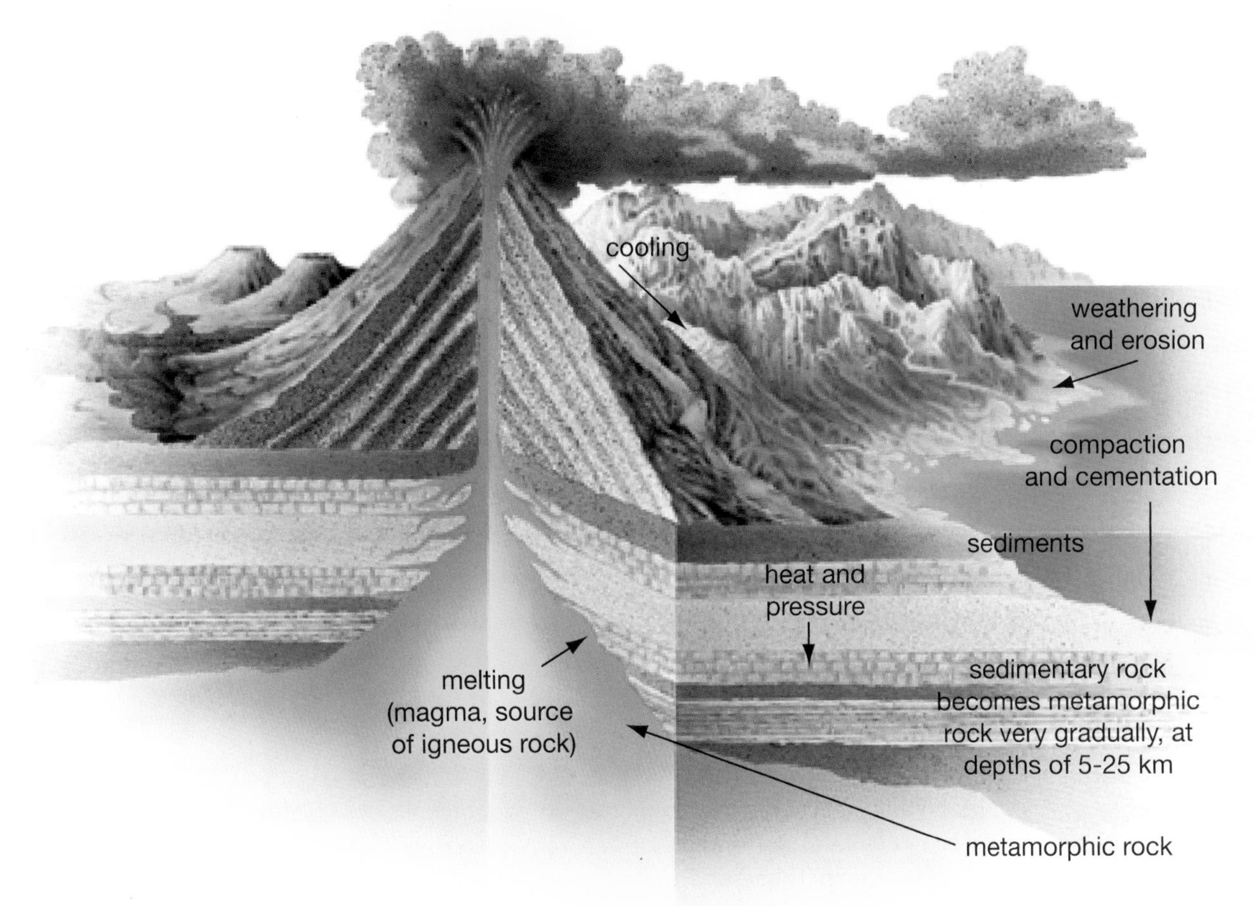

Figure 10.17 This model of the rock cycle shows the part that volcanoes play in the rock cycle. Note that the diagram is not to scale. The production of new magma occurs 25-40 km below Earth's crust.

The Rock Cycle

Much of what you have learned so far in this section suggests that rocks are constantly changing. For example, igneous rock is formed when magma or lava cool. Rock fragments and sediments can be compacted and cemented to form sedimentary rock, and both igneous and sedimentary rock can form metamorphic rock under high pressure and heat.

Does the process stop there? No, rocks continue to change in an ongoing process called the **rock cycle** (see Figure 10.17). Does the magma ever run out? As rocks sink back into the depths of Earth's crust, the heat and pressure can turn them back into magma. As well, all rocks can be broken down to form smaller rocks, fragments, and sediment. Although human activity is responsible for some of the breakdown process, most of it occurs naturally.

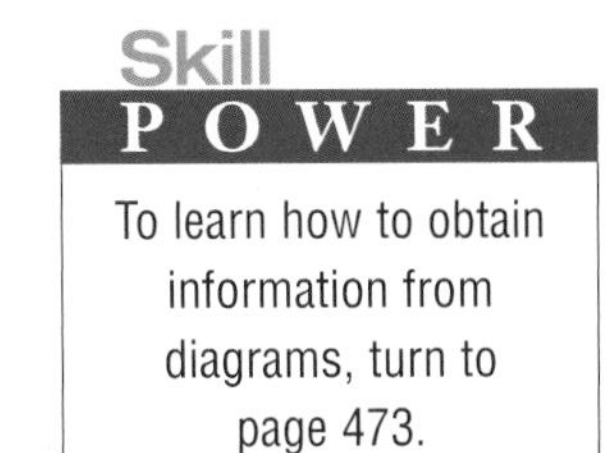

To learn how to obtain information from diagrams, turn to page 473.

Weathering

Sediment comes from larger rocks that have broken down or worn away by a natural process called **weathering**. Rocks can be weathered mechanically, chemically, or biologically.

Mechanical Weathering

Mechanical weathering is the physical break-up or disintegration of rocks. For example, gravity causes rocks to fall down a cliff and break apart. Rocks rolling down a slope or in a fast-moving stream rub and bump against each other, becoming smoother and more rounded.

Climate change can also cause mechanical weathering. In early spring or during winter warm spells, as days begin to get warm, night-time temperatures still dip

Figure 10.18 Frost wedging is caused by water that repeatedly freezes and thaws.

below freezing. This is known as the freeze-thaw period, and it can go on for several days. During the freeze-thaw period, snow and ice melt in the daytime, allowing water to seep into cracks in rock. At night, when the temperature falls below 0°C, the water freezes and expands, pushing the cracks wider apart. Each day, more water fills up the cracks, and it freezes again at night, pushing the rock pieces farther apart. Finally the rock breaks apart. If a rock has many cracks, it can seem to crumble at the end of a freeze-thaw period. This entire process is called **frost wedging** (see Figure 10.18). Materials other than water, such as crystal salts, can wedge rocks apart as well.

Pause& Reflect

Road repair crews are a common sight each spring. Use your understanding of weathering processes to explain why our roads are in need of repair. Write your ideas in your Science Log.

Another type of mechanical weathering occurs when wind and water wear away the surfaces of rocks and carry the pieces to another place, where the pieces build up. **Erosion** is the part of the process responsible for "wearing away," and **deposition** is the part of the process responsible for "building up." A dust storm and the muddy silt along a river bed are examples of erosion and deposition (see Figure 10.19A and B).

Erosion occurs more easily when small pieces of grit that are suspended in the wind and water act like sandpaper on Earth's surface. During the Ice Age, especially

Figure 10.19A Blowing dust and sand granules are carried from one place to another.

Figure 10.19B The impact of a single raindrop loosens many soil particles.

in northern lands such as Canada and Scandinavia, glaciers flowed across Earth's surface, carving and scraping and lifting away rock and soil (see Figure 10.20). The glaciers carried the sediment far away. When they began to melt and retreat, they deposited the sediment in new locations.

Figure 10.20 Which way did the glacier go? The straight-line scratches you see on the surface of the rock continue across the cracks. They are therefore not a characteristic of the rock itself, but they were caused by something heavy and large that scratched only the surface.

Find Out **ACTIVITY**

Swishing Sediment

Forces are needed to move anything. Which forces in nature can move sediment? In this activity, you will use a model to explore how sediment can move from location to location.

Safety Precautions

- Wash your hands thoroughly after completing this activity.
- Clean all surfaces.

What You Need

sand
coarse gravel or small rocks
rectangular cake pan
water
measuring cup

What to Do

1. Place a small pile of sand and gravel in one end of the cake pan.
2. Move the pile to the other end of the pan without touching the particles with your hands. Describe your method.
3. Try to move the pile in a number of different ways. Each time, describe your method, and record and draw what happens to the sand and gravel.
4. Pile up the sand and gravel in the centre of the pan. Slowly drip water onto the pile, and watch how the sand and the gravel move as the water flows down. Record and draw your observations.
5. Add a total of one-half cup of water. Record and draw where the sand and gravel end up.

What Did You Find Out?

1. How many different methods worked to move dry sand and gravel?
2. Which processes in nature did these methods represent?
3. Which human activities affect the movement of sediment in nature?
4. How might human activity prevent the movement of sediment in each of your methods?
5. Did some of your classmates use methods you did not think of? What were they?

INVESTIGATION 

10-C

Nature's Design

You have seen how models can be useful for investigating some of the processes that take place in Earth's crust. In this investigation, you will design your own model.

Challenge

Design a model to describe the origin and history of a lake or a river valley in your area or an area of your choice in Canada.

Materials

construction materials: wood, plaster of Paris, modelling clay, clay, plastic, plastic bags, rubber, paper, papier maché, cardboard, wax, sand, stones

other materials: paint, tape, glue, saran wrap, hardware (such as screws and nails), cotton batting, macaroni, glitter, beads, fabric, yarn

Design Criteria

A. Your model should be at least 0.5 m × 0.5 m, with clear labels.

B. Use at least two construction media (for example, papier maché and wood, or plastic and cardboard).

C. Choose a period of time that will be easy to recognize in a model representation of your land formation.

D. Not all models look like the system they represent. (Remember, in "Cool Crystals, Hot Gems!", you used a liquid solution in glasses and bowls rather than actual hot magma and lava.) Design a new model, for the same land formation and time period, that does not look like the model you have already made.

Plan and Construct 

1 With your group, plan how you will construct your model.

(a) Will you need to construct a sequence of small models, or will one large model be sufficient?

(b) Will your model include moving parts?

(c) Which materials will you use to best illustrate the aspects of the land formation?

(d) How will you show your labels?

2 Prepare a labelled sketch of your model. Build your model based on your sketch.

3 Demonstrate your model for your classmates.

Evaluate

1. (a) Did your model demonstrate lake or river formation effectively?

(b) If it contained moving parts, did they work successfully?

2. (a) Compare your work with that of other groups.

(b) If you were to make another model, how might you improve on your current model?

Extend Your Skills 

Design Your Own A "working model" is a model that can demonstrate how a sequence of events takes place by making the same events occur in the model itself. For example, a working model of a diverted river might involve dripping water over a cliff to demonstrate how dirt would eventually build up along one river bank, changing the river's direction. You could reset the working model by placing the sediment back on the cliff. How would you design a working model for your land formation?

Skill POWER

For tips on using models in science, turn to page 493.

Chemical Weathering

Chemical reactions can speed up the process of erosion. **Chemical weathering** involves the breakdown or decomposition of minerals as a result of a chemical reaction with water, with other chemicals dissolved in water, or with gases in the air.

An example of a chemical reaction is acid rain, which contains dissolved chemicals from air pollution. This chemical rainwater reacts with some rocks. The rock material dissolves easily in the acidic water and washes away (see Figure 10.21).

Figure 10.21 Many old and valuable marble statues and buildings in Europe and Asia have suffered the effects of acid rain and have had to be repaired. The Parthenon in Athens, Greece is shown here under reconstruction.

Biological Weathering

Biological weathering is the physical or chemical breakdown of rock caused by living organisms, such as plants, animals, bacteria, and fungi. Physical breakdown occurs, for example, when a plant root wedges into a rock by forcing its way into a crack, just as frost does. As the root grows and expands, so does the crack, and the rock is pushed apart until it eventually crumbles and breaks. As well, chemical reactions can take place between rock material and acidic fluids that are produced by plant roots, bacteria, fungi, and some insects and small animals. As the rock slowly dissolves and flows away with rainwater, cracks and crevices increase in size until the rock finally breaks apart (see Figure 10.22).

The physical environment, which includes natural processes such as weathering, earthquakes, and volcanic eruptions, has an impact on the rock cycle. If the physical environment changes, the rocks in it will also change.

Mechanical, chemical, and biological weathering work together constantly to change the landscape around us.

Figure 10.22 Tree roots work their way into cracks and, as they grow, eventually break up the rock.

Check Your Understanding

1. What is the composition of rocks?
2. Name and describe the three families of rocks. Give an example of each.
3. Describe four ways that sediment can form.
4. Describe the rock cycle, and explain how rocks may change with time.
5. **Thinking Critically** A core sample was taken from the bottom of a lake that contained first a layer of sandstone, then a layer of shale, and finally a layer of conglomerate on top. Why could these sediments not have settled at the same time? Explain what you think happened.

Pause& Reflect

Copy the rock cycle into your Science Log. Number the arrows, and write an explanation for each. If you do not remember what an arrow means, read back through the chapter to find out.

10.3 Soil

Sediment is an in-between stage in the rock cycle. The slow process of rock formation takes thousands of years to occur. What happens to sediment in the meantime? Rock sediment is commonly called dirt, but what most people call dirt is actually soil. **Soil** is a mixture of weathered rock, organic matter, mineral fragments, water, and air.

Formation of Soil

How does soil form? Earth is covered by a layer of rock and sediment. Sediment and mineral fragments do not become soil until plants and animals have lived in them and added organic matter, such as leaves, twigs, and dead worms and insects. The organic matter creates spaces that can be filled with air or water. All of these combine to form soil, a material that can support plants (see Figures 10.23A, B, and C).

Climate, the type of rock, and the amount of moisture influence soil formation. Even the slope of the land can influence it. For example, if the slope is steep, loose dirt will run off with each rainfall, leaving only large, barren rocks.

In addition to these non-living factors, the small, living creatures that invade the soil can speed up the process of soil formation. As you saw in Unit 1, soil is a complex ecosystem, where small rodents, worms, insects, algae, fungi, bacteria, and decaying organic matter all live in harmony. Most of the decaying matter is made up of dead plant matter, called **compost**. It mixes with other matter to form the dark-coloured portion of the soil called **humus**.

Humus is rich in nutrients, such as nitrogen, phosphorus, potassium, and sulfur. These nutrients dissolve in water in the soil. Plants absorb the nutrient-rich water through their roots. Humus also promotes good soil structure and helps keep the water in the soil. As worms, insects, and rodents burrow throughout the soil, they mix the humus with the fragments of rock. In good-quality soil, there are equal parts humus and broken-down rock.

A **fertile** soil is one that can supply nutrients for plant growth. Soils that develop near rivers are generally fertile. Some soils may be nutrient-poor and have low fertility, such as the eroded, rocky soil of steep cliffs and roadsides.

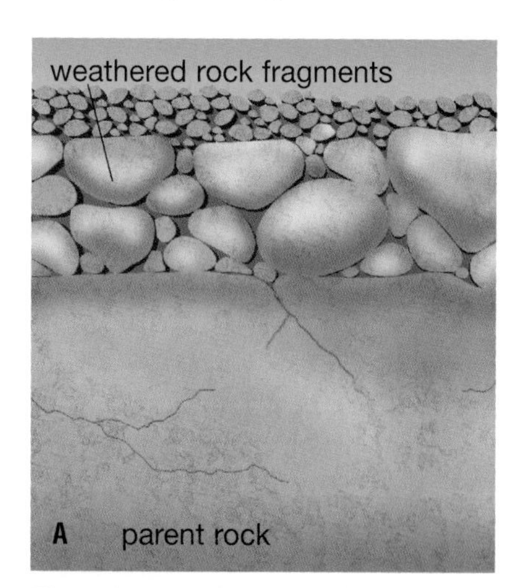

Figure 10.23A: These weathered rock fragments contain many cracks and spaces, providing areas that air and water can fill.

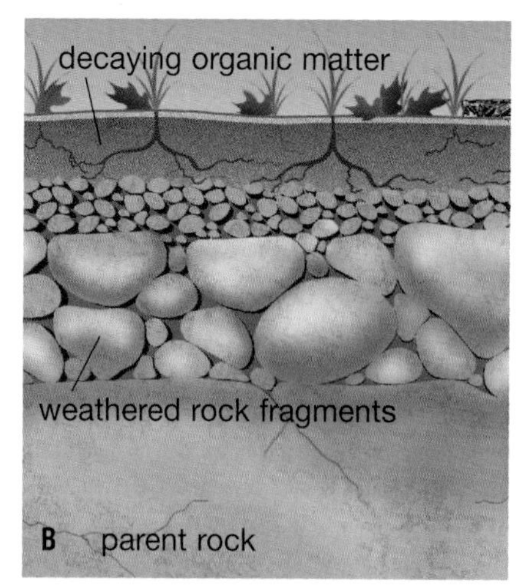

Figure 10.23B: Immature soil can support small hardy plants that attract insects and other small animals. Over time, dead plant and animal material build up, and bacteria and fungi cause them to decay. The decaying organic matter forms a layer on top of the weathered rock.

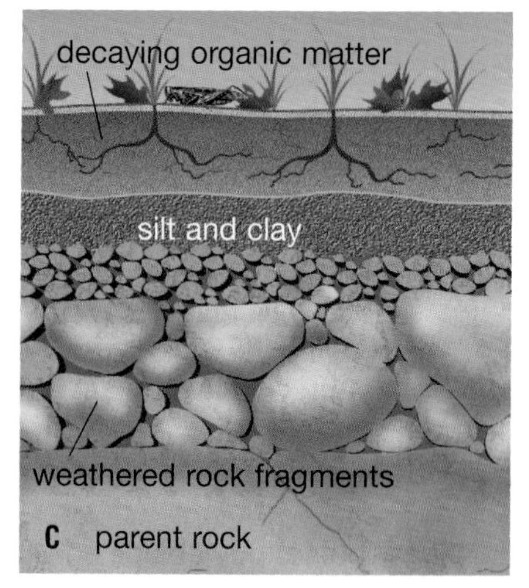

Figure 10.23C: Mature soil contains mineral-rich clay on top of weathered rock. The clay forms when water carries the minerals away from the decaying organic matter above. Above the clay, the topsoil extends to the surface and contains decayed organic matter, plant roots, and living organisms.

Soil Profiles

Soils can take thousands of years to form. They can range in thickness from 60 m in some areas to just a few centimetres in others. Soil varies in structure and appearance, depending on its depth. In examining soil, scientists have exposed layers of soil with clear differences in appearance and composition. The layers of soil make up a **soil profile**. The soil profile in Figure 10.24 illustrates how the layers are divided. They show different degrees of soil evolution.

Examine Figure 10.24. The top layer (A) is called **topsoil**. It consists of dark-coloured, rich soil that contains humus and small grains of rock. It has undergone the greatest number of changes from the underlying rock layer. In your Science Log, suggest how leaching fits into the rock cycle.

The next layer (B) is generally lighter in colour because there is little or no humus, and it contains minerals that have leached from the top layer. **Leaching** is the removal of soil materials dissolved in water. Water reacts with humus to form an acid. This acid can dissolve elements and minerals from upper layers and carry them through the spaces in the soil to lower layers. In your Science Log, suggest how leaching fits into the rock cycle.

The bottom layer (C) contains partly weathered rock and minerals leached from above. This layer most closely resembles the original, or parent, rock below and is at the beginning of the long, slow process of rock evolving into soil.

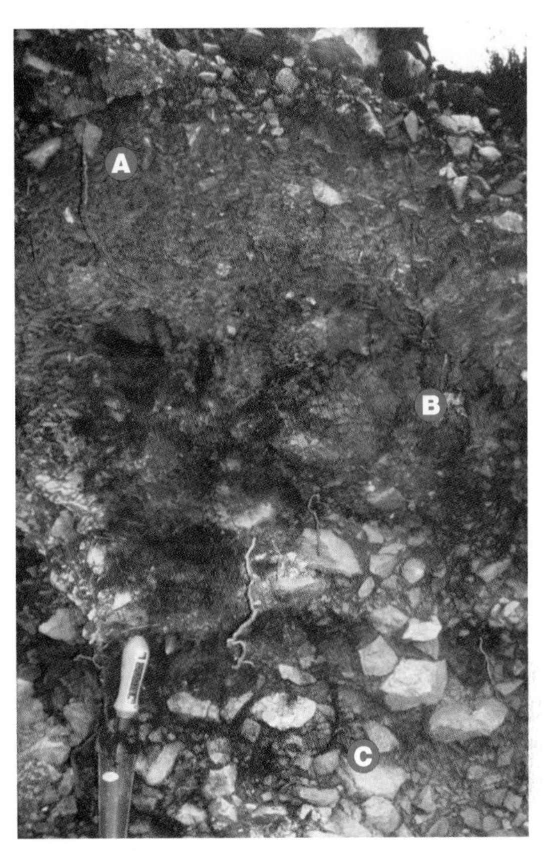

Figure 10.24 Exposed soil profile of an eroded hillside

Across Canada

Sylvia Edlund

In her first job after university, Sylvia Edlund worked as a typist. She preferred research and writing to typing other people's ideas, however. These other skills were recognized and she was soon promoted to research scientist.

Dr. Edlund joined the Geological Survey of Canada in 1974. She travelled with a team of scientists to the Arctic to study the plant and animal life. Most of her work involved "mapping," or charting, the various types of plants and showing where different kinds of plants will grow. For example, some plants can grow on bare rock while others require soil. Dr. Edlund became very interested in the area and has continued to specialize in Arctic research for over 20 years.

On Ellesmere Island, Dr. Edlund solved the puzzle of how plants can grow in northern areas that seem completely frozen or dry. She discovered that melted ice within the permafrost (permanent frozen ground) provides plants with water.

Sketching and painting the northern landscape are her other interests. Some of her paintings of wildflowers from the Arctic have appeared in a book published by Indian and Northern Affairs Canada, in Yellowknife.

Pause& Reflect

Have you ever tried to grow a flower on a rock? In your Science Log, list and explain the reasons why this would be difficult and why it would be easier if you used soil.

CONDUCT AN

INVESTIGATION 10-D

Soil Sleuth?

Can you imagine a clay beach, gravel garden, or sandy road? Sound wrong? What do you know about these types of mixtures? Look at different soils in your area. Collect a sample, and note where you obtained it. Use the following tests to determine the characteristics of your soil sample.

Problem

What are the characteristics of soils?

Safety Precaution

Be careful when using objects with sharp points or edges.

Apparatus

soil sample
sand
gravel
clay
watch
scissors
thumbtack or push pin
measuring cup or graduated cylinder
rubber bands
4 glasses or beakers (250 mL)
hand lens
ruler
cheesecloth squares

Materials

water
4 large plastic cups
4 plastic coffee-can lids
paper

Procedure

1. Make a table like the one below.
2. Spread your soil sample on a sheet of paper.
   - (a) Record the colour of the soil, and examine the soil with a hand lens. Describe the different particles. Measure the diameter of an average particle while looking through the hand lens.
   - (b) Measure the diameter of a group of soil particles. Divide the diameter by the number of particles you estimate are present along the diameter. This will give you an estimate of the size of one particle.
3. Repeat Step 2 with the sand, gravel, and clay samples.

Area Soil Sample			
	Sand	Gravel	Clay
Soil sample			
Colour			
Average particle size			
Texture when dry			
Texture when wet			
Drainage rate			

4. Rub a small amount of soil between your fingers. How does it feel? Press the soil sample together. Does it stick together or crumble? Wet the sample and try again. Record your observations in the table.

5. Repeat Step 4 with the sand, gravel, and clay samples.

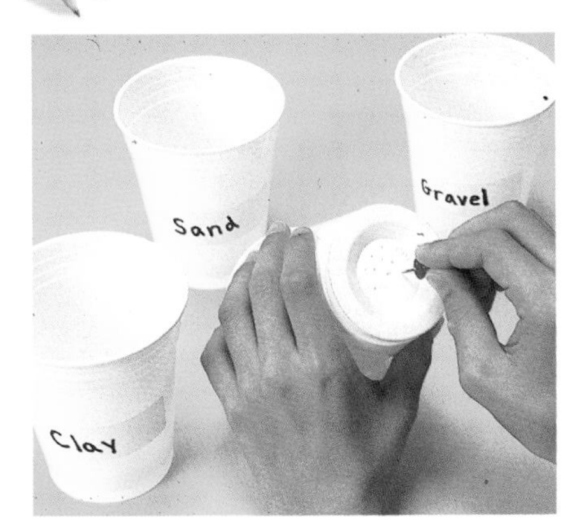

6. Test for water drainage.
   - (a) Label the plastic cups "sand," "clay," "gravel (with sand and clay)," and "area soil sample." Using the thumbtack, punch an equal number of holes around the bottom of each cup.
   - (b) Cover the area of holes on each cup with a cheesecloth square. Secure the cheese cloth with a rubber band.

7. To hold each cup over the glass or beaker, cut a hole in a coffee-can lid so that the cup will just fit inside the hole. Place a cup and lid over each glass or beaker.

8. Half-fill the appropriate cups with dry sand, clay, and your soil sample. Make a mixture of equal parts gravel, sand, and clay, and half-fill the last cup with this mixture.

9. Use the measuring cup (or graduated cylinder) to pour 100 mL of water into each cup. Record the time when the water is first poured into each cup and when the water first drips from each cup.
   - (a) Allow the water to drip for 20 min, then measure and record the amount of water in each glass or beaker.
   - (b) Wash your hands thoroughly after completing this investigation.

Computer CONNECT

Design Your Own 

Design a computer simulation that can predict the drainage rate of soil mixtures containing varying amounts of clay, sand, and gravel.

CONTINUED

Analyze

1. Describe your soil sample in as much detail as possible.

2. Compare the feel and stickiness (texture) of the sand, gravel, and clay samples. Which one is grittiest? Which one is stickiest? How does your soil sample compare to these? Is it more like one than another?

3. Now compare the drainage of the sand, clay, and gravel mixture. Which drained the most quickly? Which drained the least quickly? How did the drainage rate of the gravel mixture compare with the drainage rate of the others? Which one did your soil sample resemble the most in drainage?

Conclude and Apply

4. How does the addition of gravel and sand affect the speed at which water drains through clay?

5. What would you do to increase the speed at which water drains through your soil? What would you do to decrease its speed? Why would this be useful to know?

6. Describe three characteristics of soil. Which characteristic do you think most affects how fast water drains through soil?

7. Rank the size of soil particles from largest to smallest in clay, sand, and gravel.

8. Which types of soil would be the best to grow vegetables in your garden? Why?

Extend Your Knowledge

9. **Design Your Own** Make a hypothesis about how to change your soil's permeability, and design an investigation to test your hypothesis. What will your dependent and independent variables be? What other factors will you need to control?

10. Watch for soil profiles along roadsides and riverbeds. If you see one, try to identify the different layers. Record your observations as labelled diagrams. How could you quantify your observations?

11. Compare the soil from your neighbourhood with soil from other neighbourhoods. How can you account for the differences?

Soil Texture

Table 10.3 Classification of Soil Rock Sediment Particle Size

Name of rock sediment	Average particle size
gravel	over 2 mm in diameter
sand	2.0 to 0.05 mm in diameter
silt	0.05 to 0.002 mm in diameter
clay	less than 0.002 mm in diameter

Texture relates to how soil feels when it is rubbed between two fingers. Texture can tell us much about the soil: what is in it and what it can do.

The formation of topsoil depends largely on the particle size of the rock fragments (see Table 10.3). The **particle size** affects how gritty a soil feels when you rub it between your fingers. It also determines how large the spaces will be in the soil.

As you have just seen in the previous investigation, all these factors affect how quickly water will drain through, or **permeate**, the soil. If the soil is sandy, rain-water permeates it too quickly. This kind of soil has a low water-holding capacity. **Water-holding capacity** refers to a soil's ability to hold water. Soil with low water-holding capacity tends to be dry, or arid, most of the time.

Clay particles, on the other hand, are so small that they fill up most of the spaces in soil, leaving little room for air and water. Rainwater often sits on top in puddles and soaks in very slowly. Clay has a high water-holding capacity. To grow most plants, topsoil should allow water to drain through it at a moderate rate. Farmers and gardeners prefer a mixture of sand, silt, and clay, a type of soil referred to as **loam**.

Farming, Soil Loss, and the Environment

Soil is an important resource in Canada and throughout the world. When vegetation is removed by harvesting crops, the soil is exposed to the direct action of rain and wind. Topsoil, which contains the nutrient-rich humus, can be eroded and carried away through weathering. Also, without plants, soil development slows and sometimes stops because humus is no longer being produced. When natural vegetation is removed from land that receives little rain, plants do not return. All of these conditions can contribute to the destruction of the natural ecosystem and lead to desert formation, called **desertification**. This is currently happening on every continent.

Pause& Reflect

Where might desertification happen in Canada? Discuss this with your classmates, and, in your Science Log, propose some ideas to stop it from happening.

Career CONNECT

"Oil" in a Day's Work

Petroleum companies hire people like Jennifer Dunn to help them find the best places to dig for oil and natural gas. Jennifer is a petroleum geologist. She uses what she knows about Earth's crust to figure out where these petroleum products may be, far below the surface.

To decide if a spot may yield oil or natural gas, Jennifer needs to find out what kinds of rock are deep underground. She visits the area and examines the rocks on the surface. Then she and her co-workers use a special drill to bring up a "core sample," a thin cylinder of rock, from far under the ground. By studying the core sample, she can see the types of rock it contains and how they were formed.

After additional tests, she makes maps explaining what the whole area is like and lets the company know if she thinks there is a good chance of finding oil or natural gas there.

What kind of education do you think Jennifer needed for this job? Find out about the geology courses or programs offered by a university, college, or technical school near you. Try to find out what careers these courses or programs could lead to by talking to someone at that school or to a guidance counsellor in your own school.

While modern farming methods are capable of feeding billions of people on Earth, they can contribute to the destruction of the very environment that is needed for such large-scale farming. Productivity has increased through the use of large agricultural machinery and agricultural chemicals.

Large-scale operations can take their toll on the environment, however. Large equipment requires that the farmers plow long straight lines. This can contribute to loss of topsoil due to water erosion. Trees that are used as windbreaks often have to be cut down to make larger fields, which can expose the fields to a loss of topsoil due to wind erosion. In many cases, marshes have been drained and fields plowed right to the edge of rivers or streams. This can contribute to the elimination of different species of wildlife. The machinery used in large-scale agricultural operations relies on the combustion of fossil fuels, which adds to air pollution. Nitrogen compounds and other substances in fertilizers and manure, as well as chemicals in the widespread application of pesticides and herbicides, dissolve in water and find their way into the rivers, streams, and ground water systems. The result is the pollution of human drinking water. How can farmers minimize the damage?

Saving the Soil

Most farmers try to prevent soil erosion and to minimize the environmental impact of their farming practices. If they have had to remove windbreaks, they plant new ones in locations suitable for their fields. As well, they cover bare soils with decaying plants to hold soil particles in place. In dry areas, instead of plowing the natural vegetation under the soil to plant crops, farmers graze animals on the natural vegetation. Proper grazing management can retain plants and reduce soil erosion.

In recent years, many farmers have begun the practice of no-till farming (see Figure 10.25). Normally farmers till or plow their fields two or three times a year, exposing the topsoil to dangerous wind erosion each time. In no-till farming, plant stumps are left in the field. At the next planting, farmers seed crops without removing these stalks and without plowing the soil. No-till farming provides cover for the soil all year round and reduces soil erosion.

Figure 10.25 In no-till farming, old plant stalks are not removed.

INVESTIGATION 10-E

Where Does It Go?

Think About It

Everything we use has a life cycle, after which items must be thrown away or somehow re-used. Finding ways to cut down on our use of disposable items is one way to keep our land clean and our resources plentiful.

What to Do

1. With your group, examine the diagram below. It shows the "life cycle" of an aluminum pop can.
2. Choose a product, and think about the different steps in the product's life cycle. Discuss the following issues with your classmates, and list the pros and cons for each.
   (a) use of resources
   (b) number of people employed
   (c) type and amount of pollution (wasted resources) at each stage

Analyze

1. What kinds of changes could you suggest for the life cycle of this or another product? Where could the most beneficial changes, in terms of land use, be made? What kinds of compromises might have to be made? What groups might object to your suggestions? How might you personally be involved?
2. Write your analysis in a short report, and be prepared to present it to the class.

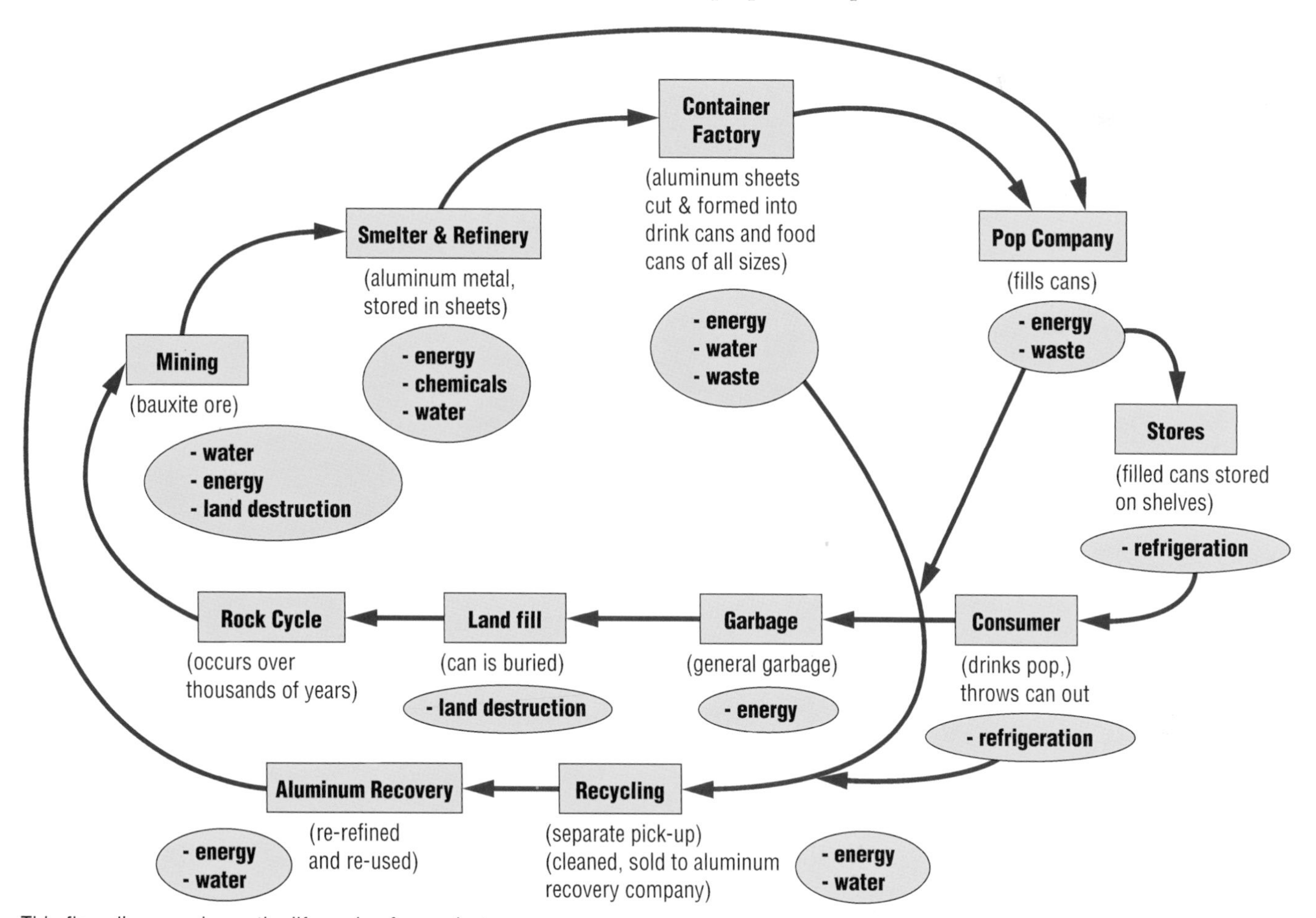

This flow diagram shows the life cycle of a product.

Landfills and Buried Waste

Figure 10.26 Hazardous waste can find its way into landfill sites.

You have seen that we use land for mining and for farming. You also know that we build our towns and cities on much of it. This "urbanization" leads to another use — landfill sites. Garbage is a part of modern life. Most garbage is brought to a **landfill site**, an area where waste is deposited (see Figure 10.26). There the garbage is dumped, sorted, and eventually buried. A great percentage of the volume of garbage brought to landfills is dry, relatively harmless garbage, or kitchen and garden waste that can decompose and contribute to the humus content of the soil below.

Although hazardous waste is supposed to go to specific sites, it sometimes ends up in landfill sites. Hazardous waste includes various types of poisonous substances that can seep into the ground and into our water. Underground gasoline tanks and septic systems that leak, as well as runoff from farm fertilizers, also flow into our water system.

Land developers have learned to be careful about choosing locations for landfill sites. They first take into account the nature of the garbage that will be buried there. Then they examine the soil and rock formations in the area. Which types of soil and rock would minimize the leaching of poisonous chemicals through the ground?

Although hard rock and clay soils can minimize leaching, this will never be a permanent solution to the storage of the more dangerous wastes. You can help by reducing the quantities of harmful materials that you use and throw out. You and your family can separate hazardous waste from the regular garbage and take it to the appropriate collection site.

Land is one of Canada's great resources. It contains the rocks and minerals we need for construction and industry. It grows our food, produces our mineral resources, and provides beautiful settings for our towns and cities. It is where you live and go to school; it is where your future lies.

Check Your Understanding

1. What are the ingredients of soil?
2. Why do soil profiles contain layers?
3. List some ways that topsoil can be lost.
4. **Apply** How would the soil profile in a rain forest be different from the soil profile in a desert?
5. **Thinking Critically** Would you support the sale of imported products made from such tropical woods as rosewood, mahogany, or teak? On one hand, you help the economy of a poor, developing country today; on the other hand, your purchase contributes to the country's long-term impoverishment. Discuss these ideas with some classmates, and write your opinions in a report.

CHAPTER at a glance

Now that you have completed this chapter, try to do the following. If you cannot, go back to the sections indicated.

Define the following: rocks, minerals, elements. (10.1)

List the properties used to identify minerals, and explain how each is used. (10.1)

List and describe uses of minerals. (10.1)

Name the three families of rocks, and explain how each type of rock is formed. (10.2)

Draw a flow diagram of the rock cycle. (10.2)

Name, describe, and give examples of the three types of weathering. (10.2)

Define and give an example of erosion and deposition. (10.2)

Explain how the three types of rocks are re-formed in the rock cycle. (10.2)

Explain the difference between soil and dirt. (10.3)

Compare compost and humus, and explain their importance in the formation of soils. (10.3)

Explain what process is occurring in the top, middle, and bottom layers of a soil profile. (10.3)

List the properties that are used to classify a soil. (10.3)

Explain the ways in which farming practices can actually damage the soil. (10.3)

Explain how a knowledge of soils and their properties can help people make decisions about land use, such as where to locate landfills and dumps for hazardous waste. (10.3)

Prepare Your Own Summary

Summarize this chapter by doing one of the following. Use a graphic organizer (such as a concept map), produce a poster, or write the summary to include the key chapter ideas. Here are a few ideas to use as a guide:

- How do different rocks form?
- What are the different processes that can break down rocks?
- Using the illustration at the right, explain how the breakdown and re-formation processes of rocks connect.
- Why is it in a farmer's best interest to take steps to reduce the impact that large-scale farming can have on the environment? Use specific examples in your answer.
- How would you decide where to locate a landfill or hazardous waste dump in your area?

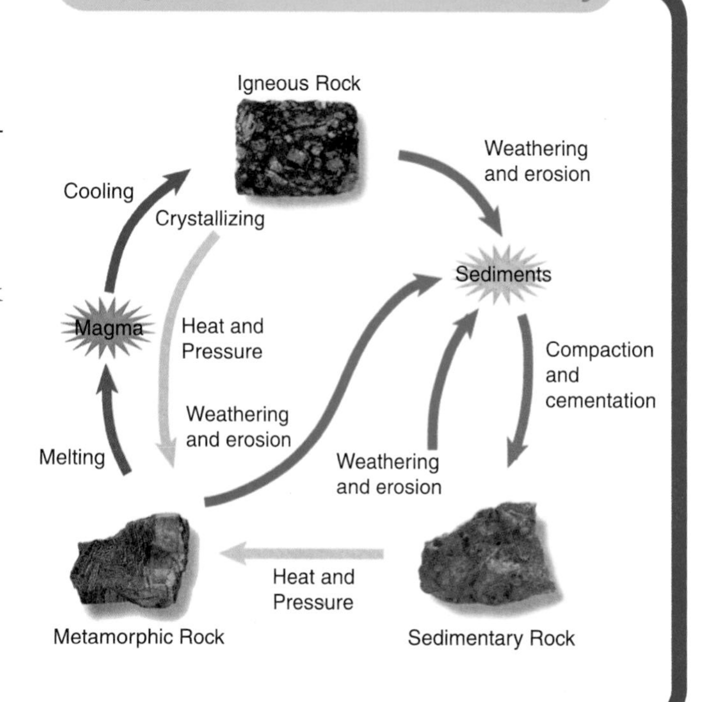

CHAPTER

10 Review

Key Terms

rock
minerals
crust
element
crystal
lustre
streak
cleavage
fracture
metallic ores
igneous rock
magma
intrusive rock
lava
extrusive rock
sedimentary rock
sediment
beds
compaction
fossils
organic sedimentary rock
metamorphic rock
parent rock
rock cycle
weathering
mechanical weathering
frost wedging
erosion
deposition
chemical weathering
biological weathering
soil
compost
humus
fertile
soil profile
topsoil
leaching
texture
particle size
permeate
water-holding capacity
loam
desertification
landfill site

Reviewing Key Terms

If you need to review, the section numbers show you where these terms were introduced.

1. Which words in the list are used in mineral identification? (10.1)
2. Which words in the list are used in the rock cycle? Draw the rock cycle. (10.2)
3. Which words in the list can be used to describe the process of sedimentation? (10.2)
4. Which words in the list are types of soils? (10.3)
5. Which words in the list refer to soil characteristics? (10.3)
6. Which words in the list refer to soil loss and environmental damage? (10.3)

Understanding Key Ideas

Section numbers are provided if you need to review.

7. What is the difference between a mineral that has cleavage and a mineral that has fracture? (10.1)
8. If a mineral belongs to the hexagonal crystal group, how many sides do its crystals have? (10.1)
9. Explain why a diamond does not leave a streak on a streak plate. (10.1)
10. Where do sediments come from? (10.2)
11. Compare magma and lava. (10.2)
12. How many soil horizons are generally present in a soil profile? Briefly describe the general characteristics of each horizon. (10.3)
13. What farming practices reduce soil loss? (10.3)
14. How can you increase the fertility of a soil? (10.3)

Developing Skills

15. Visit a nearby landfill or waste disposal site. Find out how hazardous wastes are handled: how are they collected, what is done with them, and how are decisions made about where to bury these wastes?
16. Make a mineral scrapbook by "collecting" all the minerals presented in this chapter. Organize them into a "characteristics" table, like the table you made in Conduct an Investigation 10-A. There will be many spaces left empty. Make hypotheses about what should go in these spaces. Then go to your library and look up the information in a mineral guide. How accurate were your hypotheses?
17. What hardness does a mineral have if it is scratched by glass but it can scratch an iron nail?

18. Use cardboard to make three-dimensional models of each major crystal shape.

19. In your notebook, copy and complete the concept map below.

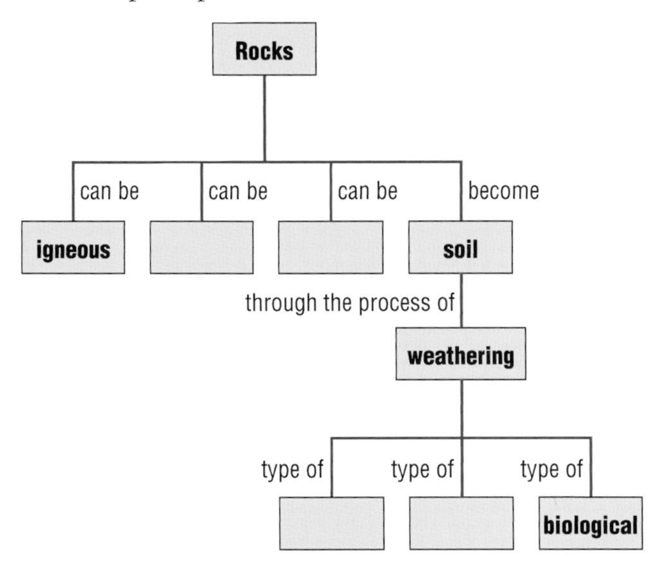

20. Plant three identical plants in three different types of soils. Why would it be important to keep everything the same in your experiment except the types of soils? Over a period of weeks, observe and measure plant growth. Then make a poster display to share your results with your classmates. The poster display should show how the soil type affected plant growth.

21. Organic farming involves environmentally friendly, alternate methods for enriching soils. Visit a nearby farm and discuss organic farming with the farmer. Find out why she or he is or is not using these methods.

Problem Solving/Applying

22. If all Joe had was a piece of paper, a steel knife, and a glass bottle, explain how he could distinguish between calcite and quartz. What other test would help him identify calcite?

23. Use the rock cycle to explain why pieces of granite and slate can be found in the same piece of conglomerate.

24. Hamida was planting flowers in the spring. When she dug into the ground, she could feel that the soil was soft, was sticky when wet, and did not let water drain very quickly. What particle type did she conclude this soil was made up of? What could she do to turn it into a loam?

Critical Thinking

25. Why do you think igneous rocks have bubbles or pores? Do you think such rocks are likely to be intrusive or extrusive? Explain why.

26. Marble is used to carve sculptures. What properties of marble make it useful for this purpose?

27. How can water be a factor in both mechanical and chemical weathering?

28. How do organisms help soils to develop? Is this mechanical or chemical weathering? Explain.

Pause& Reflect

1. Look at the development that is going on in your area. What is being built? What is being replaced? Discuss the pros and cons of land development with your parents, teacher, and other classmates. Why is land management so important?
2. Why do farmers make every attempt to keep their land as flat as possible? Write a newspaper article for your school or community newspaper, discussing why we need to care for our soils and what major problems or crises are facing our soil environments.
3. Go back to the beginning of this chapter on page 278 and check your original answers to the Getting Ready questions. How has your thinking changed? How would you answer those questions now that you have investigated the topics in this chapter?

CHAPTER

11 Earthquakes,

Getting Ready...

- Could an earthquake happen where you live?
- Why do most mountains not erupt?
- How long would it take to travel to Earth's centre?

Science Log

In your Science Log, record your thoughts about the questions above. Share them with others in your class, and find out if anyone has experienced an earthquake or a volcanic eruption. As you work through this chapter, you will find the answers to these questions.

Imagine that you were driving the vehicle on the overpass in the photograph above. Suddenly it seems as if you are in a boat being tossed on rough seas. You bring the vehicle to a screeching halt, realizing that there is a huge hole in the overpass just ahead! Somehow you scramble out of the vehicle to safety.

The photograph was taken on January 17, 1995, following an earthquake in Kobe, Japan. The earthquake caused over 5000 deaths and cost many billions of dollars in property damage. An earthquake on the same day in 1994, in Northridge, California, also caused billions of dollars in property damage and 60 deaths.

Why did these earthquakes happen? Why was there so much death and destruction? Could the damage

Volcanoes, and Mountains

have been prevented? In this chapter, you will learn why earthquakes and volcanic eruptions cause so much damage, and you will learn about the remarkable processes that result in earthquakes, volcanoes, and the formation of mountain ranges.

Starting Point

This Is Powerful!

Imagine you are the first person on Earth ever to experience an earthquake or a volcanic eruption. What would you think was happening? What would you think might be causing it?

Many civilizations have developed stories or myths about the tremendous power and forces that cause earthquakes and volcanic eruptions. Here is your chance to create your own myth!

What to Do

Write a story or a poem or create a storyboard about what is causing the event. Use your imagination. Look at the photographs on this page and elsewhere in this chapter for some ideas.

Spotlight On Key Ideas

In this chapter, you will discover

- how new technologies help scientists to study Earth's crust
- how clues in Earth's crust help scientists to predict where earthquakes are most likely to occur in Canada and around the world
- how earthquakes, volcanoes, and mountains form

Spotlight On Key Skills

In this chapter, you will

- plot locations on a map, using latitude and longitude, to find out where earthquakes and volcanoes occur
- analyze data to find patterns and trends in earthquake and volcano occurrence
- design an investigation to demonstrate types of rock movement in earthquakes

11.1 Earthquakes

DidYouKnow?

Animals have often been thought to "predict" earthquakes. Caged rabbits will hop around wildly for five minutes before an earthquake. Deep-sea fish will swim close to the ocean's surface. Some fish, such as catfish, have actually jumped out of the water onto dry land just before an earthquake has struck. Bees will evacuate their hives in a panic minutes before an earthquake, and mice will seem so dazed that they can be captured by hand.

You are in a class at school, and suddenly a bell starts to ring very loudly. The teacher asks you all to leave the room calmly and walk outside to a safe place for attendance to be taken. You are having a fire drill, just as students in schools do every year. Now imagine yourself sitting in your classroom and hearing the same bell ring. This time the teacher asks everyone to get under the desks. You are having an **earthquake** drill. Students who live in places where earthquakes can happen have earthquake drills as well as fire drills.

How do you know if your school should have earthquake drills as well as fire drills? How do people prepare for earthquakes? Where do earthquakes happen in Canada?

Figure 11.2 Seismologists study earthquakes by reading seismograms.

Figure 11.1 These students are crouched under a classroom table during an earthquake drill.

The eight dragon heads that are attached to this urn have little balls inside them. Earthquake movement shakes the balls into the toads' open mouths. The direction of the earthquake is determined by which toad swallows the ball.

On one occasion long ago, a ball fell from a dragon's mouth but no ground movement was noticed. Several days later, however, a messenger brought news of an earthquake that had happened about 650 km away.

Measuring Earthquakes

Throughout history, people have designed machines to measure earthquakes. Very old examples of these machines can be found in China. Earthquakes have been measured there for over 2000 years. The Chinese believed that earthquakes were important spiritual signs, so they wanted to find out about when and where earthquakes happened. The earthquake detector pictured on page 314 was invented in A.D. 132 by Chang Heng.

Today scientists called **seismologists** use a special machine called a **seismograph** to measure earthquakes (see Figure 11.2). Seismographs must be attached to **bedrock** (the solid rock that lies beneath the soil and looser rocks) in order to feel the vibrations that result from an earthquake. Inside the seismograph, a marking pen hangs over a rotating drum, just touching the drum. The drum is covered with paper to record the vibrations marked by the pen. When an earthquake strikes, it shakes the bedrock, causing the pen to move while the paper drum stays still. The pen point moves against the paper drum, making a jagged line. Most modern seismographs are electronic, but they are based on the same principle.

Seismologists use a method of measurement called the **Richter scale** to describe the **magnitude** (strength) of an earthquake. The scale starts at zero and can go as high as necessary. Each number on the scale represents an earthquake with about 30 times more power than the previous number. An earthquake that registered 7 would be 30 times stronger than one that registered 6, and 900 times stronger than one that registered 5. Most earthquakes that cause damage and loss of life register between 6 and 8 on the Richter scale; the Kobe earthquake registered 7.2.

Table 11.1 shows some of the numbers on the Richter scale, the effects of earthquakes of each magnitude, and their frequency.

Did You Know?

A seismograph will move only if the bedrock moves. It will not register movement even if you, and everyone else in your whole class, jump up and down right beside it. Usually several seismographs are mounted together in order to sense the different directions of movement that can happen in an earthquake. One seismograph measures up-and-down movement, while another measures side-to-side movement.

Table 11.1 Richter Scale

Richter magnitudes	Earthquake effects	Estimated number per year
< 2.0	generally not felt, but recorded	600 000
2.0 – 2.9	potentially perceptible	300 000
3.0 – 3.9	felt by some	49 000
4.0 – 4.9	felt by most	6200
5.0 – 5.9	damaging shocks	800
6.0 – 6.9	destructive in populous regions	266
7.0 – 7.9	major earthquakes, which inflict serious damage	18
≥ 8.0	great earthquakes, which produce total destruction to communities near the source	1.4

SOURCE: Earthquake Information Bulletin

Earthquake Waves

Imagine everything in your classroom suddenly vibrating for a few seconds. The vibration stops, and the teacher says, "Everyone get underneath your desks, please." Soon the floor feels like jelly and you have to hold on to the legs of your desk to keep it from moving away from you. Lights fall from the ceiling and crash onto the floor. At last, the floor stops moving. The earthquake is over. Or is it?

Find Out ACTIVITY

Shake It!

With a partner, design and make your own seismograph.

What to Do

What parts in the diagram below might present a challenge? What method could you devise, other than a rotating drum, to record movement? Draw and label your design. Write what you will need to make your seismograph and how you will go about it. If your model differs in any way from your design, draw and label your seismograph, and describe the changes and why you made them.

Demonstrate your seismograph for other students. Compare it with those designed by your classmates. Discuss your comparisons with your partner, and write down what you noticed. How might you improve your seismograph?

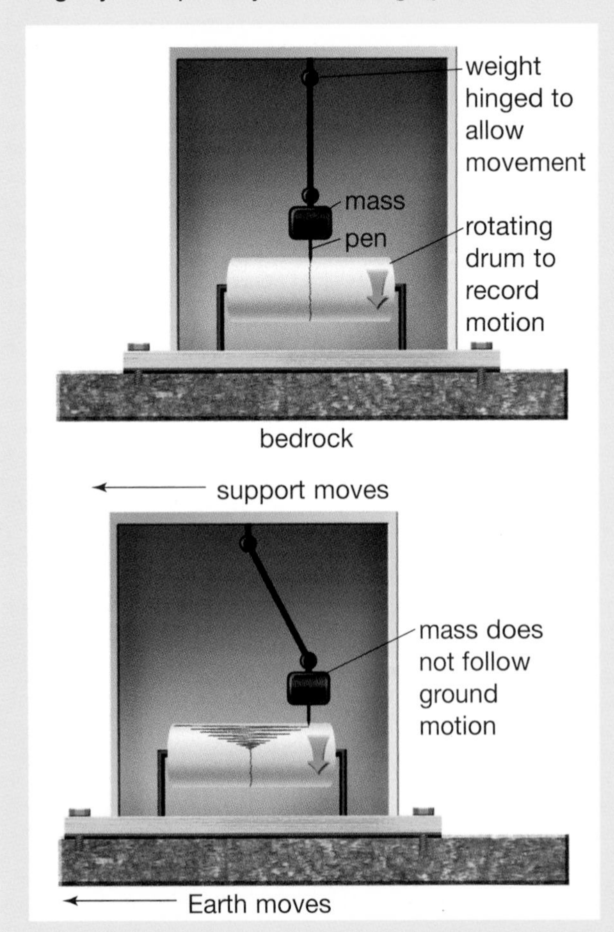

If you have never experienced an earthquake, you might think that the ground shakes only once and then the earthquake is over. There can actually be many episodes of ground-shaking movement, however, caused by **seismic waves**. These are the energy waves that travel outward from the source of the earthquake. These **aftershocks** are actually smaller earthquakes, and they produce even more ground movement. Aftershocks can cause damaged buildings to collapse. The Kobe earthquake in Japan produced over 600 aftershocks.

Figure 11.3 Lichens such as these are called pioneer organisms because they are the first to grow on solid rock. Geologists use the ages of lichens to figure out when earthquakes exposed new rock for the lichens to grow on. Such information helps geologists to predict when the next earthquake is likely to happen in these locations.

CONDUCT AN

INVESTIGATION 11-A

How Do Earthquake Waves Travel?

When an earthquake occurs, several different kinds of vibrations, or waves, travel out from the source. These waves cause differing ground movements.

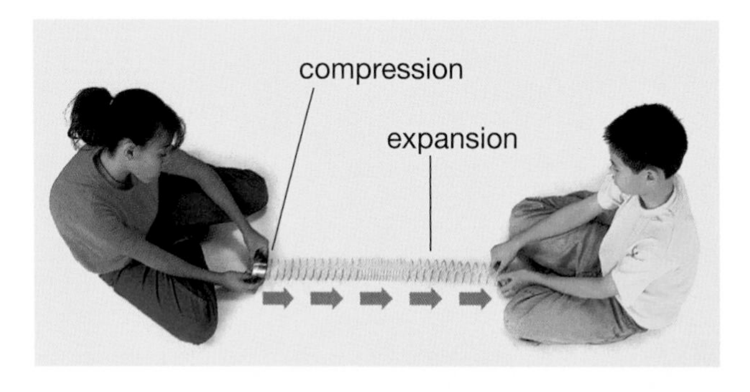

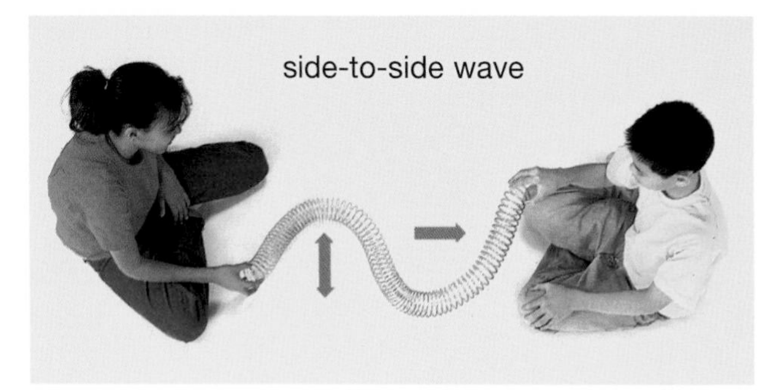

Problem

How do ground movements during earthquakes cause so much damage?

Apparatus
Slinky® coil

Procedure

1. Read steps 2 to 4. Think about the kinds of movements you will be causing in the Slinky® coil. Predict which of the "waves" will be first, second, last to travel from one end of the Slinky® coil to the other. Explain why you think so.

2. Make a table like the one below. Give your table a title.

3. Work with a partner. One person holds one end of the Slinky® coil on an uncarpeted floor so that the coils are lined up in a straight line. (Try not to stretch the coils out too far.) The other person gathers up a few of the coils at the other end and then quickly releases them. Observe what happens. Call this a compression wave.

4. One person holds one end of the Slinky® coil on the floor. The other person moves the other end from side to side. Observe what happens. Call this a side-to-side wave.

5. One person moves one end of the Slinky® coil up and down as the other person moves the other end from side to side. Call what happens a double wave.

6. Repeat steps 3 to 5, to find out which of the three waves travels the fastest and the slowest.

Type of wave	Predicted speed	Actual speed
compression wave		
side-to-side wave		
double wave		

Analyze

1. (a) How did you create the fastest wave? The slowest wave?
   (b) Check your predictions. Did any of your "race" results surprise you? Explain.
2. If you had tried your "race" a few times, might the results have been different? Explain.

Conclude and Apply

3. Which kind of wave might cause the most damage during an earthquake? Why?

Types of Earthquake Waves

What you saw in the activity represents the three kinds of seismic waves that occur in an earthquake. Seismologists have given names to them.

- **Primary** or **P waves**, represented by the compression waves of the Slinky® coil: They travel the fastest of all three types of waves and can pass through solids, liquids, and gases. They cause a slight vibration (compression) that would rattle dishes on the shelves. These waves warn people in earthquake areas that an earthquake is happening and can give people a few seconds to prepare for the movement to come.
- **Secondary** or **S waves**, represented by the side-to-side movement of the Slinky® coil: They travel more slowly than P waves and can pass only through solids, not through liquids or gases.
- **Surface waves**, represented by the double-wave movement of the Slinky® coil: They are the slowest of the three waves, but their rolling motion breaks up roads and buildings, so they do the most damage. You have probably thrown a small stone into water and watched the ripples spread out from the point where the stone entered the water. Surface waves travel through Earth in just the same way. They cause part of a building to move up while another part moves down. Rigid structures will collapse if the movement is too great.

Examine Figure 11.4. Which type of wave causes the greatest reaction in the seismograph? What does the seismogram tell you about the arrival times of the different waves?

Did You Know?

The P waves from the Kobe earthquake registered on the seismograph at the University of Manitoba. They travelled right through the centre of Earth!

Math CONNECT

Primary waves travel at about 6 km/s through Earth's crust. The distance from Toronto to Québec City is about 800 km. How long would it take for primary waves to travel between these two cities?

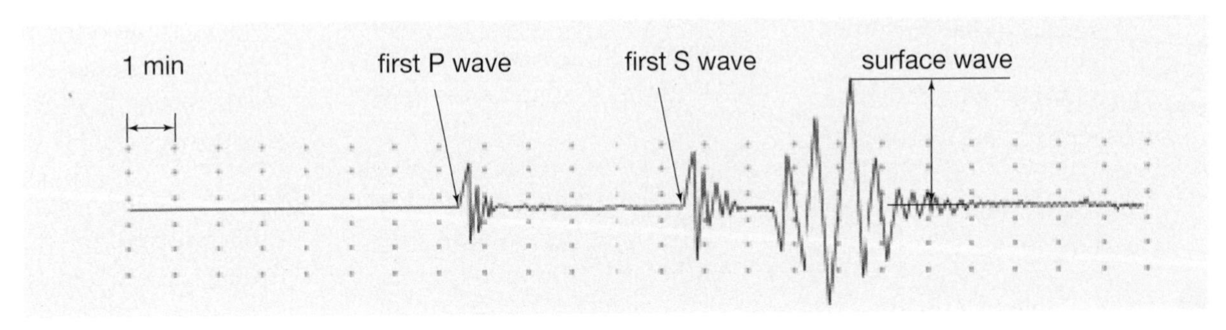

Figure 11.4 The jagged line on this seismogram represents the three different kinds of waves.

Pause & Reflect

Design Your Own 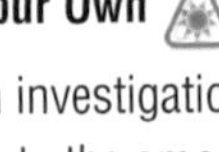

Design an investigation to demonstrate the amount of damage each type of earthquake wave might cause.

Find Out ACTIVITY

How About a Shake?

Can you watch a seismograph at work?

What to Do

Find out if there is a seismograph located somewhere close to where you live. The best place to start is the university or college closest to your home. If it is close enough to visit, check with your teacher about arranging a field trip for your class to see the seismograph.

Locating an Earthquake

During a thunderstorm, you see lightning before you hear thunder. It is possible to estimate how far away the storm is by counting the time that passes between the flash of lightning and the sound of thunder. Light travels many times faster than sound, so you see the lightning flash before you hear the thunder. The more seconds you can count between the lightning and the thunder, the farther away the storm is. If the thunder happens at almost the same time as the lightning, the storm is right above you.

You can use the same idea with earthquake locations. You know that P waves travel faster than S waves. Since this is so, it is possible to determine the location of an earthquake by the interval between the P and S waves. The farther apart the P and S waves are, the farther away the earthquake is.

Scientists have a special name for the source of an earthquake. In fact, they use two names. The place deep in the crust where the earthquake begins is called the **focus** of the earthquake. The primary and secondary waves come from the focus of the earthquake. The surface location directly above the focus is called the **epicentre.** Surface waves travel out from the epicentre (see Figure 11.5).

DidYouKnow?

The British Columbia coast is an earthquake area, as are the Atlantic seaboard, the St. Lawrence River Valley, the high Arctic, and the Yukon Territory. About one quarter of the earthquakes in Canada take place in northern regions where there are not many people to notice the vibrations. The first major Canadian earthquake on record occurred about 1535, near Québec City.

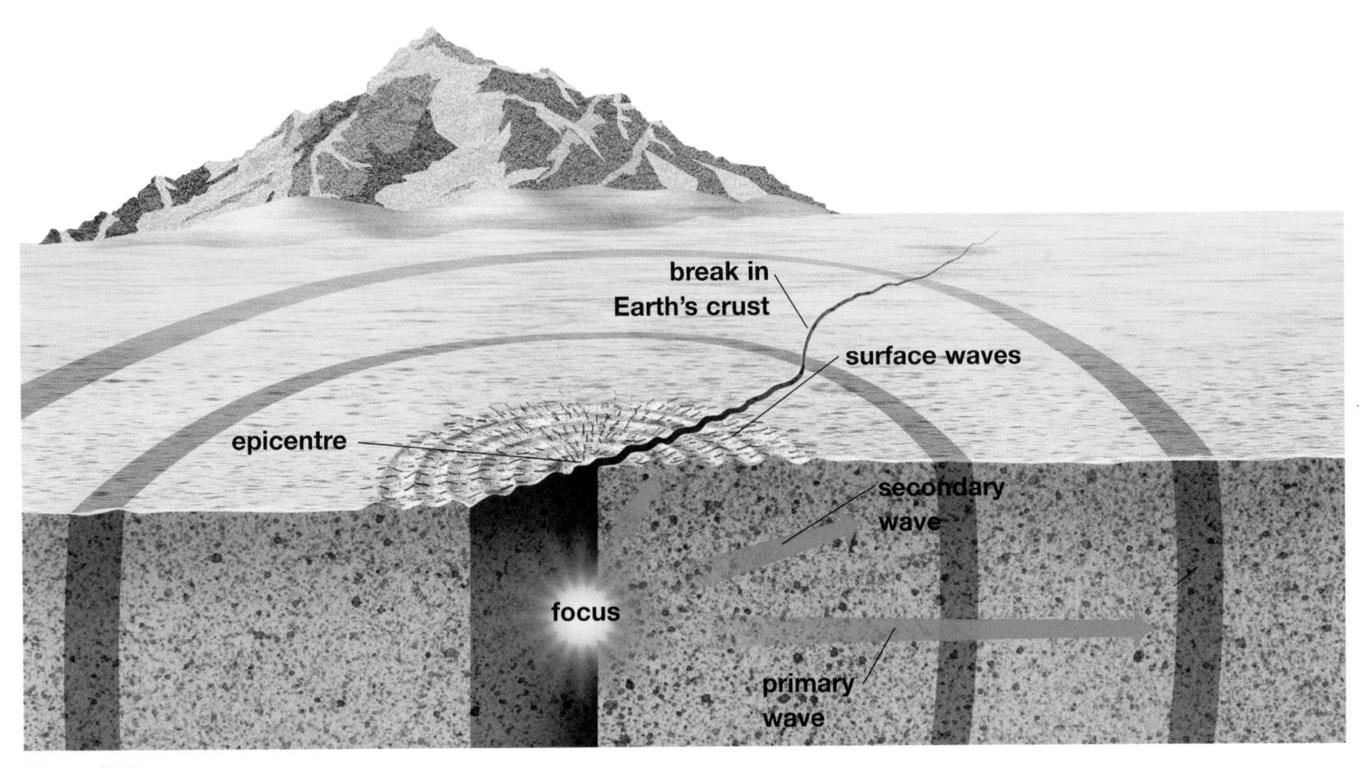

Figure 11.5

A Sudden movement in Earth's crust releases energy that causes an earthquake. The point beneath Earth's surface where the movement occurs is the focus of the earthquake.

B Primary waves and secondary waves originate at the focus and travel outward in all directions. Primary waves travel about twice as fast as secondary waves.

C The place on Earth's surface directly above the focus of the earthquake is called the epicentre. When primary and secondary waves reach the epicentre, they generate the slowest kind of seismic waves, surface waves. If you have ever floated on an inner tube in a wave pool, you have experienced waves similar to surface waves.

D Surface waves travel outward from the epicentre along Earth's surface in much the same way that ripples travel outward from a stone thrown into a pond.

Earthquake Zones

Skill POWER

For tips on using the Internet effectively, turn to page 497.

Since 1900, 4643 sizable (greater than 3.0 magnitude) earthquakes have been recorded in Canada, the United States, and Mexico. Only 17 of these earthquakes have been magnitude 8 or greater. One of these was off Canada's west coast; 8 were in Mexico, and 8 were in Alaska. Why might it be important to know when an earthquake has happened elsewhere?

Some places in the world have many earthquakes, and other places have few. Although we do occasionally experience earthquakes in Canada, we rarely feel them because most of them register low on the Richter scale.

Examine the map in Figure 11.6, and identify where earthquakes mostly occur in North America. In Chapter 12, you will look at worldwide earthquake activity and compare it to other events related to Earth's crust.

INTERNET CONNECT

www.school.mcgrawhill.ca/resources/

You can find recent information about Canadian earthquakes on the Internet from the Geological Survey of Canada, or the United States Geological Survey. Go to the above web site, then to **Science Resources**, then to **SCIENCEPOWER 7** to know where to go next. Find a list or map of recent Canadian earthquakes this year. If you were to update the map in Figure 11.6, would you include any of these earthquakes? Why or why not?

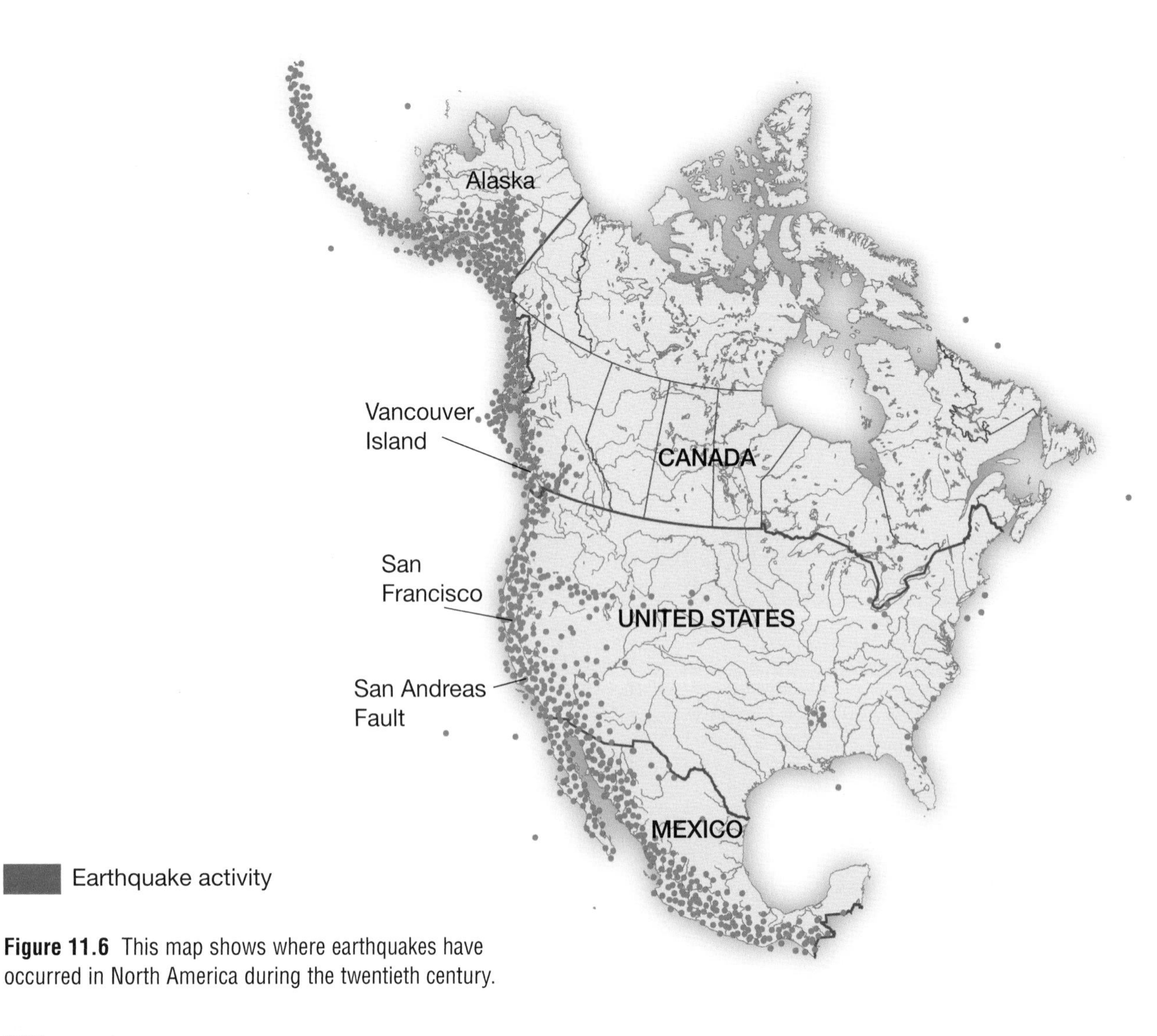

Figure 11.6 This map shows where earthquakes have occurred in North America during the twentieth century.

Types of Rock Movement in Earthquakes

As you learned in Chapter 10, the rock in Earth's crust is under pressure all the time from tremendous forces. (You will learn more about these forces in Chapter 12.) The pressure keeps the rock in constant movement. Seismologists have identified three basic kinds of ground movement that cause earthquakes (see Figure 11.7).

- Type 1: Two enormous rock surfaces are pushing together.
- Type 2: Two enormous rock surfaces are pulling apart.
- Type 3: Two enormous rock surfaces are sliding past each other.

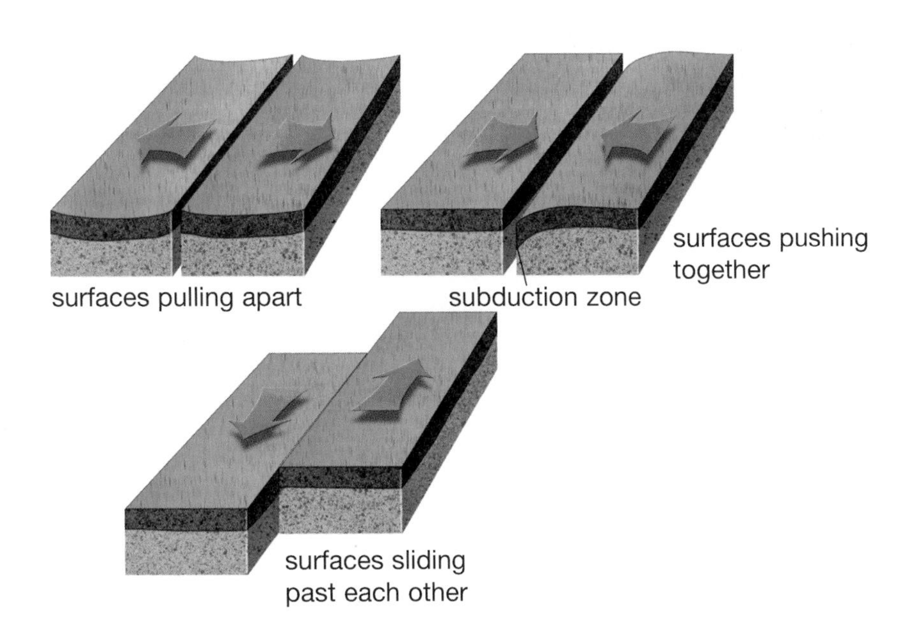

Figure 11.7

Pause & Reflect

Design Your Own

With a partner, think of some materials you could use to demonstrate types of rock movement in earthquakes. Be ready to demonstrate the various types of movement to your classmates.

Rocks That Push Together

In places where the rock surfaces push together, tremendous pressure can bend them into folds. Sometimes one piece of rock will get pushed under the other and dip deep into Earth. This is referred to as a **subduction zone.**

Scientists can tell what kind of rock movement is occurring in an area by measuring the depth of the focus of each earthquake. The deepest earthquakes are usually caused by rocks that have been pushed together, forming a subduction zone. One of the deepest subduction zones in the world is in the Pacific Ocean, in the Marianas Trench off the coast of Japan.

Rocks That Pull Apart

Earthquakes that are caused by rock surfaces pulling apart are very shallow earthquakes. The rocks pull apart when hot magma beneath Earth's crust bubbles upward to a small opening. As pressure increases, the rocks pull apart. Most of these shallow earthquakes happen on the sea floor and cause little damage. The island called Iceland in the North Atlantic Ocean experiences many shallow earthquakes. It has huge cracks on its surface where the rocks have pulled apart.

Rocks That Slide Past Each Other

In 1965 a Canadian, Tuzo Wilson, was the first person to identify the type of earthquake that is caused when rock surfaces slide past each other and lock in place until the pressure becomes too great. When the rocks shift, an earthquake happens. The area where the rocks break and move is called a **fault.** The best-known example of this type of fault is the San Andreas Fault in California. Look back at the map in Figure 11.6 to see how earthquake activity in the area around the San Andreas Fault compares with activity in other parts of the world.

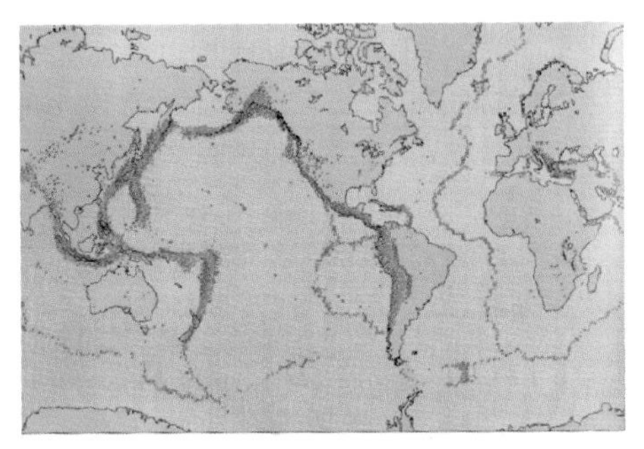

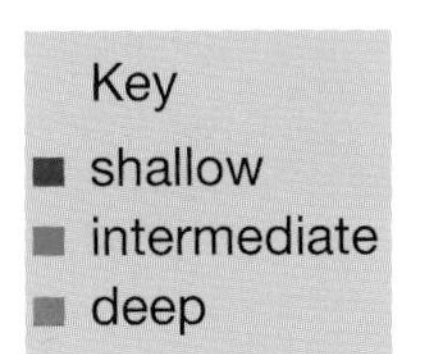

Figure 11.8 This map shows the depth of earthquakes that have occurred throughout the world.

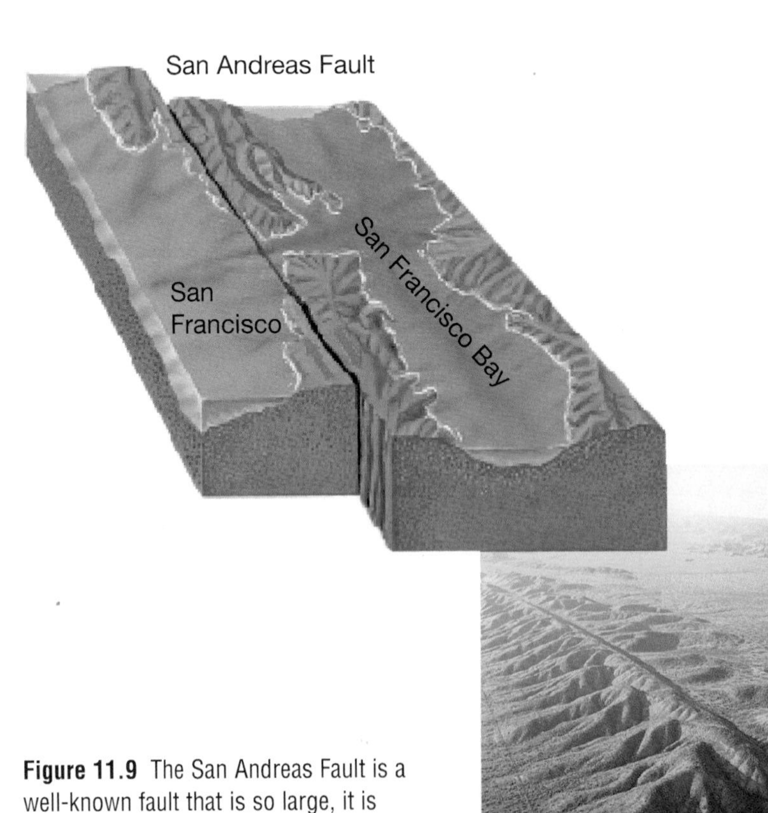

Figure 11.9 The San Andreas Fault is a well-known fault that is so large, it is visible from the air. The photograph on the right shows an aerial view of the fault, and the diagram above shows the movement of the rock surfaces that make up the fault.

INVESTIGATION 11-B

Plotting Earth's Movement

Think About It

Knowing how rock surfaces in Earth's crust are moving can help scientists predict the strength and frequency of future earthquakes.

The table below shows the location of 11 earthquakes that occurred near South America. Thousands of earthquakes have happened in this area. How can earthquake locations tell us how rock surfaces are moving?

Earthquake	Distance east (km)	Depth (km)
1	400	230
2	80	50
3	450	320
4	220	120
5	10	15
6	480	400
7	250	150
8	500	500
9	150	60
10	300	175
11	600	550

Materials

earthquake data graph paper pencil

What to Do

1. Make a copy of the graph in your notebook or on a sheet of graph paper, or obtain a blank copy from your teacher.
2. Using the table, plot the 11 earthquakes on the graph.
3. Keeping as close as you can to the pattern made by the dots, and without using a ruler, draw a smooth or curved line. (Your line might not touch all of the dots.)

Analyze

1. What do you notice about the depth of the earthquakes as you move farther from the coast underneath South America?
2. Which type of ground movement does this look like to you (Type 1, Type 2, or Type 3)?
3. Hypothesize about what might be happening to the rock below 800 km, since earthquakes do not register below this depth.

Skill POWER

For tips on drawing a graph, turn to page 486.

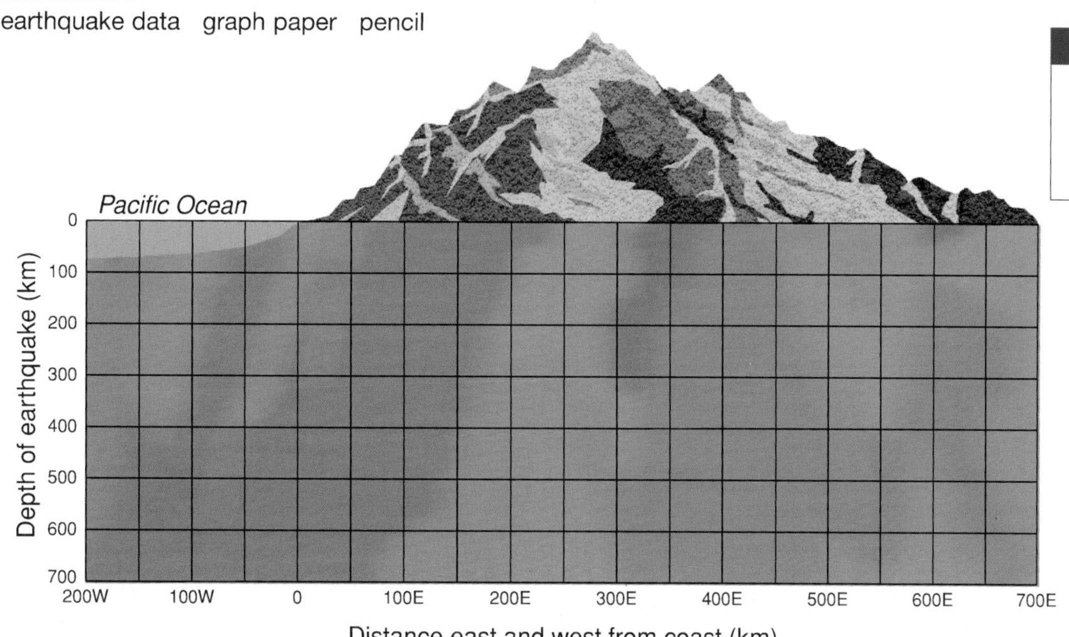

Effects of Earthquakes on People

People who live in earthquake zones learn how to prepare for earthquakes. As you know, schools have earthquake drills just like fire drills. In many homes, people attach the furniture to the walls so that it won't shift or fall over during an earthquake. They store heavier items near the floor on shelves.

Buildings and roads are constructed differently in areas that experience many earthquakes. Engineers try to make them earthquake-resistant — able to withstand the shaking of the ground that occurs during an earthquake. Rigid structures made of bricks or solid concrete break during an earthquake because they have very little flexibility. Buildings made of steel, wood, and reinforced concrete can bend a little without breaking. (You will learn more about reinforced concrete in Unit 5.)

Figure 11.10 This office building in Vancouver is specially built so that it will not collapse in an earthquake. The floors are suspended from the central core of the building by huge cables that are visible at the top. What do you think will happen to this building when the ground moves in an earthquake?

Figure 11.11 Part of this freeway collapsed during an earthquake near San Francisco. Part of it did not collapse because the concrete pillars had been reinforced to withstand an earthquake.

This system of lasers monitors movement along the San Andreas Fault. A series of 18 reflectors are positioned several kilometres away from the laser station. If a reflector's position changes, the change is measured. Movements of less than 1 mm along the fault can be detected.

At Home 

Be Prepared!

Think about how you might prepare for an earthquake in your own home.

What to Do

Consider what changes might be necessary in your bedroom to prevent you from being injured if an earthquake happened while you were sleeping. For example, are there shelves with heavy objects at the top?

Think about the items you might need in an emergency kit after an earthquake. How long might you need them? Where could you store your emergency kit?

Make a list of changes to your bedroom and another list of items for your emergency kit. Compare your lists with another student's list, and make any changes that you feel would improve your own work.

Figure 11.12 The building on the left is in the process of falling over during an earthquake in Mexico. The sediment underneath it is acting like quicksand.

Figure 11.13 Some earthquakes happen under the sea, causing huge waves called tsunamis. Tsunamis are common along Japan's coastline. This painting by artist Katsushika Hokusai shows a huge ocean wave near Japan, with Mount Fuji in the background. Tsunami is a Japanese word meaning "harbour wave."

Liquefaction

One of the most damaging earthquakes happened about 350 km east of Mexico City in 1985. When the shock waves reached the city, their size was increased by the soft sediments of the ancient lake bed on which the city is built. The sandy base turned into quicksand, and many buildings fell over (see Figure 11.12). The process of changing into a liquid-like substance such as quicksand is called **liquefaction**. The official number of deaths caused by this earthquake was over 5000.

INTERNET CONNECT

www.school.mcgrawhill.ca/resources/

Find out about the tsunami that struck Port Alberni, B.C., in March 1964 by going to the above web site. Go to **Science Resources**, then to **SCIENCEPOWER 7** to know where to go next. Write a short story, a poem such as a haiku, or a script, as though you were one of the people whose account you have just read.

Check Your Understanding

1. (a) Name the instrument that *measures* earthquakes, and explain how it works.
   (b) Name an instrument that *detects* earthquakes, and explain how it works.
2. Explain why three seismographs are required to locate an epicentre.
3. Where do earthquakes usually occur in Canada?
4. Imagine waking up in the middle of the night to find an earthquake occurring. When the shaking stops, you get out of bed to check on the rest of your family. Everyone is fine, but you notice cracks in the walls of your home. Should you stay inside or leave? Explain.
5. **Thinking Critically** What kinds of structures might suffer the least damage during an earthquake? Write your ideas, and then share them with a classmate. Check your ideas by looking at books or using the Internet.

11.2 Volcanoes

What would it be like to watch a huge volcano erupt? The photos on pages 326 and 327 give you some idea, but you have to imagine the terrific heat, the choking ash, and the streams of molten lava that threaten to overtake you. A **volcano** is an opening in Earth's crust that releases lava, smoke, and ash when it **erupts** (becomes active). The openings are called **vents**. When volcanoes are not active, they are described as **dormant**. Scientists try to predict when volcanoes will erupt so that the people living near them can avoid injury or death.

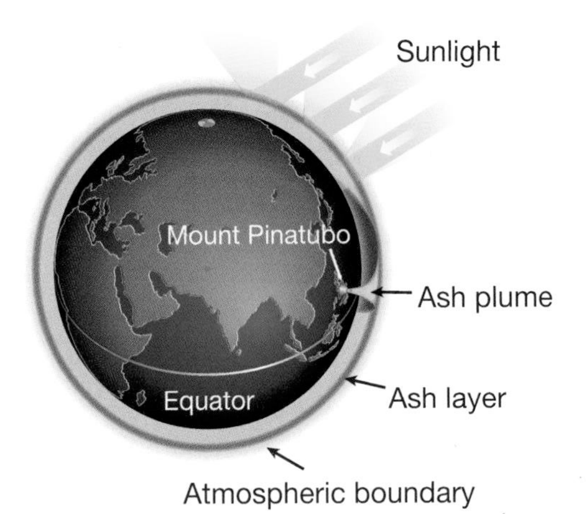

Figure 11.14A Mount Pinatubo erupted in the Philippines in 1991. The huge amount of ash blown out of the volcano formed an ash layer within the atmosphere that circled the globe and cooled temperatures around the world.

Types of Volcanoes

Imagine watching a volcano form right before your eyes. A farmer in Mexico had this experience when a hole in the ground in one of his fields suddenly sent clouds of ash, smoke, and glowing cinders into the air. The activity continued for several days. The ash and cinders cooled, fell to the ground, and eventually formed a volcano called an **ash-and-cinder cone**. The ash-and-cinder cone is the smallest type of the three main types of volcanoes. It has gently sloping sides and a large, bowl-shaped crater. Paricutin, the volcano that formed in the Mexican farmer's field, is now several hundred metres high (see Figure 11.15)

ash and rock layers

crater

central vent filled with rock fragments and magma

Figure 11.14B Ash-and-cinder cones are built from layers of ash and rock and are usually less than 300 m high.

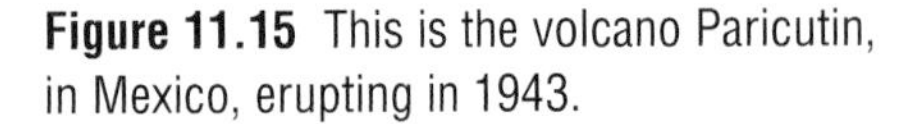

Figure 11.15 This is the volcano Paricutin, in Mexico, erupting in 1943.

Figure 11.16A A shield volcano consists of layer upon layer of magma (basalt) with no ash layers.

Figure 11.16B The wide, gently sloping structures that form the Hawaiian Islands are the largest volcanoes on Earth. The eruption in the picture is a lava fountain from Kilauea in 1983.

The Hawaiian Islands are examples of the second type of volcano, the **shield volcano** (Figure 11.16A). The shield volcano is the largest of the three main types of volcanoes. The most well-known example of this type of volcano is Mauna Loa in Hawaii. It is so huge that it is an island, growing up from the bottom of the Pacific Ocean. It is shaped like a shield that is lying face up on the ground, with gentle slopes. This type of volcano is constantly active, sending out lava most of the time. If you travel to Hawaii, you can visit the area where the volcano is active and see the lava flowing. The shield volcano does not erupt violently as the ash-and-cinder cone does.

Scientists have collected information indicating that a **hot spot** under Hawaii is responsible for the volcanic activity there (see Figure 11.17). A hot spot is a location where the temperature under the crust is much higher than elsewhere. As a result lava is forced upwards, through cracks in Earth's crust. The people who live in Hawaii see lava flowing almost continuously from Mauna Loa (see Figure 11.18).

Pause& Reflect

Write notes in your Science Log about the two types of volcanoes you have read about so far. Hypothesize about what a third type of volcano might be like if it were a combination of the first two.

DidYouKnow?

One of the worst eruptions in history happened on August 27, 1883, when the volcanic island of Krakatau in Indonesia literally blew up. The blast was heard 4800 km away. At least 32 000 people died when tsunamis reaching 30 m in height were triggered by the eruption.

DidYouKnow?

A new volcano is forming under the ocean, right beside the main island of Hawaii. It has already been named Loihi. It will continue to grow until it is another island in the middle of the Pacific Ocean. Perhaps your descendants will visit Loihi in the future!

Figure 11.17 This illustration shows the Hawaiian Island chain stretching north in the Pacific Ocean and the hot spot that has created it. The northern islands in the chain no longer have any volcanic activity. They are also much older than the southern islands. What clue might this give you as to what is happening to Earth's crust here?

DidYouKnow?

Tourists can visit the Hawaii Volcanoes National Park to see the top of the volcano Kilauea. Clouds of steam fill the air, and the odour of rotten eggs is noticeable. What do you think causes the odour? **Hint:** Some of the ground is a yellow colour!

Figure 11.18 Tourists in Hawaii can watch the lava flow into the ocean and listen to it sizzle!

DidYouKnow?

Mount Pinatubo in the Philippines is another example of a composite volcano. It was the largest volcanic eruption of the twentieth century. It had been dormant for about 600 years before it erupted in 1991.

Like earthquakes, volcanoes can be formed when rock surfaces beneath Earth's crust push against one another. The part of the crust that is pushed downward reaches very hot areas where it melts and becomes magma. Eventually there is so much magma, it is forced up through openings and erupts. The third type of volcano is called a **composite volcano**. The eruption may be violent, as in an ash-and-cinder cone, or quiet, causing lava to ooze over the cinder layers as in a shield volcano. Mount St. Helens, in Washington, is an example of a composite volcano. The rock on one side of the mountain began to bulge out in the days before the eruption that occurred in 1980. Scientists knew that an eruption would happen soon, so they had time to warn people to stay away from the area. The eruption literally blew away the side of the mountain. Figure 11.19 (below) shows how magma built up inside the volcano to cause the eruption. (You can see a cross section of a composite volcano in Figure 11.20 on page 332.)

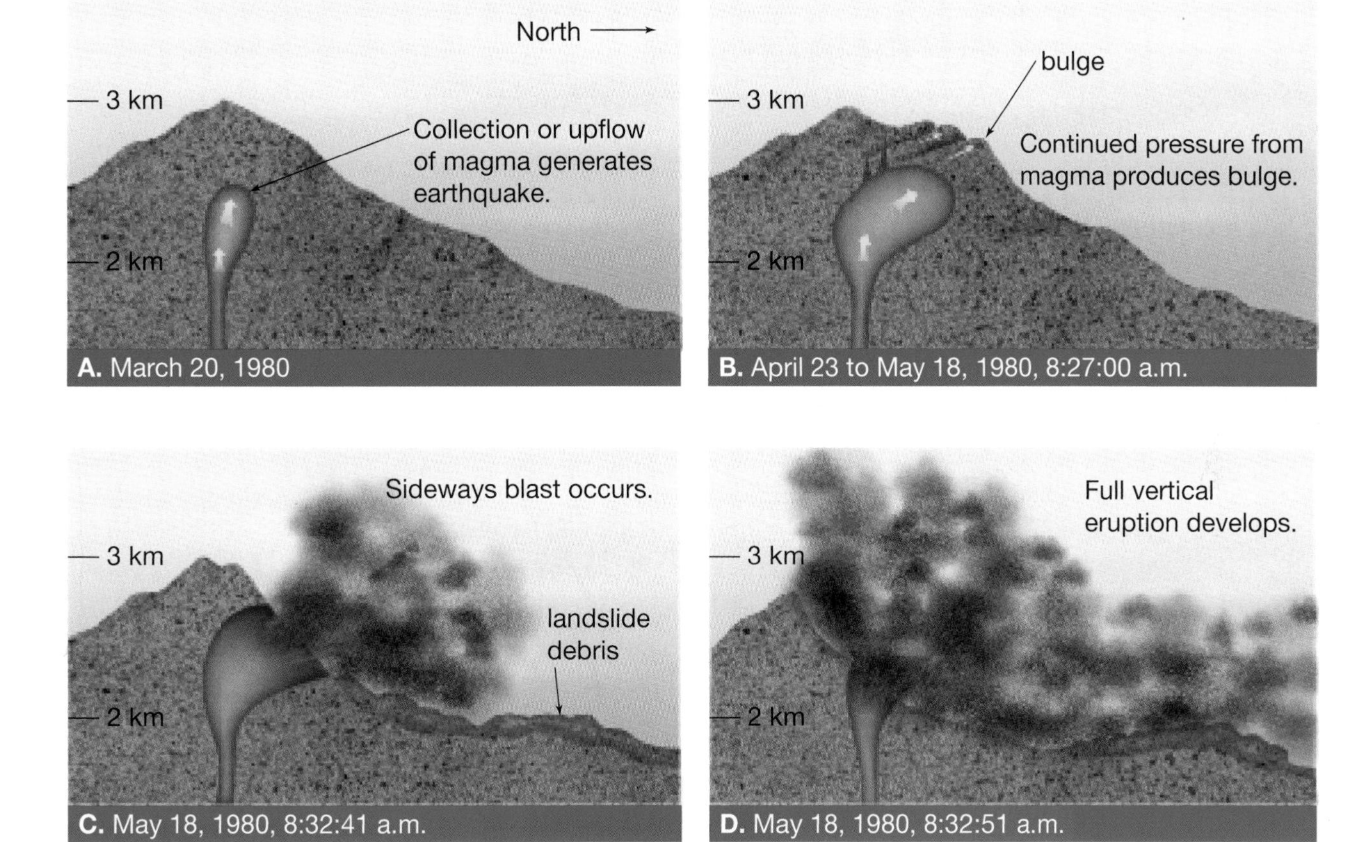

Figure 11.19 Volcanos erupt in stages, over a period of several weeks, months, or even years.

Find Out **ACTIVITY**

Myths Retold

The mythologies of Greeks, Romans, Indonesians, Japanese, Icelanders, and Hawaiians all contain accounts of gods or goddesses whose anger resulted in a volcanic eruption. For example, Hawaiians believed that the Fire Goddess, Pele, whose image is shown in the photograph, lived inside the volcano Kilauea. They believed that the volcano erupted when she became angry.

What to Do

Look for myths about volcanoes in library books and on the Internet. How do these myths compare with the story you told in the Starting Point Activity? Choose one of the myths. Tell it in your own words and illustrate it.

CONDUCT AN

INVESTIGATION 11-C

Patterns in Earthquake and Volcano Locations

One way to predict earthquakes and volcanic eruptions is to look for patterns in their occurrence. In this investigation, you will plot earthquake and volcano locations on a map using lines of latitude and longitude. You will begin to recognize a pattern of earthquake and volcano activity which suggests that something unusual occurs in Earth's crust in certain areas.

Problem

Is there an observable pattern in the occurrence of earthquake and volcanic activity?

Apparatus

world map with latitude and longitude lines
Table: Earthquakes Around the World
Table: Volcanoes Around the World
coloured pencils or markers

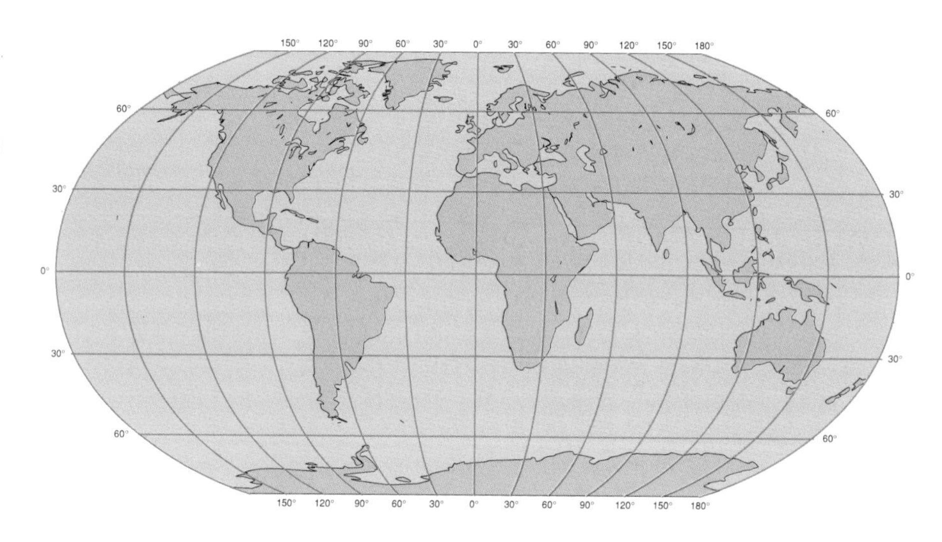

Procedure

1. Use the tables on page 331, or obtain your own earthquake/volcano list from the Internet.
2. Decide on a colour and symbol that you will use to indicate earthquake and another colour and symbol that you will use to indicate volcano locations. You might choose a blue circle for earthquakes and a red triangle for volcanoes. Put your symbols on your map under the title "Legend," indicating what each symbol represents.
3. Mark the earthquake epicentres and volcanoes on the map.
4. Keep your map in a safe place so that you can use it later in this chapter.

Skill POWER

For tips on using the Internet, turn to page 497.

Earthquakes Around the World

Longitude	Latitude	Location	Year
122°W	37°N	San Francisco, California	1906
72°W	33°S	Valparaiso, Chile	1906
78°E	44°N	Tien Shan, China	1911
105°E	36°N	Kansu, China	1920
140°E	36°N	Tokyo, Japan	1923
102°E	37°N	Nan Shan, China	1927
85°E	28°N	Bihar, India	1934
39°E	35°N	Erzincan, Turkey	1939
136°E	36°N	Fukui, Japan	1948
133°W	54°N	near Queen Charlotte Islands	1949
97°E	29°N	Assam, India	1950
3°E	35°N	Agadir, Morocco	1960
48°E	38°N	Northwestern Iran	1962
147°W	61°N	Seward, Alaska	1964
57°E	30°N	Southern Iran	1972
87°W	12°N	Managua, Nicaragua	1972
92°W	15°N	Central Guatemala	1976
118°E	39°N	Tangshan, China	1976
40°E	40°N	Eastern Turkey	1976
68°W	25°S	Northwestern Argentina	1977
78°W	1°N	Ecuador-Colombia border	1979
137°E	37°N	Honshu, Japan	1983
102°W	18°N	Western Mexico	1985
45°E	41°N	Northwestern Armenia	1988
122°W	37°N	San Francisco, California	1989

Volcanoes Around the World

Longitude	Latitude	Location
122°W	46°N	Mount St. Helen's, Washington
123°W	50°N	Garibaldi, British Columbia
130°E	32°N	Unzen, Japan
25°W	39°N	Fayal, Azores
29°E	1°S	Nyiragongo, Zaire
152°W	60°N	Redoubt, Alaska
102°W	19°N	Paricutin, Mexico
156°W	19°N	Mauna Loa, Hawaii
140°E	36°S	Tarwera, Australia
20°W	63°N	Heimaey, Iceland
14°E	41°N	Vesuvius, Italy
78°W	1°S	Cotopaxi, Ecuador
25°E	36°N	Santorini, Greece
123°E	13°N	Mayon, Philippines
93°W	17°N	Fuego, Mexico
105°E	6°S	Krakatoa, Indonesia
132°W	57°N	Edziza, British Columbia
74°W	41°S	Osorno, Chile
138°E	35°N	Fujiyama, Japan
15°E	38°N	Etna, Sicily
168°W	54°N	Bogoslov, Alaska
121°W	40°N	Lassen Peak, California
60°W	15°N	Mount Pelée, Martinique
70°W	16°S	El Misti, Peru
90°W	12°N	Coseguina, Nicaragua
122°W	49°N	Mount Baker, Washington State
121°E	15°N	Mount Pinatubo, Philippines

Analyze

1. Are most of the earthquakes located near volcanoes, or are their locations unrelated?
2. Describe the pattern of earthquakes and volcanoes in or around the Pacific Ocean?
3. Does the pattern around the Atlantic Ocean look similar to or different from the Pacific Ocean pattern?
4. Where do most earthquakes in North America occur?
5. Describe any other places in the world that appear to have a large number of earthquakes and volcanoes.

Conclude and Apply

6. What conclusion can you reach about earthquake and volcano locations, based on your observations?
7. If you were a scientist, what might you hypothesize about Earth's crust in these areas?

Extend Your Knowledge

8. Look on the Internet for a map of worldwide earthquake activity prepared by the Geological Survey of Canada and the United States Geological Survey. How does the pattern of earthquake and volcano activity on this map compare with the pattern on the map you prepared?

Famous Volcanoes

Probably the most famous composite volcano on Earth is Mount Vesuvius in southern Italy. Many scientists believe that Vesuvius, dormant since 1944, is due for a large eruption. A huge area beneath the peak is filling with magma (see Figure 11.20). The situation is even more dangerous because the opening at the peak is sealed by a rock "plug." Scientists have produced computer simulations to show that, when pressure forces the rock "plug" out, a cloud of molten rock, ash, and gas will blast about 1.5 km upwards. It will probably stay in the air, held up by heat from the volcano. Then, after several hours, it will drop to the ground and flow at about 160 km/h. This terrifying cloud is expected to reach Naples, a city of three million, five minutes after the volcano erupts. Currently there is a great deal of discussion over the appropriate measures that must be taken in case such an event occurs.

Look again at the volcano map you made. Run your finger around the edge of the Pacific Ocean, from New Zealand north to Asia and then down the west coast of North America and South America. Can you see all of the volcanoes in this circle? These volcanoes around the Pacific Ocean make up the Ring of Fire. The name comes from the circle of volcanoes that pour out red hot lava, fire, and smoke. Mount St. Helens and Mount Pinatubo are part of this Ring of Fire. Most volcanoes in the Ring of Fire occur at subduction zones.

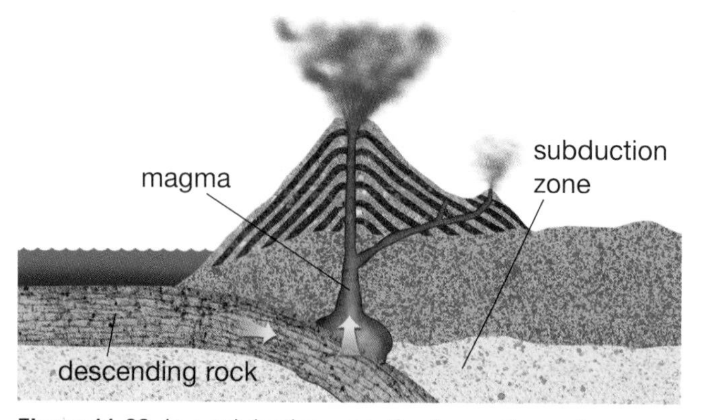

Figure 11.20 In a subduction zone, the descending rock moves deeper and deeper until it melts into magma. This magma rises up through cracks in the rock until it exerts enough pressure to cause the volcano above to erupt.

Skill POWER

To learn how to do research in science, turn to page 497.

Find Out **ACTIVITY**

A Blast from the Past

Mount Vesuvius is a volcano with a colourful past.

What to Do

With a partner, choose one of the major eruptions and find out what you can from books and from the Internet. See what you can learn about something called *pyroclastic flow*. Why is it one of the most deadly parts of an eruption?

Organize your findings in the form of an annotated poster, a storyboard, a booklet, or a demonstration (live or on video) that you can present to your class.

Key Dates in the History of Vesuvius

300 000 B.C.E.: Molten rock is forced up between two major rock surfaces to form Mount Vesuvius.

25 000 B.C.E.: Oldest dated explosive eruption occurs.

A.D. 79: First historically recorded explosion occurs. At least 3000 people are killed in Herculaneum and Pompeii.

472: Eruption causes destruction in north-eastern region around Mount Vesuvius.

1631: Coastal regions around Mount Vesuvius are destroyed. Eruption kills more than 4000 people.

1794: Lava destroys town near Vesuvius for the third time in 170 years. About 400 people are killed.

1906: Lava and explosions cause destruction of nearby towns. More than 500 people are killed.

1944: Mild eruption causes 26 deaths, due mainly to the collapse of buildings.

Extend Your Skills

Design Your Own With your team, plan the materials you would need and the steps you would take to build a "working" volcano.

Harnessing Earth's Energy

You have seen that Earth's interior contains a tremendous amount of heat. Scientists have begun to harness the energy that exerts such incredible force many kilometres under Earth's crust. The intense heat deep inside Earth has been harnessed at hot spots in more than 20 countries around the world. At such hot spots, boiling water or steam occurs naturally throughout the rock in areas where magma is close to Earth's surface. This water or steam is piped to a power plant at the surface. There it is channelled through a control system to turbines, and is transformed into electrical energy. The energy that we harness from Earth's interior is called **geothermal energy**.

Unfortunately, geothermal energy exists only where magma is close to Earth's surface or where geysers are located. Geysers are caused by water under the ground being heated by magma (see Figure 11.21). Geysers are found at various places on Earth's surface.

As well as being an alternative to our dwindling supply of fossil fuel, geothermal energy is clean and the power plants that convert it to electrical energy are reliable. Such energy would help to reduce the threat of oil spills, wastes from the mining of fossil fuels, and pollutants from their use.

Where magma is too far beneath the surface, scientists are finding other ways to use geothermal energy. A technique called *HDR* (hot, dry rock) is used to pump water into rock that has been cracked (see Figure 11.22). The heated water returns to Earth's surface as steam, where it helps to generate electricity. At present, this process is expensive, but it could be an alternative to fossil fuels in the future.

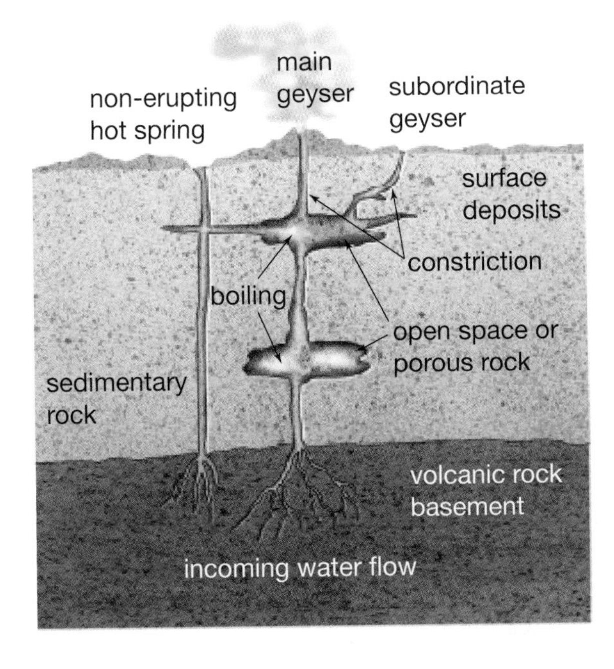

Figure 11.21 Examine the cross-section of a hot spring and a geyser. Both could be harnessed for geothermal power. British Columbia has more geothermal energy than other parts of Canada, but for a number of reasons, hydroelectric energy remains the main focus of development at this time.

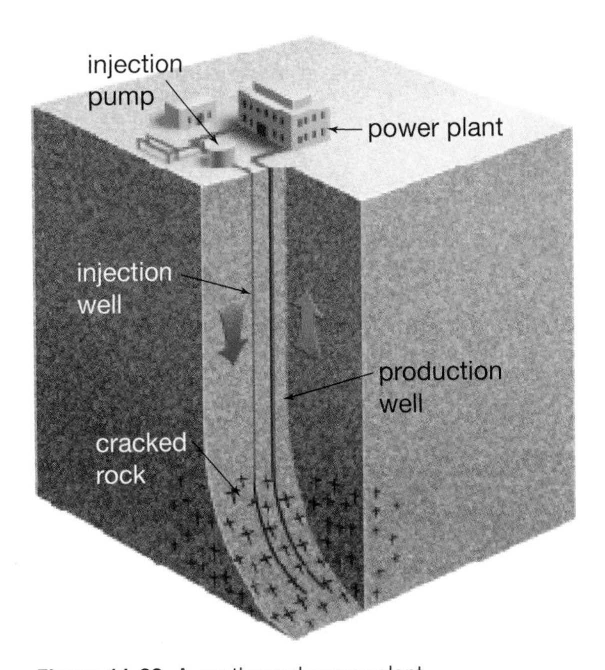

Figure 11.22 A geothermal power plant

Check Your Understanding

1. What similarities are there between the causes of earthquakes and the causes of volcanoes?
2. Where is the Ring of Fire, and how did it get its name?
3. Where might you find volcanoes in Canada?
4. **Apply** How could heat from inside Earth provide electrical energy?
5. **Thinking Critically** Why is it important to investigate sources of energy other than fossil fuels?

11.3 Mountains

"You're really growing, aren't you?" How many times has someone made a comment like this to you? Have you ever thought that the same kind of comment could be applied to a **mountain**? How do mountains form? Which ones are still growing? Why do they stop growing? These are the kinds of questions that scientists ask as they try to solve some of the mysteries related to Earth's crust.

In Chapter 10, you learned how sedimentary rock is formed. A lot of sedimentary rock has formed at the bottom of ancient oceans from shells of marine creatures. These shells have gradually settled to the bottom and, over many, many years, have compressed and built up to form layers of sedimentary rock. In other areas, sedimentary rock is made of sand, gravel, and mud.

How does sedimentary rock turn into mountains? The activity described in the margin will help you to visualize the process.

Mountain Formation

Figure 11.23 Can you see the layers in the rock of this mountain? These are exposed sedimentary layers in the Rocky Mountains of British Columbia.

Sedimentary rocks that are placed under slow, gradual pressure can either fold or break. As you have seen, rocks that are pushed against one another can cause earthquakes and volcanoes. These occur when there is relatively rapid movement. Geologists explain that rocks can fold if they are hot enough to act like a bendable plastic. It is important to remember that where a volcano exists, the rock beneath Earth's surface is hot enough to melt into magma. Where mountains are forming, the rock is only hot enough to bend.

Can you see the folded rock layers in the photograph in Figure 11.23? Study the diagram in Figure 11.24 to see the fold lines in the rock. This is what you imitated in the activity with the modelling clay. The upward or top part of the folded rock is called the **anticline**. The bottom of the fold is called the **syncline**. Over time, both of these can erode, but the folded layers still indicate what has happened. Can you identify the top of Figure 11.23 as an anticline or a syncline?

Find Out **ACTIVITY**

Make a Mountain!

How do mountains form?

What You Need

3 colours of modelling clay

What to Do

Flatten the modelling clay into circles about 15 cm across. Pile them on top of each other. Put your hands on each side of the stack and push together. What happens to the layers?

Make a hypothesis about why mountains continue growing and why they stop growing? (**Hint**: Think about what is happening in Earth's crust.)

What does a "young" mountain look like? What does an "old" mountain look like? Mountains that are jagged at the top are young; mountains that are more rounded are older. Think back to the rock cycle in Chapter 10 and try to identify another difference in appearance between a young and an old mountain range.

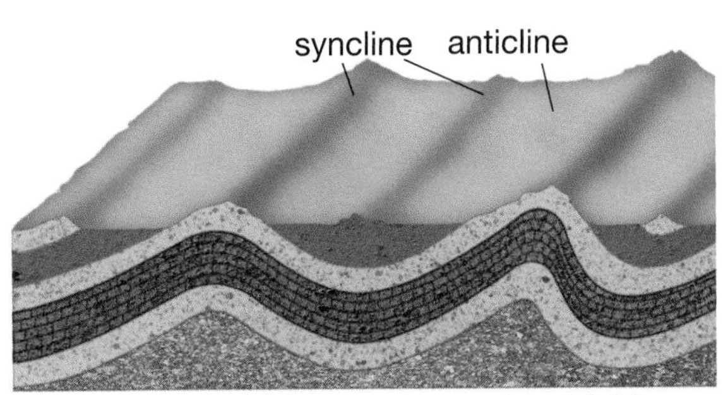

Figure 11.24 These are the main types of folding action in mountain formation.

Look again at Figure 11.23. Would you describe the Rockies as an old or a young mountain range? Some of the peaks in the Rockies are so high that they are snow-covered all year. If you think that the Rockies are a young mountain range, you are correct. They are one of many younger mountain ranges in the world, with the Himalayas in India being the youngest and highest. The top of Mount Everest in the Himalayas, like you, is still growing taller! The Laurentian Mountains in Québec are not as high as the Rockies. They are an older mountain range that is in the process of being worn down.

Sometimes the rocks in Earth's crust are too brittle to fold. When pressure is exerted on them, they break, forming a fault. Many different kinds of faults are created when the rock is too brittle to bend. As you know, the best-known example of a fault is the San Andreas Fault in California. In 1906 the San Andreas Fault shifted 6 m from side to side near San Francisco, causing a huge earthquake.

Figure 11.25 This photo of sedimentary rock layers was taken in the Rockies, near Lake Louise, Alberta.

Pause& Reflect

Find mountain ranges that are close to current earthquake and volcano activity. Now find mountains that are *not* close to current earthquake activity. The Appalachians in eastern North America are one example. These mountains are formed of much older rocks than the Rocky Mountains along the west coast of North America. What do you think might have been happening when the Appalachians and other older mountains were being formed? Write your ideas in your Science Log.

Figure 11.26 Compare this map with your map of earthquakes and volcanoes. Most of the western coastline of North and South America are areas where rock surfaces are pushing against each other.

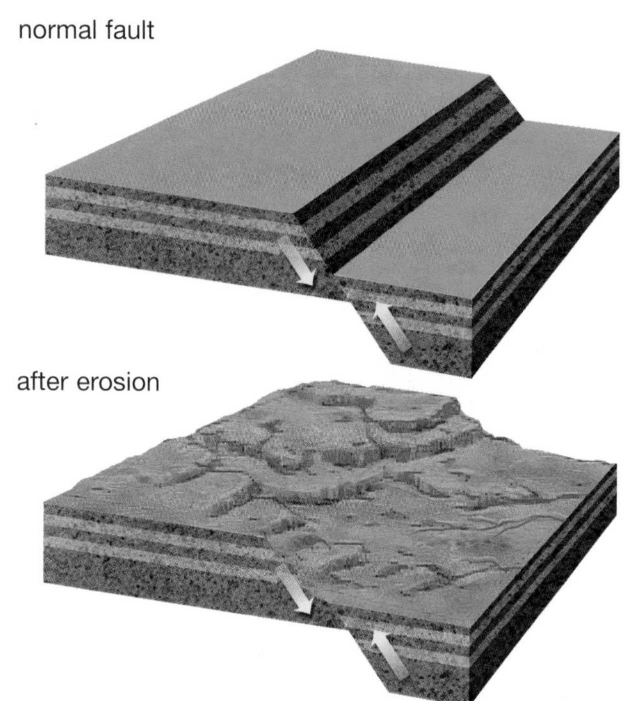

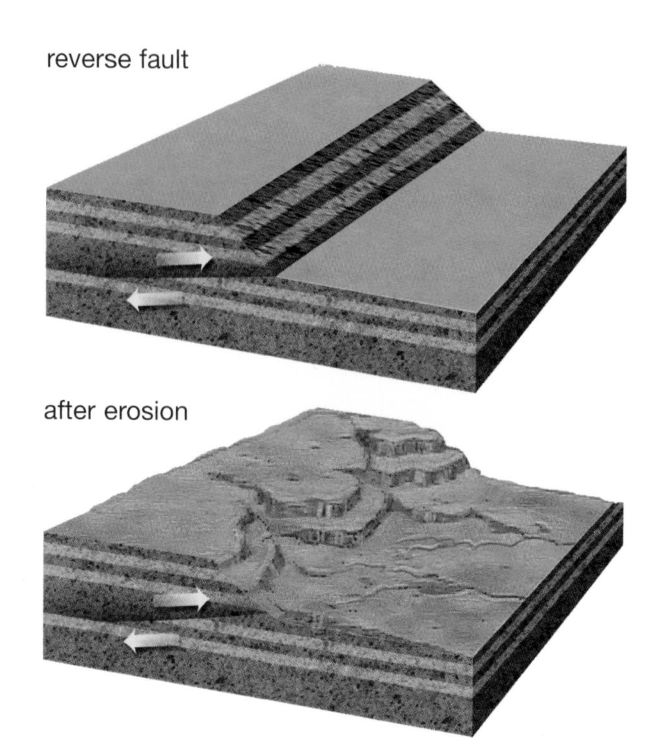

Figure 11.27 Fault movement

As well as moving from side to side, faults can move up and down, or vertically. Vertical faults are called **normal faults** if the rock moves down, and **reverse faults** if the rock is pushed up. Mountain ranges are often the result of more than one kind of action (see Figure 11.27). Mountains created as a result of both folding and faulting are described as **complex mountains**. The photos of the Rockies show folding in the rock layers and faulting in places where the layers are broken away.

Career CONNECT

Room at the Top

Scenic mountainous areas have long been popular vacation spots. In a small group, brainstorm what sorts of businesses would likely be found in mountainous areas.

What sort of business might you be interested in starting in this type of area? Write a description of your business, explaining what service it would provide, who its clients would be, and why you think it could be successful. With the help of an adult, try to arrange an interview with someone who runs a similar type of business in your own community.

Before the interview, prepare questions such as these:

- Did you begin the business by yourself or with partners?
- How much time went into planning your business before you actually started?
- Has your business grown?
- What have you learned from having your own business?

Check Your Understanding

1. Describe the two main types of mountain formation, and make a drawing of each.
2. What is an anticline? What is a syncline? What does each indicate to a geologist?
3. **Thinking Critically** Why do you think the Himalayas are getting taller and the Laurentians are getting smaller?

11.4 Layers of Earth

For a long time, people have wondered about what might be under the ground to cause earthquakes and volcanoes. Jules Verne wrote a book called *Journey to the Center of the Earth*, which described what he imagined the centre of Earth would look like. We are still not able to travel into the interior of Earth; in fact, it is too deep for us even to drill for test samples.

The earthquake waves that were discussed earlier in this chapter do give us some evidence of what might be inside Earth. The earthquake that happened in Kobe, Japan in 1995, registered on the seismograph at the University of Manitoba because P waves travel right through the centre of Earth. S waves did not register. We know that P waves can travel through liquid and that S waves cannot. Therefore, we can hypothesize that Earth's centre must be liquid. Both P and S waves actually change their speed and direction as they travel into the interior of Earth. Scientists have constructed a model of the interior based on the information from these seismic waves (see Figure 11.28).

Earth is composed of an outer layer called the **crust**, an inner layer called the **mantle**, and a central portion called the **core**. The crust is very thin under the ocean. In some places it extends only 5 km to 6 km down. Under some parts of the continents, the crust reaches a depth of 60 km. You could compare the inside structure of Earth to an egg. The crust is the egg shell. The mantle is like the white of the egg. The inner and outer core are represented by the egg yolk. The cracks that you make in the shell of a hard-boiled egg could be the faults in the rocks that are causing earthquakes to happen!

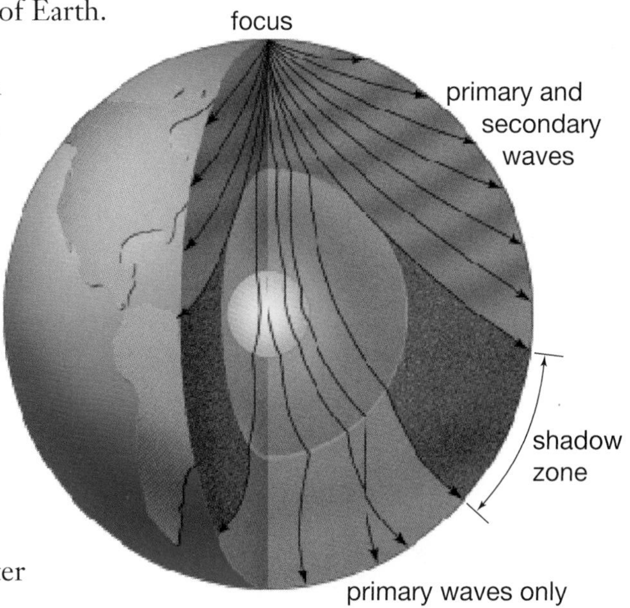

Figure 11.28 Primary waves bend when they hit the outer core, and secondary waves are stopped completely. Because primary waves are bent, there is an area where they do not come through on the other side of Earth. This area is called a *shadow zone*. Geologists have learned about the interior structure of Earth from these two kinds of waves.

Find Out **ACTIVITY**

What's Inside?

What clues help you identify hidden substances?

What to Do

Your teacher will give your group several containers, each containing a marble. There might be another substance in the container, along with the marble. Your group must examine the containers and, without looking inside, try to figure out what is inside them, besides the marble. List what you think is inside each container. Tell how you figured it out.

Pause & Reflect

Work with a partner to design a model of Earth. In your Science Log list the materials you would use.

Cool Tools

Satellites high above Earth now take pictures of Earth. These pictures are similar to X-rays because they can show what is underneath the crust. They can detect differences in temperature the same way a CAT scan can detect a tumour inside a person's body. Computers are used to interpret the pictures.

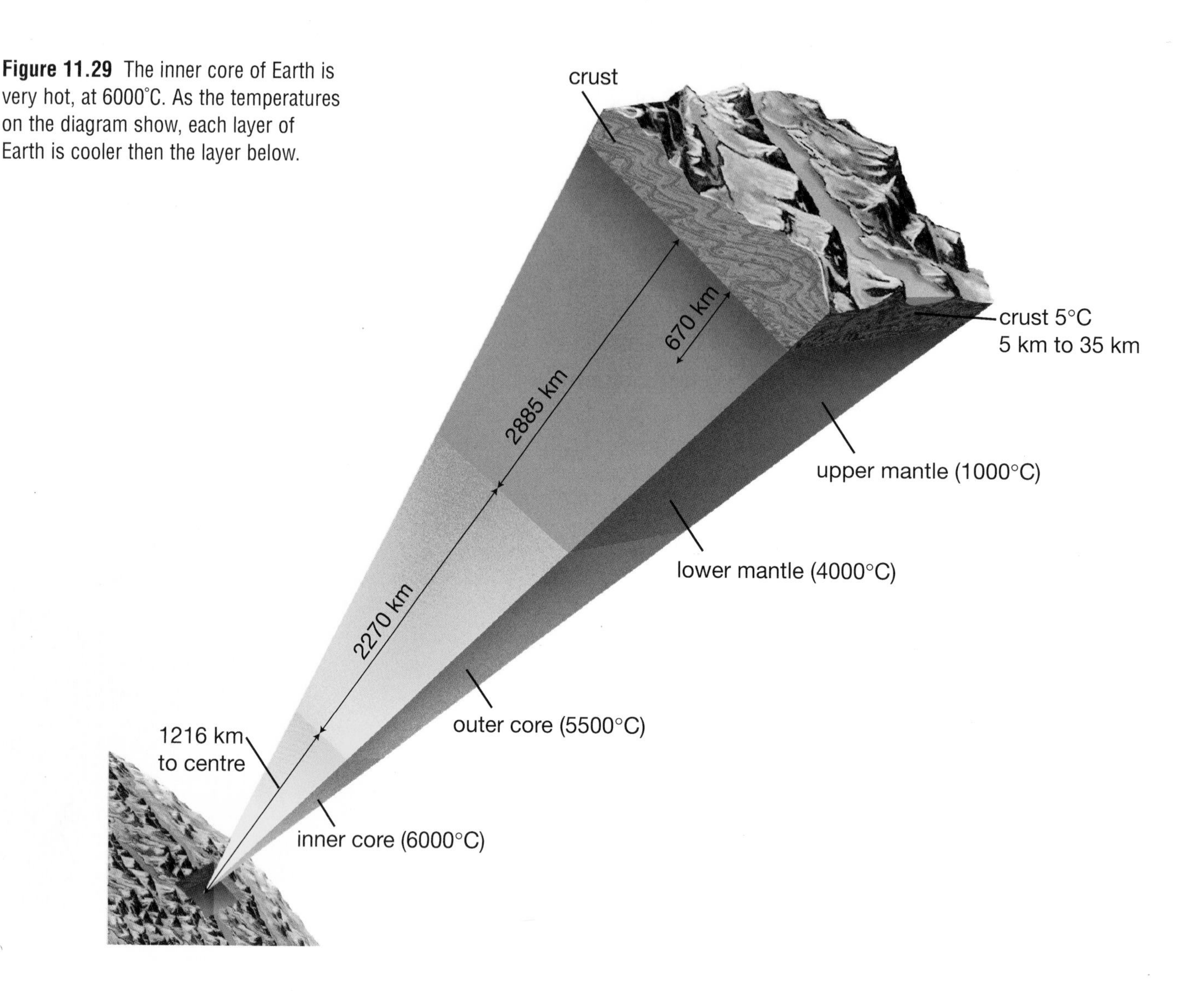

Figure 11.29 The inner core of Earth is very hot, at 6000°C. As the temperatures on the diagram show, each layer of Earth is cooler then the layer below.

Just below Earth's crust, the mantle, made of rock material, has the consistency of taffy. At a lower level, the mantle is solid. Earth's core is made of iron and nickel. The outer core, just below the mantle, is liquid. The intense pressure of all the layers about the inner core cause the inner core to be solid, in spite of its very high temperature (see Figure 11.29).

Math CONNECT

If you were travelling at 100 km/h, how long would it take you to travel through Earth's crust? How long would it take you to travel through the entire mantle?

Check Your Understanding

1. In what ways can you compare the inside structure of Earth to an egg?
2. Make a drawing of Earth and label the four main layers.
3. Describe two kinds of evidence that tell us what is inside Earth. How was the activity you did with the containers similar to the work scientists do to identify what is deep within Earth?
4. **Apply** Look at a map of Canada. Find two cities that are about the same distance apart as the distance through the lower mantle. Prepare a brief explanation for a student in an earlier grade pointing out the difficulty of exploring Earth's interior.

CHAPTER at a glance

Now that you have completed this chapter, try to do the following. If you cannot, go back to the sections indicated.

Identify three kinds of earthquake waves and the machine that is used to measure them. (11.1)

Explain how a seismogram shows the different kinds of earthquake waves and demonstrates both their speed of travel and the amount of ground movement they cause. (11.1)

Explain the difference between the epicentre and the focus of an earthquake. Describe the method used to locate the epicentre of an earthquake. (11.1)

Explain how the arrival time of earthquake waves on a seismogram relates to the location of the epicentre. (11.1)

Name the huge wave created by an earthquake in the ocean. Where does this wave often happen? (11.1)

Explain the relationship between earthquake-prone areas and the development of earthquake readiness and building specifications. (11.1)

Explain the relationship between earthquake zones and volcanic activity. (11.2)

Identify three different types of volcanoes. (11.2)

Explain the connection between volcanic activity and geothermal energy. (11.2)

Describe the relationship between earthquake, volcano, and mountain range locations. (11.3)

Describe the different processes of mountain formation. (11.3)

How does technology help us to understand whether the layers of Earth are solid or liquid? (11.4)

Prepare Your Own Summary

Summarize this chapter by doing one of the following. Use a graphic organizer (such as a concept map), produce a poster, or write the summary to include the key chapter ideas. Here are a few ideas to use as a guide:

- Explain how various technologies have provided information about earthquakes, especially where they are most likely to happen.
- Describe the impact of earthquakes that occur in heavily populated areas. What can be done to prepare for an earthquake?
- How do earthquake waves help us to understand the structure of Earth's interior?
- What is geothermal energy, and why is it not more commonly used?
- Explain how folding and faulting relate to mountains.

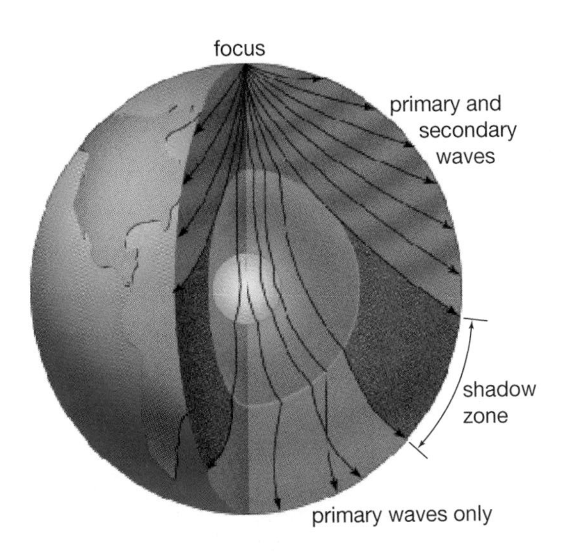

CHAPTER 11 Review

Key Terms

earthquake
seismologists
seismograph
bedrock
Richter scale
magnitude
seismic waves
aftershocks
primary (P) waves
secondary (S) waves
surface waves
focus
epicentre
subduction zone
fault
liquefaction
volcano
erupts
vents
dormant
ash-and-cinder cone
shield volcano
hot spot
composite volcano
geothermal energy
mountain
anticline
syncline
normal faults
reverse faults
complex mountains
crust
mantle
core

Reviewing Key Terms

If you need to review, the section numbers show you where these terms were introduced.

1. In your notebook, match the description in column A with the correct term in column B. Do not write in this book! Answers are found on page 498.

A

- fastest travelling earthquake wave
- surface location of an earthquake
- second type of earthquake wave generated
- bottom portion of a rock fold
- person who studies earthquakes
- most violent eruption: for example, Mt. St. Helen's
- smallest volcano: for example, Paricutin
- scale used to measure earthquake magnitude
- wave that causes the most damage
- upward portion of a rock fold
- rock break location under the ground
- huge ocean wave caused by earthquake
- largest volcano: for example, Mauna Loa

B

- epicentre (11.1)
- liquefaction (11.1)
- focus (11.1)
- P wave (11.1)
- composite volcano (11.2)
- anticline (11.3)
- seismologist (11.1)
- tsunami (11.1)
- shield volcano (11.2)
- Richter scale (11.1)
- ash-and-cinder cone (11.2)
- surface wave (11.1)
- syncline (11.3)
- dormant (11.2)
- aftershock (11.1)
- S wave (11.1)

Understanding Key Ideas

Section numbers are provided if you need to review.

2. Explain the difference between the focus and the epicentre of an earthquake. (11.1)

3. Name and describe the three kinds of earthquake waves. (11.1)

4. Describe the three kinds of rock movement that can cause earthquakes. Give an example of a place where each is happening. (11.1)

5. Name and describe the different types of volcanoes. (11.2)

6. What is a hot spot? (11.2)

7. Explain the two main processes in mountain building. (11.3)

8. How can you distinguish an "old" mountain from a "young" one? (11.3)

9. What are the names of the layers beneath Earth's crust? Which layer is thickest? Which layer is hottest? (11.3)

Developing Skills

10. In your notebook, copy and complete the following concept map on earthquakes. Use these words and connections, along with any other words you find useful: measured by, types of waves, epicentre, seismograph, Richter scale.

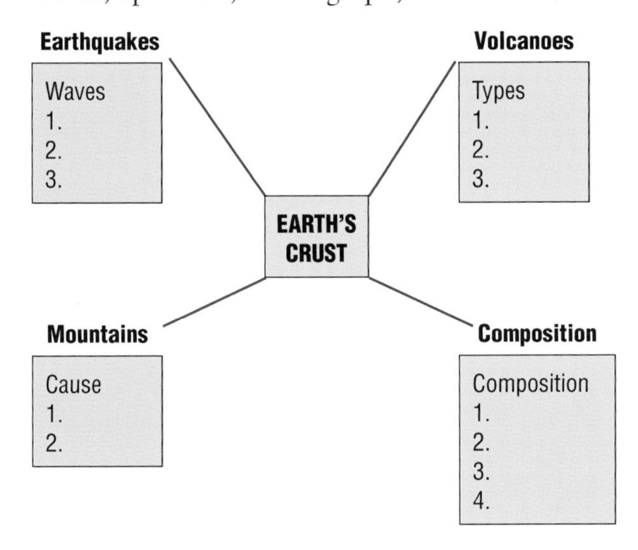

Problem Solving/Applying

11. Read the newspaper excerpt at top right, and answer the questions that follow.
    (a) The Kobe earthquake measured 7.2 on the Richter scale. How much stronger was it than the one in this article?
    (b) What inaccuracy does this article contain?
    (c) Aside from its greater strength, why do you think the Kobe earthquake killed thousands of people, while the earthquake described in this article killed no more than 60?

About 60 people were killed, many as they slept, during an earthquake in central Bolivia yesterday. A 5.9 magnitude quake struck, followed by a second one, 13 min later, with a magnitude of 6.8. The epicentre was 89 km below Earth's surface. Repeated aftershocks —up to 150 in the first 12 h alone—sent panicked residents fleeing any buildings left standing. About 30 000 people, mostly farmers, live in the area hit by the quakes. Eighty percent of houses in the community where the earthquake struck were destroyed, the hospital roof caved in, and a landslide blocked access to the town. Reports indicate that the town was almost wiped out. People gathered in main plazas after the jolts, fearing the aftershocks would bring down more buildings. Streets were cleared of rubble by tractors so that workers could assist the injured and homeless.

Adapted from *Winnipeg Free Press*, May 23, 1998

Critical Thinking

12. How do earthquakes provide us with information about the interior of Earth? How do volcanoes provide us with information about the interior of Earth?

13. What is the relationship between the locations of earthquakes, volcanoes, and mountain ranges? Which parts of the world have all three?

14. Explain why geothermal energy could be a natural energy source for the future.

15. Try to find an article in a newspaper or a magazine about an earthquake or a volcano. Bring it to school to share with the class. Write a summary of the article and keep it with the article in your journal.

Pause& Reflect

Go back to the beginning of this chapter on page 312 and check your original answers to the Getting Ready questions. How has your thinking changed? How would you answer those questions now that you have investigated the topics in this chapter?

CHAPTER

12 The Story of Earth's

Getting Ready...

- How do we know Earth's age?
- How do dinosaurs provide clues to Earth's history?
- What can magnetism tell us about Earth's crust?

Science Log

Choose one of the questions above to answer in full, or answer all of the questions in point form. Record what you already know in your Science Log. Then watch for more answers to these questions as you read this chapter.

What would it have been like to live on Earth when massive creatures like those pictured here were common? What was Earth like then? Did dinosaurs walk on exactly the same ground that you walk on today? How do we know that dinosaurs really existed? How do we know when they lived or where they lived? What clues do they provide about Earth's history?

In this chapter, you will find out how and why Earth's crust has changed over the years. You will look at the kinds of contrasting clues that scientists examine. For example, you will look at clues left by dinosaurs and clues provided by technology that let us see right into Earth's interior. By doing so, you will uncover Earth's deepest secrets and learn about the tremendous internal forces that still shake and shape the planet.

Crust

Starting Point ACTIVITY

What a Puzzle!

When you work on a jigsaw puzzle, you use certain clues that help you put it together. Think about the clues you use in this activity as you try to reassemble a torn-up picture.

What to Do

1. Choose a partner to work with on this activity. Both of you need to do all the steps.
2. Take a single page from an old magazine or newspaper.
3. Decide which side of the page to use. Glue the page to a blank piece of paper. (Do not show your partner the page you are using.)
4. Carefully tear the page into about ten pieces.
5. Trade pieces with your partner, and try to re-assemble the page. Glue the reassembled page to another blank piece of paper.
6. Write down the clues you used to help you re-assemble the page.

Spotlight On Key Ideas

In this chapter, you will discover

- fossil, rock, and climatic clues to Earth's amazing history
- how recent technology reveals clues about Earth's geological past
- current scientific thinking about the forces that have shaped Earth's crust

Spotlight On Key Skills

In this chapter, you will

- gather evidence about Earth's past on a world map
- explore and assess theories about Earth's crust
- construct a time line of the history of Earth's crust
- design an investigation into fossil formation

12.1 Early Theories about Earth's Crust

If you were the first person to observe a mountain, what clues would you use to try to figure out how it was caused and why it looked the way it did?

An early theory about our planet came from Greece. A scholar called Empedocles (495–435 B.C.E.) developed a theory that all matter consisted of four elements: earth, air, fire, and water (see Figure 12.1). Another scholar named Democritus suggested that matter was made up of tiny particles that could not be broken down further. His theory was not accepted. His contemporaries were influenced by Aristotle, a highly respected scholar who believed that Empedocles's theory of the four elements was correct. Although Democritus's theory closely resembled theories of matter that are accepted today, it took almost 2000 years for his ideas to be accepted!

Figure 12.1

In the late 1700s and early 1800s, some scientists believed that Earth was at one time completely covered by water and that all rock formation was based on sedimentation. These people were called Neptunists, after Neptune, the Roman god of the sea. Unlike the Neptunists, James Hutton and others believed that hot lava from volcanoes formed rock when it cooled. He was called a Plutonist, after Pluto, the Greek god of the Lower World.

Observation of Natural Events

As people observed natural events, such as earthquakes and volcanoes, they made these events part of the theories about Earth's development. Chinese civilizations left writings about earthquakes, volcanic eruptions, and fossil evidence of Earth's history. In the last chapter, you saw an example of an ancient Chinese seismograph from over 2000 years ago.

During the seventeenth and eighteenth centuries, people came up with an idea called **catastrophism** to explain the history of Earth. This idea stated that sudden great catastrophes, involving worldwide earthquakes and volcanic eruptions, had

Figure 12.2 This photograph shows a bust of Aristotle (384–322 B.C.E.). He was a Greek philosopher, educator, and scientist.

Figure 12.3 Hadrian's Wall was named after the Roman emperor who had the wall built in the year A.D. 120. It provided a barrier to discourage raids and revolts, and reminded tribes that the Romans were the masters of the territory. Hadrian's Wall stretches 117 km across the northern edge of present-day England. It is visible from the Space Shuttle!

created mountains and other parts of Earth's landscape. No one knew what had caused these catastrophes, but most people believed they had stopped happening. According to a popular belief at that time, Earth was created about 4000 B.C.E.

If you noticed that an ancient structure had changed very little over time, what might this suggest to you about the theory of catastrophism? In the eighteenth century, a Scottish scientist named James Hutton observed that the processes that changed the rocks and land around him were very slow. Hutton guessed that it took much longer than a few thousand years to erode mountains and build layers of rock. He could see that Hadrian's Wall, which the Romans had built 2000 years earlier, looked relatively unchanged (see Figure 12.3). He hypothesized that Earth was much older than most people thought. Catastrophism made no sense to him because he thought that whatever forces had created the mountains must still be at work. This approach to Earth's history is called **uniformitarianism**.

INTERNET CONNECT

www.school.mcgrawhill.ca/resources/
Sir Charles Lyell was a scientist who researched the history of Earth. Go to the web site above to find out more about him. Go to the **Science Resources**, then to **SCIENCEPOWER 7** to know where to go next. Write down Lyell's definition of uniformitarianism. Compare Lyell's definition with Hutton's definition.

The Shrinking Apple

A nineteenth-century explanation for Earth's mountains and valleys suggested that the world had been a hot mass that was cooling off and shrinking over time. Scientists compared the shrinking to the drying of an apple. They compared the mountains of the world to the wrinkles on the apple, and the oceans and lakes to the spaces between the wrinkles. The problem with this explanation is that apples wrinkle evenly over their entire surface, whereas Earth obviously had not, since mountains did not occur everywhere. Some scientists pointed out that mountain ranges seemed to be near the edges of continents and oceans took up more space than the

Figure 12.4 When an apple dries out, the surface becomes wrinkled. In what way does the wrinkled apple resemble Earth's surface?

Find Out **ACTIVITY**

Weather or Not!

James Hutton saw some clues to Earth's age in the weathering of rocks. You can make similar observations in this activity.

What to Do

1. With your teacher, organize a trip to a local cemetery to observe the stone markers.
2. Choose some markers. You should be able to find several types of stone; choose three or four examples of each type. Try to find the earliest examples of each type, as well as the most recent example. (If you can, identify each type of stone.)
3. Record the age of each marker and your observations of the condition of the lettering. Is the lettering clear and easy to read, or is it worn or missing? How much difference is there between different ages of markers made from the same type of rock?

What Did You Find Out?

1. Did one type of stone marker change more quickly than the others?
2. Imagine seeing three similar types of rock, all showing different amounts of weathering, at a geological site. How could you use the observations you made in this activity to learn from the rocks at the site?

spaces between the wrinkles on an apple. The theory of a shrinking Earth added nothing to the understanding of what caused mountains to form.

When scientists try to understand structures and processes that are impossible to observe, they often make models. Earth is too big to observe, and the processes that create mountains, earthquakes, and volcanoes often take too many years for one person to observe them.

You could think of an apple as a model of Earth. How much easier is it to look at an apple than to look at the entire Earth?

DidYouKnow?

A famous demonstration by Sir Charles Lyell showed how the natural forces in Earth cause rock layers to wrinkle and fold. Lyell was the first person to demonstrate the formation of folded mountains, which you learned about in Chapter 11.

Figure 12.5 Why is a globe a model of Earth? Why might you want to study a globe instead of Earth itself?

Pause& Reflect

Design Your Own 

With a partner, develop your own hypothesis about Earth's structure. Write your ideas in your Science Log.

Fixed-Continent Model

Another model of Earth's structure, which was popular until the early twentieth century, was the **fixed-continent model**. It described Earth's outer covering as a solid, rocky crust that made up the continents and the floors of the oceans. According to this model, the continents and oceans had occupied the same position since the world began. This model provided no explanation for the formation of the mountain ranges of the world, so it was no more useful than the shrinking apple model.

Check Your Understanding

1. What problems might scientists in Aristotle's time have had when they were making a map of Earth?
2. Explain catastrophism in your own words.
3. How is uniformitarianism different from catastrophism?
4. What observable features of Earth did the shrinking apple model fail to explain?
5. Describe the fixed-continent model.
6. **Apply** What problems might scientists have in making a map or a model of Earth? How do you think they might solve these problems?

12.2 Evidence for Continental Drift

The map of the world is a common sight on classroom walls. You can probably identify North America, Africa, and the Atlantic Ocean on the map. When explorers first travelled from Europe to North and South America, they produced maps that included these two continents for the first time.

In 1620 Sir Francis Bacon noticed that South America and Africa looked as though they might fit together like puzzle pieces. Examine these two continents in Figure 12.6. Can you see where the bulge of South America could fit into the indented side of Africa? Now look at Figure 12.7. It includes the **continental shelf**, a gradual, sloping edge that surrounds each continent and extends out under the ocean. When the continental shelves around all the continents are included, the fit of the puzzle pieces is even better.

This apparent fit was a mystery to scientists for a long time. If the continents were fixed in place, how could they look as though they had once been joined? One scientist who questioned this thinking was Alfred Wegener (1880–1930).

Pause& Reflect

In your Science Log, write what you might hypothesize about South America and Africa after examining Figures 12.6 and 12.7. Support your hypothesis with evidence.

Figure 12.6 This map was made by explorers during the sixteenth century. It would have been one of the maps available for Bacon to look at. Notice that the western part of North America is incomplete.

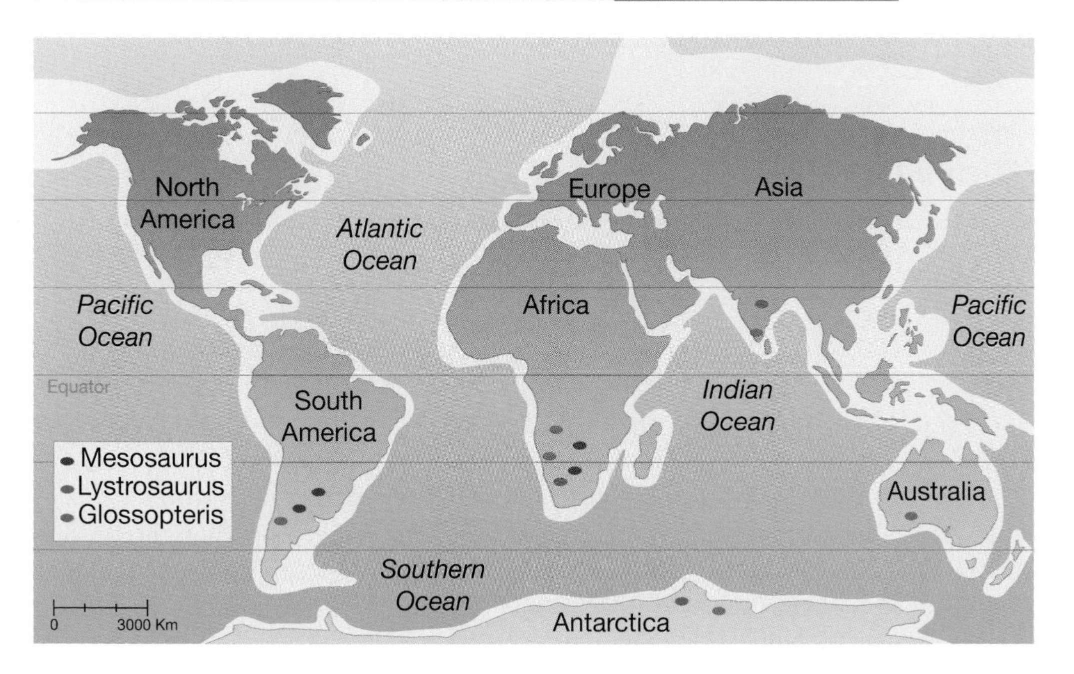

Figure 12.7 This map shows the location of three kinds of fossils that have been found on many different continents. Notice that fossils of Glossopteris have been found on Antarctica, which is totally covered with ice now.

Biological Evidence

In his readings, Wegener noticed that several fossils of similar plants and animals (like those in Figure 12.7) had been found on different continents. Mesosaurus lived in freshwater lakes, and its fossils have been found in eastern South America and southern Africa. If it was able to swim in salt water, why did it not swim to more locations? According to the evidence, it did not.

The Lystrosaurus could not swim at all, but it travelled from South America to Africa. It must have travelled by some sort of land connection.

Look at the locations where Glossopteris plant fossils have been found. What kinds of climates exist in these places today? Neither Wegener nor other scientists could understand how the same plant or animal could have developed on completely separate land masses, especially when some of the land masses now had very different climates.

Several explanations were offered to explain this **biological evidence** (evidence from plants and animals). The most popular was that a bridge of land between the continents had disappeared when Earth cooled and shrank. Another suggestion was that trees had fallen into the water, enabling animals to cross the ocean. Yet another suggestion was island "stepping stones." Perhaps, at one time, the ocean had been lower and islands had existed close enough together to allow the animals to cross.

Wegener believed that fossil evidence and the interlocking shapes of the continents indicated that the continents had been joined together when the fossil animals and plants had been alive. This meant that over thousands, maybe millions, of years, the continents had gradually moved to their present locations. Wegener called his explanation **continental drift**.

Geological Evidence from Rocks

Imagine learning something that contradicted some of your deepest beliefs. You would probably want to search for more evidence. This is just what Wegener did. He examined the observations of other scientists to see if there might be more evidence to support his idea of continental movement. He discovered that geologists had found similarities in rocks on both sides of the Atlantic Ocean. A mountain range, called the Appalachians, in eastern North America was made of the same kind and ages of rock as the mountain range that ran through Britain and Norway. These rocks looked as though they had been formed and pushed up into mountains together, but were now separated by the Atlantic Ocean (see Figure 12.9).

Meteorological Evidence from Ancient Climates

Coal provided further important information about Earth's history. In order for coal to form, there has to be rich, luxurious plant life in a tropical, swampy environment. When the plants die and are compressed under many layers of sediment, coal is formed. Thus coal beds indicate that a very warm climate existed when the plants were living. The coal beds that exist in North America, Europe, and Antarctica are now in moderate to cold climates. How did tropical plants grow there in the past? Why has the climate changed in so many places?

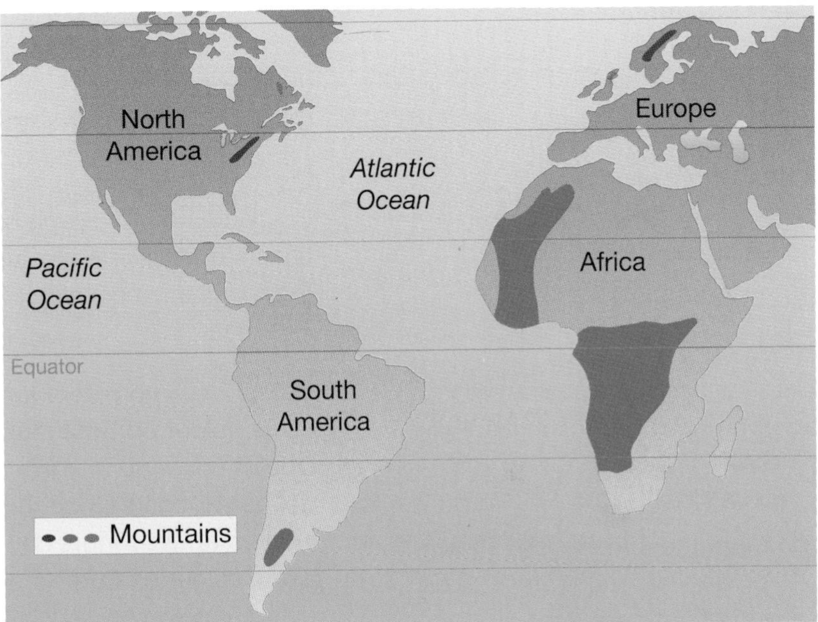

Figure 12.9 How could mountains formed from the same type of rock occur thousands of kilometres from each other across an ocean?

For Wegener, the clues provided by this **meteorological evidence** (climatic change) raised questions that had no easy answers. Since Wegener was trained as a meteorologist, he was especially interested in these clues. He found evidence of even greater climatic changes in places that had probably been covered by glaciers. Ancient glacial deposits (200 to 300 million years old) were found spread over the southern hemisphere. Layers of deposits left behind by glaciers were found in southern Africa, South America, India, and Australia. Under the deposits in some places, there were grooves in the bedrock showing the direction in which the glaciers had moved. All of these locations now had very warm climates, much too warm for glaciers. Was the whole world cold, or had these land masses moved to their present warm locations from a place nearer to the South Pole?

Figure 12.8 Trilobite fossils like this one have been found in the Himalayas. Wegener wondered how these sea creatures ended up on mountain tops.

CONDUCT AN

INVESTIGATION 12-A

Give Me a Clue!

Wegener collected all of the fossil and rock evidence that he could find and put it onto a map of the world. He imagined that all of the continents that we see on Earth today were joined together in one huge supercontinent that he called Pangaea.

To review safety symbols, turn to page 492.

Problem

What clues do fossils and rocks provide about the ancient world? Can you make the puzzle fit?

Apparatus
coloured pencils
scissors
glue

Materials
blackline master of Pangaea map
blackline master of world map

Procedure

1. Mark the three fossils shown on the map in Figure 12.8 on your blank world map.
2. Put a legend on your map, telling what each symbol represents.
3. Find other fossil evidence in other sources. Try the Internet or a CD-ROM in your library. Hint: Look for Cynognathus!
4. Mark these additional fossil locations on your map. Add the symbols to your legend.
5. Examine Figure 12.9. Add the three samples of "rock evidence" to your map. Use four different colours to shade in the locations and call each colour "same rocks" on the legend. Do not cover up your fossil evidence.
6. On the Pangaea map, write the names of the seven continents on the land masses that you can see inside Pangaea.
7. Colour each continent the same colour on both maps. Do not cover up any evidence you marked on the world map. Will India be the same colour as Eurasia or a different colour?
8. Cut out the land pieces on the world map around the continental shelf edges. Remember to cut India away from Eurasia along the tops of the Himalayas.
9. Fit the pieces of the world map together to resemble your map of Pangaea. Once the pieces are in place, glue them to a sheet of blue paper.
10. Transfer the legend to the blue paper by cutting it out or copying it.

Analyze

1. What difficulties, if any, did you experience in fitting the pieces of land together?
2. Which pieces gave you the most trouble? How might these pieces have looked 300 million years ago? How could you test your ideas?

Conclude and Apply

3. Why was Wegener's idea on continental drift a reasonable one? Why did it make sense at the time?
4. As a young child, did you have ideas that you had to change as your knowledge increased? Was it easy or hard to give up your old ideas? How might your experience be compared to the experience of scientists?

The Ancient World of Pangaea

A further clue to Earth's history came from the top of the Himalayan Mountains in India. Fossils of trilobites had been found in the Himalayas, the Swiss Alps, and other mountain ranges (see Figure 12.8). These trilobites roamed the ancient seas 250 to 500 million years ago. How did their fossils end up on the "roof" of the world? The evidence seemed to suggest that India was once a separate piece of land. Many millions of years ago, India crashed into Eurasia. The collision pushed rocks containing fossils from the bottom of the sea, up to the top of the Himalayan mountains (see Figure 12.10).

Figure 12.10 Try to name the seven biggest land masses on the map. Can you find India?

Response to Wegener

In 1915 Wegener published his findings in a book, written in German, called *The Origin of Continents and Oceans*. In the book, he stated that all of Earth's continents had been joined together in a giant supercontinent called Pangaea. Pangaea started breaking up about 200 million years ago, and the pieces began moving or drifting into their present locations. Wegener wrote, "It is just as if we were to refit the torn pieces of a newspaper by matching their edges and then check whether the lines of print run smoothly across. If they do, there is nothing left but to conclude that the pieces were in fact joined in this way."

To support his hypothesis about drifting continents, Wegener thought about what forces might be causing the movement. He proposed that the Moon might be responsible, but other scientists disagreed with him. Because Wegener could not satisfactorily explain the origin of the force that was moving the continents, the scientific community rejected his ideas on continental drift. The fixed-continent model continued to be the most widely supported idea. Wegener died in Greenland in 1930, still searching for evidence to support his theory of continental drift.

INTERNET CONNECT

www.school.mcgrawhill.ca/resources/
Find out more about Wegener by going to the web site above. Go to the **Science Resources**, then to **SCIENCEPOWER 7** to know where to go next. See if you can find any drawings from Wegener's book. Find some information about him that you did not learn in this text, and use it to prepare a brief biography.

Check Your Understanding

1. When maps of the whole world were first drawn in the seventeenth century, what was the first clue that the continents might not have always looked as they do today?

2. List the three kinds of evidence that Wegener collected to support his idea of continental drift. Give one example for each kind of evidence, and explain why the example suggested that the continents had moved.

3. **Thinking Critically**
   (a) In what ways were Democritus's and Wegener's experiences the same?
   (b) Why were others unwilling to accept their ideas?
   (c) Are people generally willing or unwilling to accept change? What does this suggest to you about scientific progress?

12.3 Evidence for Plate Tectonics

Sonar stands for **So**und **N**avigation **a**nd **R**anging. This technology is used in nature by bats to navigate around objects in the dark. Sonar works by sending out a sound and then recording the time that the sound takes to bounce back. For example, scientists can bounce a sound off the ocean floor and measure the time that it takes to bounce back. Since they know how fast the sound travels, they can calculate the distance to the bottom of the ocean.

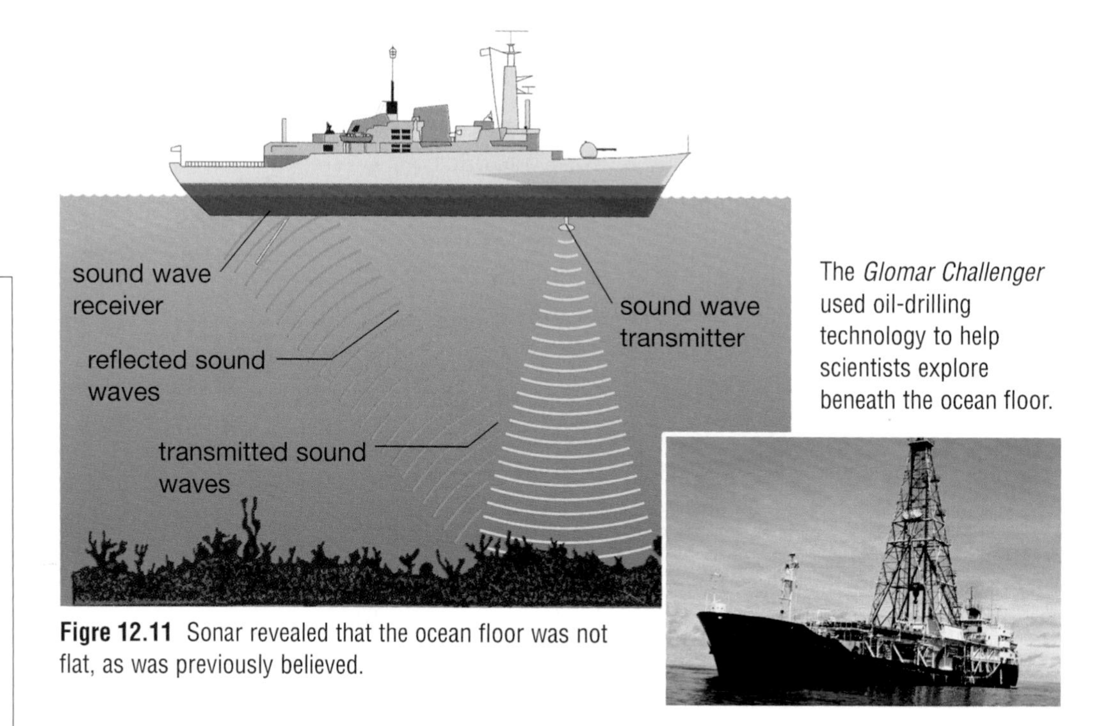

Figre 12.11 Sonar revealed that the ocean floor was not flat, as was previously believed.

The *Glomar Challenger* used oil-drilling technology to help scientists explore beneath the ocean floor.

Wegener based his idea of continental drift on evidence collected from the continents. In the 1940s, after his death, the technology needed to investigate the sea floor became available. **Sonar** (sound wave technology) and **magnetometers** (sensitive instruments for recording magnetic data) were constructed during World War II to detect submarines in the ocean. After the war, scientists used this technology to collect information about the sea floor (see Figure 12.11). They made some very surprising observations!

Figure 12.12 This map shows the mid-ocean ridges and the trenches. The long, ridged structures are the mid-ocean ridges.

Sonar

If the continents had always been in their present locations, scientists expected to find a lot of sediment on the bottom of a relatively flat sea floor. Scientists were surprised to learn from the sonar images that there was little sediment and the sea floor was not flat!

When many sonar tests from Earth's oceans were studied, the results amazed everyone. It was obvious that there were mountains on the sea floor. Moreover, there were long mountain ranges or ridges in some places, just like the mountain ranges that existed on land. Scientists identified a mountain ridge that stretched from north to south along the middle of the Atlantic Ocean. They called this the Mid-Atlantic Ridge (see Figure 12.12).

The features found on the sea floor were similar to the features found on land. What was causing these mountains to form? The answer would come from another technology.

Magnetometers

As mentioned earlier, magnetometers are electronic instruments that can detect the direction and strength of a magnetic field. Developed during World War II, they were used to locate submarines in the ocean. When they were used for research after the war, scientists found that the magnetometers recorded a magnetic field that pointed north most of the time. As the ships that carried them moved across the Atlantic Ocean, however, the magnetometers recorded a magnetic field that sometimes pointed south unexpectedly. It became clear that something on the sea floor was causing these reversals. A pattern was noticed that looked like stripes of magnetic reversals travelling along a path parallel to the Mid-Atlantic Ridge. The width and direction of the stripes on both sides of the Ridge were similar (see Figure 12.13).

Recall that igneous rock forms from magma. This magma contains iron-bearing minerals, for example, magnetite, which line themselves up with Earth's magnetic field. As the molten rock hardens at Earth's surface, the mineral particles stay in line with the magnetic field. So the stripes on the ocean floor that are lined up according to a reversed magnetic field must have formed at a different time — a time when Earth experienced a reversal of its magnetic field. If the stripes lined up with the ridges, it could mean that the sea floor was spreading; therefore new rock was being formed at the mid-ocean ridges. This is called the theory of **sea floor spreading**.

How could scientists find out exactly what was happening on the sea floor to better understand what was happening to Earth's crust?

spreading centre (new crust forming)

magnetic stripes

mid-ocean ridge

Figure 12.13 The pattern of magnetic reversals on the sea floor led scientists to the theory of sea floor spreading. As new crust forms, it takes on the magnetic polarity of Earth at the time of formation.

Find Out **ACTIVITY**

What Do the Rocks Tell Us?

What information can rocks provide about the ocean floor?

What to Do

The graph on the right shows the ages and locations of samples of rock taken from the magnetic stripes at the bottom of the Atlantic Ocean. Each dot represents a sample of rock.

1. Find the age of the oldest rock and the youngest rock on the graph.
2. State the distance of the oldest rock and the youngest rock from the Mid-Atlantic Ridge.

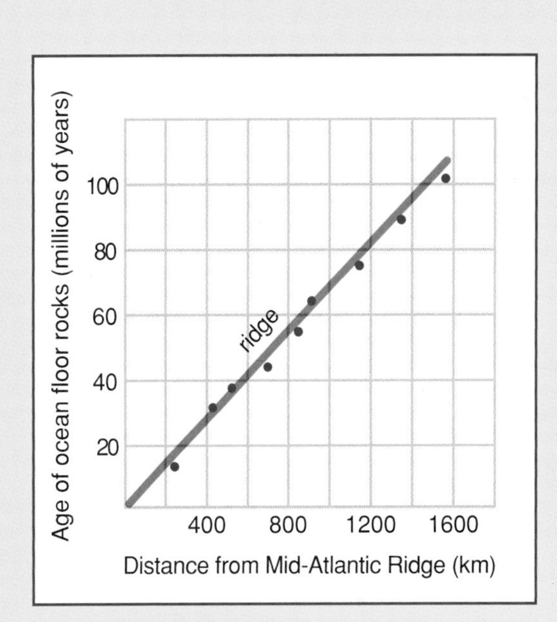

What Did You Find Out?

What does this evidence suggest was happening to the sea floor? Explain your answer.

Deep Sea Drilling and Radioactive Dating

Scientists confirmed the theory of sea floor spreading when they were able to bring up samples of rock for testing. The ship *Glomar Challenger* carried equipment that could drill deep holes into the sea floor. Rock from the holes was brought onto the ship for testing by scientists. Can you imagine the excitement of the scientists who first examined these rock samples, knowing that they were the first people in the history of the world to do so! Tests of the rock samples showed that younger rock was closer to the Mid-Atlantic Ridge and older rock was closer to the continents. Scientists found that the Atlantic Ocean is getting wider by about 2 cm every year — about the same speed that your fingernails grow!

Figure 12.14 Submersibles like *Alvin* have made it possible for us to see lava coming out of cracks in the sea floor. The lava cools so quickly in the cold water that it is called "pillow lava." Why might it have been given this name?

Even with scuba (**s**elf-**c**ontained **u**nderwater **b**reathing **a**pparatus) gear, deep-sea divers can go only a few hundred metres down into the ocean because of the tremendous pressure of the water on their bodies. Submersibles allow people to travel deeper into the ocean by protecting them from the pressure of the water. Submersibles are equipped with an air supply and powerful lights.

Submersibles

Most of the volcanic activity where rock surfaces are pulling apart happens on the sea floor, where it cannot be seen. (Iceland is one example of an area where the ridge can actually be observed.) Scientists can travel in submersibles to the bottom of the Atlantic Ocean and see lava pouring out of the underwater volcanoes that form a line from north to south in the middle of the ocean (see Figure 12.14). This is the same area where shallow earthquakes take place as rock pulls apart.

Pause& Reflect

Miners of the future may look for gold and other precious metals on the sea floor near volcanic vents called black smokers. Scientists say that the vents act like smelters. The vents dissolve metals from the surrounding rock and send them into the cold water, where they collect like snow crystals outside the vents. Use the Internet or library resources to find out about a fossilized black smoker deposit called Kidd Creek in northern Canada. Write notes about your findings in your Science Log.

Toward a New Theory

All of this evidence indicated that Earth's crust was indeed moving and not fixed in place, as most people believed. If the sea floor was moving, then the continents must be moving, too. But what force was driving the crust to move?

If you look again at your earthquake map in Chapter 11, you should see that the earthquake epicentres map out the edges of the plates. In Figure 12.17, the major plates are labelled for you. Can you see that most of the plates are named for the continent that is on the plate? Which plate has no continent on it? This plate is named for the ocean that covers it. Alfred Wegener was hypothesizing correctly about the movement of the continents.

In Chapter 11, you learned that a break in Earth's crust where rock movement occurs is called a fault. Now you can see that the rock movement occurring at a fault is the movement of the huge plates that make up Earth's crust. You also learned about the types of plate movement that can occur at a fault. The area where two plates are pushing together is called a **convergent boundary**. The area where two plates are pulling apart is called a **divergent boundary**.

Figure 12.15 Tuzo Wilson

J. Tuzo Wilson, a Canadian scientist, is one of the long line of scientists who have contributed to our understanding of Earth's crust (see Figure 12.15). During the 1960s, scientists dismissed the idea of continental drift and favoured the fixed-continent model. Wilson became a leading spokesperson for the idea of continental drift. He made an important addition to scientific observation when he developed the concept of a third kind of movement along fault lines. Instead of pushing together or pulling apart, he hypothesized that plates were sliding past each other along what he called a **transform boundary**.

Plate Tectonics

Together with the information about movement on the sea floor, Wilson's idea of transform boundaries brought about a rethinking of Earth's crust movement. It helped form a new theory that completely revolutionized this branch of science in the 1970s.

Scientists needed a name for the new scientific theory. The name "Continental Drift" was no longer appropriate since scientists now knew that the sea floor, as well as the continents, was moving. The new theory, called **plate tectonics**, states that Earth's crust is broken up into pieces, called **plates**, that are always moving around on Earth's mantle (the thick layer inside Earth between the outer core and the crust). It is important to remember that the biggest plate has no continent on it!

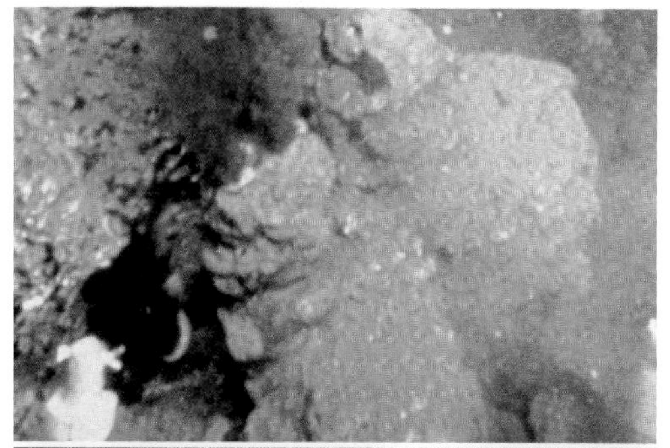

Figure 12.16 When submersibles found red-hot lava under the oceans, they also found living creatures that use the heat from the lava as their energy source! These creatures all appeared to be white in the lights of the submersible. Why do you think they have no colour?

Scientists are using satellites and lasers to measure plate movements. Plates that hold the greatest continental masses move more slowly than plates that hold smaller continental masses. The African, Eurasian, and American Plates move about 20 mm per year, in comparison with the Pacific, Nazca, and Cocos Plates, which can move up to 13 cm per year.

Figure 12.17 This diagram shows the major plates, their direction of movement, and the type of boundary between them. What is happening to the Juan de Fuca Plate where it meets the North American Plate?

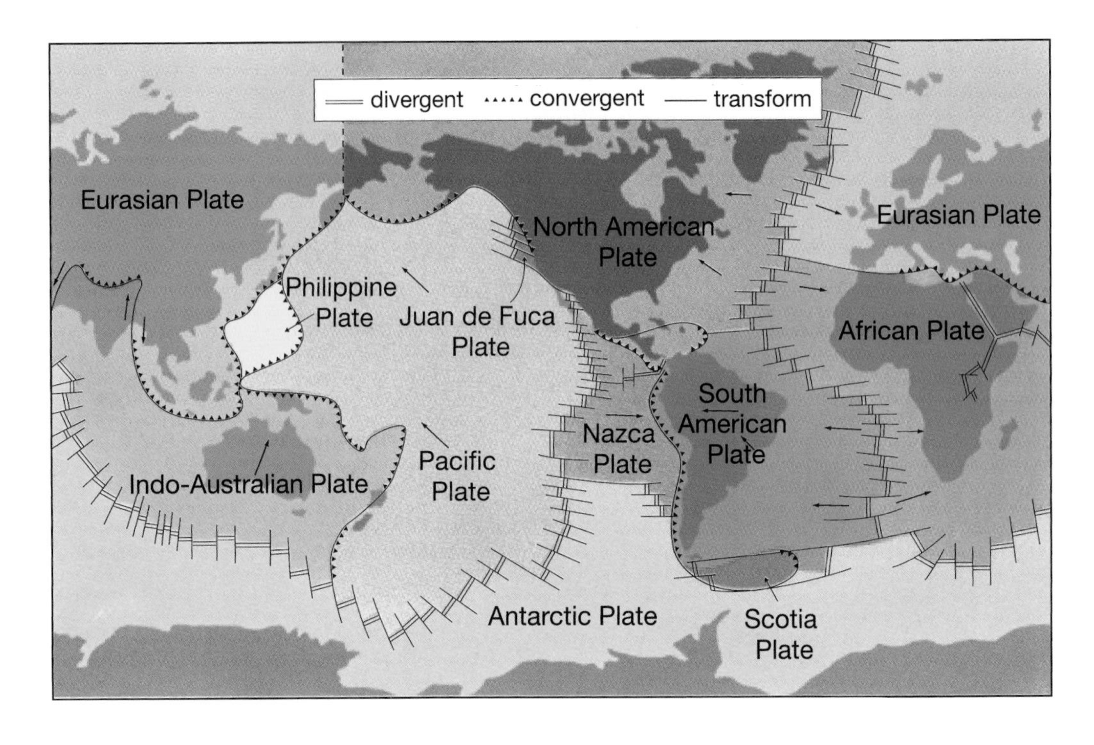

Figure 12.18 The entire cycle of heating, rising, cooling, and sinking is called a convection current.

Geologists think that plate tectonics provides the best explanation, up to now, for all of the evidence that has been gathered. Geologists are still not sure what causes the plates to move. One explanation is that convection currents in the mantle under Earth's crust move the plates (see Figure 12.18).

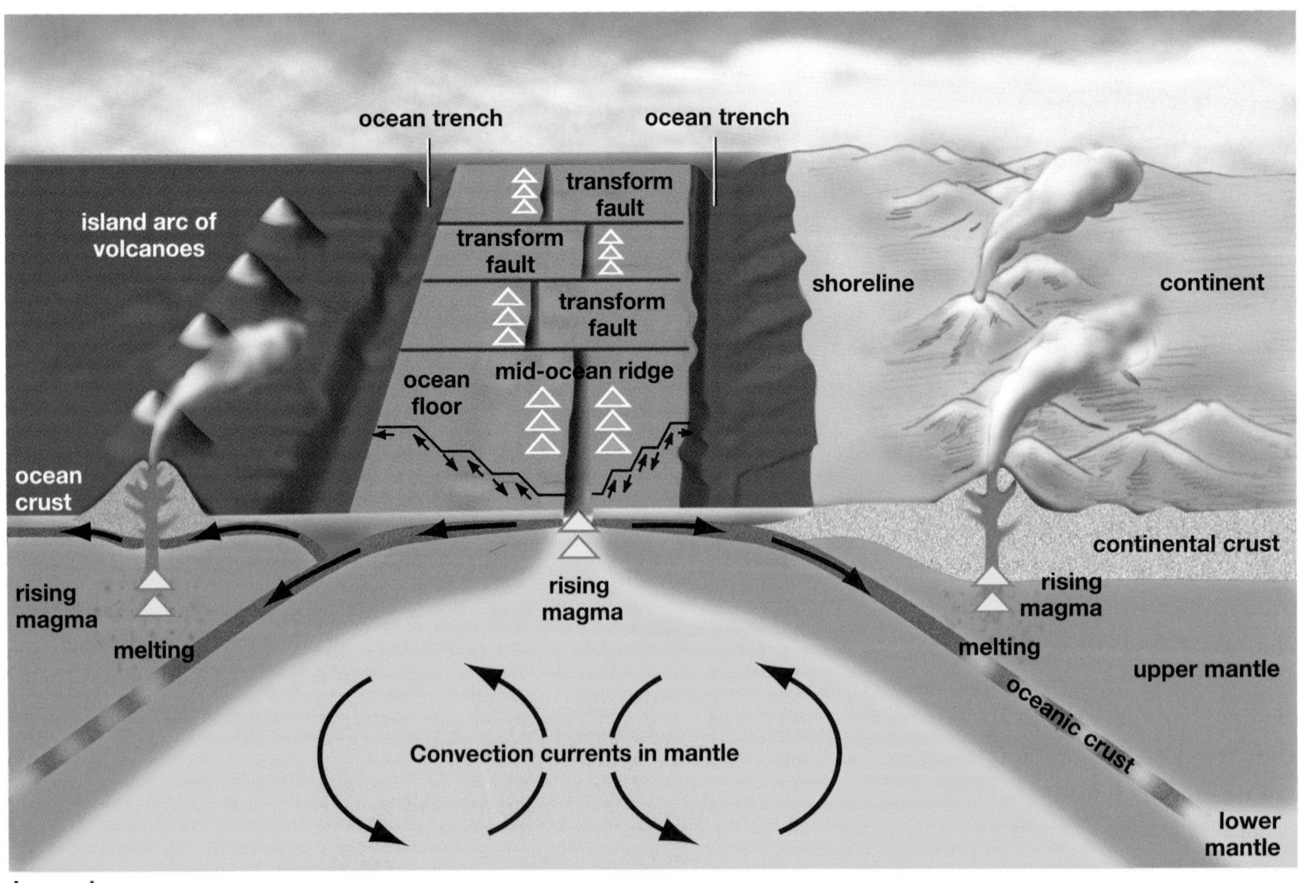

Legend:

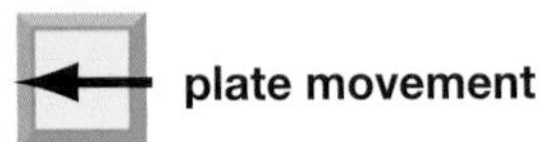

Convection is based on the idea that warm fluids move up and colder fluids move down. Convection currents cause wind and rain in weather systems. The same process might be occurring in Earth's mantle, and this is what scientists think is the force behind plate tectonics. Just as in a hot-air heating system, hot plastic-like rock in the lower mantle moves upward after it is heated by the intense heat in Earth's core. At the upper part of the mantle, the heated rock moves horizontally under the plate above it, taking the plate along as if the plate were on a conveyer belt. When the rock finally cools, it sinks down farther into the mantle. As it does so, it pulls the edge of the plate down with it, forming a deep ocean trench. The deepest ocean trench is the Marianas Trench off the coast of Japan.

The convection currents might be causing the Atlantic Ocean to widen at the Mid-Atlantic Ridge. Does this mean that Earth's crust is getting bigger? No, because while new crust is forming in the middle of the Atlantic Ocean, other crust is being pushed or pulled down into the ocean trenches and recycled back into the mantle as molten rock.

The theory of plate tectonics is called a **unifying theory**, a single theory that explains different natural events and landforms. It is our best explanation for now for earthquakes, volcanoes, and mountains. Who knows what further discoveries will be made in your lifetime!

At Home **ACTIVITY**

Convection at Home

How can you observe convection?

What to Do

Check to see if you have both warm and cold air vents in any of your rooms. When you hear that the furnace is on, put your hand near the vents. What do you feel? Record your explanation of how convection heating works in your home. Be prepared to present it to a classmate.

Figure 12.19 In 1993 new splits occurred in the Juan de Fuca Ridge. In 1997 a team investigated the area in the submersible *Alvin*. For the first time, scientists saw part of the process in which an undersea volcanic area develops new growth and life.

Check Your Understanding

1. (a) How does sonar work?
   (b) What information did sonar provide about the sea floor?
2. What evidence do we have that Earth's magnetic field has reversed over time?
3. What surprising discovery was made when rock samples from deep-sea drilling were tested?
4. What do scientists now think is causing the continents to move?
5. In what two ways is the plate tectonics theory different from the idea of continental drift?
6. (a) Why do you think that some books still use the term "continental drift"?
   (b) Why would Tuzo Wilson have used the term in a paper he wrote for the scientific community in 1965?
   (c) Give one reason why the term "continental drift" is no longer appropriate.

12.4 Time Line of Earth's History

Figure 12.20 Geologists can use the principle of superposition to determine the relative ages of these layers of rock. The top layer is the youngest, and the bottom layer is the oldest.

Suppose that on a Tuesday, you want to find an article in the newspaper from last Saturday. You probably find Tuesday's paper on the top of the pile in the recycling bin. Underneath is Monday's paper, then Sunday's, and finally Saturday's at the bottom. You knew that you would probably find the older newspaper under the more recent ones. Geologists have used this principle to infer the relative ages of different layers of rock. It is called the **principle of superposition** (see Figure 12.20).

The principle of superposition states that in undisturbed layers of rock, the oldest layers are always on the bottom and the youngest layers are on the top. As you learned in Chapter 10, sediments deposited at the bottom of bodies of water eventually form sedimentary rocks. Additional new layers of sediment continue to form over the previous layers. Unless something happens to move these layers, they stay in their original positions, like the newspapers in the pile.

Geologists use a similar technique, called **relative dating,** to find the order in which events occurred. With relative dating, scientists determine the relative age of rocks by examining their position in a sequence of layers. If a crack or a fault runs through a layer, it must have happened after the layer was in place, so the rock is older than the fault. This method tells nothing about the exact age of the rock layer. What it does indicate is that the layer is older than the rocks above it and older than the fault running through it.

Find Out ACTIVITY

Which Rock is the Oldest?

What to Do

1. (a) Examine the illustration below. The legend will help you to interpret the layers.

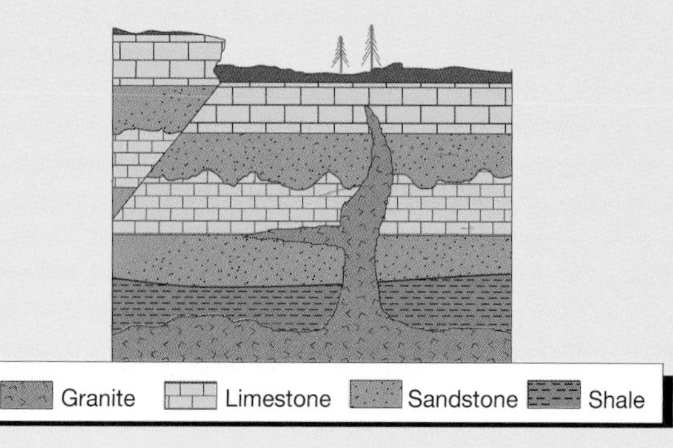

   (b) Discuss the relative ages of the rock layers and events based on what you see.

2. (a) Make a sketch similar to the illustration.

   (b) Label the relative age of each rock layer on your sketch. The bottom layer is the oldest, so mark it with a 1. Mark the next oldest layer with a 2, and so on.

What Did You Find Out?

1. Where is the fault line in the rocks?
2. How do you know that this is an old fault line, which has not moved for some time?

Clues from Technology

If a stack of newspapers is out of order, you can easily use the dates printed on them to put them in order. Unlike newspapers, rocks are not dated, so how can geologists establish their age?

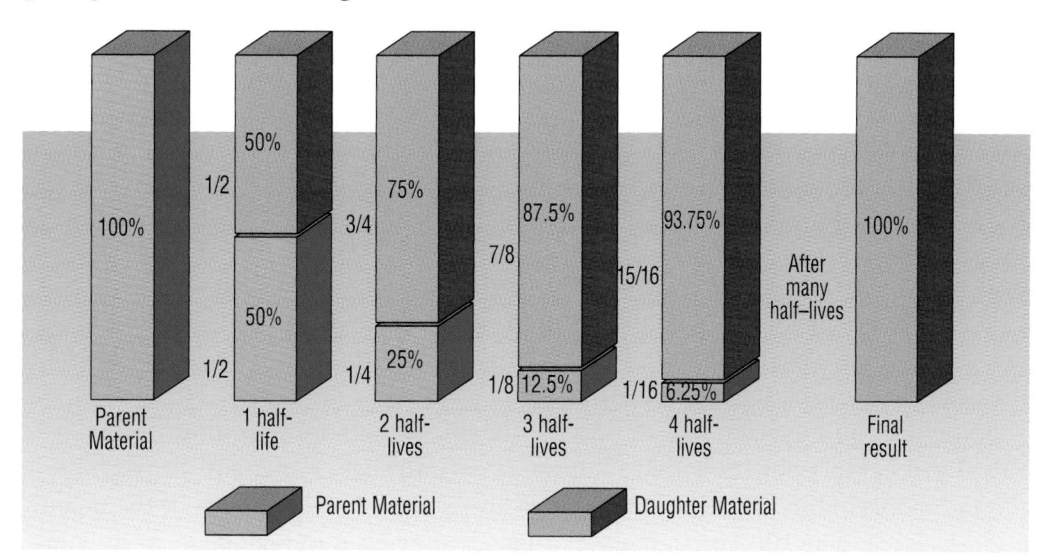

Figure 12.21 After each half-life, only one half of the parent material remains. Eventually almost all of the parent material will be gone.

The amounts of certain elements in a rock can tell geologists a great deal about its age. Over billions of years, some elements can change into others. For example, over a period of 4.5 billion years, half of the uranium in a rock will turn into lead. The lead will then undergo no further change. This time period is called the **half-life** of uranium. The uranium is called the parent element. Once half of the parent element has changed, only half of the remaining element continues to change over a similar period of time. So, in another 4.5 billion years, half of the remaining uranium will change into lead. The process will continue until such a small amount of uranium remains, it may not be measurable, as shown in Figure 12.21.

Math **CONNECT**

How old is a fossil that contains $\frac{1}{4}$ of its original carbon-14?

By measuring the amounts of such elements in a rock, and by knowing the half life of the parent, a geologist can calculate the absolute age of the rock. This process is called **radiometric dating**.

Scientists also use **radiocarbon dating**, a type of radiometric dating, to find out when recent events in Earth's history occurred. Radiocarbon dating uses carbon-14, a rare form of carbon, as its parent material, to find the age of fossils, bones, and wood that are up to 50 000 years old. After 50 000 years, the amount of carbon-14 left in the sample will be too small to measure. All living things take carbon-14 out of the environment to build cells and tissues. When they die, the carbon-14 changes into nitrogen gas in a half-life of 5730 years. The amount of carbon-14 left in the tissue allows scientists to determine the age of the remains. Radiocarbon dating is a type of radiometric dating, and both are examples of absolute dating. **Absolute dating** is a method of determining the ages of objects by using the radioactive decay of their smallest particles.

Figure 12.22 Scientists have determined that the oldest rocks on Earth's surface are the Acasta Gneiss, found in the Northwest Territories. These rocks are 4.03 billion years old. Why are these rocks not found elsewhere?

Geologic Time

Pause& Reflect

Design Your Own

With a partner, develop a procedure for making a "fossil" of your own.

Earlier you saw how fossils provided clues about continental drift and eventually about the theory of plate tectonics. Fossils also help scientists to determine the age of the rocks in which they are found.

Usually the remains of dead plants and animals quickly decay and are destroyed. When the remains are protected from scavengers and micro-organisms, however, they can become fossilized (see Figure 12.23).

If a carcass is in water and sinks to the bottom, the body can be buried by sediment. The hard parts (bones, shells, or teeth) are usually seen in fossilized remains.

When water permeates the bones of a dead animal, the water dissolves the calcium carbonate in the bones. A deposit of another very hard mineral, silica (quartz) remains, turning the bones into a petrified (rock-like) substance.

When an organism is buried under many layers of sediment, pressure and heat may build up, leaving a thin film of carbon residue on rock surfaces. This type of fossil is called a **carbonaceous film**.

Sometimes an organism falls into soft sediment, such as mud. As more sediment falls above it, the sediment below gradually turns into rock. Water and air pass through pores in the rock, reaching the dead organism. The hard parts of the organism dissolve, leaving a cavity in the rock called a mould. Other sediments or minerals may fill the hole, hardening into rock and producing a cast of the original object.

Some fossils can be used to determine the age of the layer of rock in which they are found. If the creature that became fossilized was on Earth for only a short period of time, it can give us very accurate information about the age of the rock in which it is found. This type of fossil is called an **index fossil**.

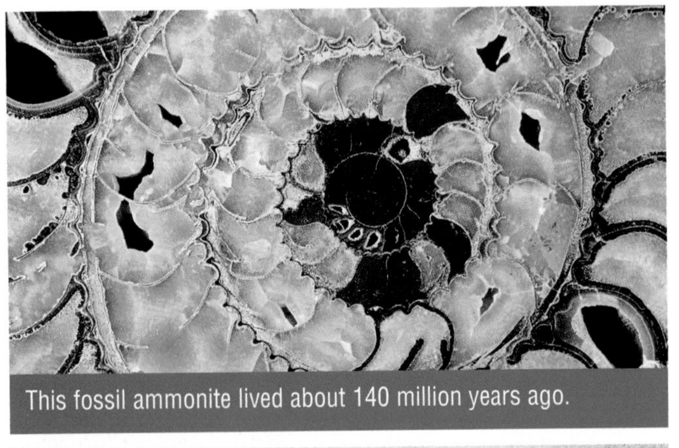

This fossil ammonite lived about 140 million years ago.

The hard parts of this leaf helped to preserve it.

This eurypterid is preserved as carbon film.

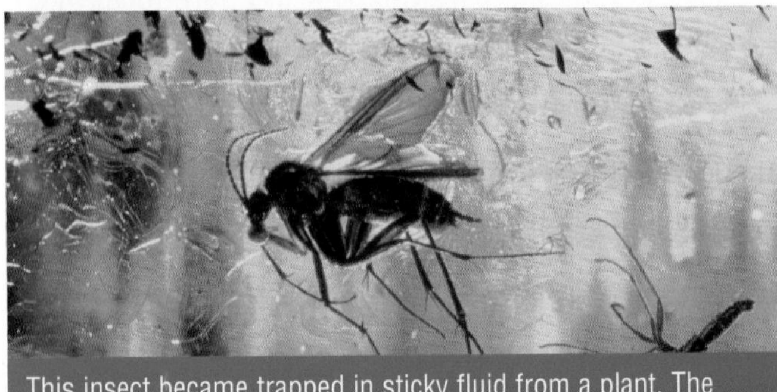

This insect became trapped in sticky fluid from a plant. The fluid crystallized over time, becoming a substance called amber, and preserved the insect.

Figure 12.23 There are several ways in which dead plants and animals can become fossilized.

Tell-Tale Layers

What can fossils tell us about Earth's age?

What to Do

1. Examine the figure at the right. It shows three layers of sedimentary rock with the fossils found in each layer.
2. Match each fossil with a fossil in the table, showing when it was alive.

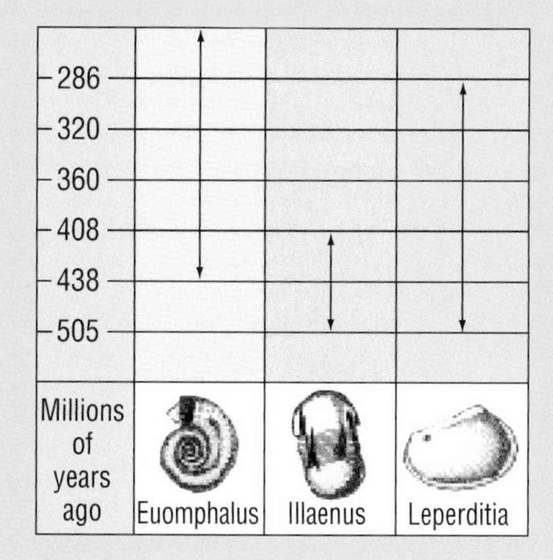

3. How can scientists infer that the middle layer of rock was formed between 438 and 408 million years ago?

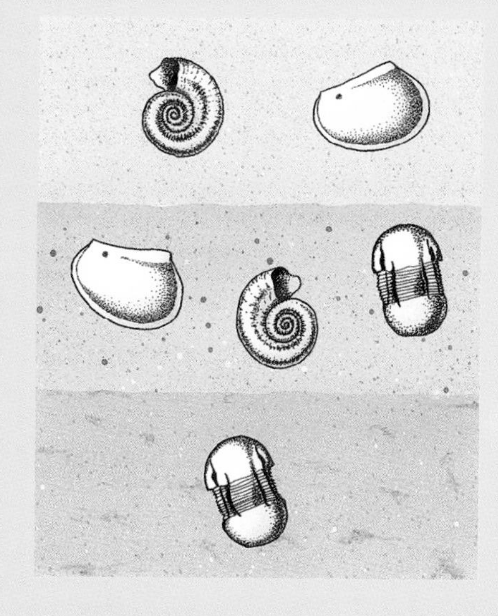

Geologic Time Scale

The geologic time scale is a division of Earth's history into smaller units, based on the appearances of different kinds of life forms in the fossil record. It starts 4.5 billion years ago when Earth was formed. The divisions in the time scale are called **eras** and **periods**. Eras are the four largest subdivisions in the geologic time scale. Periods are subdivisions in this scale.

For the first 4 billion years, there is little fossil evidence. This vast expanse of time is referred to as the **Precambrian Era**. Scientists think that the earliest supercontinent, **Rodinia**, formed during this time, about 1.1 billion years ago. Rodinia split apart 750 million years ago, forming the ocean basins. There is evidence that simple forms of life, such as bacteria, algae, fungi, and worms, lived on Earth during the Precambrian Era. Because the bodies of these creatures were soft, they have left very little fossil evidence of their existence. The three eras that are rich in fossil evidence began 570 million years ago: the **Paleozoic Era** (ancient life), the **Mesozoic Era** (middle life), and the **Cenozoic Era** (recent life).

Pangaea, the second supercontinent to form, came together during the Paleozoic Era about 350 million years ago, and broke up about 180 million years ago during the Mesozoic Era. The dinosaurs dominated Earth in the Jurassic Period, which was 200 million years ago during the Mesozoic Era. The fossil evidence indicates that Pangaea first split into a northern portion called **Laurasia** and a southern portion called **Gondwanaland** (see Figure 12.24). Look on a modern map to see which continents have replaced Laurasia and Gondwanaland.

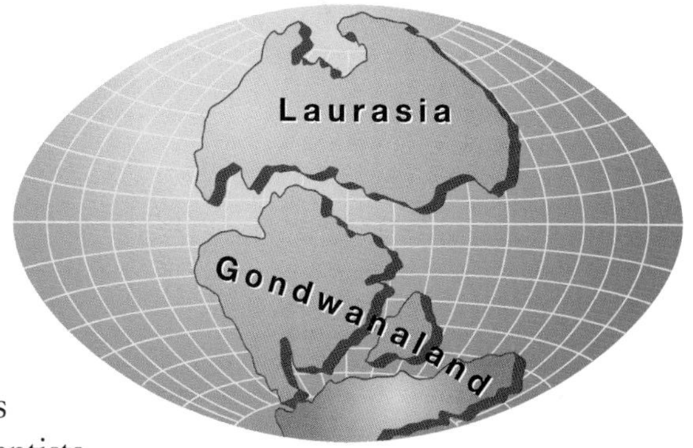

Figure 12.24 Present-day North America was once part of Laurasia. The split into Laurasia and Gondwanaland was the first break-up of Pangaea.

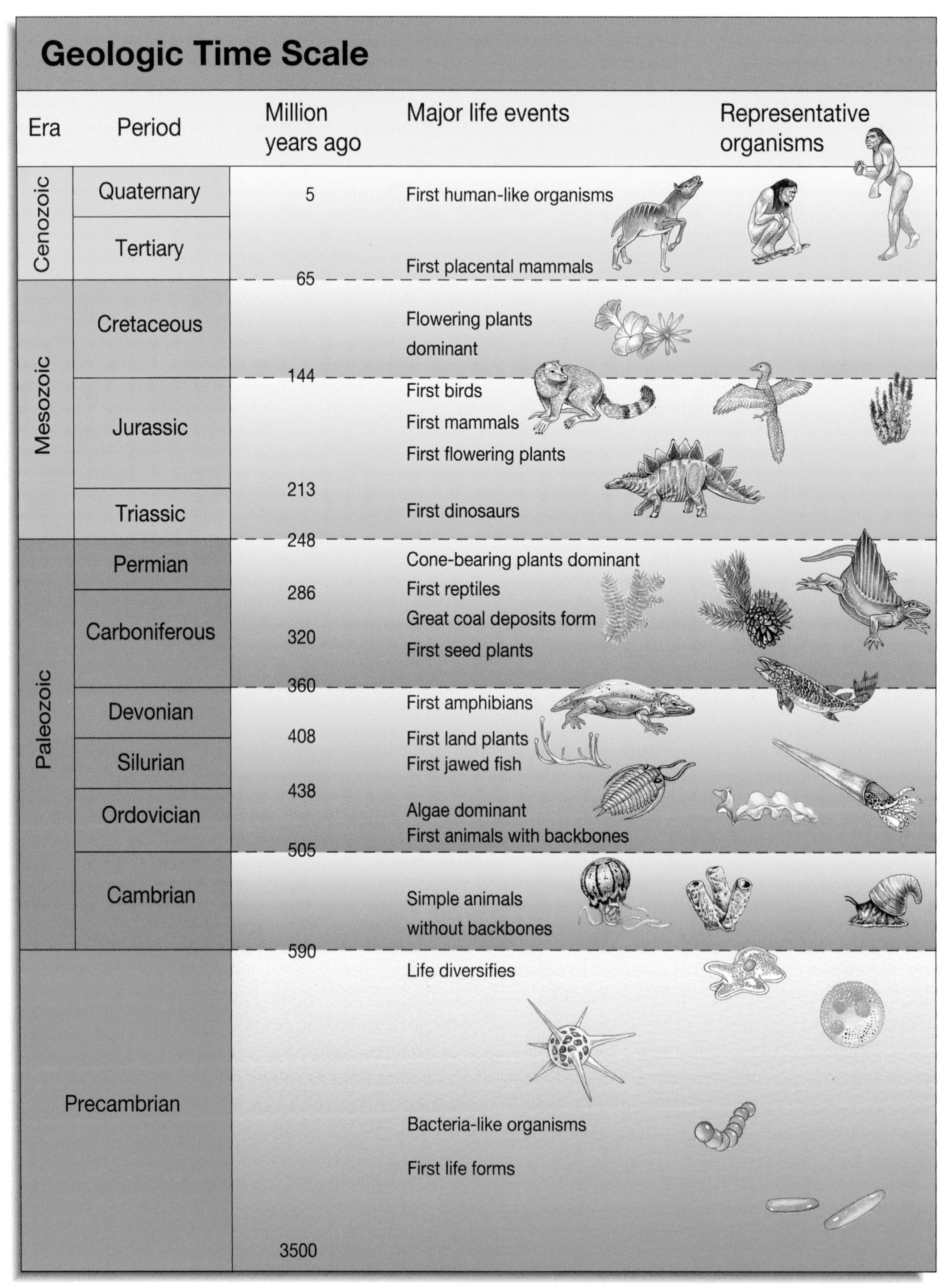

Geologic Time Scale

Era	Period	Million years ago	Major life events
Cenozoic	Quaternary	5	First human-like organisms
	Tertiary		First placental mammals
		65	
Mesozoic	Cretaceous		Flowering plants dominant
		144	
	Jurassic		First birds First mammals First flowering plants
		213	
	Triassic		First dinosaurs
		248	
Paleozoic	Permian		Cone-bearing plants dominant
		286	First reptiles
	Carboniferous		Great coal deposits form
		320	First seed plants
		360	
	Devonian		First amphibians
		408	First land plants
	Silurian		First jawed fish
		438	
	Ordovician		Algae dominant First animals with backbones
		505	
	Cambrian		Simple animals without backbones
		590	
Precambrian			Life diversifies Bacteria-like organisms First life forms
		3500	

Figure 12.25 This geologic time scale provides even greater detail about geological events than you have just studied. It divides the eras you have learned about into a number of periods and shows the approximate dates that scientists now have for the first appearance of each general group of organisms on Earth.

Call That Old?

You are a geologist. You must organize the information you have about the history of Earth's crust and construct an informative time line.

What You Need

scissors
adding machine tape
metre stick or ruler
pencil
set of coloured pencils

What to Do

1. Look through the chapter and note when various events in Earth's history occurred and when various ideas about Earth's history were suggested.
2. Look back over Chapters 10 and 11. Decide which events from those chapters you want to include on your time line. If you have time, you might want to research other events you could also include.
3. Decide how long a piece of adding machine tape you need for your time line. If you have ideas about other materials you might use to construct a time line, have them approved by your teacher.
4. Decide how you will indicate events on your time line.
5. Complete your time line.

What Did You Find Out?

1. Where on the time line do most events occur? Why might they occur here?
2. Compare your time line with your classmates' time lines. Did your classmates include any events that you feel you should have included in your time line?
3. If you were to do it again, how might you improve your time line?

Extension

4. Choose a period of time on your time line, and research it. Try to find out about other major events that you could add to your time line.

Across Canada

Helen Belyea was one of the first women to work in the field of geology in Canada. She was the only woman who went out into the field before 1970. Dr. Belyea worked in the oil fields of Alberta.

Before Dr. Belyea found work as a geologist, she worked as a teacher and served in the Canadian navy. She joined the Geological Survey of Canada in 1945. She started out as a technologist, was promoted to technical officer, and then to geologist. Dr. Belyea made these jumps in her career in just two years!

In 1956 Dr. Belyea received the Barlow Memorial Medal from the Canadian Institute of Mining and Metallurgy (the science of working with metals). In 1976 she was named Officer of the Order of Canada — a great honour for a Canadian.

Helen Belyea

Figure 12.26 A paleontologist studies fossilized dinosaur footprints in an ancient lake bed in the Andean highlands of South America. The area is now one of the most important in the world for dinosaur study.

INTERNET CONNECT

www.school.mcgrawhill.ca/resources/

Go to the web site above to find out more about fossils. Go to the **Science Resources**, then to **SCIENCEPOWER 7** to know where to go next. Why is "dinosaur" or "terrible lizard" no longer a suitable name for all of the fossils of animals found from the Jurassic Period? Write a short article about your findings.

It is very difficult for us to comprehend the vast expanse of 4.6 billion years. If we compare this period of time to a 24 h day, for example, humans have been here for only the last second, and the ancient civilizations in Egypt, Greece, and Rome took place in less than one second! (It has taken you much more than a second just to read the last three sentences.) In your time line you decided how best to represent 4.6 billion years of history. By doing so, you helped yourself to understand the time involved and the slowness of the processes that shape our Earth.

Check Your Understanding

1. What is the principle of superposition? How does it help geologists to determine the ages of rocks?
2. Compare absolute dating with relative dating. What do they indicate?
3. What does "half-life" mean? Give one example.
4. List four different kinds of fossils. Give examples and explain how each one is formed.
5. How old is Earth? Describe Pangaea, Laurasia, and Gondwanaland. When did they exist, and what evidence do we have about them?

Career CONNECT

Digging out the Facts

Take a look at the following occupations:

mineralogist	miner	archaeologist	gardener
museum curator	paleontologist	landscape architect	surveyor
oil rig operator			

With your group, produce a booklet or a slide presentation about one of the occupations. Make sure that you describe what it has to do with the study of Earth's crust. You can use photographs and illustrations from magazines or from the Internet. If possible, interview someone working in the occupation and include your interview as an article. Try to present the occupation in as inviting and attractive a way as you can, so that people will become interested in it.

CHAPTER at a glance

Now that you have completed this chapter, try to do the following. If you cannot, go back to the sections indicated.

Identify and explain one example of a model used in the past to explain the formation of the continents on Earth. (12.1)

Describe the first clue that made people think the continents may have moved. (12.2)

Explain Wegener's theory of continental drift and two forms of evidence that he collected to support his theory. (12.2)

Describe Pangaea, and explain why Wegener thought that it must have existed. (12.2)

Explain why the scientific community did not accept Wegener's ideas. (12.2)

Describe how magnetometers, sonar, deep-sea drilling, and radioactive dating were used to collect information about the sea floor after World War II. (12.3)

Give two examples of how new evidence from technology provided scientists with support for sea floor spreading and the theory of plate tectonics. (12.3)

Compare continental drift and plate tectonics. (12.3)

Differentiate between relative and absolute dating. (12.4)

Identify and explain three different methods of fossilization. (12.4)

Describe how analysis of fossils has contributed to our knowledge of life in the past and provides evidence of geological change. (12.4)

State the age of Earth, and explain how scientists learned this information. (12.4)

Describe some of the major eras and events in Earth's history. (12.4)

Describe how science-related technologies have led to a better understanding of Earth's crust. (12.3, 12.4)

Prepare Your Own Summary

Summarize this chapter by doing one of the following. Use a graphic organizer (such as a concept map), produce a poster, or write the summary to include the key chapter ideas. Here are a few ideas to use as a guide:

- Explain the continental drift idea proposed by Alfred Wegener and the evidence he used to support it.
- Name several forms of technology that were used after World War II to gather information about the sea floor. Explain how this information led to the idea of sea floor spreading.
- Use the diagram at the right to describe the theory of plate tectonics. Explain how this theory helps to explain catastrophic events, such as volcanoes and earthquakes.
- Explain different types of fossilization. How do fossils help us to understand the geological history of Earth's crust?
- Briefly describe the main eras and events in the history of Earth, including Rodinia and Pangaea.

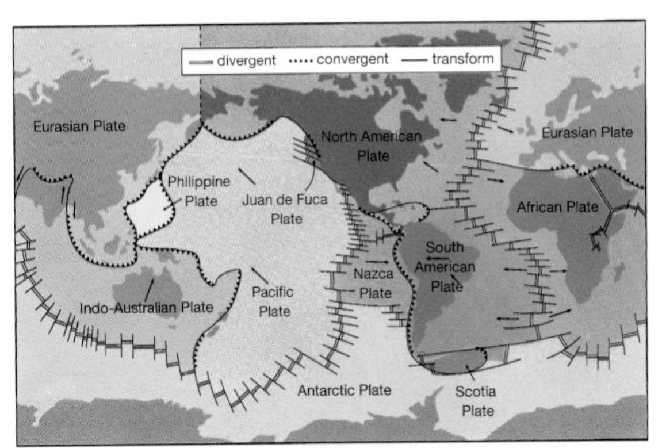

CHAPTER 12 Review

Key Terms

catastrophism
uniformitarianism
fixed-continent model
continental shelf
biological evidence
continental drift
meteorological evidence
sonar
magnetometers
sea floor spreading
convergent boundary
divergent boundary
transform boundary
plate tectonics
plates
unifying theory
principle of superposition
relative dating
half-life
radiometric dating
radiocarbon dating
absolute dating
carbonaceous film
index fossil
eras
periods
Precambrian Era
Rodinia
Paleozoic Era
Mesozoic Era
Cenozoic Era
Pangaea
Laurasia
Gondwanaland

Reviewing Key Terms

If you need to review, the section numbers show you where these terms were introduced.

1. In your notebook, copy and complete the crossword puzzle on the following page, using Key Terms.

Understanding Key Ideas

Section numbers are provided if you need to review.

2. Explain one example of an early idea of how Earth's crust had formed. (12.1)

3. Name the three forms of evidence collected by Wegener, and give an example of each. (12.2)

4. Why was Wegener unable to convince others of his theory of continental drift? (12.2)

5. Name two forms of technology used after World War II to investigate the sea floor. What evidence did scientists find to indicate sea floor spreading? (12.3)

6. How do we know how old Earth's crust is? (12.4)

Developing Skills

In your notebook, copy and complete the following flowchart.

7. **(a)** Clues noticed by Alfred Wegener.
   1. ______
   2. ______
   3. ______
   4. ______

↓

(b) Technology used to study Earth's crust after World War II.
   1. ______
   2. ______
   3. ______
   4. ______

↓

(c) What two kinds of observations led to the most recent theory?
   1. ______
   2. ______

↓

(d) Define and describe the most recent theory.

______
______
______

Problem Solving/Applying

8. You have just discovered a fossil common to your province in the ground at a construction site. Find out how old the fossil could be and infer what this tells you about the ground that it was found in. What can you hypothesize about the ancient climate and environment of the province?

Critical Thinking

9. Imagine that you are designing a display of dinosaurs for the local museum. Why would you not put any people into the display? How would you explain this to the museum staff?

10. Laser beams are used today to measure plate movement. How would they need to be set up and monitored to give you information?

Pause& Reflect

1. Are there any questions that you are still not sure how to answer? Why might this be? Why is it acceptable to have "unanswered" questions?
2. Go back to the beginning of this chapter on page 342 and check your original answers to the Getting Ready questions. How has your thinking changed? How would you answer those questions now that you have investigated the topics in this chapter?

Across

1. remains of ancient living things that have turned into rock (12.2)
2. the northern part of the supercontinent that included North America and Eurasia (12.4)
3. Canadian geologist who developed the concept of the transform boundary (12.3)
4. term meaning "ancient life" (12.4)
5. term meaning "Sound Navigation and Ranging" (12.3)
6. idea that the forces that shaped Earth's crust are still at work today (12.1)
7. dating: process used to find the age of rocks by measuring the amount of parent material that has changed (12.4)
8. process used to find the approximate age of rocks based on the younger rock layers on top and the older rock layers underneath (12.4)

Down

1. change that relates to meteorological evidence (12.2)
2. magnetometer revealed magnetic stripes ______ to the mid-Atlantic Ridge (12.3)
5. principle demonstrated by newspapers piled in order of date (12.4)
11. evidence provided by fossils (two words) (12.2)
12. continents moving (two words) (12.2)
13. term meaning "recent life" (12.4)
14. first supercontinent to form about 1 billion years ago (12.4)
15. the idea that the Earth's crust was formed by catastrophic events (12.1)
16. southern part of the supercontinent that contains South America, Africa, Australia, Antarctica, and India (12.4)
17. a device that reads the magnetism of rocks at the bottom of the ocean (12.3)
19. the longest time period in the history of the Earth's crust, with little fossil evidence (12.4)
20. term meaning "middle life" (12.4)

4 Ask an Expert

Is it possible to study rock that is many kilometres below Earth's surface? Charlotte Keen would answer with a resounding, "Yes!" Charlotte is a senior research scientist with the Bedford Institute of Oceanography and the Geological Survey of Canada. For 28 years, she has been studying the rock under the ocean floor. In 1995, she won the J. Tuzo Wilson Medal for her outstanding contributions to geophysics in Canada.

Q What exactly is a geophysicist?

A A geophysicist is a scientist who uses knowledge of physics — the interactions of matter and energy — to study parts of Earth that are beneath its surface.

Q What part of Earth do you study?

A I look at the continental margin off the east coast of Canada that runs from Baffin Bay right down to south of Nova Scotia. This is the area where the North American continent, of which Canada is part, once separated from neighbouring continental plates. Continental margins can tell us a lot about how Earth is changing, and the rock underneath them holds great potential for natural resources: oil, gas, and valuable minerals. I study the rock that lies up to 40 or 50 km below that margin.

Q How can you study rock so far below Earth's surface?

A I look at the way vibrations from sound waves travel through that rock. My colleagues and I plan experiments for the area we want to study. We assemble a team of nearly 100 people, gather the necessary equipment, and then head out to sea. We lower very sensitive recording devices onto the ocean floor of the area we will study. After that, the ship travels away from the devices, firing an air gun which makes a very loud sound. The sound waves go down through the water and into the rock. Some waves bounce up off layers of rock, others go deeper before they bounce back. When each sound wave reaches the device, it is recorded.

Q How the sound is recorded tells you something about the rock?

A Yes. Sound waves travel through different types of materials at different speeds. The amount of time it takes for each sound wave to be recorded helps me figure out what kind of rock the sound wave may have passed through. We keep firing the air gun from many different distances, keeping track of exactly where we were and the time at which it was fired. Usually, we are at sea for two to four weeks to record everything that we need. Once we're back on land, I analyze the data from the recording devices to figure out what type of rock exists in each place and how thick the layers of rock are.

Q Why did you decide to study the rock under the ocean, rather than somewhere else?

A When I graduated from physics at Dalhousie University in Halifax, research in oceanography was expanding in Canada. It was an exciting time. I think my love of the sea had a lot to do with my decision too. I was born in Halifax and my whole family enjoyed sailing.

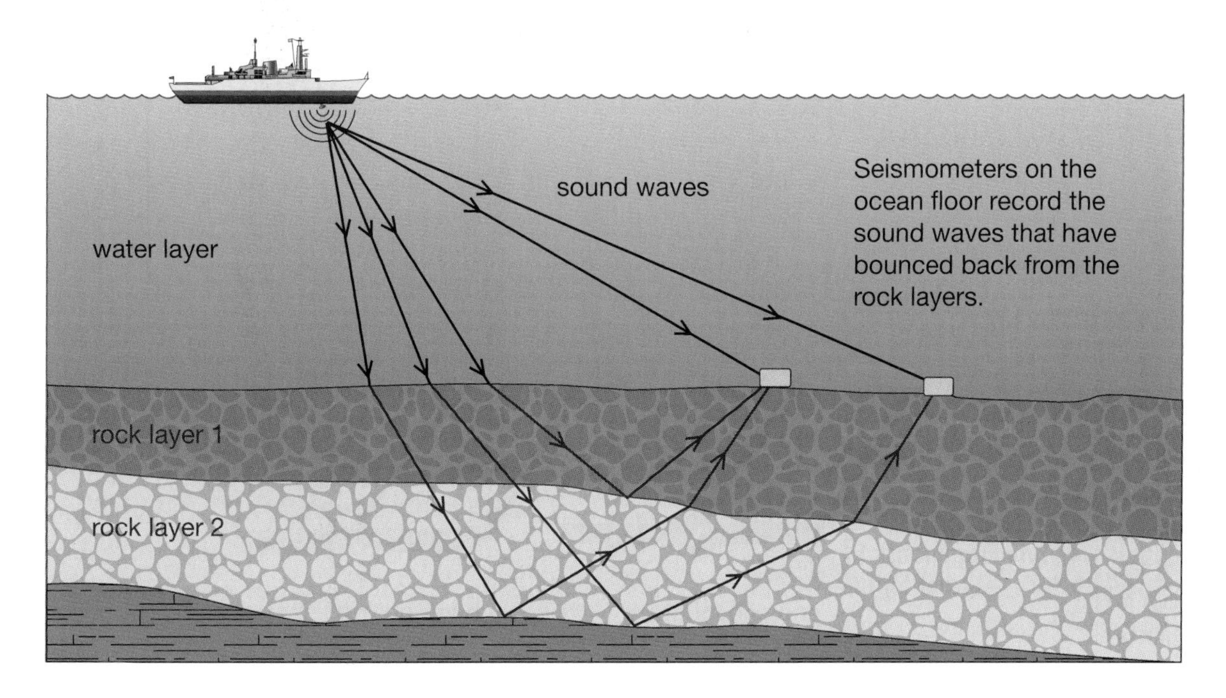

Q What can the rock under the ocean tell you that other rocks cannot?

A Rock under the ocean provides information about ocean floor spreading. This is something we have been investigating since technology developed during World War II became available to us.

Q Does the fact that the sea floor is spreading mean that Earth is getting bigger?

A No, it doesn't. As new crust is produced in one area, it disappears in another. There are deep trenches in certain parts of Earth's oceans. At these trenches, a plate will slip past another one, down into the mantle where it melts and turns back into magma. This is a continuing process.

Q Why is the study of rock under the ocean "special"?

A The technology is fairly recent, and the latest theories about what is occurring on and beneath Earth's crust are still quite new. The theory of sea-floor spreading was only developed in the 1960s. It is exciting to be able to explore areas where new observations can be made. Who knows what the ocean floor still has to tell us about Earth's crust?

EXPLORING Further

The Speed of Sound

Geophysicists like Charlotte Keen study how vibrations travel through different materials. Try this experiment to compare how sound vibrations travel through wood, glass, and metal.

Hold a ticking watch by its band and place the back of the watch against a large glass window. Put one ear on the window 1 m away from the watch and cover your other ear with your hand. Can you hear the ticking? Now, try the same thing with the watch 1 m from your ear on:

- a metal vacuum cleaner hose or water pipe
- a wooden table or board.

How did the sounds compare? Which materials transferred the ticking sound best?

UNIT 4

An *Issue* to Analyze

A SIMULATION

A Mine Just Outside of Town

Think About It

Put yourself in the place of a person in a northern town. The town is small, and there are few local industries; thus job opportunities are limited. You have just received information in the mail from your local town council about the proposed construction of a mine just outside the town. You ask your friends what they think. Some feel the new mine will be a lifesaver! It will provide lots of job opportunities, and the mine may even bring more people into the area. This means more business for shops and restaurants. Others view the mine as a disaster. It will disturb, perhaps even destroy, part of the natural environment. Jobs are needed, but mines have created eyesores and major environmental problems in other places.

Mineralite Mining has presented a proposal to the town council for an open-pit mine to be developed in the area 5 km north of the town. An open town council meeting is going to be held to decide whether to allow the mine to be built. Start preparing for the meeting by considering the Background Information.

Background Information

The following information provides background on what will be involved in constructing and operating the mine:

- Mining involves two major kinds of *surface extraction* (the act of taking a mineral out of the ground). One is strip mining, and the other is open-pit mining. Strip mining does the least amount of damage, but it can be used only when a mineral lies in horizontal layers close to Earth's surface. Open pit mining is necessary when the deposits of minerals are close to the surface but are irregularly shaped. This is the case for the proposed mine outside your town.

- The pit that will be required will be more than 1000 m in diameter and 100 m to 200 m deep. It will be cut into solid rock in a series of terraces or steps. Trucks can use these steps as roads when they haul minerals and equipment in and out of the mine.
- It is expected that up to 3 to 4 tonnes of waste will be removed for every tonne of mineral that is extracted. What will happen to the waste?
- In a process called *beneficiation*, extractions will be crushed and ground in a mill to separate the tailings (unwanted rock materials) from the ore that contains the valuable minerals. This process will require huge amounts of water. The waste water, containing the tailings, will be transported to settling ponds, where the tailings will settle to the bottom. It is expected that the mine will recirculate treated water from the settling ponds back to the mill for re-use. When the settling ponds fill up, the waste will have to be moved to another site, or new settling ponds will be required. Mineralite Mining has promised to do everything within its power to ensure that chemicals and minerals from the tailings are not washed into streams and lakes by rainwater.
- After beneficiation comes the smelting process. *Smelters* are furnaces that reach very high temperatures and turn the ore concentrates into metals. The metals are then refined in refineries. Mineralite Mining recognizes that smelting and refining can cause serious pollution. Waste rock, called *slag*, is sometimes dumped in heaps on mine property, and fumes from smokestacks go into the air. Because many minerals contain sulfur, smelters and refineries are major contributors to acid rain. Mineralite Mining believes that these problems can be prevented or lessened. They propose to install state-of-the-art, tall smokestacks. Fumes will be blown out of the area before falling to the ground. As well, Mineralite Mining promises that when the area has been mined and the mine closes, they will reclaim the land. They will bury the slag, replace the topsoil, and turn the area into parkland.

Plan and Act

1. Plan to attend the town council meeting. These are some of the people who have asked to make a presentation at the meeting:

   Mineralite Mining executive

   town councillor

   Aboriginal representative

   unemployed town citizen

2. Consider these questions: What point of view do you think each of these people might have before the meeting? What are some concerns and considerations they might want to discuss? Do you think anything could change their points of view?

3. Your teacher will give your group the role of one of these people with additional information to help you plan your presentation. As a group, research your role, and be prepared to argue your case at the town council meeting.

4. At the town council meeting, your task, as a class, will be to decide whether to accept the mining company's proposal. Your teacher will provide you with a blackline master to show you the correct *Procedure* for a Public Hearing.

Analyze

1. What was your decision?

2. Were all points of view well researched and presented? If not, how do you think they could have been improved?

3. Were all participants fairly treated? If not, how could this be improved?

4. How did your understanding of science and technology help you make your decision?

UNIT 4 PROJECT

A Creative Crust

This unit has shown how Earth's crust is always moving and changing. Some changes take many millions of years, as you saw in the time line on page xxx. Other changes occur in a few short, terrifying hours, as you saw in the photographs of the eruption of Mount St. Helen's on page xxx. In whatever way they occur, the changes in Earth's crust result in events such as the following:

- Minerals form when magma cools.
- Rocks form in various ways and are classified as igneous, sedimentary, or metamorphic.
- Soils form when rocks weather and organic matter is added.
- Weathering continues, causing soil to erode.
- Mountain formation is caused by convection currents under Earth's tectonic plates.
- Earthquakes occur when pressure at fault lines becomes too great.
- Volcanoes erupt when magma that has built up inside a mountain pushes upward, usually due to plate movement.
- Fossils indicate when various layers in Earth's crust formed.

Challenge

Design a model of Earth's crust that shows events like those listed above.

Materials

You will brainstorm about the materials that will be most appropriate for your model. Here is a partial list. You may have additional ideas.

household goods, such as food colouring, salt, flour, baking soda, water, cornstarch, modelling clay
gravel, stones, small rocks
cardboard
plywood
paper, tin foil, wax paper, construction paper
tape, glue
tacks, small nails

Safety Precautions

- Do not mix chemicals without your teacher's knowledge and approval.
- Wash your hands after completing this project.

Design Criteria

A. Include at least four of the eight listed events in your model.

B. Your model should be no smaller than your textbook and no larger than your desktop.

C. You must have a completed model ready to demonstrate to your class in the time frame provided by your teacher.

D. Your model and demonstration must be accompanied by a clear summary of each person's contribution to the model and demonstration.

Plan and Construct

1. With your group, brainstorm ways to set up your model. How can you build it so that all (or most) of the events described above are clearly displayed? How can you demonstrate how much time is involved in each event? Which events will you choose to include? Are there other events, not mentioned in the list, that you would like to include?

2. Are there areas where you need to do some research or advance planning? For example, if you want to include a working volcano in your model, you need to find out how to make it work. Maybe you want to include some earthquake-resistant structures. Chapter 11 touched on these briefly; however, you will need to do

some research and practise building such structures before you add them. Perhaps you plan to have fossils in sedimentary layers to demonstrate superposition. These must be researched and prepared before you build the model.

3. Divide all stages of the work among the group. All the group members must be clear about their role, what they need to do, and when their work must be ready.

4. Decide how you will actually build your model. Will you build multiple components and then assemble them, or will you build one basic structure that includes all of the components?

5. Brainstorm the materials list. Be as creative as you can. Your first ideas may not be the best ones, so be open to other ideas until the group agrees on the best possible materials for the model.

6. Prepare a plan. Include a materials list and a neat, labelled diagram of your model. Present it to your teacher for approval.

7. Begin the building process. Parts of your plan may not be successful, so your group should discuss what is not working and come up with an alternate approach. Be prepared to modify your plan as you go along.

8. Try out the working parts of your model. If an earthquake or a volcanic eruption are to occur, make sure they work. Make any demonstrations of erosion and weathering as realistic as you can.

9. Prepare an explanatory poster to accompany your model. Display it on the wall beside your model.

Evaluate

You assessed your model as you built it, and you might have modified it during the process. At this stage, think about how well the model worked as a whole. Did any other model in the class work better than yours did? If so, why did it work so well? Were there ideas you would like to incorporate into your own model? What might you change if you were to start from scratch and build the model again?

Design and build a model of an earthquake-resistant building. As a class, develop the design criteria for your models, and determine in advance how they will be evaluated.

UNIT 5

Structural Strength and Stability

Have you ever been on an amusement park ride that is similar to one shown in this photograph? Did you find yourself wondering about the safety of its various parts — the cables, the towers, or the concrete on which the structure rested?

All of us trust our lives to some very strange and wonderful structures. Some structures, such as dams and skyscrapers, contain such enormous amounts of materials that their weight distorts the ground beneath them. Other structures, such as amusement park rides and gymnastics apparatus, have carefully arranged parts that hold them up and support whatever they must carry. All structures, no matter how they are built, must keep their shape, stay in place, and do a particular job.

Small or large, all structures are designed using similar principles. Whenever you construct models, handicrafts, or clothing, you need to answer the same questions and solve the same problems as professional architects and designers. By learning the principles that designers use, you can do a better job of designing and building your own projects. As well, you will have a better understanding of how a ski lift or an amusement park ride can be safe, and why you can depend on the apparatus in gym class!

Unit Contents

CHAPTER

13 Describing Structures

Getting Ready...

- What is the largest human-made structure that you know of?
- What is the strongest natural structure you have ever seen?
- What is the most unusual thing you have ever built?

Science Log

Think about the questions above. Record your thoughts in your Science Log. Then compare your ideas with a partner's thoughts. What new ideas arose as a result of the comparison? Add them in your Science Log.

Skill POWER

For tips on how to make and use a Science Log, turn to page 476.

Think of a hobby or craft you enjoy. You have probably learned a lot about your special interest from reading, trial and error, and talking with other people. Potters learn about clay and how to shape it. Knitters learn about different kinds of wool and stitches. Artists, crafters, and tradespeople find out about the materials and methods — the technology — of their trade. They know from experience what works.

Scientists and engineers find out how to make things in a different way. They carefully investigate the behaviour of materials. They measure the forces that hold things together and break things apart. They develop general theories and models so that they can predict how a building or machine will behave before it is built.

In this chapter, you will explore both methods of thinking. You can find out how things work and why they work. Then when you design and build your own projects, you can use both scientific and technical skills to do the best possible job.

and Their Construction

Starting Point ACTIVITY

All the Details

Use what you already know to describe and analyze each structure you see on these two pages.

What to Do

1. Match each of the small photographs on these two pages to the best description in the list below.
   - a hard clay container for soil and small plants
   - an ancient Egyptian stone sculpture whose purpose is unknown
   - carefully arranged layers of silicon and metal that control electrical signals
   - a tourist attraction in Paris, built to show that iron frames could support large buildings
2. Describe the important features of each structure by answering the following questions. Guesses are allowed!
   (a) Function: What is it designed to do?
   (b) Material: What is it made of?
   (c) Strength: How is it made strong enough to do its job?
   (d) Assembly: How is it held together?
   (e) Special feature: How is it unusual?
3. Choose one structure in or around your home. Give the name of the structure, and then describe it by answering questions (a) to (e) in step 2.

Spotlight On Key Ideas

In this chapter, you will discover

- why we classify structures as mass, frame, or shell
- how the function of a structure affects its design
- how the choice of building materials, shapes of parts, and methods of fastening parts together all affect the performance of a structure

Spotlight On Key Skills

In this chapter, you will

- identify examples of mass, frame, and shell structures, and describe key features of each group
- design structures and explain why you chose particular shapes, building materials, and fastenings
- design your own investigation of the strength of a frame structure

13.1 Types of Structures

Have you ever made a sand castle or created a snow sculpture? What about building an igloo or assembling a tent? Perhaps you have woven a basket or baked a cake. All of these objects are examples of structures: things that have a definite size and shape. We can classify structures as **manufactured** (made by people) or **natural** (*not* made by people).

Structures are also classified according to what they do. All manufactured structures have a purpose or **function**. To perform this main job, every part of the structure must resist forces that could change its shape or size. For example, you might build a brick wall around your backyard. The wall must be able to stand up to the force of the wind. The bricks at the bottom must support the weight of the bricks above. If a person climbs on top of the wall, the bricks must support that **load** (the weight carried or supported by a structure) as well.

A third classification of structures is based on how they are built. There are three basic designs: mass, frame, and shell structures. Each design uses a particular type of construction, with its own set of advantages and problems.

Figure 13.1 The pattern in these bricks is called a "running bond." It is used for strength.

Mass Structures

To build a sand castle, you start by making a big pile of sand. If you want the castle to be larger or last longer, you start with a bigger pile. Sand castles and other things built by piling materials up are called **mass structures**. Snow sculptures, dams, and brick walls are mass structures. So are natural structures like mountains and coral reefs, and foods like omelettes, cakes, and breads.

Making something from a lot of building materials has advantages. The structure is held firmly in place by its own weight. If small parts are worn away or broken, this usually makes very little difference. Mass structures like Hadrian's Wall in England have been eroding for thousands of years without being destroyed.

Pause& Reflect

You will learn a lot about functions, forces, and loads in this unit. What do these words mean to you now? In your Science Log, write a brief definition of each word, using your own words, then write a sentence that uses each word correctly.

A Layered Look

All around you there are mass structures made of carefully arranged pieces. Have you ever noticed the pattern of bricks in a brick wall? The centre of each brick is usually placed over the ends of two bricks in the row below, as Figure 13.1 shows. Bricks and concrete blocks are often arranged in other ways, however. Compare the patterns used for several outside and inside walls made of bricks or blocks. Look around doors and windows to see if the arrangement is different there. If you can, ask a bricklayer or a mason to explain when particular patterns are used.

Skill POWER

To find out how to do this kind of research, turn to page 497.

DidYouKnow?

What is the largest amount of material ever used in a dam? The dam that was built to create the tailings pond at the Syncrude Oil Sands project in northern Alberta contains over 540 million m^3 of material! Use print or Internet resources to find out what the Syncrude factories produce, why the tailings pond needs to be so big, and where the water in the pond comes from.

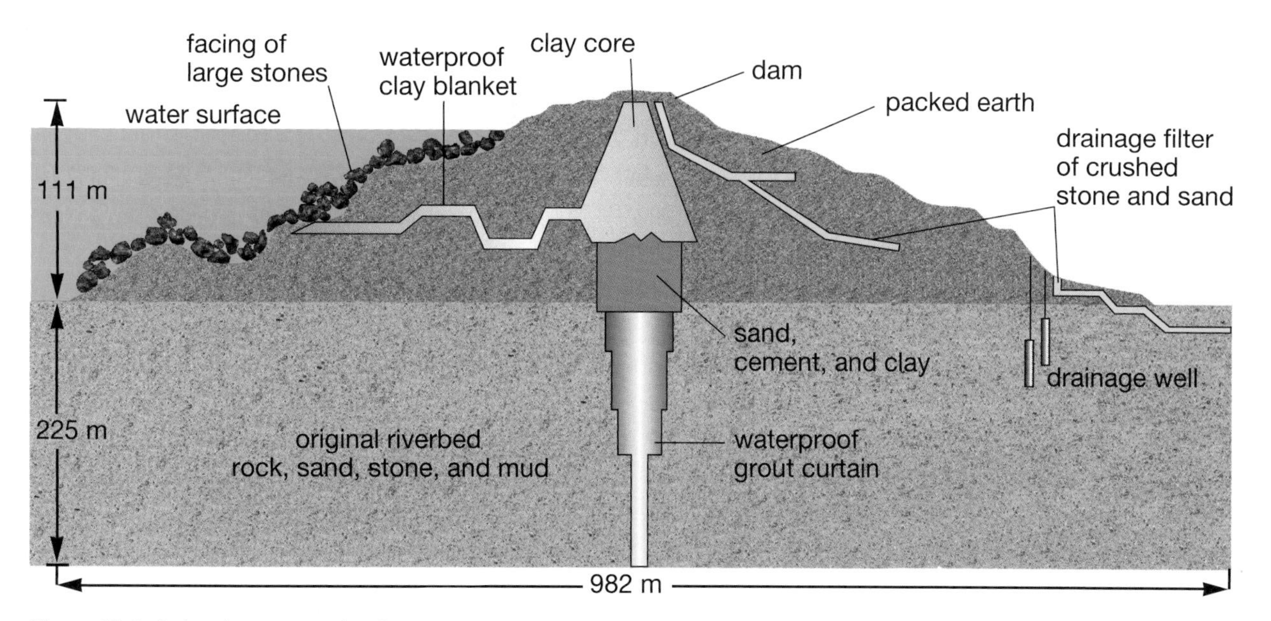

Figure 13.2 A dam is an example of a mass structure.

Take a close look at the dam in Figure 13.2. Try to guess the purpose of each type of material.

Mass structures are not always solid. Inside many power dams are enormous rooms that hold electric generators. Bricks and concrete building blocks are hollowed out so that wires and pipes can pass through them.

Because of their large size and weight, mass structures must be very carefully designed. Think of a wall of sandbags holding back a flooding river. There will be big problems if the wall fails! There are four main ways that this can happen.

- The wall may not be heavy enough to stay in place. The whole structure is pushed out of place by the force of the water against it.
- The wall may be so heavy that the earth beneath it is pressed down unevenly. The structure becomes unstable and tips over or falls apart.
- The wall may not be thick enough or fastened tightly together, so parts of it are pushed out of place. Then the whole structure breaks apart.
- The structure may not be anchored firmly to the ground. If there are very large forces pressing against the top, the whole structure may tip over.

The Great Pyramid in Egypt contains passages and rooms that once held everything that Ancient Egyptians believed the dead Pharaoh would need as he journeyed to the spirit world.

DidYouKnow?

Flying passenger aircraft across the oceans seemed almost impossible in the 1930s. The trip required too much fuel. Canadian inventor Frederick Creed designed and built models of huge floating islands where planes could land and refuel in the middle of the ocean. During the Second World War, another inventor, Geoffrey Pyke, suggested building these artificial islands or even aircraft carrying ships out of ice! A trial project in Alberta found that a frozen mixture of water and wood pulp was strong enough to build a refueling island or a ship. Building with new material was as expensive as using steel, however, so the project was dropped.

Frame Structures

Human dwellings and office buildings are not usually mass structures made by hollowing out piles of building materials. Like the houses in Figures 13.3 and 13.4, many buildings are **frame structures**. They have a skeleton of very strong materials, which supports the weight of the roof and covering materials. Most of the inside of the building is empty space. Extra partition walls can be built to separate different rooms, but they do not need to be particularly strong because the load-bearing framework supports the structure and everything in it. Can you identify the load-bearing walls and the partition walls in the diagrams of the house?

Career **CONNECT**

Many different people work together to construct a building. Getting the house in the photograph built to this stage required the work of an architect, a contractor, a surveyor, an excavation team, and foundation builders. What was the responsibility of each member of the team? Which members of the team were responsible for the house's stability? Brainstorm the types of workers who would be involved in the next stages of construction for the house. Choose one, and find out about the kind of work this worker does.

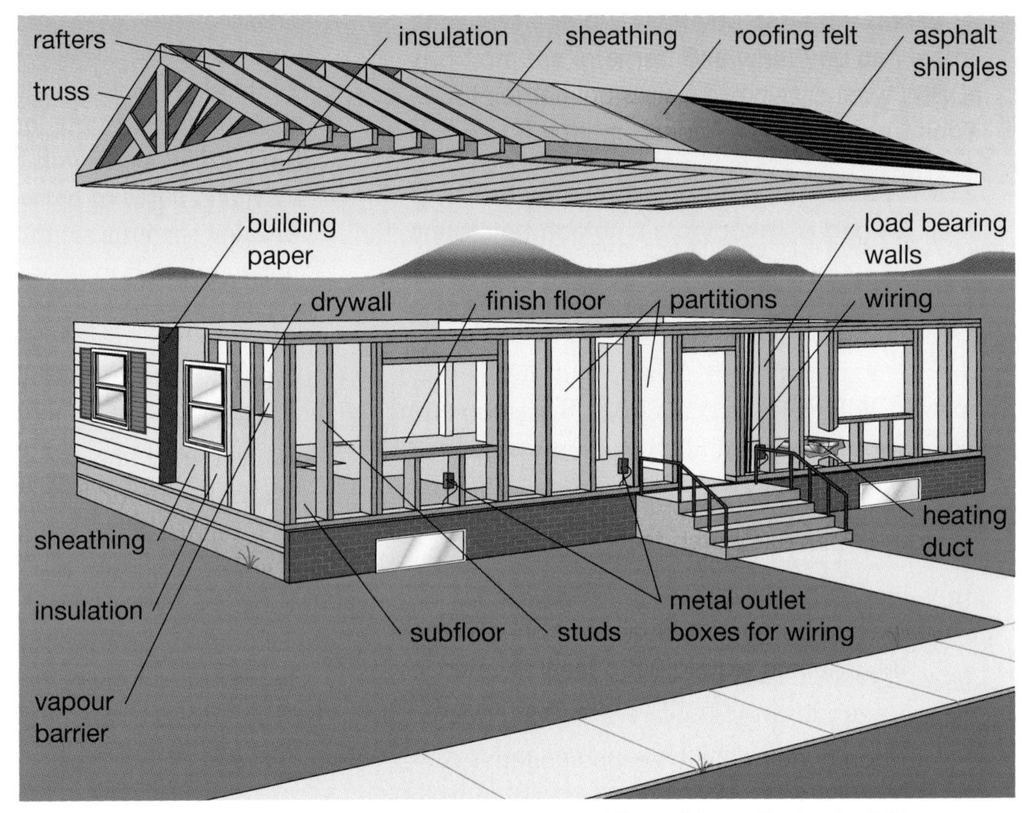

Figure 13.3 Load-bearing walls hold up a frame structure, while partition walls simply divide rooms.

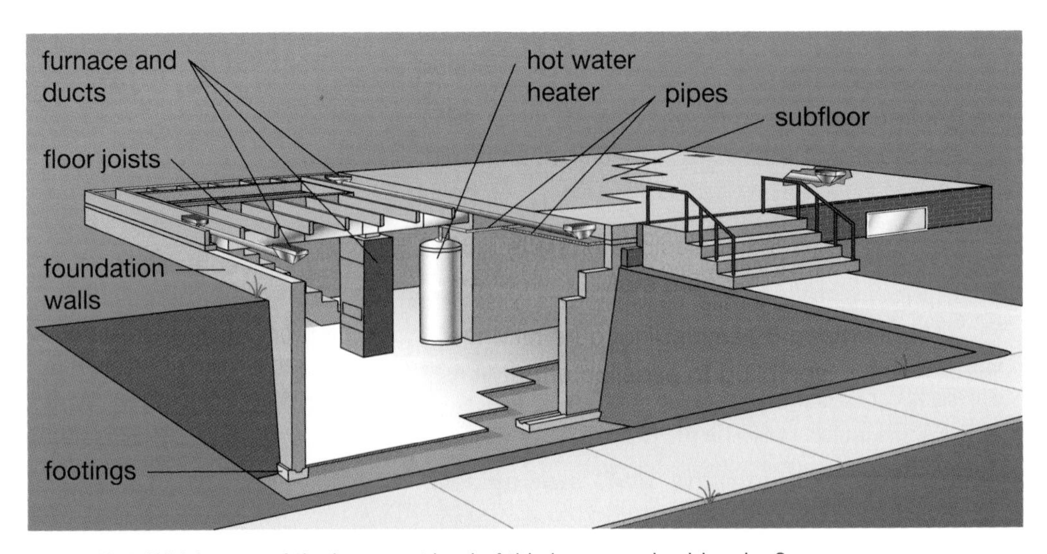

Figure 13.4 Which parts of the basement level of this house are load-bearing?

Frame structures are very common and are designed in many different ways. Some objects, such as ladders, snowshoes, and spider webs, consist of only a frame. More complex objects may have other parts added to the frame, such as the pedals, gears, and brakes of a bicycle. The frame may be hidden beneath covering materials (as in umbrellas, automobiles, and boats) or left exposed (as in drilling rigs and steel bridges). Whether simple or complex, hidden or exposed, all frames must overcome similar problems.

At Home **ACTIVITY**

Examine a Frame

What to Do

1. Choose a frame structure, such as a bicycle, that you can examine closely at home. You may pick any other structure approved by your teacher.

2. Sketch the shape of the frame in the centre of a piece of poster paper. Leave space for a title and for notes all around your sketch. Work neatly, and design your poster so that it can be seen and read easily from a distance.

3. Find at least one place or part on your structure that illustrates each of the features below. Circle the location on your sketch and add a brief note on your poster explaining how this feature is designed on your particular structure.

   (a) rigid joint: fastens parts of the frame together so that they cannot move

   (b) mobile joint: holds parts of the frame together but allows them to move or turn

   (c) brace: strengthens a joint or another part that must support a heavy load

   (d) rigid shape: will not collapse or change shape even when large forces push or pull on it

   (e) thin, lightweight material: does not have to be an especially strong part or place

   (f) part that uses extra material for strength

   (g) part that must resist a pulling force

   (h) part that must resist a pushing force

   (i) part that must resist a twisting force

   (j) place where the material has been carefully finished

   (k) one other interesting part of the frame structure: your choice

Do you remember building frame structures, like towers or bridges, in other science classes? How did you fasten the parts together? How did you make your frames strong without using too much material? How did you shape or brace them so that they would not bend or collapse?

Certain kinds of frame structures present special design challenges. Tents and other lightweight structures do not have enough mass to stay in place without some type of anchor to fasten them securely to the ground. Very tall frame structures, such as communications towers, can easily become unstable unless they are carefully braced. Large, complicated projects, such as buildings and bridges, have many parts that all have to fit together perfectly when they are finally assembled at the building site. This can happen only if every detail of the design is calculated in advance.

DidYou**Know**?

Alexander Graham Bell, the inventor of the telephone, was also very interested in flying machines. Bell experimented with many different kinds of kite frames, trying to find one that would be strong enough and light enough to lift a person and a gasoline-powered motor. The most promising kites used bamboo or aluminum frames made of many tetrahedral (pyramid-shaped) sections or cells covered in silk. In 1905, at his summer home in Baddeck, Nova Scotia, Bell demonstrated a 1300-cell kite named *Frostking*, which could lift a person into the air in only a light breeze!

Bell was certain that a practical flying machine could be built from his kite designs, but he could not do it alone. He and his wife Mabel Hubbard gathered a small troop of skilled helpers who called themselves the Aerial Experiment Association, shown above. They were one of the first modern research groups.

By the end of 1907, the Association had completed a giant kite named *Cygnet*, which carried pilot Thomas Selfridge to a record height of over 50 m. *Cygnet* was very strong and light, but the kite had far too much air resistance to be powered by a lightweight engine. In fact, it had to be towed behind a powerful boat. So Bell and the Aerial Experiment Association stopped working on people-carrying kites and started designing aircraft more like those we use today.

INVESTIGATION 13-A

The Windproof Wonder

Moving air can be very powerful. It has enough energy to turn a windmill (see the windmill "farm" on the right). If it is as strong as a tornado or a hurricane, it can blow down buildings. One of the tasks of an architect is to design buildings that can withstand high winds. Can you design a frame structure that uses the power of the wind but is not blown over by it?

Challenge

Build a frame to support a sign that stands 50 cm or more above the ground, contains at least one moving part, and cannot be blown over by a hair drier.

Safety Precautions

Handle electrical equipment and sharp objects with care.

Materials

scissors
2-speed hair dryer or fan
small model airplane propeller (optional)
fasteners/connectors (paper clips or pins)
other apparatus approved by your teacher
20 plastic drinking straws
index card (20 cm × 13 cm)
10 cm masking tape
2 small sandbags (plastic sandwich bags filled with 200 g of sand)

Design Criteria

A. The free-standing frame must support the top of an index-card sign at least 50 cm above ground.
B. The sign must contain at least one wind-powered, attention-getting, moving part.
C. The sign can be taped to the straws, but the straws must be fastened together without tape.
D. Only the sandbags can be used to hold the frame to the ground.
E. After creating your initial design, you have 25 min to build, test, and modify your structure.

Plan and Construct

1. With your group, brainstorm ideas for your sign. Make sure your design includes all of the design criteria. Plan how to prevent your sign from being blown over.
2. Use the most practical design ideas as the basis for a neat labelled sketch of your design. Then list the materials you need. After getting your teacher's approval, gather the materials and start building.
3. You may test your frame once before judging, but only with the fan or hair drier set to half power. If you modify your original design, sketch how your structure actually looks when it is complete.

Evaluate

1. (a) List the ways you tried to make your structure strong and stable.
   (b) Record how well your structure met the specifications. Then describe at least three specific changes you could make to improve the design or construction of your sign.
2. With your group, evaluate your structure. How might you modify its design to eliminate weaknesses and improve its performance?

Figure 13.5 The dome of the Taj Mahal in Agra, India, is one of the most famous shell structures in the world.

Shell Structures

Think igloo. Think egg. Think cardboard box. All of these objects are strong and hollow. They keep their shape and support loads even without a frame or solid mass of material inside. Egg cartons, food cans and bottles, pipes, and clay pots are other examples of **shell structures**: objects that use a thin, carefully shaped outer layer of material to provide their strength and rigidity. Flexible structures, such as parachutes, balloons, and many kinds of clothing, are a different type of shell. Even the bubbles in foams and cream puffs can be thought of as shell structures.

Shell structures have two very useful features. They are completely empty, so they make great containers. Because they have only a thin outside layer, they use very little building material.

You might think that the material in a shell structure would have to be extremely strong, but this is not always the case. The shape of a shell spreads forces through the whole structure. Each part of the structure supports only a small part of the load, and the complete structure can be amazingly strong.

Figure 13.6 When polarized light (light waves of only one direction) passes through transparent materials, such as this cassette tape, the stressed areas appear as coloured fringes.

Constructing strong shell structures can be tricky. Builders face problems like these:

- Tiny weaknesses or flaws like scratches on a glass jar can cause the whole structure to fail. Bubbles pop, balloons burst, and glass seems to explode when forces are not resisted equally by all parts of the shell.
- If a shell is formed from hot or moist materials, such as melted plastic or clay, uneven cooling or drying can cause some areas to push and pull on nearby sections. Strong forces (stresses) build up inside the shell, as in Figure 13.6. If any extra force, even a small one, acts on the shell, the stressed places may break unexpectedly.
- Flat materials, such as sheets of plywood, are not easily turned into the rounded shape of a shell structure. Imagine building a plywood igloo! Each piece would need to be shaped and fitted into place individually, so construction would be slow and difficult. If you were paying a builder, the cost would be higher than for a frame structure.
- Assembling flexible materials into a shell is also tricky. Garment pieces need to be pinned into position before sewing. Afterward, the fabric edges must be specially finished so that the cloth will not pull apart along the seams.

Canoes are shell structures that are built from birch bark, aluminum, or fibreglass. Did you know they can also be built from concrete? Some engineers believe that concrete shells could be used to build the floating ocean platforms used for offshore oil drilling. They believe that concrete shells would perform better than the steel construction now used.

www.school.mcgrawhill.ca/resources/

Find out more about concrete canoes by going to the web site above. Go to the **Science Resources**, then to **SCIENCEPOWER 7** to know where to go next. Write a short newspaper report telling who builds concrete canoes and why they do it.

Across Canada

As a child, Marie-Anne Erki often took apart her toys. "I remember, with some sadness, my little white electric poodle that I dismantled immediately upon receipt, never to be put back together again." Now she is a professor of civil engineering at the Royal Military College of Canada in Kingston, Ontario. These days, she is more concerned with putting structures together than taking them apart!

Dr. Erki works with high-strength fibres embedded in a polymer (plastic). They are called Fibre Reinforced Polymers (FRP). FRPs are as strong as steel but much lighter. "My field is making existing structures like bridges and buildings stronger using carbon tape. It's just like taping up a hockey stick," she says. In Canada and other cold-weather countries, road salt eats away at the steel bars that reinforce concrete bridges. FRP tape can patch and wrap these weak, rusted bridge columns and beams in order to strengthen them. Also, new concrete bridges can be reinforced with FRP bars instead of steel reinforcements so they will never rust.

Dr. Marie-Anne Erki

In her work for the Canadian Forces, she is developing materials to make lightweight bridges that can be transported by airplane. This technology will be helpful when there is a natural disaster, such as a hurricane or a flood. It will enable washed-out bridges to be quickly and inexpensively replaced so that people are not cut off from food or medical aid.

When she is not developing improved materials for construction, Dr. Erki helps university students with their research projects. "I deeply enjoy working with students and getting them interested in structural design. Their enthusiasm for the topic energizes me."

CONDUCT AN

INVESTIGATION 13-B

Crush It!

You can find out about the strength of a structure by placing it on an ordinary bathroom scale and pressing down. A spring in the scale pushes back, and the scale reading shows the size of the force. Since the spring's force just balances your downward push, the scale reading also shows how hard you are pushing. If you push until you break the structure, you can find what engineers call the *ultimate strength* of the structure.

Problem

What is the ultimate strength of several common shell structures?

Safety Precautions

- Wear safety goggles and an apron during this investigation.
- Work on newsprint or a plastic drop cloth to simplify clean-up.

Apparatus

bathroom scale

flat board, the size of a paperback book

Materials

newspaper; 2 or 3 shell structures to test: ice-cream cone (flat bottom); Styrofoam™ cup; section of a cardboard or foam egg carton; paper muffin tin liner; small section of celery stalk; half an orange, eggshell, or ping-pong ball; plastic blister packaging

Procedure

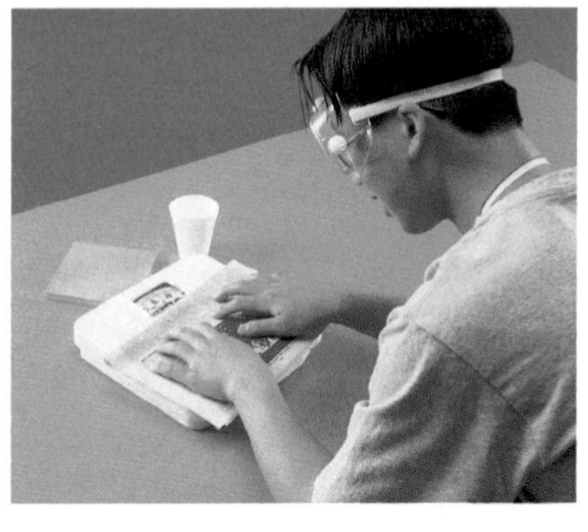

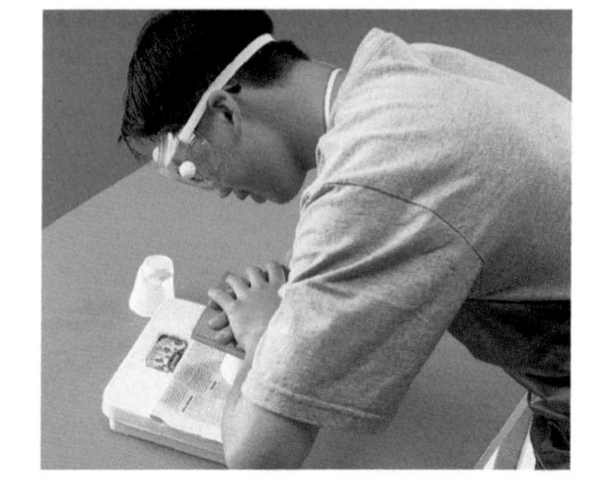

1. Put a sheet of newspaper on the bathroom scale. Then place the object you are testing on top of the newspaper.
2. Use the flat board to press gently and evenly on the top of the test object. Slowly and carefully increase your force on the object while watching the scale. Notice and record
   - the scale reading when the structure first begins to fail (Watch and listen!)
   - how the object fails (Does it bend first or just crack? Do small cracks appear and then spread, or does it crumble all at once? Where does it fail first?)
3. If possible, test more than one sample of each structure.
4. Clean up your work area. Wash your hands thoroughly after completing this investigation.

Analyze

1. Make a neat data table to summarize your observations. If you tested more than one sample of each structure, add a column for the average force needed to break the samples.
2. Rank the objects from weakest to strongest.
3. Draw a graph that clearly shows the strengths of the different objects. A bar graph would show these data clearly.

Conclude and Apply

4. Think of a shell structure in which strength would be
   (a) very important
   (b) not particularly important
5. What might have happened if you had suddenly pressed hard on the test object, or put all your force on a tiny part of it? How can the way in which you apply force affect the results?
6. Describe the bursting forces that each shell structure pictured here must resist.

Space suit   Water pipe   Basketball

Math CONNECT

Stand on a bathroom scale yourself. How large a downward force is created by gravity? Calculate how many ice-cream cones (or egg cartons or Styrofoam™ cups) would be needed to support your weight. If you have enough materials, try this out to check your calculations. (Remember to distribute your weight evenly over all of the objects.)

Math CONNECT

Forces are measured in newtons, but most scales are marked in kilograms or pounds. Each kilogram represents a force of 9.8 N (newtons). Each pound represents a force of 4.45 N. If necessary, multiply kilograms by 9.8 or pounds by 4.45 to change your scale readings to newtons.

Check Your Understanding

1. What type of structure (mass, frame, or shell) is each object listed below?
   (a) a backpacking tent
   (b) a pop can
   (c) a concrete barrier in a parking lot
   (d) the "jewel box" in which a CD is packed

2. Name the type of structure that is most likely to fail because
   (a) the material it was built from has small cracks or weaknesses
   (b) the weight of the structure caused the ground underneath it to shift
   (c) the outside walls were tilted slightly by an earthquake

3. **Apply** Model airplane wings are sometimes built by shaping solid pieces of foam plastic with a hot wire. They can also be made from many pieces of lightweight balsa wood glued together and covered with paper, plastic, or cloth. Full-sized, modern aircraft wings have a metal frame, but the frame is quite weak until a metal covering is stretched over it. Then the wings become strong enough to support the weight of the plane, as well as the weight of its fuel, which is carried in tanks inside the wings.
   (a) Name the type of structure represented by
   - a foam plastic wing
   - a balsa wood wing
   - a real airplane wing

   (b) If a model airplane crashed, which type of wing would be easier to repair?
   (c) Why do modellers not build wings as shell structures?

4. **Design Your Own** Choose a small frame structure to test, perhaps one that you have already designed and built. Make a hypothesis about the ultimate strength of the structure, and design an experimental procedure to test your hypothesis. If your teacher agrees that your chosen frame structure is safe to test, carry out your procedure.

Designers need to be able to visualize how the parts of a project can be arranged to create the final structure. Imagine that you are designing a small orchard. The owner wants the trees to be arranged in certain patterns.

1. How can ten trees be arranged in five rows, with four trees in each row?
2. If nine trees are arranged in rows of three, what is the largest possible number of rows you can make?

13.2 Describing Structures and Their Features

Imagine being a TV reporter checking the plans for a new bridge with its designer.

You: What an interesting looking structure you have planned! But why don't the drawings show any details at this middle part?

Designer: Well, I'm not exactly sure what to do there. I thought we'd just build the first part and then figure out the next step.

You: I see. What about the entrance? Could a large truck make that sharp turn?

Designer: Hmm. I didn't really worry about different kinds of vehicles. I just made it narrow to save materials and money.

You: That would reduce the strength, too, wouldn't it?

Designer: Who knows? If it breaks, the repair crew can always brace it a bit more when they fix it.

Architects and engineers do not, of course, work this way! Every detail of each design needs to be figured out in advance. The size and strength of each part must be found in advance by calculating or building computer or scale models. The quantity and cost of the parts must be estimated as accurately as possible, or the builder may lose money.

You can work like an architect or engineer. If you plan carefully at the very beginning of a project, your final design will be improved. There are four especially important features to consider: function, shape, materials, and joints.

Figure 13.7 This trestle bridge in Lethbridge, Alberta, was the longest structure of its height in the world when it was built in 1909. It carries trains 97 m above the valley of the Oldman River, for a distance of 1.6 km.

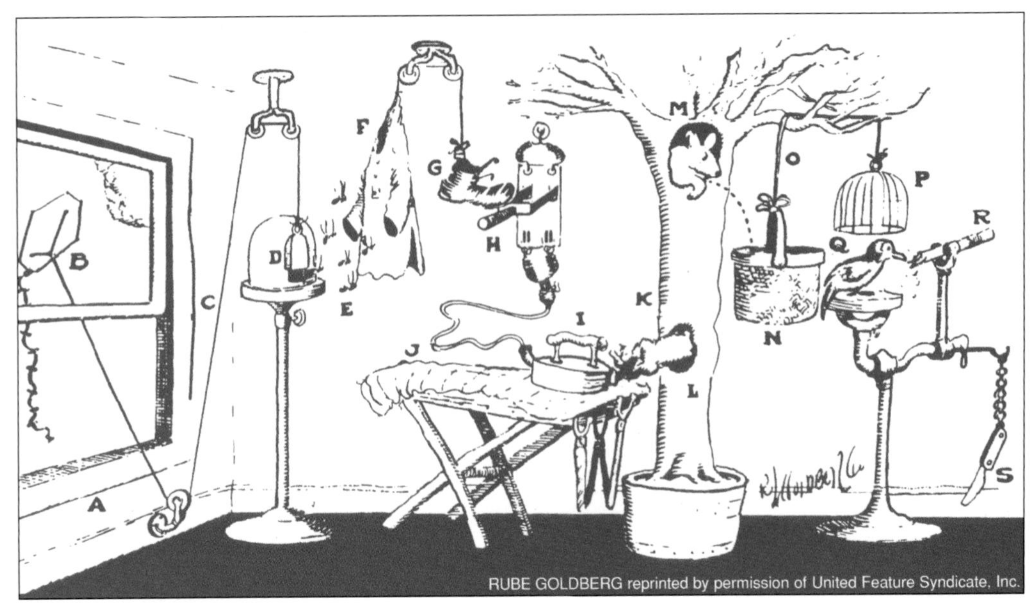

Figure 13.8 Fantastic devices like this were made famous by a cartoonist named Rube Goldberg.

Function

What is this thing supposed to do? The answer to this question will guide all of your design decisions. Simple? Not really. Most structures have several functions. Think of a bridge. Its job is to support . . . what? Vehicles and people, of course, but the steel and concrete the bridge is made from can weigh many times more than the cars and trucks travelling across the bridge. Thus one very important function of any structure is to support its own weight.

Structures do more than just support loads. For example, a running shoe, such as the one in Figure 13.9, grips the ground or gym floor and cushions your foot bones from the impact of running. But it has many other functions, too. The words on the left may suggest some of them.

Designers have a hard time creating structures that perform all of their functions equally well. Plastic-covered running shoes certainly keep water *out* when you run through a puddle. But they also keep perspiration *in*, so your feet soon get hot and sweaty. Most runners have rubber and plastic soles and cloth uppers. This compromise does not let much puddle water in but does let most perspiration out. The shoe does an acceptable job most of the time without having any serious disadvantages.

Word CONNECT

Here are a few words that describe common functions of structures:

- containing
- transporting
- sheltering
- supporting
- lifting
- fastening
- separating
- communicating
- breaking
- holding

Examine Figure 13.9, and then use the list of common "function words" to list as many functions of a running shoe as you can. Make sure that each statement is clear and detailed.

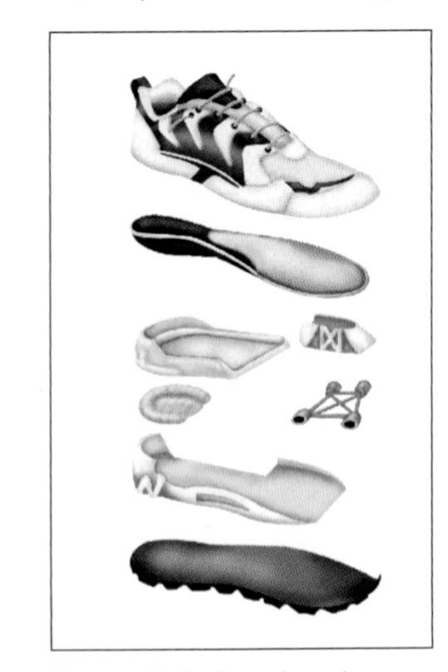

Figure 13.9 Running shoes have much more to them than meets the eye.

INTERNET CONNECT

www.school.mcgrawhill.ca/resources/

To find out more about Rube Goldberg, go to the web site above. Go to the **Science Resources**, then to **SCIENCEPOWER 7** to know where to go next. Think up your own Rube Goldberg device, and write a brief description of it.

How well must a structure perform its functions? Designers work to a set of **criteria** or **specifications** that give precise, measurable standards their structure must meet. Specifications for a running shoe might include criteria like these:

- sole can flex 100 000 times without cracking
- materials do not contain chemicals that could irritate the skin

One very important design criterion (the singular form of "criteria") is seldom written down. The best designs look good. Designers refer to good-looking designs as "aesthetically pleasing." (**Aesthetics** is the study of beauty in art and nature.) Think of an attractive car, building, or Internet page. It might use shapes that are repeated or carefully arranged to give clear lines. There might be interesting textures and colours that are carefully chosen to be harmonious or contrasting. Check with your art teacher for other principles of design that can help your projects look attractive.

Above all, architects and engineers try to keep their designs simple. Clean, designs look better than over-complicated, busy ones. So remember to keep it simple!

Pause& Reflect

Imagine that you are planning to go into the business of manufacturing running shoes. In your Science Log, write at least five specifications that your shoe design must meet.

Figure 13.10 The Museum of Civilization in Hull, Quebéc echoes a seashell in its design.

Shape

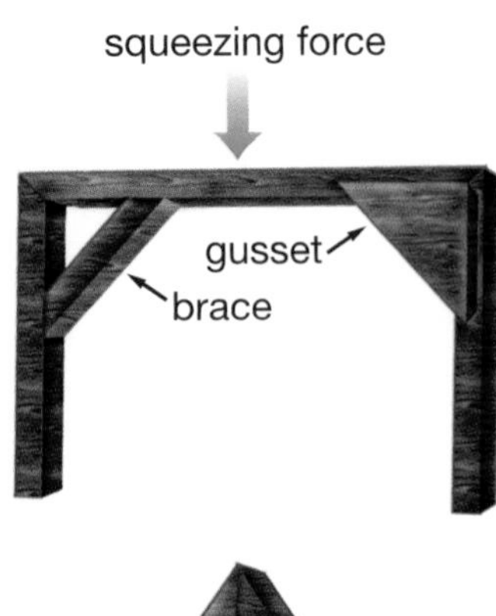

Figure 13.11 Solution to Problem 1. Triangular shapes give a structure strength.

Square and rectangular frames squish. When a load presses on them, enormous forces develop at the joints, and the frame is likely to collapse.

Rectangular frames are very easy to build, however, so designers have found ways to overcome most of their weaknesses.

Problem 1: Rectangular frames can easily be pushed and pulled out of shape.

Solution: Use triangles, as shown in Figure 13.11. **Braces**, **ties**, and **gussets** turn rectangular corners into rigid triangles, able to resist large forces. Whole buildings can be made with triangular sides, creating a pyramid shape.

Problem 2: Frames made of vertical **columns** and horizontal **beams** are weak in the middle. Pressing down on the middle of a beam causes it to bend and push sideways. If the supporting columns spread apart, the structure will collapse.

Solution 1: Brace the middle of the beam, as shown in Figure 13.12. Some of the load forces are carried toward the bottom of the supporting columns, which do not spread apart easily. The triangular shape also makes the frame more rigid.

Solution 2: Support the load with an **arch**, as shown in Figure 13.12. The arch carries load forces all the way to the ground. Domes use this principle to spread forces evenly through the whole structure and down to the ground.

Solution 3: Support the load with an arch, as shown in figure 13.12 (see Figure 13.12). A **double cantilever** is a particularly strong design with a very strong central column and braces that support beams on either side.

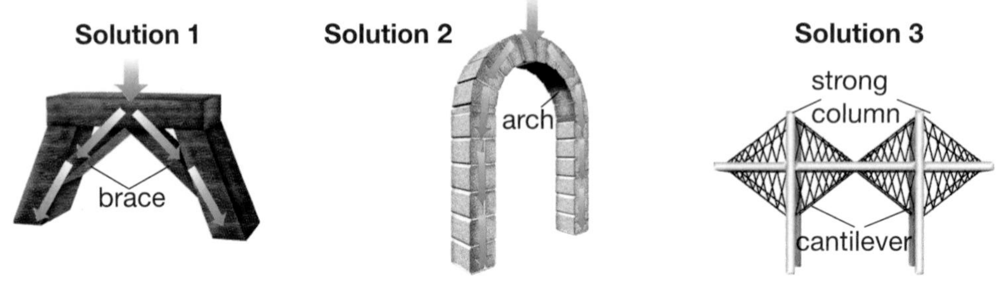

Figure 13.12 Solutions to Problem 2. Carrying load forces to the bottom of a supporting column strengthens the column.

Find Out **ACTIVITY**

A Living Machine

Think of your home as a machine that does things for the people who live inside it. Exactly what does the machine you live in do, and how does it work?

What to Do

1. Think of all the things your home-machine does for you. Make a note of each one.
2. Now list the parts of the home-machine that perform each function.
3. Finally, write a short description of how a very advanced home-machine, built 50 years from now, might perform the same function. Use your imagination, but be realistic.

What Did You Find Out?

Write a paragraph describing one job that your home-machine performs well and how it does this job. In a second paragraph, describe one function that your home-machine does not do very well and how its performance could be improved.

Problem 3: Solid, four-sided beams use a lot of building materials and are very heavy.

Solution 1: Use cylindrical columns, as shown in Figure 13.13. Though harder to build, they use less material.

Solution 2: Make the beam or column thinner in places that carry less load. It could be shaped like a capital letter I, L, or T. It could even be hollow, like the stem of a bamboo plant or a bird's bones. (see Figure 13.13)

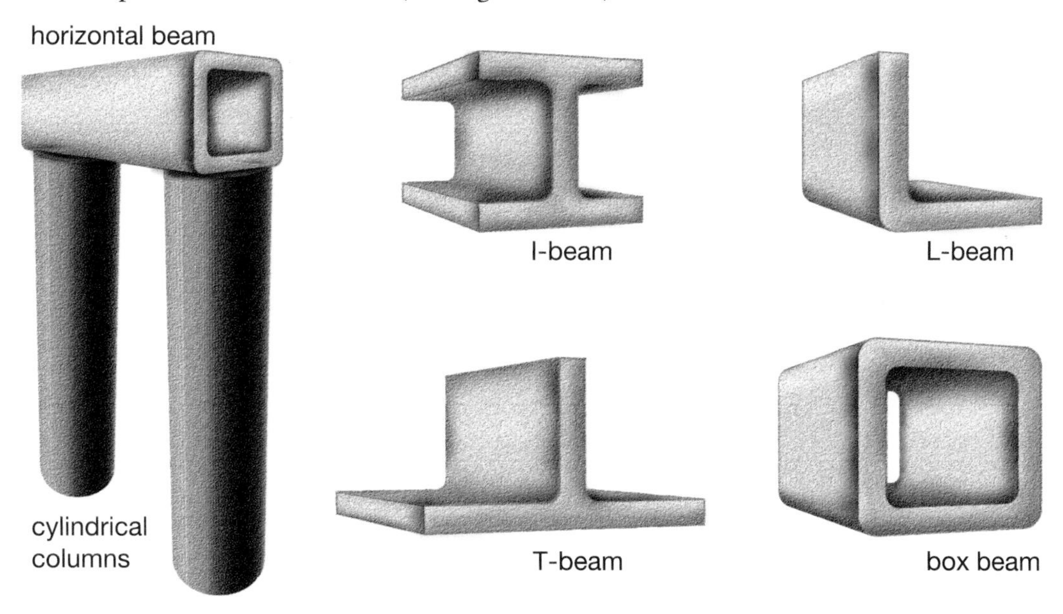

Figure 13.13 Various beam designs can be used to add strength to a structure.

Problem 4: Plywood, cardboard, and many other building materials are made in flat rectangular sheets. When a load presses on the middle of the sheet, or when the sides are pushed together by compression forces, the sheet bends out of shape.

Solution: Make the material of layers (**laminations**) that are glued or pressed together. Arrange the layers so that they strengthen each other. In plywood, the **fibres** in different layers point in different directions. Cardboard has a middle layer that is folded into a wavelike or **corrugated** shape, as shown in Figure 13.14.

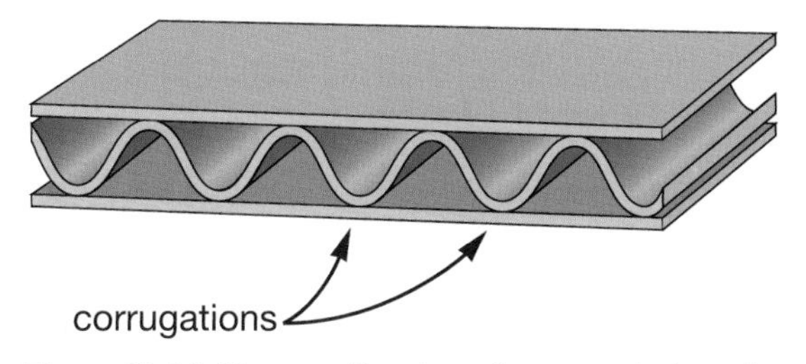

Figure 13.14 The wavelike shape in corrugated cardboard gives the cardboard strength.

Career **CONNECT**

With the help of an adult, contact an architect or building contractor in your area. Interview her or him to find out how architects are trained and what their work is like. Ask the architect to think of a particular building that was challenging to build. What were the functions and specifications for this building, and how did they influence its design?

Vancouver Public Library

The Paper Olympics

Architects and designers are sometimes quite limited as to the materials they can use, but they still need to design the sturdiest structure possible.

Challenge

How can you use your knowledge of shape and strength to turn flimsy paper index cards into load-bearing columns, beams, and flat panels?

Safety Precautions

- Use sharp objects such as scissors with care.
- A glue gun is hot and the glue remains hot for several minutes.
- Wash your hands thoroughly after completing this investigation.

Materials

scissors
templates for beams (optional)
bathroom scale
spring scale (or ice cream bucket and masses)
5 index cards (12.5 cm × 20 cm)
glue gun, white glue, or glue stick
flat plywood (20 cm × 20 cm)
string

Design Criteria/ Plan and Construct

A. Use one card to build a circular column that is the full height of the card (20 cm).

B. Cut and fold a second card into a strong beam that will span a 15 cm gap. Use the diagrams in Figure 13.13 (on page 393) to decide what type of beam to make. Then design your own pattern.

C. Use the last three cards to make a strong panel, larger than 10 cm × 12 cm. The panel must have at least three layers that are glued together, and it must be at least 1 cm thick.

D. Let the glue harden overnight.

E. Check each structure to make sure that it meets the specifications. Disqualify any structures that are too small or do not match the specifications.

Judging

1. In each category, the strongest structure receives 100 points, the next strongest gets 90 points, and so on. Your teacher might award extra points for that all-important factor, aesthetics (attractive design), and for careful construction.
2. Stand the column upright on a bathroom scale. Place a flat piece of wood on top. While watching the scale reading, press down slowly until the column fails. Record (in newtons) the largest compression force that the column can support. (If the scale reads in kilograms, multiply its reading by 9.8 to find the force in newtons. If the scale reads in pounds, multiply by 4.45.)
3. Place the beam across a 15 cm gap between two desks. Put a loop of string around the middle and attach the string to a spring scale or an ice cream bucket. Pull down on the spring scale, or add masses to the ice cream bucket, until the beam breaks, twists out of shape, or bends more than 1 cm. Record (in newtons) the largest force that the beam can support.
4. Place the panel on its edge on the bathroom scale. Watch the scale reading, as you press down on the panel until it collapses. Record (in newtons) the largest compression force supported by the panel.

Evaluate

Sketch the design your group used for each structure, and record its strength. Record any improvements that you could make or ideas that might be useful in your next construction activity.

Materials

What should we use to make this structure? Choosing building materials is another very important design decision. The **properties** or characteristics of the materials must match the purpose of the structure. Different materials can be combined or carefully arranged to give the exact properties you need.

Did You Know?

Imagine driving a car made of foam. The Karmann company in Germany has developed a layered material with thin sheets of aluminum covering a core of hard metal foam. Car body panels made from the metal "sandwich" are ten times stiffer than steel panels and much lighter, because the foam is filled with empty spaces. The foam is made by forcing gases through hot, liquid metal, making bubbles that harden as they cool.

Composite Materials

There are different kinds of strength. Steel rods and cables can support very strong tension (pulling) forces, but they bend and twist if you compress them. Concrete is just the opposite: it resists large compression forces, but breaks if it is pulled or twisted. The workers in Figure 13.15 are pouring concrete around steel rods and mesh. This produces something called a **composite** material, reinforced concrete.

As you saw in Chapter 11, inflexible substances, such as rock and concrete, are often stressed to the point where they break during earthquakes. If concrete breaks, the steel rods help to support the structure. Other composites have different properties. Fibreglass cloth embedded in rigid plastic is molded into boat hulls and other strong waterproof structures with complicated shapes. Graphite fibres in the shafts of composite golf clubs and frames of composite tennis racquets increase their strength and flexibility. Flexible plastic formed around a nylon mesh is used in lightweight garden hose to strengthen it against the pressure of the water.

Figure 13.15 Reinforced concrete can withstand both tension and compression. It is used to make buildings, bridges, and other large, strong structures.

Layered Materials

Use a fingernail to separate the edge of a TetraPak™ beverage container. Can you see the thin sheets of paper, plastic, and aluminum foil that make the container waterproof, airtight, lightweight, easily transported, and inexpensive? Layers of different materials, pressed and glued together, often produce useful combinations of properties. Inside the safety glass of car windshields is a plastic film that helps the glass resist shattering. Drywall panels on the walls of a room, and tiles or linoleum on the floor, contain layers of different materials. Even layers of one substance can be more useful than a single thick piece. Examine Figure 13.16 to find out how layers of wood are laminated into plywood. Then compare the diagram to a real piece of plywood. What differences can you find?

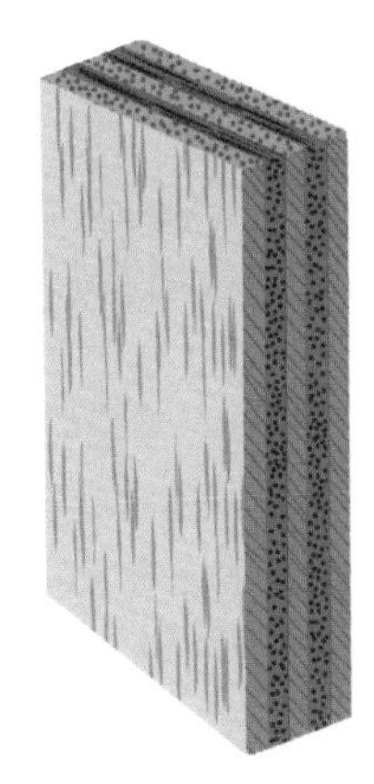

Figure 13.16 How do plywood laminations add strength to the substance?

Woven and Knit Materials

Use a magnifying glass to examine a piece of cloth. Can you see the hair-like fibres that have been spun (twisted together) into long, thin strings called yarn? A loom, such as the one in Figure 13.17, is used to weave two or more pieces of yarn together in a crisscross pattern to make cloth. Yarn can also be looped and knotted together to make knit materials. Figure 13.18 shows how each section of the knit is interlocked with many other sections so that tension forces are spread over the whole fabric. Knit materials stretch in all directions, so they fit well over complex shapes, such as human bodies. If you have any knit clothing, however, you know that the whole fabric can unravel when any piece of yarn breaks.

Weaving and knitting are not the only ways to make flexible materials. Paper and felt are made from fibres that have been pressed and matted together. Aluminum foil and plastic wrap are made by melting and dissolving a substance and then letting it harden into thin solid sheets. No matter how they are made, materials that can be folded or rolled are extremely useful for lightweight structures that must be easily transported and stored, such as clothing, tents, and parachutes.

Find Out ACTIVITY

Sneezeproof Strength

Look closely at a facial tissue. Use the information in the previous section to infer why such thin material has enough strength to resist sneezes.

What You Need

facial tissue
magnifying glass
pencil

What to Do

1. Partly separate a facial tissue into its layers. Look at the fibres through a magnifying glass. Are they pressed, woven, or knit together?
2. Put one layer of the tissue flat on a desk. Gently pull the two top corners apart. Notice how much force is needed to tear the tissue.
3. Now turn the tissue 90 degrees, and try pulling apart the same layer again. Is the layer easier or harder to pull apart in this direction?
4. Mark the grain of the layer along the direction that pulls apart more easily.
5. Repeat steps 2 to 4 for the other layer.

What Did You Find Out?

1. Are the layers arranged with their grains pointing in the same direction or in different directions? Which arrangement would be stronger? Which arrangement would be easier to tear apart? Include sketches to make your answers clearer.

Extension

2. Facial tissues are made from wood pulp. So are writing paper, newsprint, paper towels, and paper napkins. Find out if these other products have a grain and, if they are arranged in layers, whether the grains of different layers are aligned or not.

Figure 13.17 Yarn is twisted and woven to make cloth.

Figure 13.18 Interlocking yarn in a knit material spreads the tension forces throughout the fabric.

Find Out **ACTIVITY**

Putting It All Together

What to Do

1. At home, find a small device with several parts that you can bring to school. Look for something that will be unfamiliar to your classmates, so they will have trouble guessing its function. A device made of unusual materials or fastened together in an interesting way would also be a good choice.

2. Print your name on one side of a small index card. On the other side, list the following features of the device:
   (a) name
   (b) function
   (c) material(s) from which it is made
   (d) how it is fastened together

3. Bring your device to class, and display it as your teacher directs. Put the index card beside the device with your name facing upward.

4. Examine each device on display. Observe what it is made of and how its parts are fastened together. Guess its function and what it is called.

5. At your teacher's signal, compare your ideas with the information on the hidden side of each object's display card. Give yourself one point for every correct answer, so you can earn up to four points per device.

What Did You Find Out?

Choose one device, and write a short paragraph analyzing why its particular materials and fastening methods were selected. What advantages do they have over other possible materials and fastening devices? What disadvantages do they have?

Solid metal can be cut into complex shapes by milling machines like the one in the photograph. Other materials are formed and finished with different kinds of specialized equipment. You can find out more about the cool tools used with materials like clay, plastics, rubber, leather, rock, or glass by talking to a crafter or materials teacher, or by doing library or Internet research.

Joints

How should we fasten this structure together? That is a critical decision because structures are often weakest where their parts are joined together.

Mobile joints are joints that allow movement. Door hinges, elbows, and the pins in a bicycle chain are examples of mobile joints. They hold parts together while still allowing some movement. Their complicated shapes are tricky to make, and they must be coated with a lubricant (a slippery substance) so that they move smoothly. Without lubrication, door hinges squeak, bicycle chains wear out, and human joints, as in Figure 13.19, develop arthritis and similar painful diseases.

Rigid joints fasten parts firmly together, but they, too, can be weak points in a structure. Chair legs and table legs become wobbly as wooden parts shrink, glue ages, or nails, screws, and bolts loosen. Flexible materials are even harder to join securely. Clothing comes apart at seams where it is sewed together. Tape comes loose. Zippers and Velcro™ stop working.

People have invented an amazing number of rigid joints, but they all fit into just five groups: fastners, ties, adhesives, melted joints, and interlocking joints.

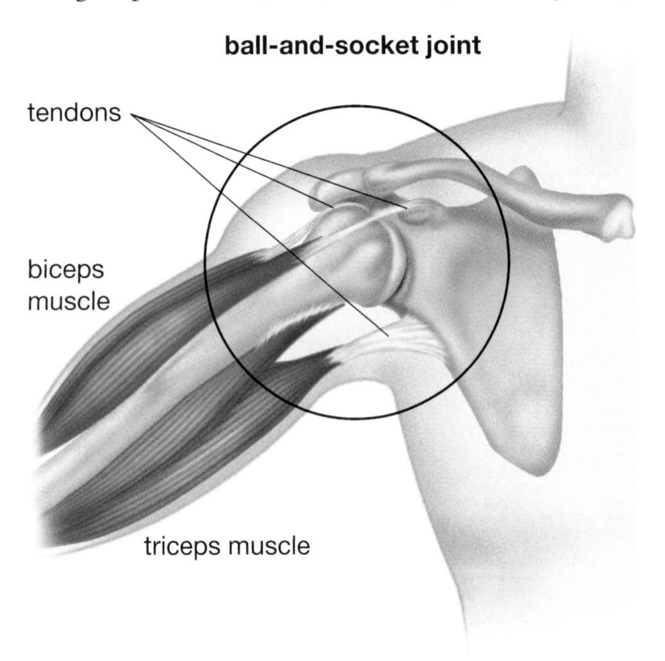

Figure 13.19 This ball-and-socket joint holds parts of the body together while still allowing movement. Such joints are lubricated by a substance called synovial fluid.

Fasteners

Nails, staples, bolts, screws, rivets, and dowels (shown in Figure 13.20) are used to hold many structures together. Unfortunately the holes that fasteners make also weaken the materials they fasten. Staples and nails are usually forced into the parts they join, and this can crack and separate the material. Drilling holes for bolts, screws, and dowels does not weaken the material as much, but the holes are time-consuming to position and cut.

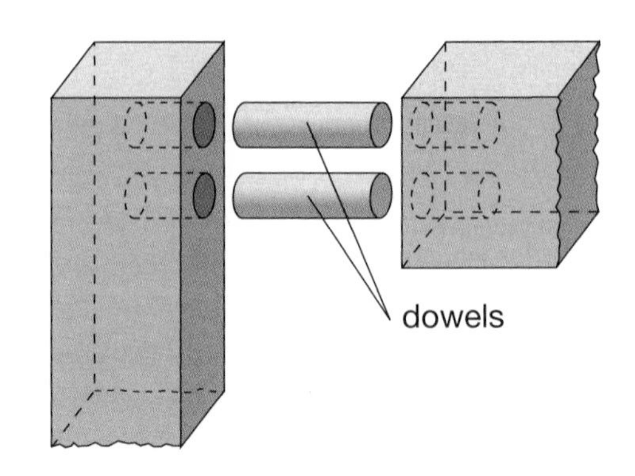

Figure 13.20 Dowelled joints are used by cabinetmakers to hold pieces of furniture together.

Attaching parts with just one fastener allows the parts to twist around it when they are pushed or pulled. Several fasteners make a more rigid joint, but the extra holes weaken the material more. So making a strong joint does not mean simply pounding in a bunch of big nails! Both the kind and number of fasteners must be carefully planned.

Ties

Thread, string, and rope can also fasten things together. Shoes are tied with laces. Jacket hoods are tightened with drawstrings. Seams in clothing are "tied together" with a sewing machine. Figure 13.21 shows how the needle and bobbin thread are intertwined to tie the seam in place.

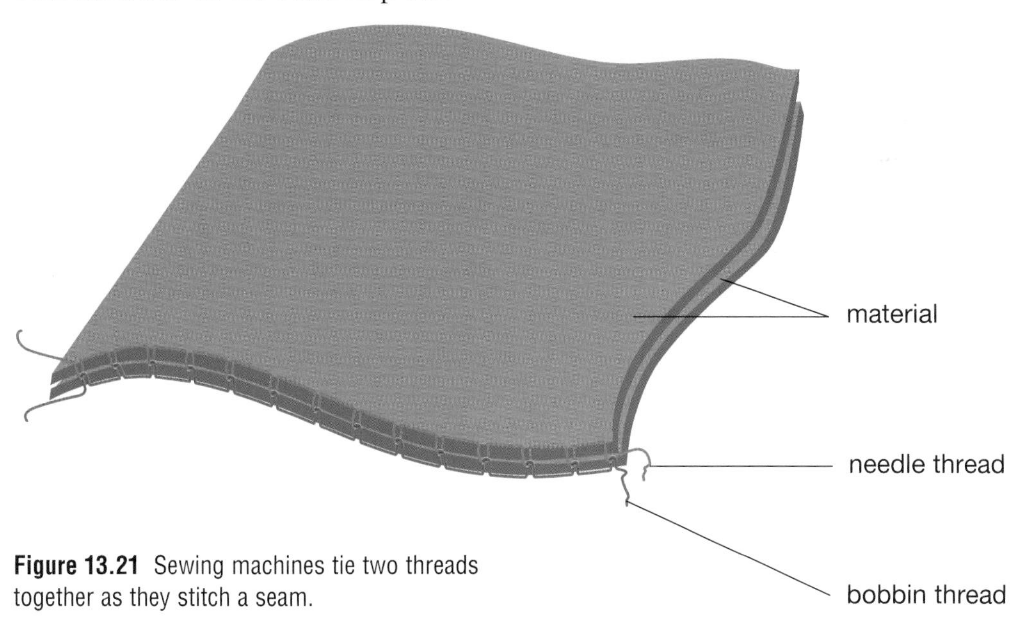

Figure 13.21 Sewing machines tie two threads together as they stitch a seam.

Adhesives

Sticky substances, called **adhesives** or glues, can hold things together. Figure 13.22 on page 400 shows how glue flows into tiny rough areas on the surface of the pieces it joins. When the glue hardens, it locks the pieces together. *Thermosetting glues* like those used in glue guns harden when they cool. *Solvent-based glues* harden as they dry out. The strongest glues also create a special kind of force between the tiniest particles of the pieces being joined. Because of these forces, epoxy resins and super glues are strong enough to hold pieces of car bodies together.

Even the strongest glued joints fail under extreme conditions. Glues may soften in water or under very hot conditions. If a glue is stronger than the substances it joins, the material next to the joint may break.

Adhesives can be a health hazard. Some glues start to harden as soon as they touch moisture. If a drop gets on your fingers or in your eye, it can stick your fingers or eyelids together almost instantly. Other glues, such as those used to make plywood and particleboard, release powerful chemicals into the air as they harden. These gases collect between walls, in basements, and in other places that have poor air circulation. If people with asthma and allergies live or work in these areas, the gases may trigger breathing difficulties, skin problems, headaches, and other health problems. Some people experience very sudden, serious reactions, but it is more common to feel just a little bit ill as long as exposure to the chemicals continues.

DidYouKnow?

Chemist Spence Silver was supposed to develop a super powerful glue. He tried many recipes. One recipe made a not-very-sticky substance that definitely did not meet specifications. It held papers together temporarily, but the joint could easily be pulled apart. Silver kept experimenting for almost ten years, sure that his glue could be useful somehow. Finally, in 1974, a co-worker, Arthur Fry, suddenly thought of a use for the temporary adhesive — sticky bookmarks that would not fall out of place! Post-It™ notes were the result, and they were an instant success.

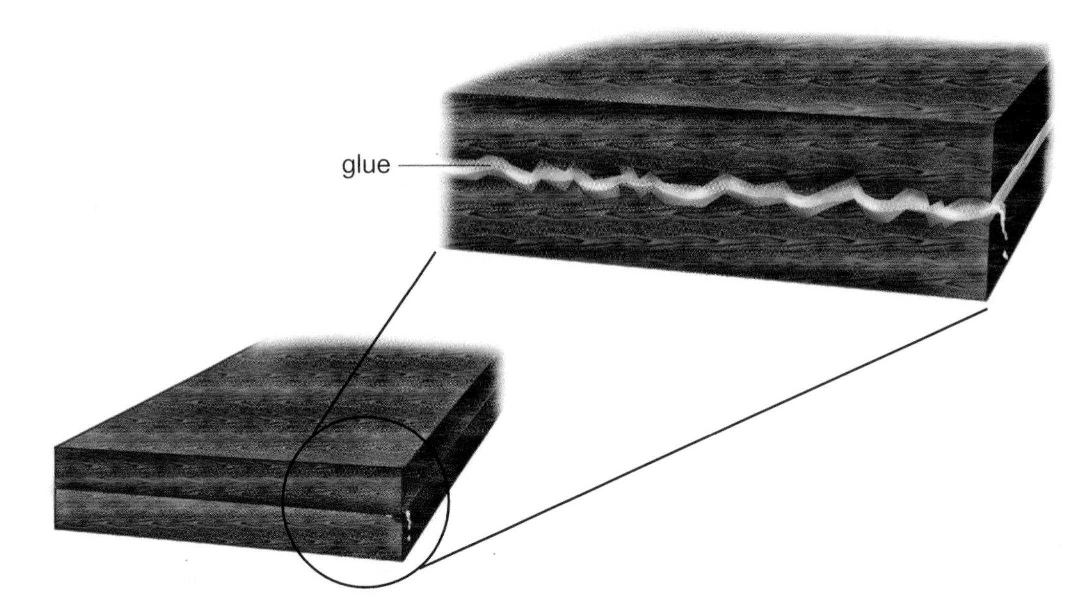

Figure 13.22 Glue creates a bridge between two surfaces and locks them together.

Melting

Pieces of metal or pieces of plastic can be melted together. **Welding** melts the pieces themselves. **Brazing** and **soldering** surround pieces with a different melted material, which locks the pieces together as it cools and hardens. To increase strength, the pieces to be joined may be twisted or folded together. No matter which process is used, the pieces must be carefully cleaned before joining, and the melted material must be cooled slowly and carefully or the joint will be brittle and weak.

There are many ways of melting materials to make welded joints. Torches use a hot flame. Arc welders and spot welders use heat from an electric spark. Plastics can be melted and welded together with strong chemicals and even with sound waves.

Figure 13.23 Welders use a dark mask to protect their eyes and to see white-hot joints clearly.

Interlocking Shapes

Carefully shaped parts can hold themselves together. Lego™ bricks and some paving stones fit together and stay together because of their shape. The fronts of wooden drawers are often locked to the sides with dovetail joints. Dentists shape the holes they drill in teeth to keep the filling material in place.

The joints in flexible materials are also carefully shaped. The sheet metal in the furnace and heating ducts of your home is overlapped or folded to strengthen the places where it is joined. Folded seams protect the cut edges of pieces of cloth and give a neat, finished appearance to the joints in clothing. Figure 13.24 shows some of the different kinds of interlocking joints.

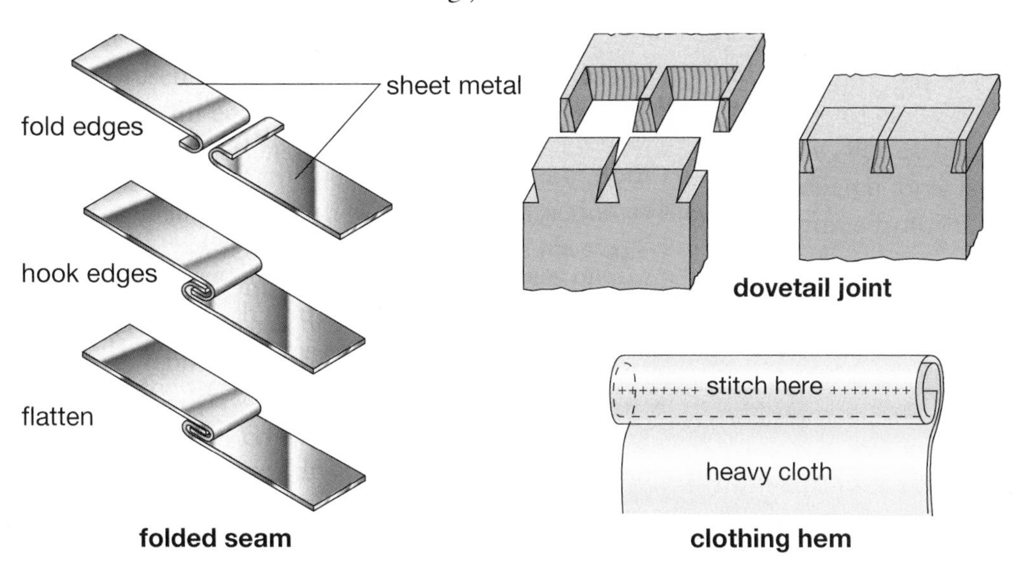

Figure 13.24 Various kinds of interlocking joints can be used to strengthen different types of structures.

Builders need to avoid waste when cutting large pieces of building materials. How can you make one straight vertical cut and one straight horizontal cut to divide the cross shape below into four pieces that can be reassembled into a square?

Skill POWER

For tips on library and Internet research, turn to page 497.

A Better Mousetrap?

What to Do

1. Use library or Internet resources to find out how a common product or invention was developed. Choose your own product, or investigate a product in the list at the right.

- Velcro™
- Super Glue™
- soft drinks
- zippers
- ball-point pens
- bubble gum
- nylon stockings
- automotive air bags
- telephones

2. Prepare a short paragraph, poster, or oral presentation to tell the story of the product.

Check Your Understanding

1. Imagine that you are stranded on a tropical island with very few supplies. Among your supplies is a box of drinking straws. Describe five possible functions they could have. Be creative!

2. Look around the room where you are working right now. Find an example of
   (a) a woven material
   (b) a laminated material
   (c) a composite material
   (d) a glued joint
   (e) a structure that is tied or sewn together
   (f) something that is welded together

3. Sketch a small diagram of each structure listed below.
   (a) an arch
   (b) a column and beam gateway
   (c) a double cantilever bridge

4. **Apply** Your foot is a wonderful, complex structure. Use what you have learned in this section to analyze the structure of your foot and answer the questions below.
   (a) Name three functions of your foot.
   (b) Identify three materials that make up your foot. For each material, describe one particular function that it has and one property that allows it to perform this function.
   (c) What type of structure is your foot (mass, frame, or shell)?
   (d) Imagine sketching the outline of the inside of your foot, from the heel to the first joint on your big toe. What shape does this part of your foot have? Why?
   (e) What type of fastening method holds the parts of your foot together? How many parts does your foot have? (You may have to do some extra research to answer this!)

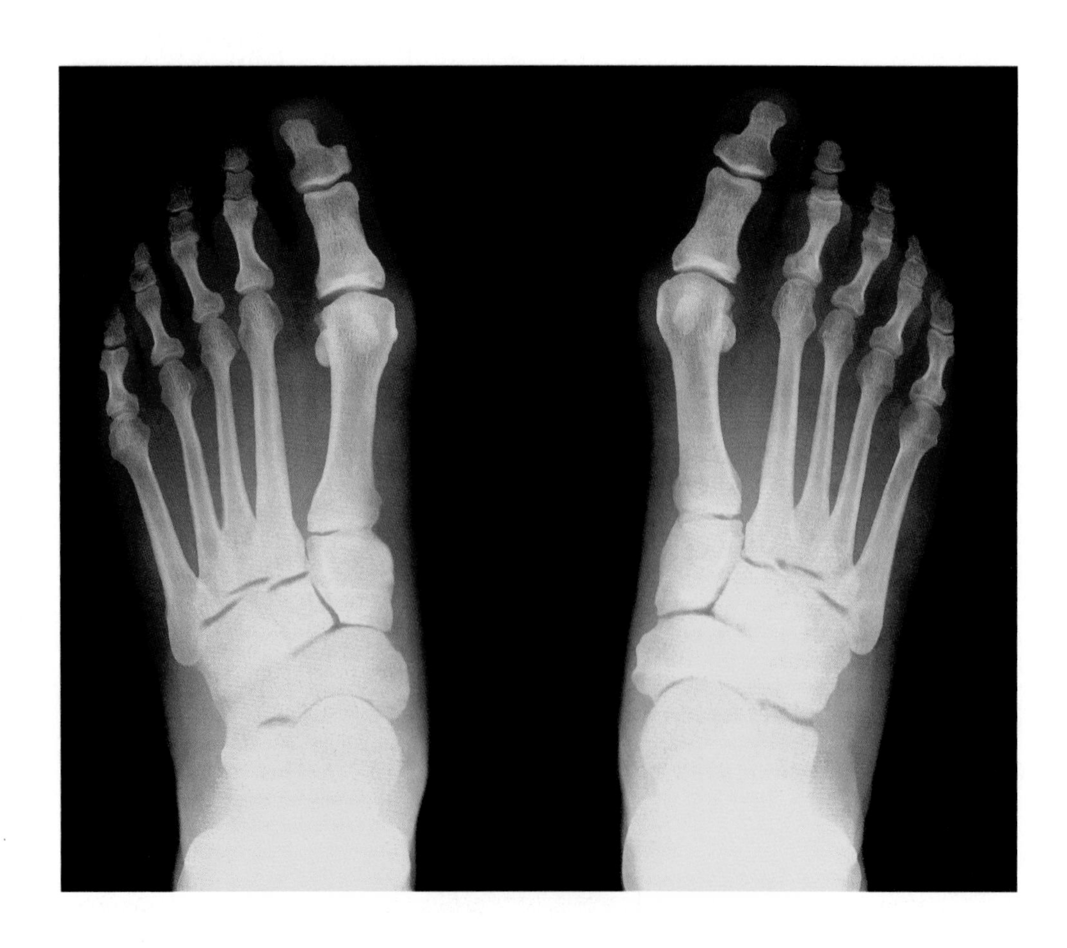

CHAPTER at a glance

Now that you have completed this chapter, try to do the following. If you cannot, go back to the sections indicated.

Give an example of each type of structure (mass, frame, and shell). (13.1)

Pick a common structure that has several parts. Identify the material it is made from and how its parts are fastened together. (13.2)

Give two specifications that a designer who is planning a new ballpoint pen would probably have to satisfy. (13.2)

Identify the shape of frame that is the most rigid. Sketch two ways of using this frame to strengthen a stepladder. (13.2)

Classify these materials as composite, layered, knit, or woven. Identify a particularly useful feature of each type of material. (13.2)

- denim in a pair of jeans
- corrugated cardboard in a storage box
- screen on a window

Identify the fastening method that holds the paper pieces together in each structure: telephone book, volume of an encyclopedia, jigsaw puzzle. (13.2)

Describe how to test the compressive strength of a small wooden column. (13.2)

List three ways to increase the strength of the concrete used to build a patio. (13.2)

Prepare Your Own Summary

Summarize this chapter by doing one of the following. Use a graphic organizer (such as a concept map), produce a poster, or write the summary to include the key chapter ideas. Here are a few ideas to use as a guide:

- What are the key features of each type of structure (mass, frame, and shell)?
- Examine a tape cassette closely, then answer the following questions.
  - **(a)** How does its function affect its design?
  - **(b)** What properties must its building materials have?
  - **(c)** What is an advantage of the particular method used to fasten its parts together?
- How is each of the following types of material made: composite, layered, woven, knit? Give an example and a typical use of each one.
- What are the five basic ways of fastening structures together. Give an example and a typical use of each one.
- Why does using more nails in a joint often weaken it?

CHAPTER

13 Review

Key Terms

manufactured structure
natural structure
function
load
mass structures
frame structures
shell structures
criteria
specifications
aesthetics
braces
ties
gussets
columns
beams
arch
double cantilever
laminations
fibres
corrugated
properties
composite
mobile joints
rigid joints
adhesives
welding
brazing
soldering

Reviewing Key Terms

If you need to review, the section numbers show you where these terms were introduced.

1. Draw labelled sketches that show the difference between
   - (a) a beam and a column (13.2)
   - (b) a cantilever and an arch (13.2)
   - (c) a laminated material and a corrugated material (13.2)

2. Think of examples of
   - (a) four mass structures whose names begin with the letters m, a, s, and s (13.1)
   - (b) five frame structures whose names begin with the letters f, r, a, m, and e (13.1)
   - (c) five shell structures whose names begin with the letters s, h, e, l, and l (13.1)

3. What is the difference between a welded joint and a joint held together with adhesives? (13.2)

Understanding Key Ideas

Section numbers are provided if you need to review.

4. Think about a glass filled with orange juice. (13.1)
   - (a) Is the juice a structure? Explain your answer.
   - (b) Is the glass a structure? Explain your answer.

5. Describe one property of glass that makes it
   - (a) a good material for making beverage containers (13.1)
   - (b) a poor material for making beverage containers (13.1)

6. People have invented many different types of foot gear. Three examples are wooden shoes, soft leather moccasins, and downhill ski boots moulded from strong, rigid plastic. (13.2)
   - (a) What is the main function of each of these types of foot gear?
   - (b) What type of structure is each type of foot gear?
   - (c) How is each type of foot gear fastened together?

7. You are stranded on an island covered in vegetation. To escape, you decide to build a raft out of layers of reeds and small branches. (13.2)
   - (a) Would it be stronger to line up the reeds in each layer and tie them together, or to criss-cross them and weave them together?
   - (b) Would it be stronger to line up the reeds in each layer in the same direction as the layer below, or should the reeds in each layer point in different directions?

8. Describe how the material in a structure can be weakened when pieces are held together by (13.2)
   - (a) ties, such as shoelaces
   - (b) welded joints

Developing Skills

9. Draw a concept map to summarize section 13.1. Make sure that it has three sections, one for each type of structure. Include examples, advantages, and problems for each type of structure.

10. Make a spider concept map to summarize section 13.2. Make sure that it has five sections, one for each of the four design features and one for other important ideas.

Problem Solving/Applying

11. Suggest three specifications that need to be met by a successful design for

(a) an emergency flashlight

(b) a toothpaste tube

12. You need to store some winter clothes so that they stay clean, dry, and free from insects over the summer.

(a) Describe the simplest structure you can think of to perform this function.

(b) Describe the most complicated structure you can think of for this function.

13. Polyester is a type of plastic that can be made into fibres and woven into cloth. Polyester clothing is strong, but it traps perspiration and is not very warm. To overcome these problems, clothing is often made with a blend of wool and polyester fibres. Pure wool is very warm and attractive but it is not strong, especially when it is wet, and it shrinks unless it is washed very gently in cold water. Some people find that woollen clothing is itchy and irritates their skin.

(a) Check the labels on some of your clothing to find a garment that is mostly wool or polyester. Which of the properties described above does it have?

(b) To make warm winter clothing, would you use a cloth that was mostly polyester or mostly wool? Why?

(c) To make comfortable indoor clothing which could be machine washed, would you use a lot of wool in the blend or a little? Why?

Critical Thinking

14. Some chairs are built from pieces of metal or wood covered with softer material. Some are made from solid pieces of foam plastic, and some are made of single pieces of moulded plastic.

(a) Classify each type of chair according to its type of structure.

(b) Give one advantage and one design problem for each type of design.

15. Very large structures can cause serious environmental problems. A lot of materials are used in construction, and a large area is affected. Suggest at least two specific problems that designers had to overcome when planning each project described below.

(a) The "Chunnel" is a 50 km tunnel under the sea between England and France. Three tubes, each large enough for railway trains to pass through, had to be dug through soft, water-filled rock.

(b) The Confederation Bridge between Prince Edward Island and New Brunswick is a 12.9 km reinforced concrete structure. Icebergs and high winds are common in the area, especially in the winter, and important fishing grounds are nearby.

Pause& Reflect

1. Review the activities you completed in this chapter. Describe two things you learned from them that might be useful in your everyday life.
2. Go back to the beginning of this chapter on page 376 and check your original answers to the Getting Ready questions. How has your thinking changed? How would you answer those questions now that you have investigated the topics in this chapter?

CHAPTER

14 Strength, Forces, and

Getting Ready...

- What would happen if a tall building were hit by a low-flying airplane?
- Can a tall building, made of steel and concrete, sway in a strong wind?
- What is the strongest material on Earth?

Science Log

In your Science Log, identify and describe a structure you have seen that collapsed or broke apart. How did its designers try to make it strong? What forces were so powerful that the structure could not resist them? Which parts of the structure failed first?

Skill POWER

To find out the proper way to use a balance, turn to page 481.

Making structures that stay together is not easy. Buildings and bridges are blown about by the wind, weighed down by rain and snow, and twisted by Earth's movements. Boxes, cars, and sporting equipment are battered by sudden impacts. Glues and fasteners loosen over time, and even the strongest materials rot or corrode. So even before designers start to work on a project, they can be certain that it will eventually fall apart.

Scientists have learned how to measure and calculate many of the forces that act on a structure. This scientific knowledge helps to solve the technological problem of building things that are efficient and strong enough to be useful.

In this chapter, you will identify some of the forces that affect structures and you will learn how these forces are measured. You will also experiment with common methods of strengthening structures so that they can resist destructive forces for their intended lifetime.

Efficiency

Spotlight On Key Ideas

In this chapter, you will discover

- how to measure the mass that a structure must support
- what structural efficiency is, and how to express it
- how to measure and describe forces
- how external and internal forces affect structures

Spotlight On Key Skills

In this chapter, you will

- calculate the structural efficiency of a load-bearing structure
- design your own investigation of the strength of modelling materials
- design and test ways to strengthen a load-bearing structure
- analyze how a new product is designed, marketed, used, and recycled

Starting Point ACTIVITY

Super Strength Straws

Can you use as few materials as possible to design a straw framework strong enough to support a full cup of water at least 5 cm above the ground?

What to Do

1. Your teacher will provide ten straws, connectors (for example, paper clips, pins, and tape), and a plastic cup with a lid.
2. You have 15 min to design and construct your straw framework. Imagine that the straws cost hundreds of dollars each, so you must use as few as possible.
3. You may test your design as you build it, but you may use only half a cup of water. Look closely to find any weaknesses, so that you can correct them before the official test with a full cup of water. Make a neat sketch of your final design. Seal the lid securely for each test.

What Did You Find Out?

What effects did the weight of the water have on your frame when you tested it? Did parts of your frame twist, bend, or flatten? Explain how you tried to make your frame strong.

14.1 Mass and Efficiency

Not so many years ago, animals were not treated with appropriate respect, and circus elephants were trained to balance on tiny stools. Of course, the stool had to be very strong or it would collapse under such an enormous load. If an elephant were taken to the Moon, even a flimsy stool would support it. The mass of the elephant — the amount of material that it is made of — would be the same in both places. The load force that the elephant exerted on the stool would be very different, though.

If you understand how mass causes a load force on a structure, you will find it easier to make strong, efficient designs.

DidYouKnow?

The metric system of measurement was created in 1795 in France. It replaced many traditional systems of measuring with just one system. The modern International System of Units (in French, *Le Système International d'unités* or SI) is based on the original metric measures. Standard SI units were set up in 1960 by an international committee called the General Conference on Weights and Measures. Experts meet regularly to review and revise SI.

Figure 14.1 If you took an elephant to the Moon, what would stay the same? What would change?

Mass

The **mass** of an object is the measure of the amount of material in it. Measurements of mass tell something about the tiniest particles in the object. An elephant is made of a very large number of particles, so it has a large mass. An egg-sized lump of lead contains fewer particles, so it has less mass than an elephant. The lead particles are packed very tightly together, however, so the egg-sized lump of lead has more mass than a real egg, a baseball, or other objects of the same size. A piece of Styrofoam™ or a soap bubble is mostly air. Air is made of even fewer particles, spaced widely apart. So Styrofoam™ or a soap bubble has a very low mass.

When the metric measuring system was first designed, scientists decided to measure mass by comparing objects to a particular small cylinder of metal. They called this cylinder the **primary standard** of mass, and the amount of material in it was called one **kilogram (kg)**. Exact copies of the primary standard kilogram are kept in different countries, including Canada.

Skill POWER

To find a summary of basic SI units, turn to page 479.

The standard kilogram has about the same mass as 1 L of water, milk, or juice. Some other common masses are shown in these photographs.

1000 kg (1 t)

250 kg

100 kg

10 kg

Smaller masses are usually expressed in grams (g). "Kilo" always means "thousand," so one kilogram (1 kg) is just another way to say one thousand grams (1000 g).

100 g

25 g

5 g

1 g

Even smaller masses are usually expressed in milligrams (mg). "Milli" means "one thousandth," so a milligram is one thousandth of a gram (0.001 g). How many milligrams would be needed to make one gram (1 g)?

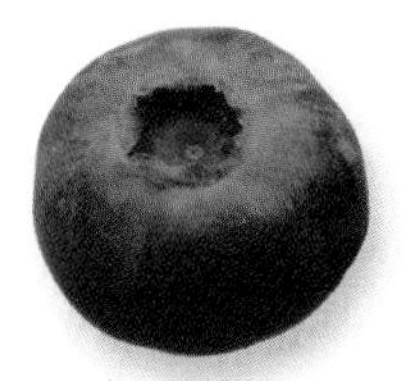
500 mg (half a gram)

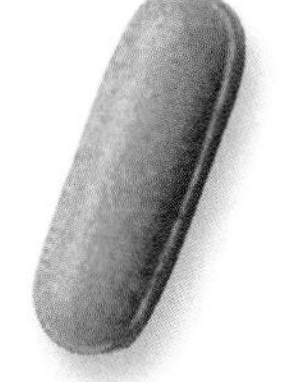
100 mg ($\frac{1}{10}$ of a gram)

20 mg

A **balance** is the most common type of measuring instrument for mass. Most balances compare the pull of gravity on the object being measured with the pull of gravity on standard masses. If the pull of gravity is equal, the masses must be equal, too. Examine Figure 14.2 carefully for clues on how to use different types of laboratory balances.

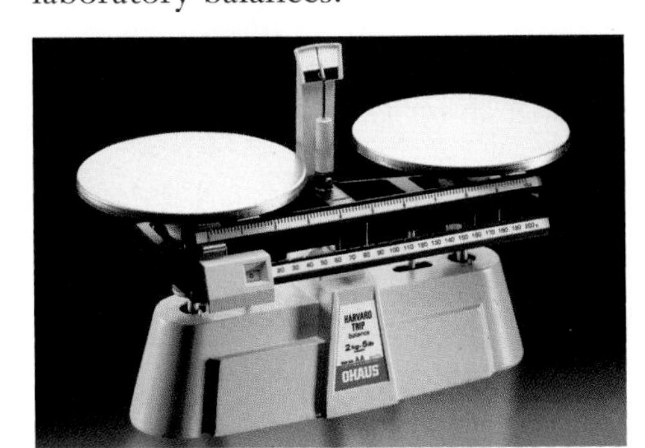

Figure 14.2A Equal-arm balance

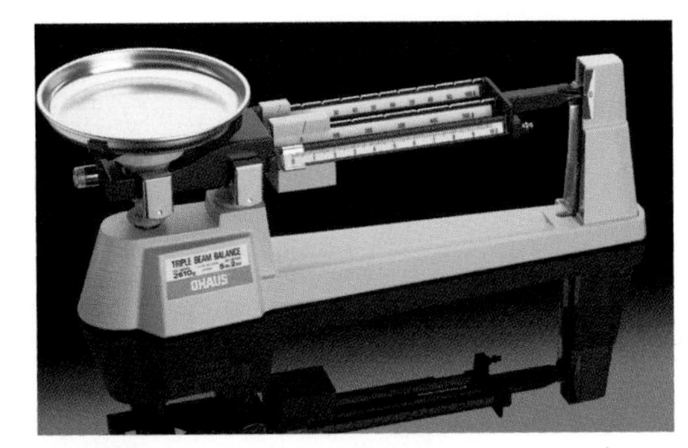

Figure 14.2B Triple-beam balance (low form)

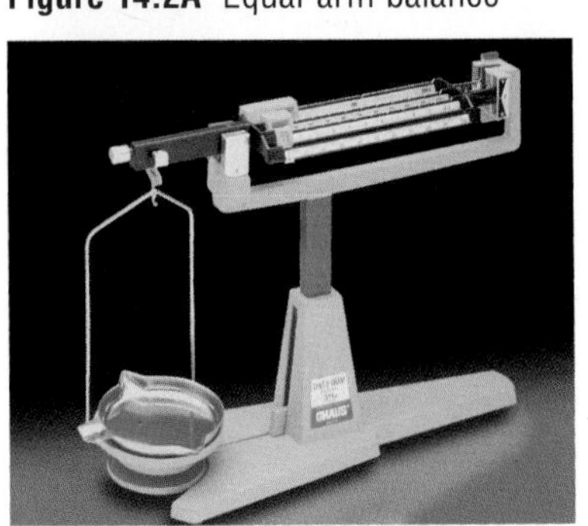
Figure 14.2C Triple-beam balance (high form)

Figure 14.2D Electronic balance

Mass is a very useful property to measure because it stays the same no matter where an object is located. An elephant has a mass of about 5000 kg whether it is on Earth, on the Moon, or in space. Its mass will change only if it gains extra matter (by eating a large meal, for example) or if it loses matter (perhaps by converting some of its body fat into energy through exercise).

Structural Efficiency

Look closely at the bicycle in the photograph. Has the bicycle been strengthened only by using bigger, thicker parts and lots of bracing? What would it be like to ride a bicycle with all that extra weight? Would it be easy or hard to balance, pedal, and stop?

Bicycles and many other structures work best if they can support a load without having too much mass themselves. It is easier to compare different designs if you calculate their structural efficiency.

High structural efficiency is a sign of a strong, relatively light design. A paper bag is surprisingly efficient because it can support loads many times greater than its own mass. If the bag gets wet, however, both its strength and its efficiency will drop. Low efficiency structures like beaver dams can sometimes be strong, but their strength comes from extra mass.

Figure 14.3 A bicycle is an example of a well-designed structure.

Did You Know?

You can calculate Earth's mass by using the distance from Earth to the Moon and the amount of time it takes for the Moon to orbit Earth.

Off the Wall

Earth has an estimated mass of about 6 000 000 000 000 000 000 000 000 kg! This figure was first calculated in 1798 by a British scientist, Lord Cavendish. He made careful measurements of the tiny gravitational force between two small lead balls in his laboratory. Then he calculated how much more mass would be needed to create the stronger gravitational force that makes falling objects move as fast as they do. That is the mass of Earth!

Structural efficiency is a single number that compares the *mass* of a structure with the *load* it supports.

$$\text{Structural efficiency} = \frac{\text{Maximum mass supported}}{\text{Mass of structure}}$$

Example

Saraj used 200 g of materials to build a model bridge that supported an 8.0 kg load. George's bridge had a mass of 150 g and supported a load of 5.8 kg. Who won the prize for the more efficient structure?

Start by finding the structural efficiency of Saraj's bridge.

$$\text{Structural efficiency} = \frac{\text{maximum mass supported}}{\text{mass of structure}}$$
$$= \frac{8.0\ \text{kg}}{200\ \text{g}}$$

Before calculating, you need to express both masses in the same units. One way to do this is to convert 8 kg to grams. Each kilogram is 1000 g, so you can think

$$= \frac{8.0\ \text{kg} \times \frac{1000\ \text{g}}{1\ \text{kg}}}{200\ \text{g}}$$
$$= \frac{8000\ \text{g}}{200\ \text{g}}$$

Therefore, structural efficiency = 40

To complete this example, follow the Math Connect opposite.

Math CONNECT

Find the efficiency of George's bridge. Compare the efficiency of Saraj's and George's bridges. Which bridge won?

DidYouKnow?

High structural efficiency is not important for space stations or satellites because their mass does not need to be supported — gravity has almost no effect on the structure of objects in orbit. These structures need a framework with enough strength to resist other forces, however. Parts twist if the station rotates. Parts expand and contract during every orbit as the station heats up in sunlight and cools in Earth's shadow. The frame must resist impact with supply rockets and other moving objects, such as astronauts. Both the Hubble Space Telescope and the Mir Space Station had to be repaired when they could not withstand these kinds of forces.

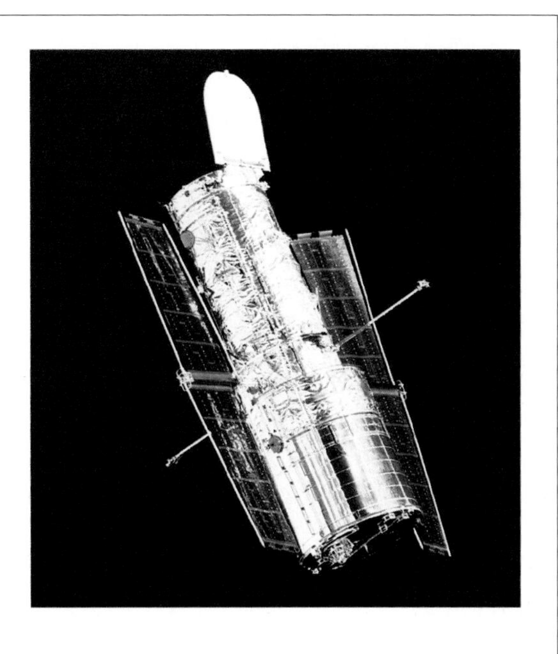

Word CONNECT

Physiologists study the ways that living creatures perform functions necessary for life, such as obtaining oxygen or digesting foods. The term "physiologist" comes from the ancient Greek words *physike*, which means "science of nature," and *logos*, which refers to written or spoken words. So a physiologist studies and tells others about the science of nature. What other English words have the same Greek roots?

To increase a structure's efficiency, you need to reduce its mass without weakening it. Designers know many ways to do this. For example, if you can identify parts that do not need to be particularly strong, they can be made thinner and lighter. Beams can be designed with less material along their **neutral axis** (the direction in which they carry the least load). Stronger building materials can reduce the need for braces and very thick, heavy parts.

With less material, you might think that an efficient structure would also be more economical, but this is not always true. Just check the prices of strong, lightweight racing bicycles! Stronger materials are not cheap, and carefully shaped pieces cost a lot to produce. Efficiency can be expensive.

Check Your Understanding

1. For many years, the youngest world-record weight lifter was 16-year-old Naim Suleimanov of Bulgaria. In 1983 Naim lifted and held 160 kg overhead. At the time, his own mass was 56 kg. Calculate Naim's structural efficiency.
2. The strongest insect is believed to be the rhinoceros beetle. For many years, the *Guiness Book of World Records* stated that this insect could support loads of 850 times its own body mass.
   (a) According to this figure, what load could a 75 g beetle support?
   (b) What is the beetle's structural efficiency?
3. In 1996 Dr. Roger Kram, a physiologist at the University of California, reported that he had actually tested the strength of rhinoceros beetles. According to his results, a 65 g beetle could support a load of 6.5 kg when standing still. With a smaller load of 1.95 kg, the beetle could even walk. Calculate the structural efficiency of
   (a) a standing beetle (b) a walking beetle
4. Use your results from questions 2 and 3 to decide if the *Guiness Book of World Records* correctly stated the beetle's strength.
5. Draw a graph that compares the structural efficiencies you calculated in questions 1, 3(a), and 3(b). The numbers come from different situations, so a bar graph would show the information most clearly.

To learn how to draw a bar graph, turn to page 486.

A standard set of laboratory "weights" contains only a few metal masses, but they can be combined to balance many different objects. For example, there is no 30 g mass, but you can use a 20 g mass with a 10 g mass to balance a 30 g object.

1. What masses need to be included in a set to make every whole-number mass from 1 g to 10 g?
2. A student set of "weights" usually includes a 10 g mass (the smallest) and a 500 g mass (the largest). Without using the largest mass, you can combine the other masses in the set to make all the multiples of 10 g (10 g, 20 g, 30 g, and so on) up to 500 g. What masses must be included in the smallest possible set that can do this?
3. If you used all the masses from the set in question 2, including the 500 g mass, how massive an object would they balance?

14.2 Forces

Structures break when they are not able to resist **forces** — pushes or pulls — that act on them. How can you decide which parts of a structure need extra strength to stand up to forces that tug and twist them? To start with, you need to find the size of the forces. The standard SI unit of force is called a **newton (N)**. One newton (1 N) is only a small force, just enough to stretch a thin rubber band a bit. Some other examples are shown in Figure 14.4.

1 N: force that can lift a flashlight battery (D-cell)

10 N: force that can lift a 1 L carton of milk

200 N: force of a hard-thrown baseball hitting your hand

400 000 N: pushing force (thrust) of a jet engine; also written as 400 kN (kilonewtons)

10 000 000 N: thrust of a rocket as it lifts off; also written as 10 MN (meganewtons)

Figure 14.4 Some forces of different sizes

DidYouKnow?

In everyday talk, "mega" means "really big." In science, "mega" means a particular very large number: one million. Whenever you see the prefix "mega," you can replace it with the word "million" or the number 1 000 000.

DidYouKnow?

Isaac Newton (born on December 25, 1642) was a British mathematician and scientist. By the time he was 24, Newton had begun to develop differential calculus, a powerful type of advanced mathematics, and had used it to analyze the gravitational forces between objects. In 1687 Newton published a book called *Principia Mathematica*. In this book, he showed how to use mathematics to analyze and predict the motion of any object and the forces that caused the motion. In 1699 Newton became Master of the Mint and was in charge of producing coins for the British government. Newton was disagreeable to work with, but his books and ideas revolutionized science and mathematics. He died in 1727 and is buried in Westminster Abbey, a famous church in London, England.

Study Figure 14.5 to find out how to use a **spring scale**, the most common laboratory instrument for measuring forces. Spring scales are not very accurate, but they are less expensive and more sturdy than electronic sensors, which use the force you are measuring to bend or twist a tiny crystal. Forces that are very large or otherwise difficult to measure, such as the strength of a rocket engine, can often be calculated by observing their effect on the motion of an object.

To completely describe a force, you need to determine both its direction and its size. To lift a box, for example, you might have to exert a force of 50 N upward. A book falling on your toe might exert a force of 15 N downward.

Figure 14.5 A force can be measured by seeing how far it stretches the spring in a spring scale.

Weight

The very important force called **weight** was carefully investigated by Isaac Newton in the seventeenth century. According to a famous (and probably untrue) story, Newton once sat under an apple tree and began to wonder why the apples always fell down, toward Earth. They never fell up into the sky or just floated in mid-air. Newton realized that there is a force between any two objects, anywhere in the universe, that tries to pull them together.

Using his mathematical skills, Newton analyzed the size of this force, which he called gravity. He found that the **gravitational force** between two objects depends on the masses of the objects and the distance between them. The force of gravity is very small between ordinary-sized objects. You do not notice gravity pulling you toward trees or other people as you pass by them, but the force is there. If objects are very massive, however, the gravitational forces near them become much larger.

Earth is big enough for the gravitational force between it and nearby objects to be important. In everyday language, we call this force "weight." Instead of saying that there is a gravitational force of 10 N between a 1 kg mass and Earth, we usually say that the mass has a weight of 10 N. If you have a mass of 40 kg, there is a gravitational force of about 400 N pulling you toward Earth. Your weight is 400 N! This might sound strange, but in the SI system, weight — the force of gravity — is properly expressed in newtons, just like any other force.

Precision balances are basic tools in modern laboratories. You will use them a lot in later courses. If your school has balances and spring scales, you may be able to practise measuring small objects with them now. Estimate each object's mass and weight first. Then check your guesses by using a balance (carefully!) to find mass and a spring scale to find weight. Your teacher may have more specific directions for you to follow.

Because gravitational force depends on the distance between objects, an object's weight changes depending on where it is. In an airplane or on a high mountain, where you are farther from the centre of Earth, your weight is a little bit less. Gravitational force also depends on the mass of an object. On the Moon, your weight would be about one sixth what it is on Earth, because the Moon's smaller mass exerts less gravitational force than Earth's mass. If you could travel far down beneath Earth's surface, your weight would also be reduced, because there would be less matter between you and the centre of the planet. In space, very far from any matter, and at the centre of Earth, you would be weightless!

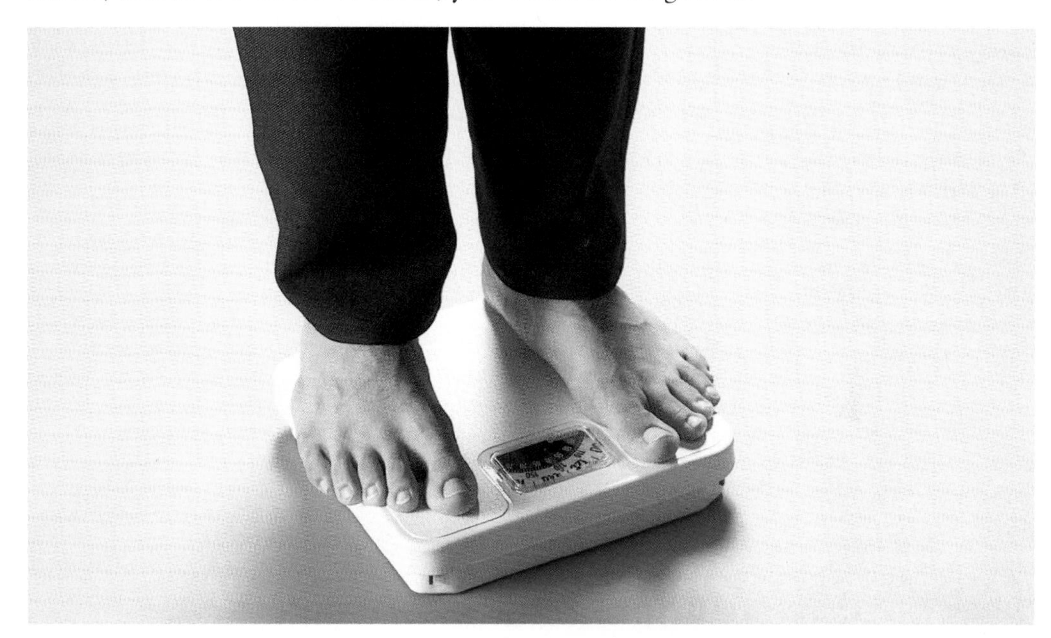

Figure 14.6 The reading on a bathroom scale depends on the gravitational force between you and Earth.

Find Out **ACTIVITY**

How Forceful!

What to Do

Stretch a rubber band between two fingers. Do you feel a pulling force on just one finger, or on both?

What Did You Find Out?

Now think about gravity. Like the rubber band, it acts between two objects. So — does it pull on both objects, or just one? If your weight is 400 N, is that just the force pulling you toward Earth? Is there also a force of 400 N pulling Earth toward you?

Why do we always see things falling down toward Earth? Why is Earth not pulled up toward them? Or is it? These are the kinds of questions Newton tried to answer 300 years ago. Write a short paragraph to explain your ideas about them.

Picturing Forces

Math CONNECT

Very precise measurements show that the gravitational force at Earth's surface is a bit smaller than 10 N on each kilogram of matter. The actual figure is close to 9.81 N. Try recalculating the predicted weight of each object you measured in *Cool Tools* on page 415, using 9.81 N as the force on each kilogram instead of 10 N. Are these predictions closer to your measurements than your original predictions?

A **force diagram** is a simple picture that uses arrows to show the strength and direction of one or more forces. As you can see in Figures 14.7A and 14.7B, a circle or rectangle stands for the object on which the forces act. Each force is shown by an arrow. The length of the arrow shows the size of the force: a longer arrow represents a larger force. The direction of the arrow shows the direction of the force. The arrow is usually drawn pointing away from the place where the force is acting, like a rope pulling an object.

Diagrams are especially useful to find the combined effect of several forces acting on the same object. A neat sketch is often enough to solve a simple problem. Even when many forces are acting together, mathematicians have found ways to use exact scale drawings and calculations to predict what will happen.

A science student is studying her textbook a week before a unit test. She reviews every page from the top of page 100 to the bottom of page 203. How many pages does she study? (Assume that each sheet of paper is printed on both sides, so that it contains two pages.)

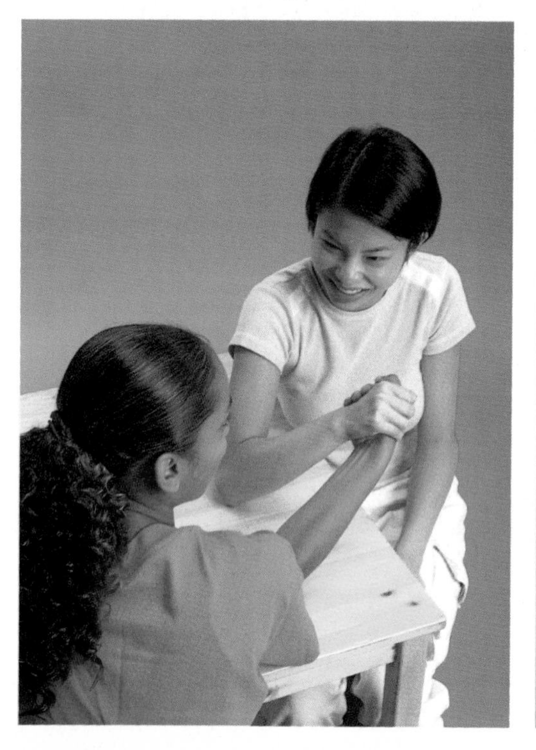

Figure 14.7A The arrows show that the forces in the photograph above are unequal.

Figure 14.7B The arrows show that the forces in the photograph above are equal.

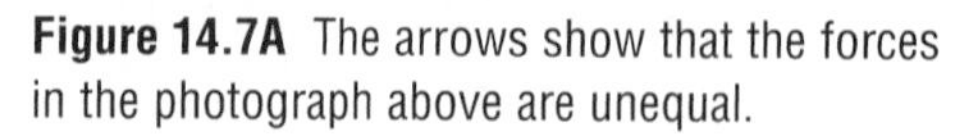

A farmer is planning to build a fence around a square field that measures 33 m on a side. How many fence posts will be required if there must be a post every 3 m? Prove your answer with a careful sketch.

INVESTIGATION 14-A

Forces on Display

Think About It

Whenever objects change their shape or motion, forces are acting. You cannot see forces, but you *can* see what they do to objects. Powerful forces might also be present even when nothing is moving, as in a tug of war with equally matched teams. Analyze the forces in the situations and diagrams below.

What to Do

1. Draw a force diagram to show the forces on a baseball in each of the following situations. Make your diagrams simple. Just draw a small circle for the ball, and arrows for the forces. Label the force arrows neatly.
   - **(a)** The pitcher is throwing the ball straight and level across your page to the right. The pitcher's force on the ball is much larger than the gravitational force pulling it down or the force of air resistance (three forces).
   - **(b)** The ball is speeding toward the plate. The pitcher is not pushing on it any more, but gravitational force is pulling on it, and air resistance is slowing it down (two forces).
   - **(c)** The ball hits the catcher's mitt and stops moving (two forces).
2. Look at the forces on the aircraft in the diagrams below. For each diagram state whether each numbered force is:
   - **(a)** force from engine
   - **(b)** gravitational force
   - **(c)** lifting force on wings

Analyze

What is the combined result of all the forces in each of the diagrams below?

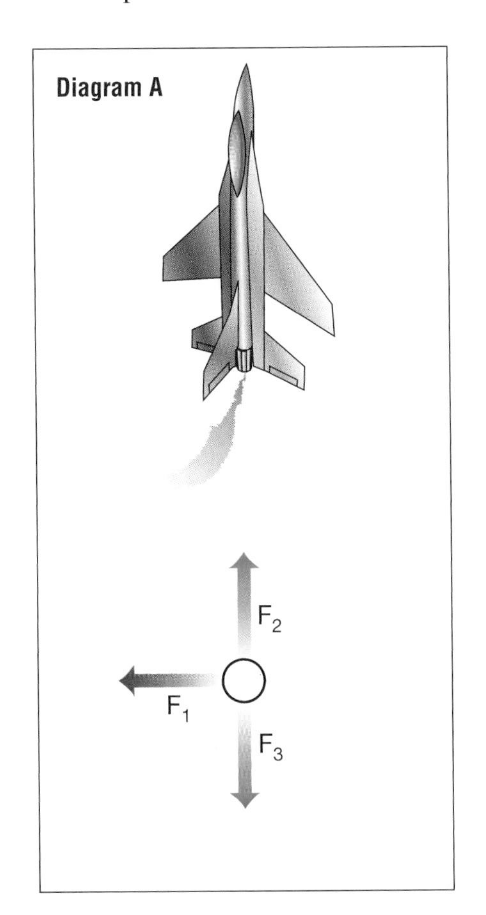

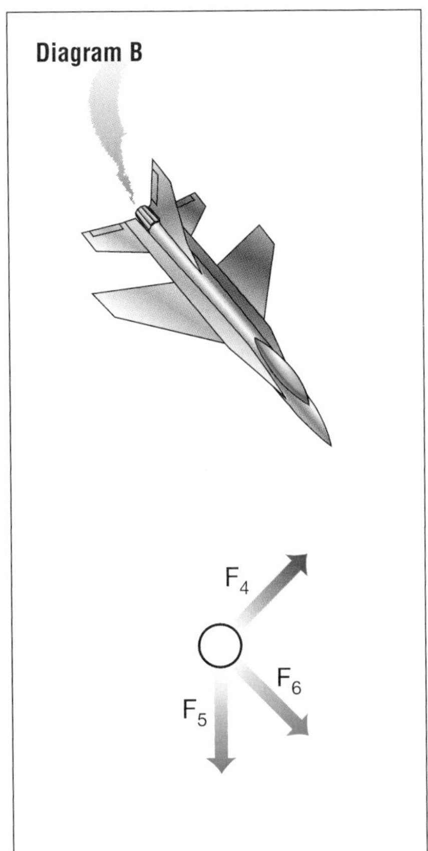

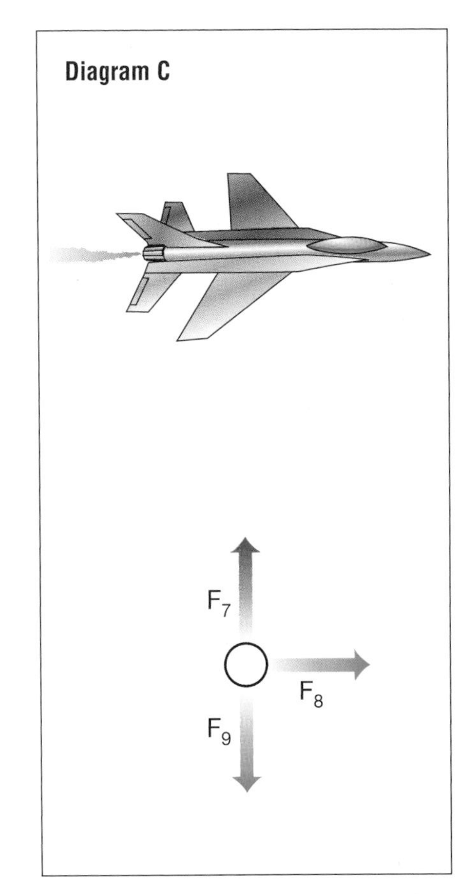

Career CONNECT

Producing Results

Industrial designers, such as Deanna Glen, create the products we use every day. They create everything from clothing to car bumpers to lamp posts. They need skills and training in both art and science to ensure that each product they develop is suited to the customer. Deanna explains, "The engineer knows whether or not a product can be built and what are the best building materials to use. The designer makes sure the product looks right and feels right."

When designing the gardening cart shown here, Deanna had to consider a gardener's needs. She included a waste bucket, hooks for tools, and space for soil and pots. The cart is on wheels so that gardeners can easily move their materials around the yard with them.

Think about an interest you have: maybe fashion, music, a sport. Now think of a product related to your interest: a winter coat, a CD package, a racket cover. What factors did the designer have to consider when designing this product?

Check Your Understanding

1. Suppose you want to measure the weight of a pencil.
   (a) What measuring instrument should you use?
   (b) What units should you use for your answer?

2. In a lab report, two students reported that they had applied a force of 6.5 N to a brick. Their answer received only half marks. What extra information should they have reported?

3. (a) What is shown on a force diagram?
   (b) What is not shown on a force diagram?
   (c) When would a force diagram be useful to draw?

4. **Apply**
   (a) Express the mass of a 125 g tube of toothpaste in kilograms.
   (b) Calculate the weight of the tube of toothpaste.

14.3 Forces in Structures

Figure 14.8 A train derailment produces a number of effects on the cars that collide with each other.

You do not have to witness a train crash to know that it creates dangerously large forces. Look closely at Figure 14.8. What effects did the forces in this collision have on the colliding objects? List as many effects as you can.

Now imagine a smaller "collision," such as kicking a soccer ball. When your foot applies a force to the ball, or any small object that is free to move, three things can happen. The object's motion can speed up, slow down, or change direction. When you kick something larger, such as a building, it does not usually move, but the force still has an effect. Your push on the outside of the building (an **external force**) creates forces inside the building material (**internal forces**). If these **stresses** (forces exerted on an object) become large enough, the shape or size of the building may change very suddenly. To design a strong structure, engineers need a good understanding of the different external forces that can act on it and the internal forces that can build up inside it and that may cause parts to fail.

External Forces

Engineers know of many forces that may affect buildings. They divide these forces into two groups.

Live load includes the force of the wind and the weight of things that are in or on a structure (people, furniture, and snow and rain on the roof). Impact forces, caused by objects colliding with the structure, are another type of live load. Most structures are designed to withstand forces at least two or three times larger than their expected live load. Sometimes, though, live loads become extremely large for a short time, as in a storm or a collision, and the structure can be damaged.

Dead load is the weight of the structure itself. Over time, this gravitational force can cause the structure to sag, tilt, or pull apart as the ground beneath it shifts or compresses under the load.

When you act as a live load on a teeter-totter, you create forces that spread through the whole apparatus. Your weight pushes down on the seat and the bar to which the seat is fastened, but the opposite seat is lifted up. The centre of the teeter-totter twists around its pivot. One external force (your weight) creates several internal forces. These stresses affect different parts of the structure in different ways. Study Figure 14.9 to learn about four of the most important internal forces.

DidYouKnow?

Although the mass of the CN Tower in Toronto is over 130 000 t, this hollow concrete structure is still flexible. In 190 km/h winds, which are thought to occur only once or twice every 100 years in Toronto, the glass-floored observation deck near the top of the tower would move 0.46 m (about the width of this open textbook) off centre. Instead of making the tower unstable, this movement would press the specially shaped foundation even more firmly into the ground.

INTERNET CONNECT

www.school.mcgrawhill.ca/resources/

Find out more about the Leaning Tower of Pisa by going to the web site above. Go to the **Science Resources**, then to **SCIENCEPOWER 7** to know where to go next. Or submit "**Pisa**" to an Internet search engine.

INTERNET CONNECT

www.school.mcgrawhill.ca/resources/

For information about how the CN Tower was designed and built, and some of the records it has set, go to the web site above. Go to the **Science Resources**, then to **SCIENCEPOWER 7** to know where to go next. Or submit "**CN Tower**" to an Internet search engine.

DidYouKnow?

The Leaning Tower was designed to be a beautiful bell tower for a church in the Italian town of Pisa. Construction began in 1173, but after the first three storeys were built, the ground beneath the heavy stone building began to sink unevenly. Even before the 55 m tower was completed (around 1370), it had developed a noticeable tilt. By 1990, when the tower was closed to the public, the edge of the top storey was about 4.4 m outside the edge of the foundation and the tilt was increasing by about 1.3 mm each year. There have been many attempts to stop or reverse the leaning to keep the tower from collapsing. Recently, engineers have been able to straighten it about 10 mm — still not safe enough to be reopened.

Internal Forces

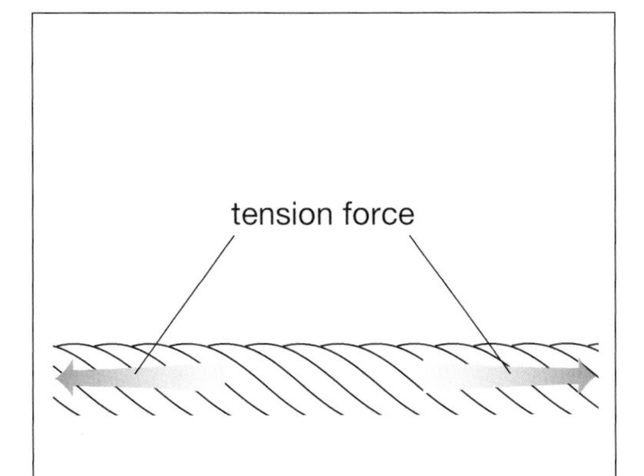

Figure 14.9A

Tension forces stretch a material by pulling its ends apart. **Tensile strength** measures the largest tension force the material can stand before breaking.

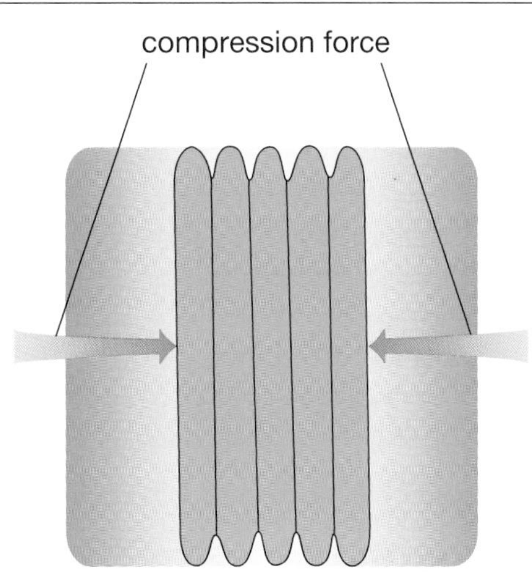

Figure 14.9B

Compression forces crush a material by squeezing it together. **Compressive strength** measures the largest compression force the material can stand before losing its shape or breaking into pieces.

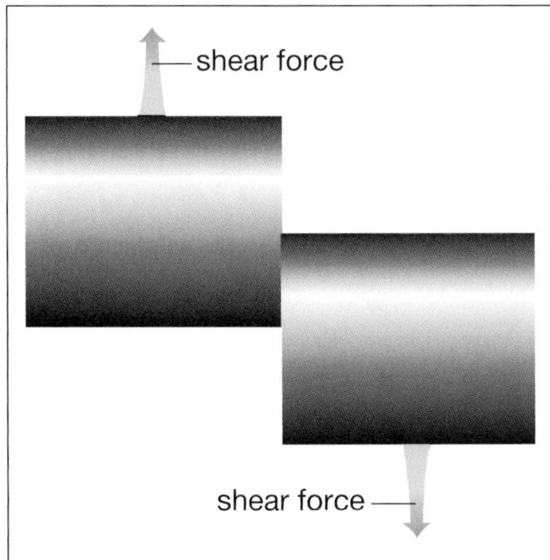

Figure 14.9C

Shear forces bend or tear a material by pressing different parts in opposite directions at the same time. **Shear strength** measures the largest shear force the material can stand before ripping apart.

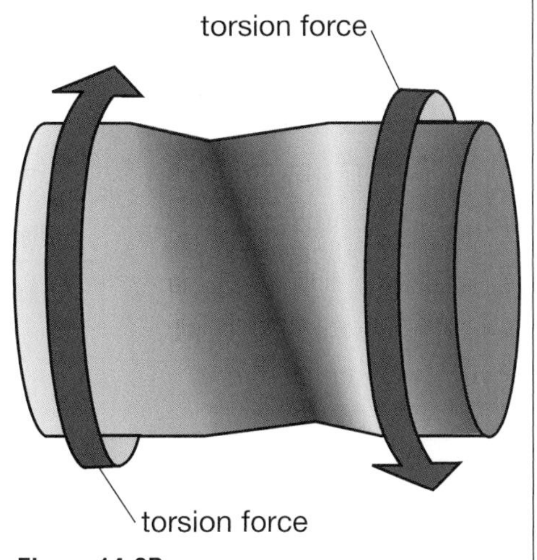

Figure 14.9D

Torsion forces twist a material by turning the ends in opposite direc tions. **Torsion strength** measures the largest torsion force the material can stand and still spring back to its original shape.

Pause& Reflect

In your Science Log, write a short paragraph explaining the difference between external forces and internal forces. To illustrate your explanation, give an example of each type of force when a strong wind blows on a flag.

Word CONNECT

The same words that name internal forces are used in other situations.

- How is a metal pair of cutting *shears* similar to a *shear* force?
- How is a *tension* headache similar to a *tension* force?
- How is a *compressed* computer file similar to a *compressive* force?

CONDUCT AN

INVESTIGATION 14-B

Examining Forces

If you know the types of internal forces that stress part of a structure, you can design that part with the exact kind and amount of strength it needs to support the forces efficiently. In this investigation, you will identify the forces acting in four different structures and the effect each force has.

Problem

What forces create stress in various structures, and where does the stress occur?

Part 1
Bend That Bike!

A spokes: stretched slightly by the weight of the rider

B seat post: pressed down by the rider's weight

C handlebars: twisted and turned by the rider

D pedal bolt: pressed down by the rider but held up by the pedal crank

E tire: pressed down by the rider's weight, up by the ground, and outward by air particles inside

Part	Letter name	Force acting	Type of strength needed

Procedure

1. Set up a data table like the one above to record your observations for five parts of a bicycle. Give your table a title.

2. Study the diagram of the bicycle.
   - (a) Read the descriptions of the forces acting in each part of the bicycle.
   - (b) Fill out one row of the data table for each labelled part of the bicycle.

3. Identify one more part of the bicycle that might fail because it has too little
   - tensile strength
   - compressive strength
   - shear strength
   - torsion strength

Part 2
Stress That Straw!
Procedure

Materials

unbent pleated drinking straw
deck of cards

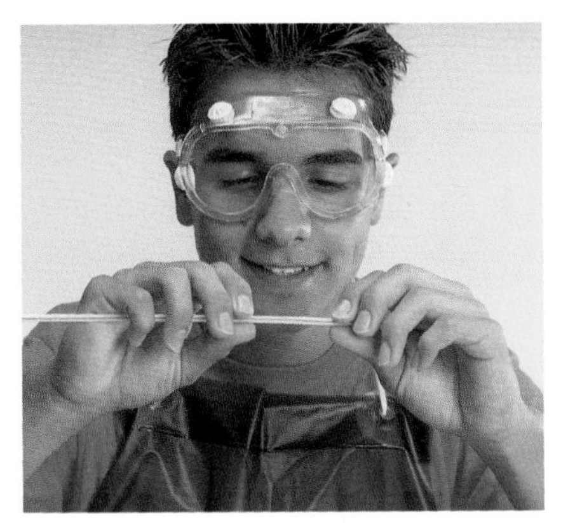

1. Make a data table with four sections, each one large enough for a neat labelled sketch of the straw or deck of cards. Give your table a title.
   - **(a)** Put the unbent straw on your desk and look closely at the pleats. They should be identical all the way around the straw. Make a sketch similar to the one below, and label it "no stress."
   - **(b)** Gently pull the ends of the straw apart about 1 cm. What is the effect on the pleats?
   - **(c)** Gently push the ends of the straw together about 1 cm. What is the effect on the pleats?

2. In the second section of your data table, make two small sketches that clearly show what the pleated part of the straw looked like after step 1. Put the labels "compression" and "tension" beside the correct sketches.

3. Watch the pleats as you bend the straw slowly until it forms a right angle.
   - **(a)** Make a third sketch that shows any difference between the top and bottom pleats. Label it "bending stress."
   - **(b)** Label one side of the straw "compression" and the other side "tension." Use your sketch from step 2 to decide where each label should go.

4. Straighten the deck of cards until the edges are square. Do the edges of the card stack stay square as you gently bend the middle down?
   - **(a)** Sketch the edge of the deck of cards before and after bending. Draw arrows to show the direction of the forces that caused any movement of the top and bottom of the deck.
   - **(b)** Label the sketch with the type of internal force that was acting.

CONTINUED

Part 3
Twist That Towel!

Apparatus

small towel or washcloth
basin of water
pail or pan to catch drips
graduated cylinder
cloth or sponge for clean-up
rubber gloves

Procedure

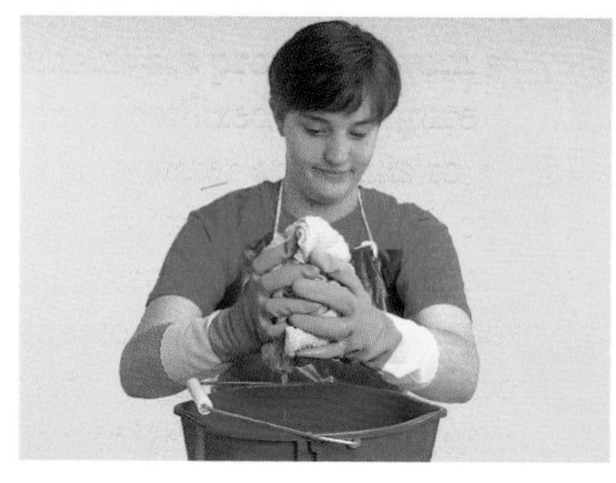

1. Soak the towel in water. Squeeze but *do not twist* the towel. Squeeze as much water out of the towel as possble.
2. Use the graduated cylinder to measure the volume of the water you squeezed out. Record your answer.
3. Repeat step 1, but this time twist the wet towel tightly to squeeze out the water. How much extra water was squeezed out when you twisted the towel? What type of force were you applying? Record your answers.

Part 4
Tug That Thread!

Apparatus

20 N and 50 N spring scale
strong metal rod or metre stick

Materials

30 cm lengths of dental floss, sewing thread, fine fishing line, and knitting wool or crochet cotton

Procedure

1. Tie each sample into a loop. Make sure that the knot is tight and does not slip.
2. Put the loop of one sample over the rod. Place the rod between two desks. Hook your spring scale to the loop. Slowly pull down harder and harder. Record the force needed to break the sample.
3. Repeat step 2 for each fibre sample you have. Record the type of fibre you tested and the force needed to break it.
4. Braid three pieces of thread or floss together, and test the braided sample. Record the force needed to break the braided sample.

Analyze

1. Use your observations to decide if each statement below is true or false. In your answers, tell which part of the investigation gives evidence for your decision, and use it to give a reason for your decision.
   (a) Only one force can act on one part of a structure at one time.
   (b) Torsion forces reduce the size of the spaces in a substance.
   (c) A piece of yarn or rope made by twisting several fibres together has much higher tensile strength than a single fibre.

2. The behaviour of beams in the ceiling or floor of a building, when they bend under a load, is similar to the behaviour of your pleated straw. Use your observations from Part 2 to answer these questions.
   (a) If the top of a bending beam is pushed together by compression and the bottom is pulled apart by tension, what is probably happening to the middle of the beam (its neutral axis)? How much force is stressing it?
   (b) Look back at the diagrams of different types of beams on page 393. Why can the neutral axis of the beam be made of less material than the top or bottom?
   (c) Nails, screws, and rivets often go right through the end of a beam to fasten it to other parts of a structure. What can the forces at the end of the beam do to a fastener if the beam starts to bend?

Extend Your Skills

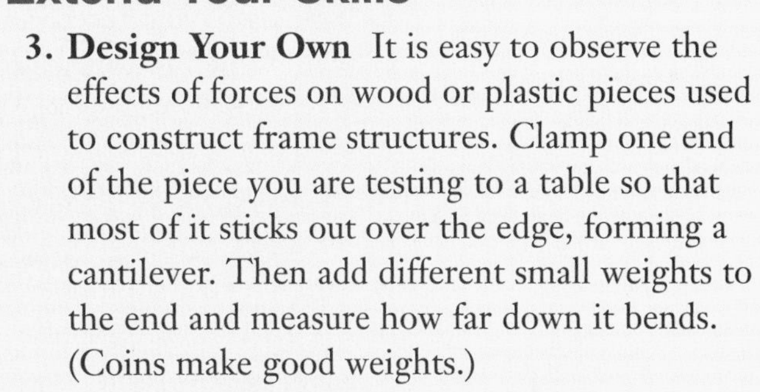

3. **Design Your Own** It is easy to observe the effects of forces on wood or plastic pieces used to construct frame structures. Clamp one end of the piece you are testing to a table so that most of it sticks out over the edge, forming a cantilever. Then add different small weights to the end and measure how far down it bends. (Coins make good weights.)

   Design an experiment that uses this method to measure the stiffness of different sizes or types of materials used for model building. Balsa and bass wood pieces, and plastic beams of different shapes, are available from hobby shops. Have your teacher approve your experimental procedure.

INTERNET CONNECT

www.school.mcgrawhill.ca/resources/

Find out more about the most recent efforts to manufacture fibres as strong as spider silk by going to the web site above. Go to **Science Resources**, then to **SCIENCEPOWER 7** to know where to go next. Or submit "**spider silk**" and "**strength**" to an Internet search engine.

Computer CONNECT

If you have access to a computer, you could use it to prepare tables and graphs of the data you collected in Part 4 of the investigation. Other computer programs can use the results of this experiment to predict the strength of trusses and bridges built from the material you test.

Did You Know?

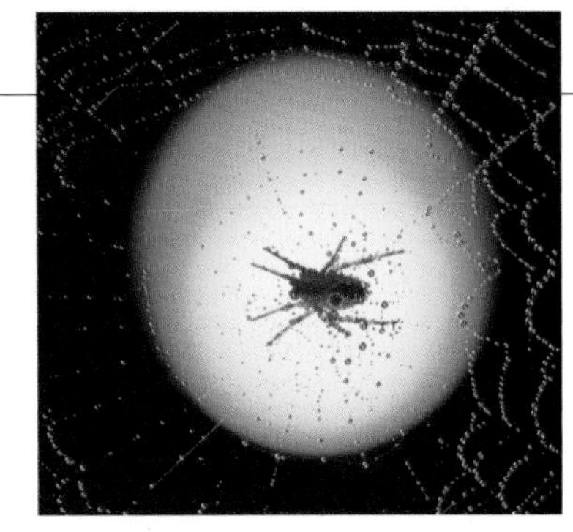

One kind of spider silk is the strongest material known. Dragline fibres are spun by spiders when they make the frame of a web and when they drop from high places. Because the fibres have high tensile strength and are also very stretchy, it takes a great deal of energy to break them. The web of the golden silk spider is strong enough to trap a bird! Measurements of the dragline fibres show that they are at least five times stronger than an equal mass of steel. That is even stronger than Kevlar™, the fibre used to manufacture bulletproof vests. If a strand of this spider silk were as thick as a pencil, it could stop a speeding 747 passenger jet!

Strengthening Structures

DidYouKnow?

Symmetrical shapes can be turned or folded to fit exactly on top of themselves. Most people's faces are quite symmetrical, as are many other natural and manufactured structures. Because symmetry looks very pleasing, it is a powerful element of good design. Symmetrical parts, braces, and decorations on an object help it look attractive. If you start watching for symmetry in structures you observe every day, you might find ways to use this principle in your next design project. Can you find the axis of symmetry — the line that divides the butterfly into two parts with almost identical shapes?

In science fiction stories, you can read about wonderful imaginary materials that stand up to almost any force. Real materials are more limited. As you saw in Chapter 13, concrete and mortar have very high compressive strength if they are made according to the correct recipe. Concrete is quite weak if it is pulled or sheared, however. Similarly, most other materials have one kind of strength but not another. That is why engineers must analyze structures in great detail to find what types of internal forces are stressing each part. They can then choose materials and shapes with the strength to withstand each force. Even a simple swing needs to be designed in this way (see Figure 14.11).

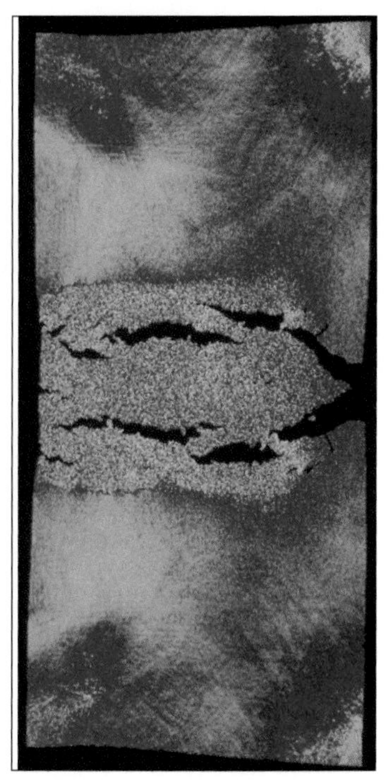

Figure 14.10 Stress cracks in metal can lead to structural failure.

Shear forces were a big problem for early railways. Tiny cracks inside the rails often weakened them enough that the weight of a loaded train would shear a rail in half, causing a serious accident. But the cracks could be detected only after the rails broke. In 1932 a Canadian metallurgist, J. Cameron Mackie, discovered that the cracks formed when the rails cooled too quickly during the manufacturing process. Mackie tried putting red-hot rails in a covered steel box where they could cool more slowly. He found that this eliminated the cracks completely. Within ten years, Mackie's process was being used by steel companies all over the world to produce strong, crack-free rails.

Figure 14.11 Different structural stresses in a garden swing

A Lots of tension here. Use rope or chain for high tensile strength.

B Brace gets pushed and pulled if the frame wiggles. It needs high tensile and compressive strength. Use wood or steel.

C Joint gets twisted as the swinger moves back and forth. Make sure it has high torsion strength and it is not brittle.

D Anchor needs compressive strength to hold the weight of the apparatus. Concrete is good, and it will not rot if the ground is wet.

INVESTIGATION 14-C

The Stable Table

Designers must identify the forces that will act on an object and then try to figure out the most efficient way to build the object so that it can withstand these forces.

Challenge

Use your knowledge of forces and structural strength to design and build a cardboard table that is efficient, good looking, and strong enough to support the smallest member of your group for one minute.

Materials

corrugated cardboard
glue
tape
metal fasteners
string
rope
other household materials

Design Criteria

A. The table must be at least 30 cm square and 30 cm high.

B. The main parts of the table must be built only of corrugated cardboard. Fasteners can be made of any other materials approved by your teacher.

C. The table must support its load for 1 min to demonstrate good structural efficiency. If it is also pleasing to look at, you have achieved a solution to the challenge set for you.

D. As a class, you might decide to add another requirement. For example, you could require the table to collapse or come apart for storage.

Plan and Construct

1. Decide what part of the project will be the responsibility of each group member.
2. Review Chapter 13, section 13.2, and then brainstorm possible shapes and designs for your table.
3. Choose the most promising design and make a list of materials and a detailed drawing showing the size of different parts. It might help to make a paper model before building the final table.
4. Build the table, following your design. Make sure it is carefully finished and looks pleasing.
5. Very carefully, do a trial test of the table to see if will support the necessary load.
6. If necessary, modify the table and do more trial tests until it is strong enough.
7. Before doing a final test of the table, show it to the class and explain the following details of your design.
   (a) what type of strength each part needs
   (b) how you have made each part strong enough
   (c) any special design features of your table
   If you wish, prepare a poster to illustrate your explanation.
8. Find the mass of the table and the load it must support. Calculate its structural efficiency.
9. Follow your teacher's directions to do a final test of the table. You will need at least one independent observer to certify your results.

Evaluate

Write a brief evaluation that tells

(a) whether your table met the design criteria
(b) the structural efficiency of your table
(c) specific ways to improve your table

INVESTIGATION 14-D

The Design Process

Think About It

Real products are designed using the principles you have been studying, but real designers face additional challenges. The strongest material might be too expensive to use. Bracing might look ugly even if it does add strength. People must want to buy the product, and it must be sold for a price that they are willing to pay. There must be a way to recycle the product or dispose of it safely. Making a usable design is more challenging than it might seem.

Analyze

Imagine that you are the person in the following story. Whenever you come to a ✎, do the task or write an answer to the question. Support your ideas with reasons or examples.

The events in the story are fictional, but the invention is real.

Plastic Battery Project

I guess someone heard me complaining about how boring this job is getting. Today I was assigned to a new project, and is it neat! A battery made out of plastic! Somebody has invented one, and our company has to decide if it is worth producing. Lots of people are already gathering information, and I am supposed to talk with them, put it all together, and recommend what to do. To start out, I am going to spend a day on each part of the product's life cycle, trying to find out what questions must be answered. I'll write brief notes every day to make sure that I leave nothing out of my final report.

Day 1: Needs

You know, this plastic battery could be really useful. People in the engineering department have met with the inventor and checked out the demonstration model she made. Even the prototype is very impressive! It is a lot lighter than ordinary batteries. Because it is a solid piece of plastic, there is nothing inside to leak out. And the plastic can be made in any shape you like — even a thin, flexible sheet. It could be put in spaces where an ordinary battery would never fit. It can be recharged, too, hundreds of times. Maybe even thousands of rechargings are possible if we change the design slightly.

✎ *I think I need to make a list of products that could use a battery with these features.*

Day 2: Production

More talks with engineers today! We have a major decision to make. Our whole factory is designed to make ordinary batteries with metal cases. There is no way the machinery can be modified to work with plastic. Anyway, we need to keep producing and selling normal batteries. Profit from them is needed to pay for developing the plastic battery.

So there are two choices. We could build a new factory. That would be really expensive. The company would have to borrow a lot of money, and paying it back would reduce our profits. In the end, though, the payments would be finished and we would own the factory. On the other hand, we could hire a plastics company to manufacture the battery. Our staff would do the design, advertising, and selling. Our factory might be able to do the final labelling and packaging. We would have to pay the manufacturing company for every battery, however, and that would cost us money forever. The accounting department is trying to estimate all of these costs so we can compare them.

I think I will try to list all the possible advantages and disadvantages of building our own factory compared with hiring another company to make the batteries.

Day 3: Sales and Marketing

The sales department is already getting inquiries from other manufacturers who could use a lightweight, flexible battery in their products. The sales staff have interviewed ordinary consumers and have heard lots of complaints about the mess batteries make when they leak and the way they stop recharging properly after only a few uses. People even say they would pay extra for a battery without these problems. Looks as if a plastic battery could be a big seller!

Our advertising agency is using this market research to create names and slogans that could be used if we decide to produce the battery. Our industrial designer is working on a distinct package design and colour, one that does not look like any other battery. Someone is even finding out how much we would have to pay a famous athlete or TV personality to try our product and maybe recommend it in commercials.

Just for fun, I'm going to make up a magazine ad for the plastic battery. My ideas cannot possibly be any stranger than some of the ideas I heard today!

Day 4: Disposal

Wow! Today I found out that people in North America buy over four billion batteries every year, and over half of these batteries are just thrown out after they go dead. What a waste, because the metals in rechargeable batteries can be recycled. They make pretty dangerous garbage if they are thrown out and really should be treated as hazardous waste.

The plastic battery is made from much safer materials. Even better, it could be collected for recycling using the system that battery manufacturers have already set up. That might save people money in countries like Denmark, which have extra taxes on rechargeable batteries that are not recycled.

What I don't know is how dangerous the battery manufacturing process might be. Our legal department is checking to see if there are any special government regulations about handling, storage, or disposal of the chemicals we would need to use, or if the factory workers might need protective equipment. Any of these things could make this battery pretty expensive to manufacture.

I wonder how many batteries there are in all the toys, appliances, and gadgets in my house? I'm going to count them and estimate how many batteries I buy in a year. I wonder how much they all cost?

Extend Your Knowledge

Choose a product that interests you, and find out about its development and manufacturing. Analyze the different parts of its life cycle and imagine some of the questions that were asked when businesses were deciding whether to manufacture it.

Possible topics could be jogging shoes, air bags in cars, cellular phones, cassette tapes or video tapes, CDs, tubeless tires, disposable cameras, freeze-dried foods, or anything else that interests you. Report on your research as your teacher directs.

Resisting Stress — The Inside View

What determines the strength of a material? Scientists trace strength and many other properties, to forces between the tiniest particles of the material. (Recall that you learned about the attraction between particles in Chapter 5.) Study the examples below to learn what scientists have been able to infer about particles that are far too small to see.

- Steel has high tensile strength. It must have strong forces pulling its particles together. A very strong tension force is needed to separate the particles and break the material.
- Graphite (a form of carbon) has low shear strength. Its particles are arranged in layers, but the forces between the layers are relatively weak. Because the layers slide over one another easily, graphite is slippery and makes a good dry lubricant. As you saw in Chapter 10, the layers of graphite in a pencil "lead" rub off and leave a mark on the paper when you write.
- Rubber has high torsion strength. Each particle is attracted in all directions to the other particles around it. The particles hold together even when a piece of rubber is twisted out of shape.

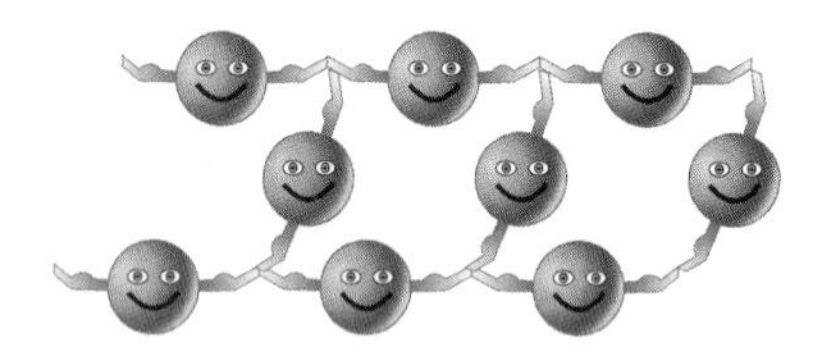

Each metal particle attracts a few other particles very strongly. The forces are quite directional, so the particles form a regular arrangement in space.

B graphite

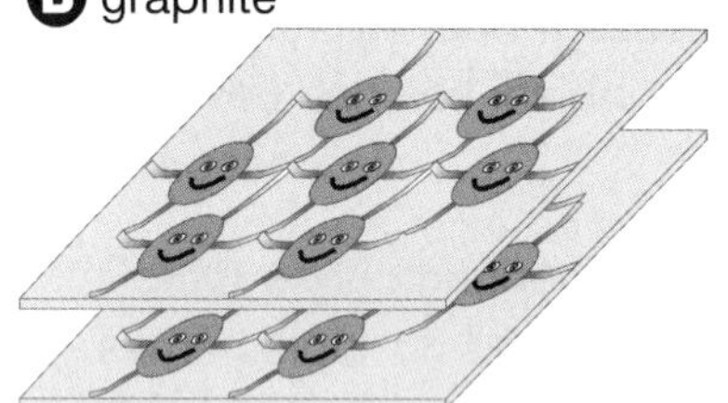

Graphite particles attract strongly in some directions, but hardly at all in other directions.

C rubber

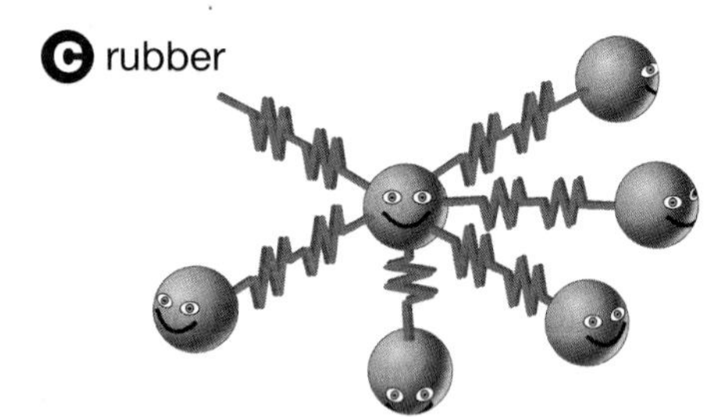

Each rubber particle attracts many other particles in all directions.

Check Your Understanding

1. Name three effects that a force can have on the motion of an object.
2. Name three types of stress that a force can cause inside an object.
3. (a) What is the difference between a live load and a dead load?
   (b) Classify as live load or dead load:
   - wind blowing against a tree
   - the weight of the tree
   - the weight of a bird that is sitting in the tree
4. Identify the type of strength that is shown by
   (a) the chain that connects a ship to its anchor
   (b) a piece of dried meat that is too tough to bite through
   (c) a bolt that is being tightened with a wrench
   (d) the legs of the chair you are sitting on
5. What can you infer about the forces between the particles in a piece of bubble gum from each fact below?
   (a) The dry stick of gum snaps easily between your fingers.
   (b) After chewing, the gum stretches in all directions so you can blow a bubble.
   (c) Moist gum sticks to many surfaces.
6. List four stages of a product's life cycle. Identify two decisions that must be made at each stage.

CHAPTER at a glance

Now that you have completed this chapter, try to do the following. If you cannot, go back to the sections indicated.

Measure the mass of an object using a triple-beam balance. (14.1)

Measure the weight of an object using a spring scale. (14.2)

Calculate Kofi's weight, if his mass is 45 kg. Calculate Amanda's mass, if her weight is 390 N. (14.2)

Give an example of a manufactured structure that

- must have high structural efficiency
- can have low structural efficiency (14.1)

Draw a force diagram showing a spacecraft that is

- being pushed to the right by a small force
- being pulled to the left by a large force
- hovering in one spot because the upward force from its engines is just balancing the downward force of gravity (14.2)

Identify one external force and one internal force acting on the piece of furniture you are sitting on right now. (14.3)

Name a live load and a dead load on your desk when you are doing homework on it. (14.3)

Write directions for creating each of these internal forces in a marshmallow: compression, tension, shear, torsion. (14.2)

Draw a ruler that has been bent in a curved shape. Label the side under compression, the side under tension, and the neutral axis. (14.3)

Prepare Your Own Summary

Summarize this chapter by doing one of the following. Use a graphic organizer (such as a concept map), produce a poster, or write the summary to include the key chapter ideas. Here are a few ideas to use as a guide:

- Identify key features of mass and weight: measuring instrument, proper SI units, on what each quantity depends. (14.1)
- Give the mathematical formula used to calculate structural efficiency. (14.1)

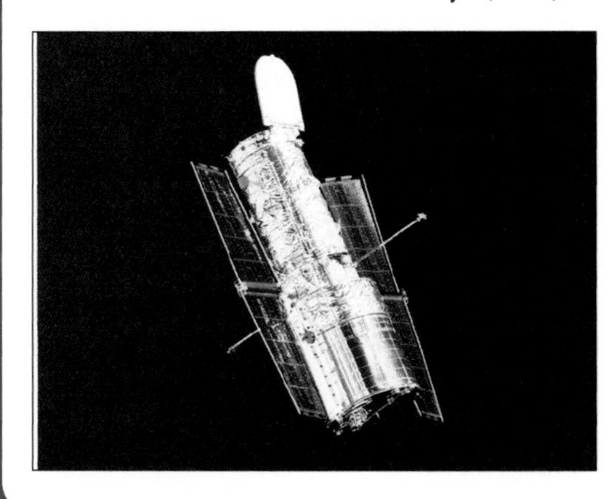

- Identify at least two differences you would notice between a device with high structural efficiency and one with low structural efficiency. (14.1)
- List the four main types of internal forces. Then give an example of a material with the type of strength to resist each force. (14.3)
- The neutral axis of a beam has several special features. Explain
  (a) where it is located
  (b) how the forces on it are different from the forces on other parts of the beam
  (c) what practical effect all this has on the shape of beams in buildings (14.3)
- What causes different materials to have different types of strength? In your answer, give reasons why a material might have
  (a) high tensile and torsional strength
  (b) low shear strength
  (c) high compressive strength (14.3)

CHAPTER 14 Review

Key Terms

mass
primary standard
kilogram (kg)
balance
structural efficiency
neutral axis
force
newton (N)
spring scale
weight
gravitational force
force diagram
external force
internal force
stress
live load
dead load
tension force
tensile strength
compression force
compressive strength
shear force
shear strength
torsion force
torsion strength

Reviewing Key Terms

If you need to review, the section numbers show you where these terms were introduced.

1. Copy the puzzle below into your notebook. Do not write in the textbook. Use the clues to complete each line. The number of blanks gives the number of letters in the word. If your answers are correct, the letters in the box will spell the name of an important force.

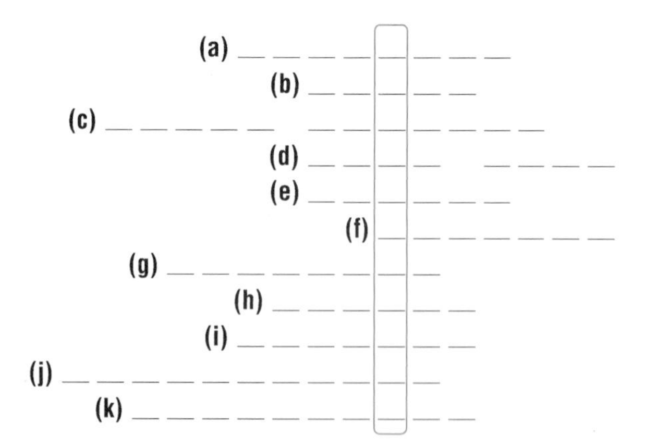

(a) primary standard of mass (14.1)
(b) push or pull (14.2)
(c) picture of forces on an object (14.2)
(d) force that causes extra stress on a structure (14.3)
(e) force that pulls you against Earth (14.2)
(f) force that is needed to turn off a water tap (14.3)
(g) forces that resist outside forces (14.3)
(h) standard unit of force (14.2)
(i) kind of strength that spider silk has (14.3)
(j) forces that develop when you squeeze something (14.3)
(k) Structural ■■■■■ is the measure of a structure's ability to support a load. (14.1)

Understanding Key Ideas

Section numbers are provided if you need to review.

2. Classify each statement as referring to force (F) or mass (M). (14.1, 14.2)
   (a) measured in newtons
   (b) stays the same no matter where the object is located
   (c) measured with a balance
   (d) your weight

3. Identify the main type of stress (internal force) produced in each situation below. (14.3) Then draw a force diagram of each object showing the external force that is producing the stress. Draw a simple box, square, or rectangle to stand for the object. Use arrows to represent the forces. (14.2)
   (a) stepping onto a tin can to crush it for recycling
   (b) twisting a piece of licorice candy to break it in two
   (c) cutting a piece of cardboard with scissors

4. Describe two ways to increase the structural efficiency of a beam in a bridge. (14.1)

5. What two factors affect the gravitational force between two objects? (14.2)

6. In a force diagram, state what is shown by
   (a) the length of the force arrows
   (b) the direction of the force arrows (14.2)

7. In the diagram below, name
   (a) the type of force stressing the top of the bookshelf
   (b) the type of force stressing the bottom of the bookshelf
   (c) the shelf to which the "x's" are pointing (14.3)

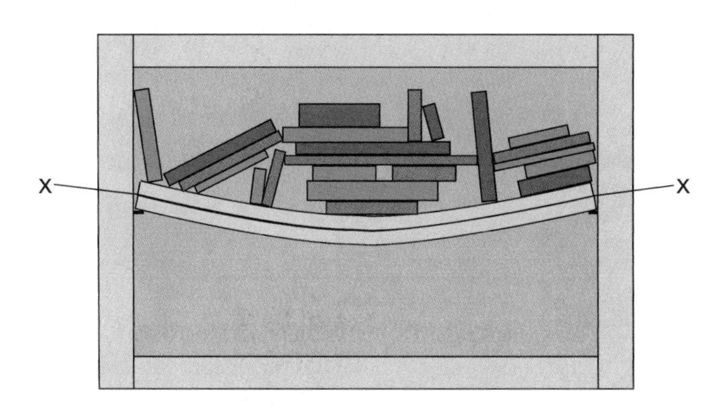

Developing Skills

8. In your notebook, copy and complete the following concept map. The following words will help you get started: kilogram, constant, balance, structural efficiency. Then draw a similar concept map for force and for stress.

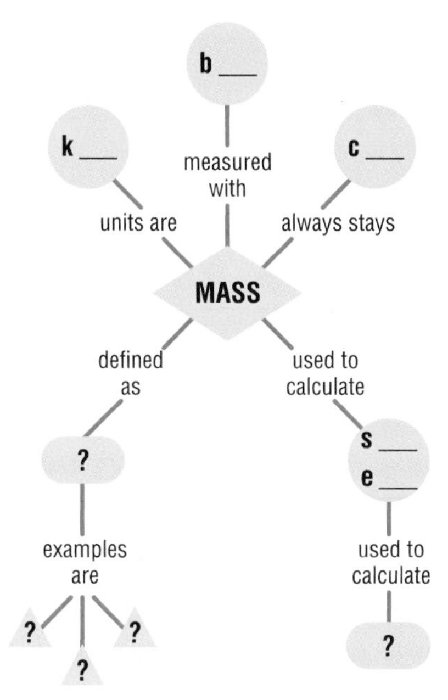

Problem Solving/Applying

9. Elephants are believed to be the world's strongest animals. A 6.0 t elephant is said to be able to carry as much as 1500 kg. Calculate the elephant's structural efficiency. (Express all masses in the same units before calculating.)

10. In 1985 Dr. William Gonyea tested the strength of house cats by training them to lift weights using a special apparatus. He found that a 2.8 kg cat could lift as much as 1.96 kg. Calculate the cat's structural efficiency.

Critical Thinking

11. Use your answers to questions 9 and 10 to compare the structural efficiency of the two animals. Which would you say is stronger?

12. Laboratory balances will not work in "weightless" conditions far from Earth. Why? A diagram might help explain your answer.

13. In your notebook, copy the four situations below in order of increasing gravitational force.
   (a) two small objects close together
   (b) the same two small objects far apart
   (c) a large object close to a small object
   (d) the same large object equally close to another large object

Pause& Reflect

Go back to the beginning of this chapter on page 406, and check your original answers to the Getting Ready questions. How has your thinking changed? How would you answer those questions now that you have investigated the topics in this chapter?

CHAPTER

15 Stability and Centre

Getting Ready...

- How can designers be sure that very large structures will not tip over?
- Why are bicycle helmets made from rigid plastic foam instead of soft material that would cushion your head in a crash?
- Why is it easier to balance a moving bicycle than a bicycle that is almost standing still?

Science Log

Think about the questions and photographs on this page. Try to identify some differences between structures that remain stable and structures that do not. Record your ideas in your Science Log. You will discover more ideas as you work through this chapter.

You can learn a lot by examining why something did not work. Russian and American scientists stayed aboard *Mir*, the first space station run by humans, long after its designed lifetime of three years. Fires, collisions with supply rockets, and electrical and computer failures could not make the astronauts abandon *Mir* as it slowly wore out. Why? They were learning so much about what goes wrong with structures in space and how to fix them.

What caused the problems in the other structure pictured on these pages? What forces cause strong metals to rip and tear? What have scientists learned by studying broken, collapsed structures? In this chapter, you will use what you know about forces to find out how to keep structures strong and stable.

# of Gravity

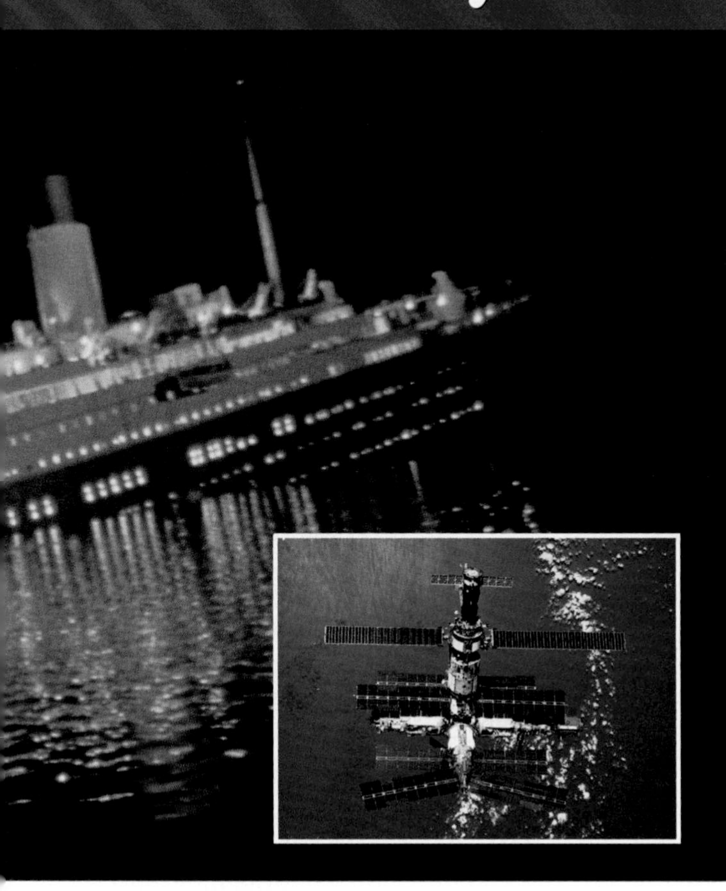

Spotlight On Key Ideas

In this chapter, you will discover

- how forces on structures cause different parts to shear, twist, and buckle
- what the centre of gravity is, and how its position determines whether something stands or falls
- how shape, spin, and foundations affect a structure's stability

Spotlight On Key Skills

In this chapter, you will

- predict the effects of forces on different parts of a structure
- balance objects by finding their centre of gravity
- draw thrust lines, and use them to predict the stability of structures

Starting Point ACTIVITY

Stand Tall!

Do you have photos or videos that show how you learned to walk? As difficult as it was, after a few years, you do not even think about it. How *do* people keep from falling over?

What to Do

Do these activities away from obstacles, so that you do not hit anything if you fall. Do not strain your body! Try each activity to see what happens, not to prove that you can do the impossible!

1. Stand so that your right cheek, arm, and foot are against a wall. Try to lift your left foot. Can you do it? What happens to the rest of your body?
2. Sit straight in a chair with your feet flat on the floor. Try to stand without using your hands or leaning forward. Can you do it? Which part of your body feels the strain?
3. With your feet together, bend over and grab your toes with both hands. (It is okay to bend your knees.) Now try to hop forward. Can you do it? What has to happen, but cannot, in order for you to hop?

What Did You Find Out?

In each part of this activity, you started in a stable position. You did not move because forces pulling you in one direction were balanced by forces pulling you in other directions. Suggest possible answers to these questions:

1. If you are in a stable position, how does your body position have to change before you can start to move?
2. How does this change affect the forces that are acting on your body?

15.1 Forces and Failure

Figure 15.1 How does the person pictured here create such a strong force?

Large enough forces will break the parts from which any structure is made, and then the whole structure can become unstable and collapse. You often create forces as large as the force pictured here, even if you are not a karate expert. Rip solid metal? No problem. It happens every time you open a pop can with a pull tab; and with just a small wrench, you can tighten bolts on a bicycle so much that you damage the threads or twist the bolt head right off. Yank nails out of solid wood? Just exert the force through a claw hammer or a crowbar. Small forces applied in particular ways can damage well-made parts, weaken joints, and destroy structures.

Levers

Pull tabs, wrenches, and crowbars develop large forces because they are levers. From earlier grades, you might remember the key features of a lever: a long rigid bar that can pivot around a balance point or **fulcrum**; a place where an effort or **input force** is applied; and another place where an **output force** acts on a load. Many common levers resemble the one in Figure 15.2, but other levers have their fulcrum at one end.

The output force can be different in strength and direction from the input force. Even small input forces can develop very large stresses in the load, large enough to damage materials or destroy a structure. Study Figure 15.3 to understand how this works.

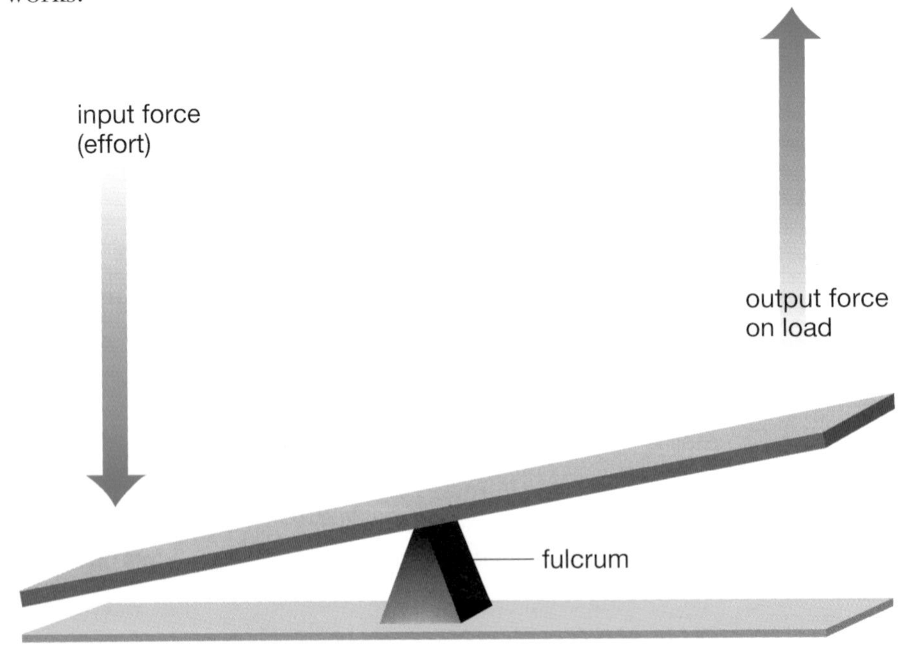

Figure 15.2 Some levers, such as a teeter-totter, have their fulcrum between the load and the effort. Can you think of a lever that has its fulcrum at one end instead?

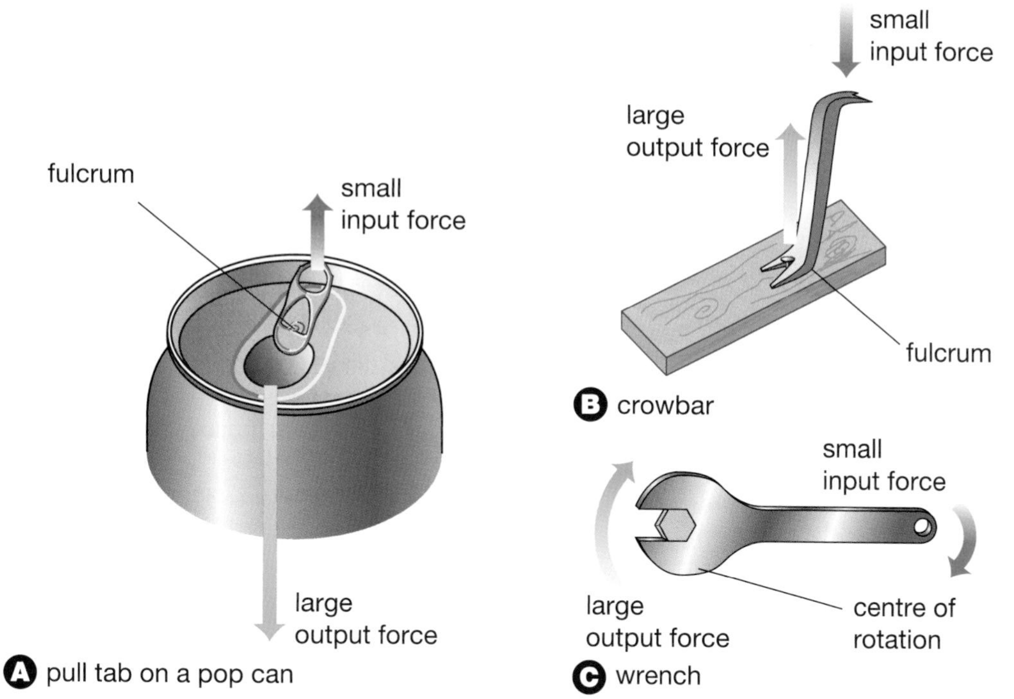

fulcrum
small input force
large output force
A pull tab on a pop can

small input force
large output force
fulcrum
B crowbar

small input force
large output force
centre of rotation
C wrench

Figure 15.3 Many everyday devices develop very large forces because they act as levers.

Forces developed by levers are a special problem in frame structures. Frames are made of long, rigid pieces. If they twist, bend, or move even a little, the pieces can act as levers. Large forces can then develop in unexpected places. For example, the force of the wind on a flagpole or sign is transmitted through the pole to the base. Even though the original force was horizontal, one side of the base can be pulled *up*. Bolts holding the pole to its concrete foundation are put under tension, and the metal around the bolt experiences large shear stresses. If the bolt or the surrounding metal holding fails, the next gust of wind may tip over the unstable structure.

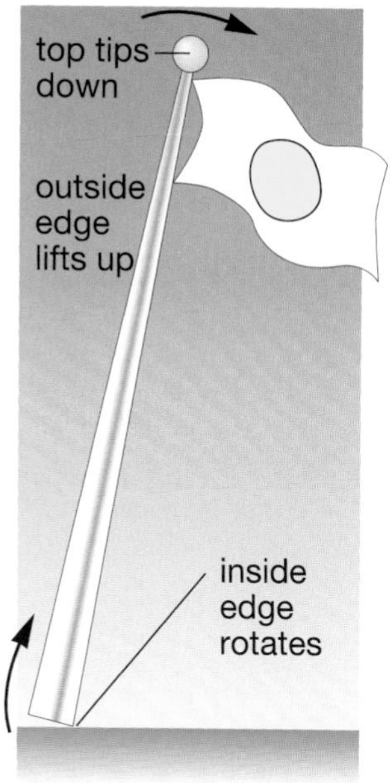

top tips down
outside edge lifts up
inside edge rotates

Figure 15.4 Strong winds can tip a flagpole.

How Materials Fail

Tension forces are forces that stretch materials. They always cause materials to fail in the same way. They snap. Compression forces are also forces that cause materials to fail, but this can happen in two different ways:

- *Shear* Solid materials always contain tiny cracks or weaknesses. When a solid material is compressed, one section may **shear** (slide over another section along a weakness). This type of failure can happen in the soil beneath a heavy building, such as the Leaning Tower of Pisa. The weight of the building compresses the soil, causing some of it to slip sideways. The ground beneath the building sinks, and the building tilts or collapses. Crushing rock, glass, or similar materials is another example of shearing under compression.

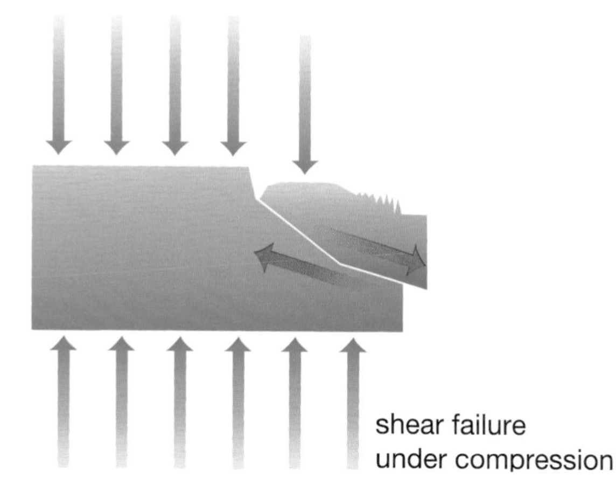

Figure 15.5 Compression causes sections of material to slide sideways, causing shear failure.

Enormous shear forces are also created naturally in Earth's rocky crust as it slowly rises, falls, and twists. As you saw in Chapter 11, if the rock fails under these forces, there is a sudden release of energy — an earthquake. Can you find evidence of rock that has sheared in Figure 15.6?

Figure 15.6 Shear forces in solid rock can cause movement along weaknesses called faults. You learned about what can happen along fault lines in Chapters 11 and 12. Some of Canada's largest cities, including Vancouver and Toronto, are built above faults.

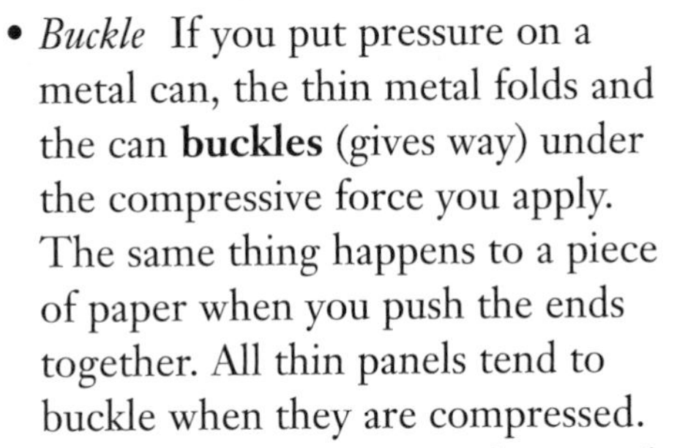

- *Buckle* If you put pressure on a metal can, the thin metal folds and the can **buckles** (gives way) under the compressive force you apply. The same thing happens to a piece of paper when you push the ends together. All thin panels tend to buckle when they are compressed. Shell structures that use thin panels to support their entire load, such as boats and aircraft, are reinforced to prevent buckling. Examine Figure 15.7 to see some common methods of reinforcement.

Torsion forces can cause material failure, too. Brittle structures, such as dry spaghetti and plastic cutlery, often shear when they are twisted. Sections of the structure slide past each other, and the structure cracks or breaks in two. Very flexible structures, such as rubber bands, hoses, and electrical cords, shear less easily. Instead, torsion forces make them fold up and **twist** into tangles and knots. Although the structure is unbroken, it has lost its shape, which is a form of failure.

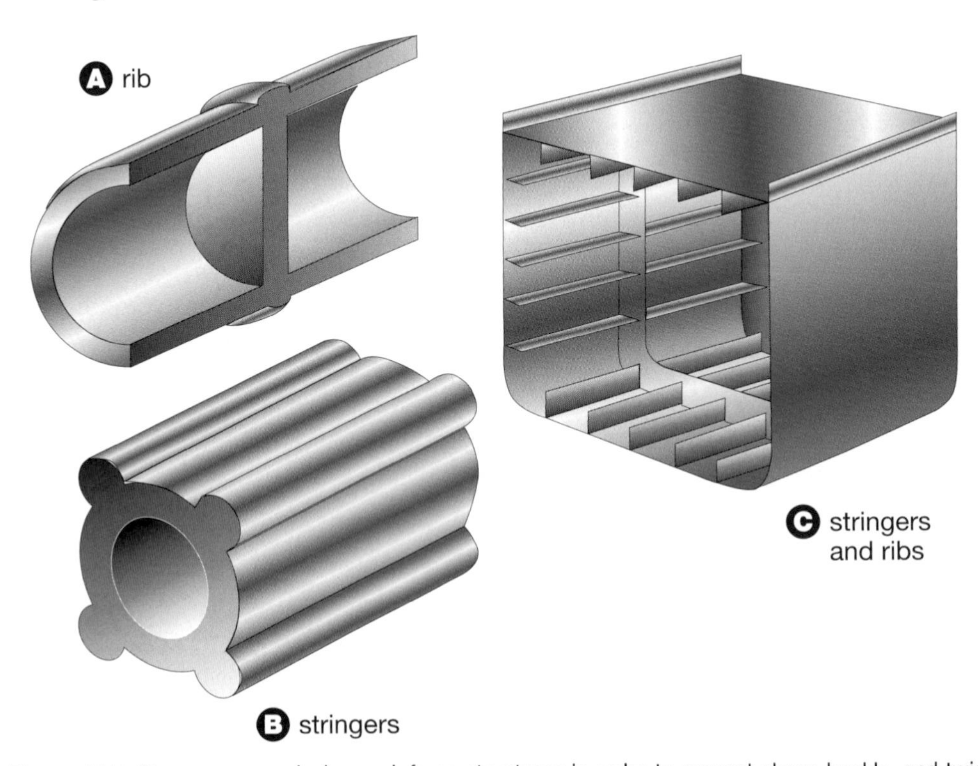

Figure 15.7 Three common designs reinforce structures in order to prevent shear, buckle, and twist failures.

Metal Fatigue

Early railway cars often developed a serious problem after they had been used for several years. Their solid metal axles broke and their wheels fell off, even though the parts were designed to be much stronger than necessary. About 150 years ago, a German railway official identified the problem. Metals weaken when they are bent or twisted over and over again.

You can explain this loss of strength using the particle theory of matter, which you learned about in Units 2 and 3. In a bent or twisted part of a metal structure, the arrangement of particles is changed. Where particles move apart, the forces holding them together become weaker. If enough particles are affected, small cracks develop. Eventually the material may fail under only a small stress, one that it could easily resist when it was new. Engineers call this weakening **metal fatigue**, and it is still a problem, especially in lightweight, flexible structures such as aircraft. Metal fatigue has caused many plane crashes, including the crash of the world's first jet airliner, the *Comet*, in the 1950s.

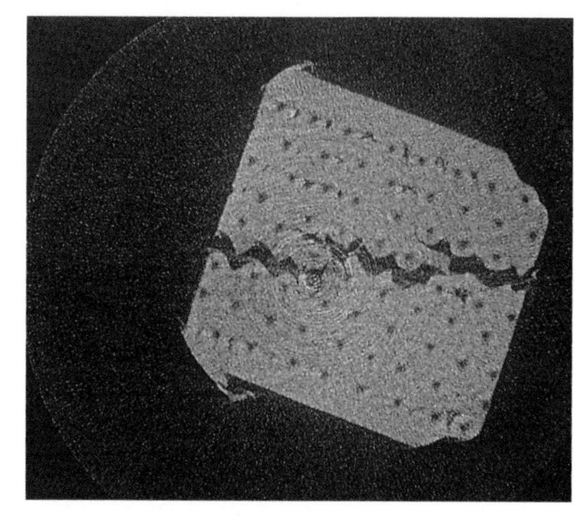

Figure 15.8 The material around this rivet shows signs of metal fatigue.

Pause& Reflect

There is a lot of information in this section. In your Science Log, write a point-form outline to summarize the main ideas you have studied so far.

Bend and Break

What You Need

2 silver-coloured metal paper clips

What to Do

1. Straighten one paper clip, and touch it to your inner forearm to check its temperature.
2. Bend the paper clip back and forth quickly ten times. Can you feel it beginning to weaken? How many bends are needed to produce noticeable metal fatigue? Immediately touch the paper clip to your forearm again. How has its temperature changed?
3. Repeat step 2 until the paper clip breaks. Keep track of how many bends are needed.
4. Straighten the other paper clip. This time, twist the ends of the paper clip back and forth. Count the number of twists needed to produce noticeable metal fatigue. Count the number of twists needed to break the metal.

What Did You Find Out?

Answer the following questions:

1. Which type of stress seems to fatigue the metal sooner?
2. Did you notice any change in temperature after bending or twisting the paper clip? If so, what provided the energy to warm or cool it?

CONDUCT AN

INVESTIGATION 15-A

The Wrecking Crew

Wrecking crews need to apply knowledge of forces and leverage to break structures apart without causing unnecessary damage. Can you predict the effect of a wrecking ball on a model of a brick wall?

Problem

What effect does a swinging mass have when it hits different parts of an object? (You will use a mass at the end of a string to represent a wrecking ball and wooden blocks to represent the bricks of a brick wall.)

Apparatus

3 wooden blocks, 20 cm x 5 cm x 10 cm

500 g or 1 kg mass

string

metre stick

ruler

Safety Precautions

- Keep the swinging mass away from people or objects that it might damage.
- Stand clear of the impact area.

Procedure

1. Set up a data table like the one below to record your observations. Give your table a title. Tie about 1 m of string to the mass.

Number of blocks	Placed	Hit	Observation
1	on edge	middle	
		end	
1	on end	top	
		middle	
		bottom	
2	on edge	bottom brick at middle	
3	on edge	bottom brick at end	
		middle brick at middle	
		bottom brick at middle	

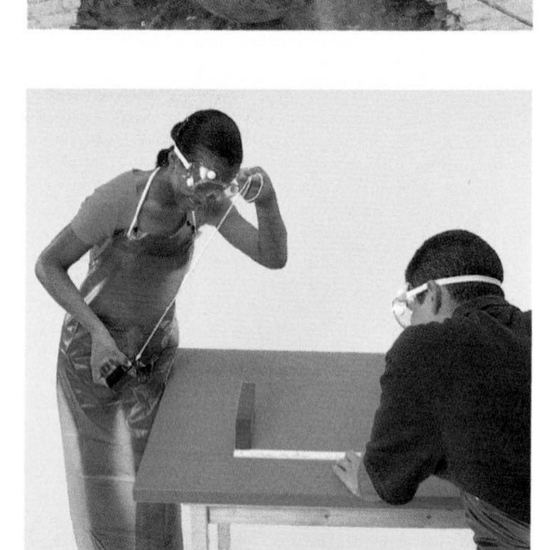

2. Set one block on its longest edge. Place the metre stick so that it touches the block. Rest your arm on the metre stick and hold the wrecking ball (the metal object you are using) level with the middle of the block and about 30 cm away. Have your partner pull the ball back about 40 cm and let it swing directly toward the block. Experiment to find a launching point that moves the block a few centimetres across the floor but does not knock it across the room. Launch the ball from this same position throughout the investigation so that it will deliver the same force each time it hits.

3 Launch the ball toward the *middle* of the block.

(a) Observe
- how far the block is pushed
- whether it moves straight or twists
- whether the impact had any effect on the structure of the block

(b) Repeat your test to find out if your observations are consistent.

(c) Record your observations. Describe the effect of the force on the block. Did it shear, twist, buckle, or do something else?

(d) Repeat parts (a), (b), and (c), but this time launch the ball so that it hits the *end* of the block.

4 Stand the block on end and repeat step 3, parts (a), (b), and (c). Try hitting the block at the top, in the middle, and at the bottom. Observe and record the effect in each position.

5 Stack two blocks on edge to make a wall.

(a) Observe and record the effect on the wall when the wrecking ball hits
- the middle of the bottom brick
- the end of the bottom brick

(b) Stack three blocks on edge to make a wall. Observe and record the effect on the wall when the ball hits
- the middle of the middle brick
- the middle of the bottom brick

CONTINUED ▶

Analyze

1. Think about the following conditions of the collisions you observed: the point where the ball hit, the force the ball created, and the damage it did.
   (a) Which condition did you try to keep the same throughout the whole investigation? (Conditions that are not allowed to change during an experiment are called controlled variables. Use this term in your answer.)
   (b) Which conditions did you change during the investigation? (Conditions that you change are called independent variables. Use this term in your answer.)
   (c) Which condition did you study to discover how it was affected by your actions? (Conditions that change as a result of what you do are called dependent variables. Use this term in your answer.)

2. Does applying force to the end of a block have a different effect than pushing the middle of the block? If there is a difference, describe it.

3. Does the force of the impact have the same effect when the block is placed in different positions (on edge and on end)? Describe any differences.

4. In which position did the block act like a lever, increasing the effect of the impact? Give evidence for your answer.

5. In which conditions did you observe shearing between the blocks? How did you recognize that shearing was occurring?

6. When the wall collapsed, did the blocks move only in the direction that the ball pushed them? If not, describe their motion in detail.

Conclude and Apply

7. Draw a diagram that shows one way of arranging the blocks to increase the stability of a three-layer block wall.

8. Demolition experts can collapse a building using only a small number of carefully placed explosive charges. Based on this investigation, where do you suppose the charges are placed: near the top, middle, or bottom of the building?

9. When a wall is hit by a wrecking ball, the bricks do not start to fall until the impact force from the ball has already acted. What force is actually pulling down the wall?

Extend Your Skills

10. Draw a vector diagram that shows the forces acting on a block in a wall just as it is hit by a wrecking ball.

11. Demolishing old buildings can be a dangerous business, not just because something might fall on you. The building materials themselves may have health risks that were not known when the buildings were constructed.

    At the library or on the Internet, find out about a material that was used in the construction industry and later found to be hazardous to health (for example, asbestos fibres, pictured below). Prepare a brief report, telling about the material's benefits in construction and how its dangers were identified.

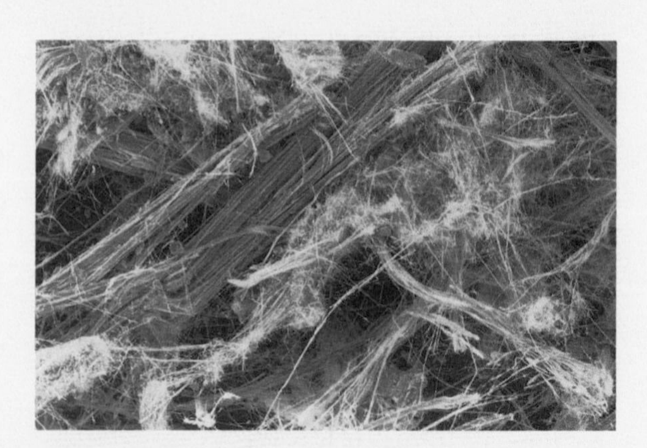

Making Use of Stress

Snap, twist, buckle, and shear: that is what materials and structures do when they fail, as you saw in the investigation. These same behaviours can also be put to good use. Here are some examples.

- *Buckle* Car bumpers and sheet metal can be designed to buckle in a collision. As the metal deforms, it absorbs some of the energy of the impact. The car may be badly damaged, but the occupants are less likely to be seriously hurt. Blades of grass on a sports field buckle as athletes land on them, absorbing some of the energy of the impact. Certain kinds of panels, such as the bellows in an accordion or the flaps on a cardboard box, are designed to buckle in specific places when you press on them.
- *Shear* In a boat's outboard motor, the propeller is held to the drive shaft with a shear pin. If the propeller becomes tangled in weeds, the shear pin will break, allowing the motor and the gears to spin freely instead of being damaged. The clutch and automatic transmission of a car are designed so that shear forces allow parts to slip past each other, speeding up or slowing down gradually until they are moving at the same speed. This produces a much smoother ride than connecting the engine directly to the wheels.
- *Twist* Spinning wheels twist cotton or wool fibres tightly so that they lock together. The twisted yarn is much stronger than a bundle of straight fibres and is long enough to knit or weave into cloth. If the fibres are twisted too much, they tangle and shorten unless you keep pulling on them. That's one way to make stretchy fabrics. Controlled twisting turns hair into braids, string into ropes, and wires into cables.

Computer CONNECT

Make one stem-and-leaf plot to combine the class data for bending from the Bend and Break activity on page 439. Make another stem-and-leaf plot for the twisting data. Then prepare graphs of the results. If you have access to a computer, use it to prepare your graphs. (Why would it not be fair to combine the data if different types of paper clips were used?)

Skill POWER

For help with stem-and-leaf plots and graphs, turn to page 486.

Across Canada

One day, Sandford Fleming (1827–1915) invited some governors to have lunch under a bridge. Fleming was chief engineer on the construction of the railroad across Canada. He was trying to convince the governors that iron was the best material for building train bridges because it would stand up well to fire, moisture, and cold. Everyone else wanted to use wood, fearing that iron beams would crack from the weight of the trains.

When a heavy train thundered across the bridge during lunch, the governors shook with fear, and some ran away from the table. Of course, nothing happened; the bridge was safe. Fleming went on to build more iron bridges across Canada.

Solving problems was always one of Fleming's passions. In his late teens, in 1845, he sketched a design for a type of rollerblade! Later he created the first Canadian postage stamp. It featured a beaver and was issued on April 23, 1851. Hard work and determination helped Sandford Fleming become one of the most respected scientists in Canadian history.

Sandford Fleming

Career CONNECT

There are many different types of engineers. Using a dictionary and logic, try to guess what type of engineer each of the people below is. Note that there may be more than one person for some branches of engineering.

- chemical engineer
- electrical engineer
- mechanical engineer
- aerospace engineer
- civil engineer
- materials science engineer

Stanley Arthurson: designs custom tools for assembly line industries

Tom Cardinal: developed a new metal alloy for a lighter, faster bicycle frame

Josh Cohen: oversees smelting operations at a large mining company

Tony Chung: designs and modifies equipment to lower emissions of pollution from a large processing plant

Susan Erickson: develops circuitry for stereo components

Bob Gonzales: designs and oversees construction of roads and overpasses.

Karen Ouimette: oversees the operation of the equipment needed to process petroleum into its components

Sasha Salinsky: creates new wing designs for a small airplane manufacturer

Find a partner and compare your matchings. Then, as partners, devise a plan to check your accuracy. Perhaps you will look in your telephone book for information, or your local college or university course calendar. The Internet or the guidance office in your school could provide some clues as well. Keep a journal of the steps in your plan and what information you gain at each stage. Share the journal, as well as your original and revised matchings, with the other partnerships in your class.

INTERNET CONNECT

www.school.mcgrawhill.ca/resources/

Find out more about engineering by going to the web site above. Go to **Science Resources**, then to **SCIENCEPOWER 7** to know where to go next. Can you discover why many Canadian engineers wear a special ring, what metal it is made of, and what famous disaster it commemorates? Write a short article about your findings.

Skill POWER

To review researching on the Internet, turn to page 497.

Check Your Understanding

1. Draw a sketch that shows how a wheelbarrow acts as a lever. Label the fulcrum. Mark the input force and the output force with labelled force arrows. (Remember that the arrows must point in the direction of the force.)
2. (a) Which type of structure (mass, frame, or shell) is most likely to be damaged when its parts act as levers and create very strong forces?
   (b) Why are the other two types of structures not also weakened by lever action?
3. Name four ways that materials fail, and identify the type of internal force that causes each kind of failure.
4. Which type of material failure occurs when you
   (a) leave a trail of footprints in a carpet?
   (b) sprain your ankle in a soccer game?
   (c) accidentally hit a baseball through a window?
   (d) crinkle a new $5 bill as you stuff it into your pocket?
   (e) twist the lid of a partly-opened tin can back and forth until it breaks off?
5. Write a technical description of a time when you knocked something down or demolished a structure. Use the terms you learned in this section and in the two previous chapters. For example, to describe kicking over a sand castle, you might begin, "I applied an external force to a small mass structure. As a result, . . . "

15.2 Balance and Centre of Gravity

There is more than one way to collapse a structure. Nothing has to break to make an athlete, a stepladder, or a bicycle rider fall. They may just lose their balance and tip over. The force of gravity pulls them down — but not always. Almost all structures can lean a bit without falling down. A bicycle rider who is moving fast enough can even lean a lot without losing balance. To design **stable** (less likely to tip) structures, engineers need to know what features of a leaning object determine whether it will tip over or stay balanced.

Figure 15.9 Why do these structures not fall over?

The Empire State Building in New York City is a steel frame building, 102 storeys tall. It has a mass of 365 000 t. After it was built, many people worried about the building's safety because strong winds made the top storeys sway back and forth slightly. On a foggy day in July 1945, a U.S. Air Force bomber hit the Empire State Building between the 78th and 79th floors. Fourteen people in the airplane were killed, but the building was not seriously damaged. Doubts about its safety disappeared.

INVESTIGATION 15-B

Tip It!

Think About It

If you can stand or walk, your body has learned a lot about balance. Examine the pictures and answer the questions to put your experience into words.

What to Do

Figure A

1. To find one key to stability, examine the photographs in Figure A and answer these questions.
   - **(a)** Which person is in a more stable position?
   - **(b)** What difference in their positions creates the difference in stability?
   - **(c)** Based on your observations, suggest a hypothesis to explain why an opened stepladder is more stable than the same stepladder with its two legs folded together.

Figure B

2. To find a second key to stability, examine the photographs in Figure B.
   - **(a)** Which athlete is in a more stable position?
   - **(b)** A large part of your body mass is in the area around your hips. What difference in the position of body mass puts one athlete in a more stable position than the other?
   - **(c)** Explain how the same principle makes balancing on stilts much harder than balancing on your feet.

Figure C

3. Football players are coached to keep their stance "wide and low" (Figure C).
   - **(a)** Explain how this advice uses both keys to stability that you have discovered.
   - **(b)** Why is it so hard to balance a pencil on its point (Figure C)? Use the ideas from steps 1 and 2 in your answer.

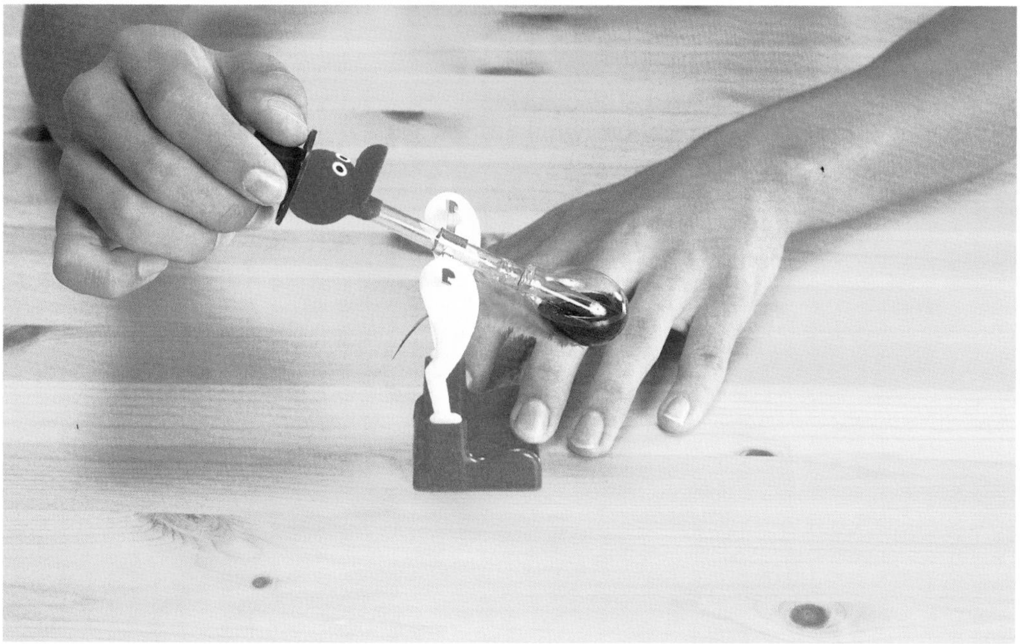

Figure 15.10 Where is this object's centre of gravity?

Centre of Gravity

In the last investigation, did you conclude that objects are more stable if they rest on a large area and have most of their mass close to the ground? These are useful general principles, but they are not precise enough to ensure that a particular structure, such as a bridge or a building, will be stable. Engineers need to calculate exactly how large a foundation is necessary or the best place to put heavy heating and air conditioning machinery. They also need to design structures, such as aircraft and rockets, that have to be stable even when they are not resting on the ground.

The key to stability is an idea developed by Isaac Newton when he was analyzing the force of gravity between Earth and the Moon. He knew that the force depended on the distance between them, but exact calculations were too difficult because some parts of Earth are always closer to the Moon than others. Newton wondered if he could pretend that all the gravitational force on an object acted from one point — the **centre of gravity**. He supposed the centre of gravity would be in the exact centre of a large object like Earth or the Moon, far below the surface. Calculations using Newton's idea agree very well with observation of gravitational force.

If you can find an object's centre of gravity, you can predict how gravitational forces will affect it: for example, whether they will cause it to tip over. Unfortunately, if something is not solid and spherical like a planet, calculating the exact position of its centre of gravity can be difficult, even using computers. It is often easier to find the centre of gravity by experimenting with the object or a model of it.

Word CONNECT

The idea of a centre of gravity helps scientists to analyze the effects of gravitational forces. A similar idea, the centre of mass, simplifies other calculations, especially predictions of how objects balance, rotate, or move after they collide or break into pieces. For normal-sized objects on Earth, the centre of gravity and the centre of mass are at the same place, so the two terms are often used as if they had the same meaning. In certain situations, however, the two centres are at different places, so scientists must be careful to use the correct term to describe their observations. Try to think of another pair of scientific words with slightly different meanings.

Pause& Reflect

Look at the balancing toy in Figure 15.10. Make a list of all the construction techniques you can identify that help to make it stable.

CONDUCT AN

INVESTIGATION 15-C

Standing on a Slant

When an object is balanced on one support, the downward force of gravity is being resisted by an equal upward supporting force. The two opposing forces and the object's centre of gravity are always in a straight vertical line. Study the diagrams on the next page, showing three pennies taped to a disk. Imagine what happens if the opposing forces do not line up with the centre of gravity. If the object is able to rotate around the support, the force of gravity pulls the centre of gravity down until it is directly under the support.

Problem

How can you use the last fact above to find the position of the centre of gravity of many small objects?

Safety Precaution

Apparatus

several small masses (pennies or small pieces of modelling clay)
pin
ruler
short wooden dowel

Materials

disposable plate or pie pan, about 30 cm in diameter
tape
cardboard tube (paper towel roll)

Procedure

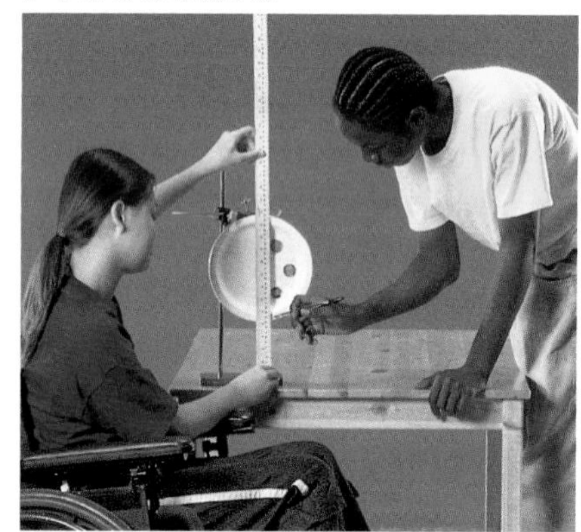

1. Use tape to fasten two or three masses close to one another on the plate.
   (a) Put the pin through the weighted plate near the edge opposite the masses. Hold the pin, and let the plate hang freely. The centre of gravity will be pulled down until it lies directly below the pin.
   (b) Draw a line across the plate straight down from the pin. The centre of gravity will be somewhere along this line.

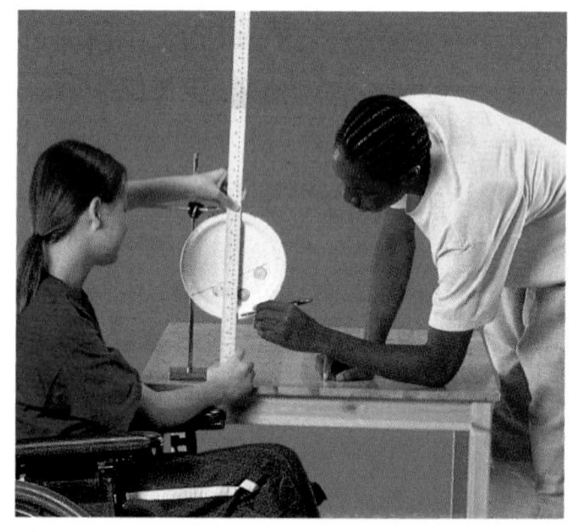

2. Repeat step 1 with the pin in a different position near the edge of the plate and about one third of the way farther around.

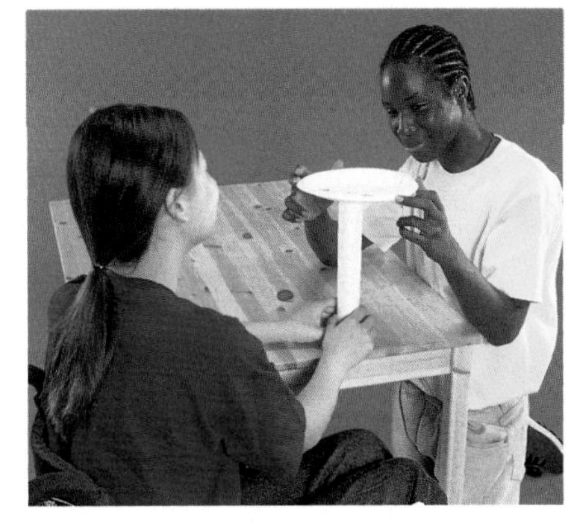

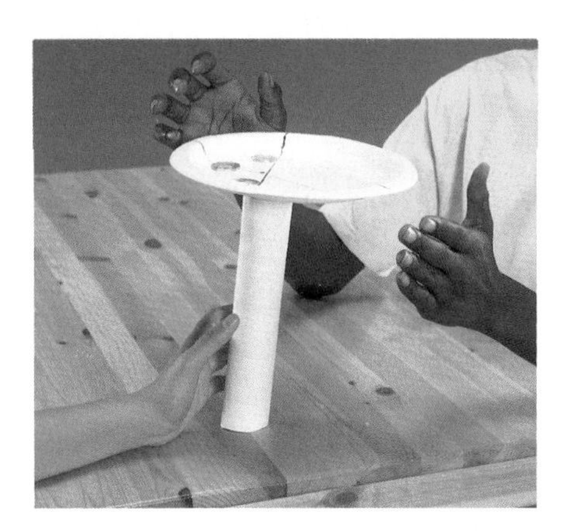

3 The centre of gravity of the plate is near the point at which the lines from steps 1 and 2 cross. Support the plate at this point with the dowel. If it will not balance, move the dowel slightly until it does. Mark that exact position as the centre of gravity.

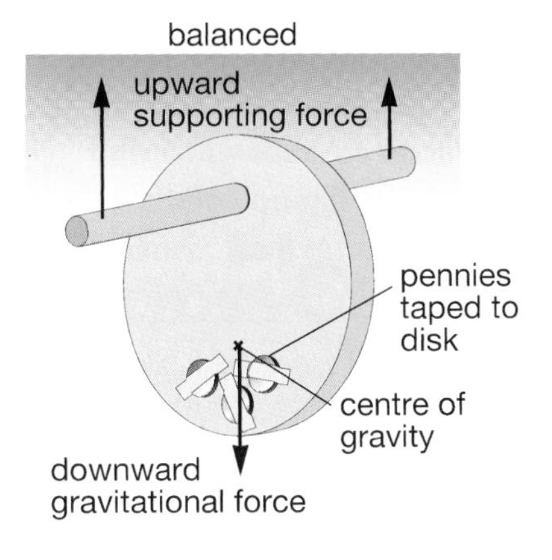

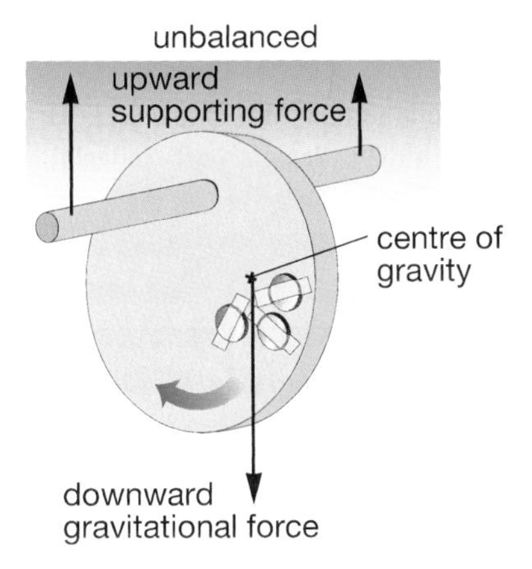

4 Rest the plate on the top end of the cardboard tube so that its centre of gravity lies inside the tube. Does it balance?

(a) Move the plate so that its centre of gravity lies outside the cardboard tube. Does it still balance?

(b) Tape the plate on the cardboard tube at the centre of gravity you found in step 3. Carefully squeeze the bottom end of the tube flat, and cut it at a slight angle.

5 Open the bottom of the tube, and try to stand the structure on a level surface.

(a) Imagine a line going from the centre of gravity of the plate straight down through the tube. Where does this line hit the surface, inside or outside the edge of the tube?

(b) Cut the end of the tube at greater and greater angles, testing it after each cut, until it will no longer stand upright. Where does the line you imagined hit the surface now?

Analyze

Engineers use the position of the imaginary line through an object's centre of gravity to predict whether or not a structure will tip over. Use your results from step 5 to suggest a rule that tells what positions are most stable and what positions are unstable.

Conclude and Apply

Summarize three or four ideas you learned or remembered while doing this investigation.

Thrust Lines

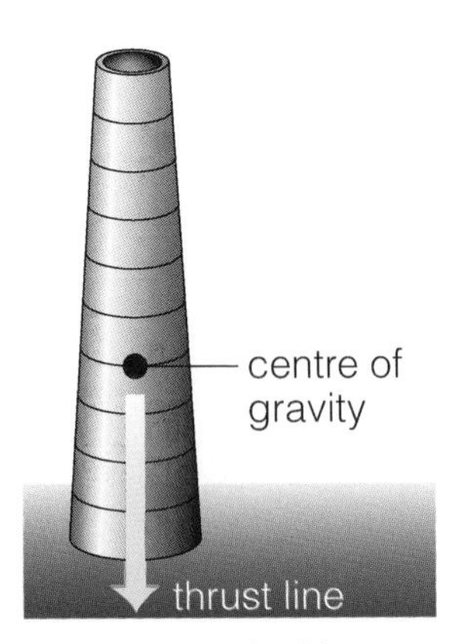

Figure 15.11A In this position, the chimney is stable. The thrust line is inside the foundation.

Did you discover that your tower was most stable when its centre of gravity was directly above its base? The line you imagined passing vertically through the centre of gravity is called a **thrust line**. The thrust line shows how the tower's weight presses straight down from its centre of gravity. When the thrust line passes inside the base, then all sides of the base help to support the weight. As long as the material below the structure does not fail under the compression force, the tower will not tip. It is stable.

If a tower's centre of gravity is not over its base, the thrust line meets the ground outside the base. The nearest edge of the base acts as a fulcrum, so when gravitational force pulls the centre of gravity *down*, the far side of the base is pushed *up*. This is what happens when a tree falls over and the roots come out of the ground, or when a skater "wipes out" after catching a skate in a crack in the ice.

To prevent this type of instability, most tall structures do not just rest on top of the ground. They are usually fastened to a foundation that cannot rotate because it is set firmly into the earth. Even with a foundation, however, a structure may be in danger if it leans over. The thrust line now points across the structure, as well as down. The structure's weight acts as a shear force, sliding one section across another. The base may hold firm, but the structure can crack in the middle, letting the top part break away from the bottom part and fall. Fence posts, flagpoles, and well-rooted trees often break in this way. So do skiers' legs if their safety bindings do not release and let their feet rotate in a fall.

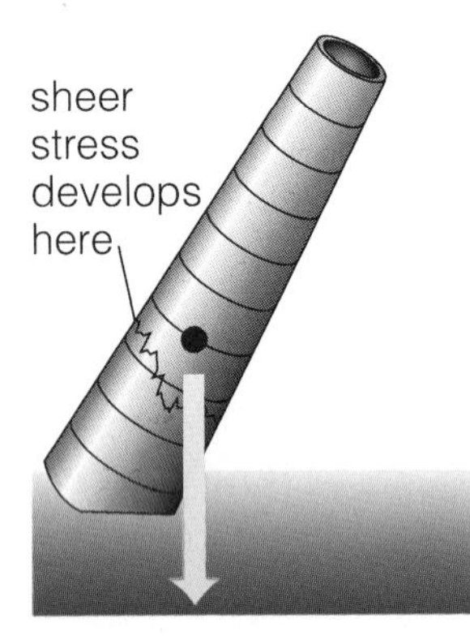

Figure 15.11B In this position, the chimney is unstable. It will tip or break apart as a result of the large shear forces near the arrow.

Figure 15.12 Where is the thrust line in this chimney?

Check Your Understanding

1. In each pair, choose the condition that is more stable.
   (a) low centre of gravity/high centre of gravity
   (b) narrow base/wide base
   (c) thrust line inside base/thrust line outside base
2. What problem did Isaac Newton simplify by developing the idea of centre of gravity?
3. What does a thrust line show about the gravitational force on a tower?
4. Write a rule (scientists would call it a "scientific law") that describes where to support an object so that it will balance. (If you were the first person to state this law, it would probably be named after you. Scientists still speak of Newton's law of universal gravitation and laws of motion.)

15.3 Principles of Stability

Balancing a loaded dinner plate on your arm, or your body on a high wire, is easier than it looks. Just make sure that the support is directly under the centre of gravity, and do not lean! Keeping buildings, bridges, and other very large structures from falling down is more complex. Also, rotating objects, or mechanisms with heavy rotating parts, can use a completely different method of stabilization.

Figure 15.13 What principle does the waiter use in order to carry this load successfully?

Principle 1: Build on a Firm Foundation

What is happening to the ground in the photograph on the right? How could anyone build a stable structure on it?

Solid ground is not always firm and stable, especially if it is moist. Water near the surface can freeze and expand (swell) in the winter, lifting the ground. In warmer weather, the melted ice water drains away, shifting tiny soil particles and leaving spaces that collapse under pressure. The soil compacts, and potholes appear. Some clay soils act in the same way when they absorb water and then dry out. Larger sections of soil can slip sideways over moist layers underground, causing sinkholes and landslides. Worse yet, very moist soil sometimes flows like a thick liquid when it is shaken or vibrated. When this happens in an earthquake, the weight of a road or a building can easily squeeze the soil out from under it.

Figure 15.14 Why has this bridge collapsed?

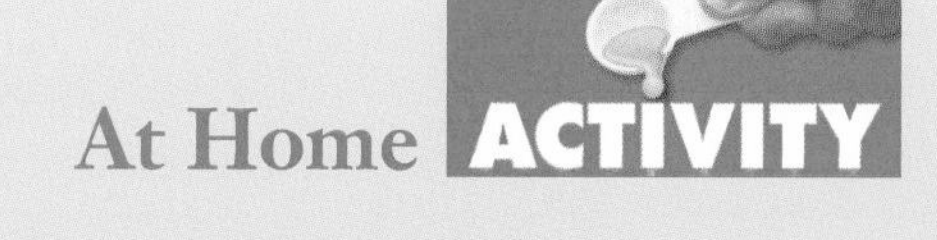

Looking Below

What to Do

Find out what type of foundation was used for the building you live in, and for two or three nearby structures such as a garage, driveway, tree, or street light. Examine each structure and the soil near it for signs of stability and instability. Record your findings as your teacher directs.

CONDUCT AN

INVESTIGATION 15-D

Sink the Stick

Soil testing is an early step in designing any large structure. Knowing how the soil reacts to forces helps an architect or engineer decide what type of foundation will be necessary.

Problem

How does loose sand react to impact forces over different areas? (Read through the procedure before you begin. Predict which dowel will have the greatest impact and which will have the least impact.)

Safety Precautions

- Do this experiment on newspaper, and have a broom handy to sweep up scattered sand.
- Handle all sharp objects with care.
- Handle the hammer with care, following your teacher's instructions.

Apparatus

small tray or plastic container
hammer
at least 4 wooden dowels of different diameters
eating fork (for loosening the sand)
sharp pencil
ruler

Materials

500 mL of sand
1 cm graph paper

Procedure

1. Make a data table with four columns, headed "Object," "Base area (cm^2)," "Depth in loose sand (mm)," and "Depth in packed sand (mm)." Give your table a title.

2. Sort your dowels according to the size of their base. Calculate the size of each base, and record it in your table. Test the smallest one first.

3. Loosen the sand with the fork. Then level the surface.

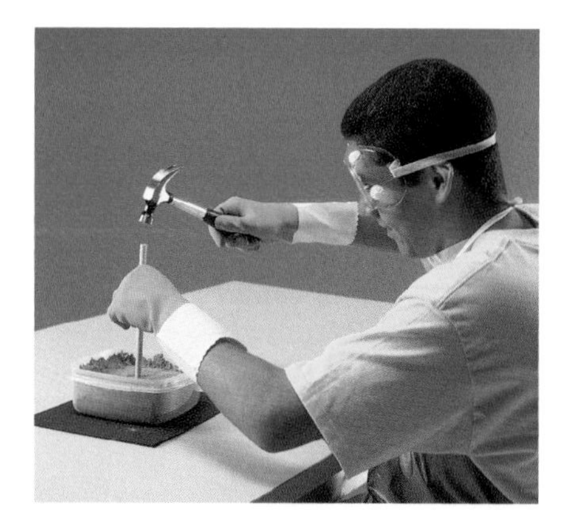

4. Hold the dowel gently on the sand so that it is upright but can slide through your fingers. Hold the hammer about 20 cm above the top of the dowel. While still holding it, let it fall by its own weight and hit the dowel. If you force the hammer down, it will deliver a different force each time, so your tests will not be fair.

5. Mark the dowel at the level of the sand surface.

Skill POWER

To find out when to use each type of graph, turn to page 486.

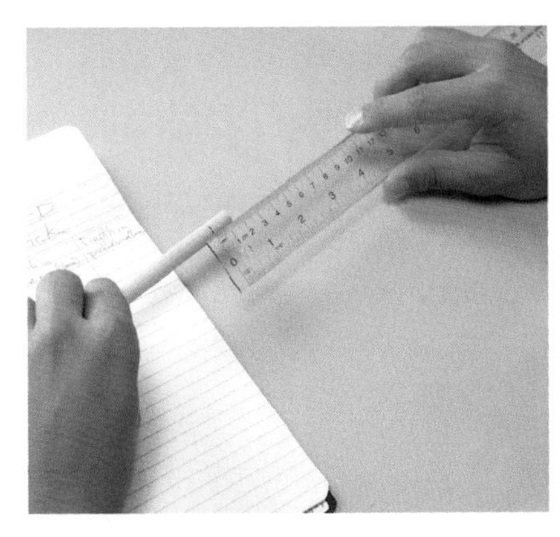

6. Measure and record how far the dowel is pushed into the sand.

7. Repeat steps 3 to 6 with each dowel.

8. Repeat steps 3 to 6, but this time pack the sand firmly before each test.
   - (a) Record the results in the proper column of your data table.
   - (b) If you have time, moisten the sand and repeat steps 3 to 6.

Math CONNECT

To find the area of the base of the first dowel, use one (or both) of these methods.

1. Stand the dowel on the graph paper, and trace around its base. Count the number of centimetre squares it covers. Group any half or quarter squares into whole squares, and include them in the count.
2. If you know the mathematical formula for finding the area of a circle, make the necessary measurements and calculate the area. Round your answer to one decimal place.

Analyze

1. What type of graph (line, bar, or circle) would be most useful for predicting what would happen if you experimented with other dowels of different sizes?
2. Draw the most suitable type of graph to show the results of your tests. Use different colours, or solid and dotted lines, to show what happened in the loose and packed sand. Make a key that explains the meaning of each colour or style of line.
3. Was the hammer blow an external force or an internal force on the sand?
4. Did the sand have to support a live load, a dead load, or both? Explain your answer.
5. What type of stress did your hammer blow create in the sand?
6. Did you observe any signs of shearing in the sand beside the dowel? If so, describe them.

Conclude and Apply

7. Do your tests support the idea that spreading a load over a larger area reduces the effect of the force on the soil? Explain how you know.
8. If a load is applied to a smaller surface, what happens to the effect of the force under the surface?
9. Why are nails, drill bits, and sometimes fence posts sharpened on the bottom? In your answer, use the ideas and words you learned in this section.
10. When might a builder need the sort of soil information that you found out in this investigation?
11. Even if every group used the same dowels, sand, and hammer, the results would probably not be exactly the same. Can you think of some reasons why?
12. If there is a lot of variation in your class results, you can study this variation by making separate stem-and-leaf plots of the results for each size of dowel. Then use the stem-and-leaf plots to draw an appropriate type of graph for your class results.

How do builders construct a stable structure on such a shifty substance? They start by creating a firm **foundation**, using one of these strategies.

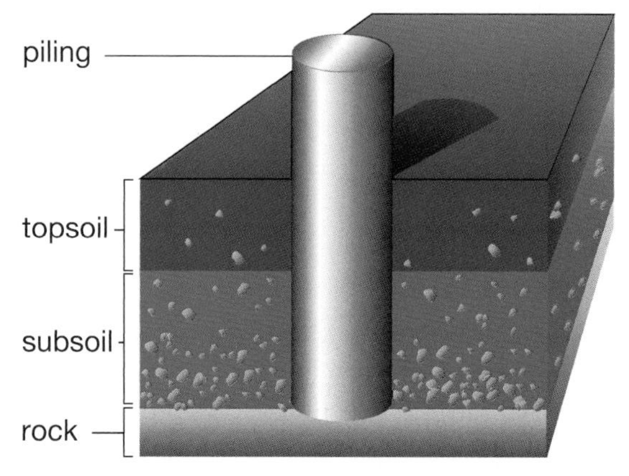

Figure 15.15A Pilings support the weight of many buildings and bridges.

Find Something Solid

Below the soil lies solid bedrock. If the loose surface soil is not too deep, builders can dig it out completely and build a stable foundation directly on the bedrock. If the loose soil is too deep, they can sink **pilings** (large metal, concrete, or wood cylinders) through the loose soil until the pilings rest on bedrock. Then the structure can be constructed so that its weight is carried by the pilings to the rock beneath. To support a structure such as a garage or a house, which is not really heavy, builders might not need to dig to bedrock. In many parts of Canada, foundation walls about 1.5 m deep reach firm layers of soil that give enough support and never freeze.

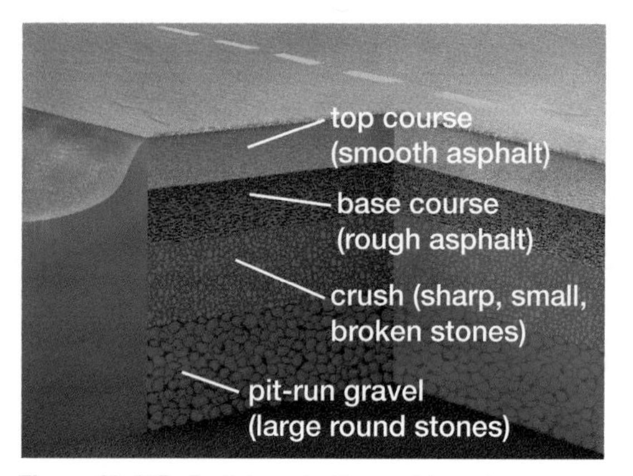

Figure 15.15B Each layer in the road base has a specific function.

Make a Solid Layer

Road builders always pack loose surface soil before paving to create a solid base for the asphalt or concrete. Later on, if the pavement cracks badly, repair crews dig out the soil and replace it with a solidly packed layer of gravel. They then repave on top of the more solid material. Packed gravel foundations are also used for dams and other mass structures.

Spread the Load

If the weight of a structure is spread over a large area, any particular part of the ground supports only a small part of the weight. This is why buildings are often constructed on many shallow pilings rather than on a few. Even if the pilings do not reach bedrock, the soil beneath each one is strong enough to carry its part of the load. This is also why the **footings**, the concrete foundation beneath house basement walls, are wider than the walls themselves. Spreading the weight of the walls over a larger area reduces the stress on every part of the soil beneath them.

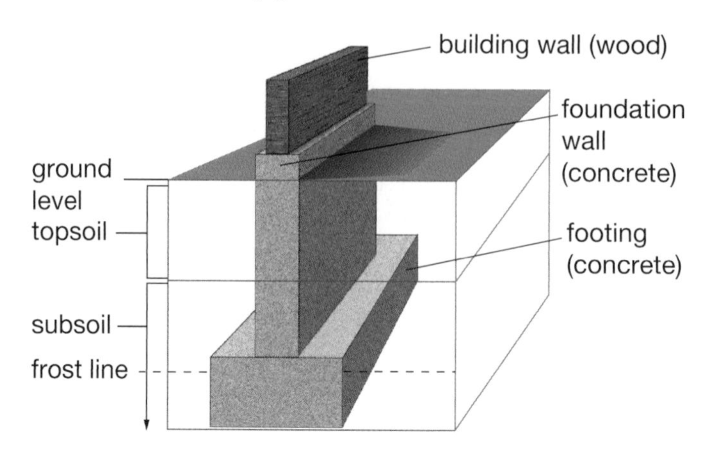

Figure 15.15C Footings beneath a foundation wall reduce stress on the soil.

Principle 2: Balance Forces

Arm wrestlers and tug-of-war teams strain every muscle, but their opponents do the same. The forces are **balanced** if they are equal in size, opposite in direction, and pointing along the same straight line. Balanced forces do not change an object's motion. Counterweights, guy wires, and "hiking" sailors use this principle. A **counterweight** is a mass that balances another mass. **Guy wires** are ropes or cables attached to an object to keep it stable. Counterweights and guy wires create a balance by pulling in the direction opposite to a dangerous force. If the structure is not pulled apart under tension, the balanced forces will stabilize it. If you use this principle in your projects, be sure to use materials with high shear and tensile strength. You know what happens in a tug-of-war if the rope is not strong enough!

Figure 15.16A A heavy mass acts as a counterweight to keep the crane stable.

Figure 15.16B The cables on this boat keep the masts stable.

Figure 15.16C A sailor can act as a counterweight to balance a small boat.

Pause& Reflect

Each photograph shows balanced forces stabilizing a structure. In your Science Log, draw a vector diagram with labelled arrows to show

- the force that is making each structure unstable
- the balancing force

For each diagram, write a sentence telling what produces each force.

Figure 15.17 The roots of these mangrove trees act as buttresses to stabilize the trees, which grow in swampy ground.

Principle 3: Keep Thrust Lines Vertical

Many bridges are a series of arches supported on a series of columns. Building roofs rest on several walls. Chairs rest on four legs. In all of these structures, the centre of gravity is somewhere in the middle, not directly over a support. The weight of the structure pushes outward on its supports, as well as down (see Figure 15.18). The supports can be tipped over, as well as collapsed, by the weight of the structure resting on it.

Builders solve this problem by adapting designs from nature or inventing new ones. Study the diagrams and the explanations below to find out how different designs work.

- **Half-circle arches** or domes distribute their weight to press straight down on their foundations. Planetariums, and some mosques and orthodox churches, are built using half-circle domes. They do not span a very large space. Can you imagine what a domed football stadium would look like if its roof were the shape of a half circle?
- **Buttresses** are slanted braces outside a wall that prevent the roof from pushing the walls outward. The slanted thrust line from the roof carries the weight along each buttress to the ground. Some old European churches have elaborate systems of buttresses.
- **Trusses** are arches with their ends tied together by a strong beam. The outward force on the support is balanced by the inward tension force in the **tie beam**, so the support does not tip. Steel frame bridges are often built with **truss girders**, and many houses have roof trusses.

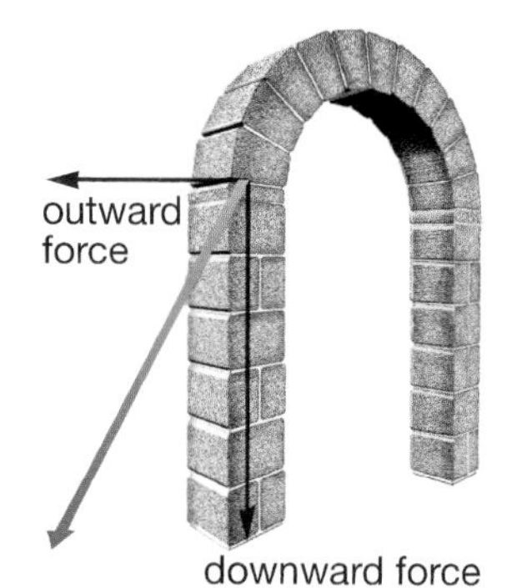

Figure 15.18 How do builders prevent the collapse of structures whose centre of gravity is not over a support?

Figure 15.19A A domed structure presses straight down on its foundation.

Figure 15.19B Buttresses prevent walls from tipping over.

Figure 15.19C Trusses balance tension so that supports do not tip.

Principle 4: Use Rapid Rotation

Every bicycle rider knows that it is harder to balance while moving slowly. The faster the wheels spin, the more the bicycle resists being tipped (see Figure 15.20A). If most of the mass of a wheel is located far from the centre, the stability is even greater. **Gyroscopes** (see Figure 15.20B), devices with heavy outer rims, can be built to spin tens of thousands of times per minute. When balanced on its axle, a gyroscope keeps pointing in the same direction, even if the device that is carrying it turns. Navigation equipment can use this property to detect when a ship or an airplane is wandering off course and to correct its heading (direction).

Spin stabilization, the principle demonstrated by the gyroscope, is especially useful for objects that do not rest on a solid foundation. Space satellites need to keep their antennas pointed back to Earth and their instruments facing the proper location in space. Football players want the ball to travel in a stable, predictable path, so they practise for hours to throw or kick perfect spirals. Rotation also stabilizes objects that are supported at only one end, such as a twirling figure skater and the tub in a washing machine. If you have ever spun a toy top, a yo-yo, or a Frisbee™, you have used spin stabilization.

Pause& Reflect

Look around your home and community, and find at least one structure that uses each type of support discussed in this section. In your Science Log, describe or sketch each structure.

Figure 15.20A Rapidly spinning wheels stabilize a bicycle.

Figure 15.20B Spin stabilization keeps a gyroscope pointing in the same direction.

Pause& Reflect

Imagine that you have been chosen to be interviewed by the education reporter for a local newspaper or radio station. The reporter wants an explanation, in your own words, of what your class has been studying in science for the past few weeks and why it is useful in everyday life. There is space for a paragraph of about 150 words. In your Science Log, write what you would say.

Check Your Understanding

1. For each pair, select one item that is likely to be more stable than the other.
   (a) wet soil/dry soil
   (b) bedrock/loose soil
   (c) unbalanced forces/balanced forces
   (d) arch without a tie beam/arch with a tie beam
   (e) rapidly spinning wheel/slowly spinning wheel
2. Identify one possible cause of each event.
   (a) soil heaving up and tilting a sidewalk
   (b) soil sinking down and collapsing a driveway
3. List three different ways to prevent the wall under an arched gate from being tipped over by outward pressure from the arch.
4. **Apply** Early settlers built corduroy roads over muskeg by placing logs side by side over the swampy ground. Why would this provide a stable foundation for the road?
5. **Thinking Critically** What problem could arise in a cold climate if house foundations were not dug down deep enough to reach soil that never freezes?

Builders and designers must become very skilled at using flat drawings to represent three-dimensional solid objects. Try to solve the problem shown below by studying the drawings. Then test your answer by making a paper or block model.

Problem

How many cubes would be needed to make each stack below? Assume that all the blocks rest on another block when there is more than one layer.

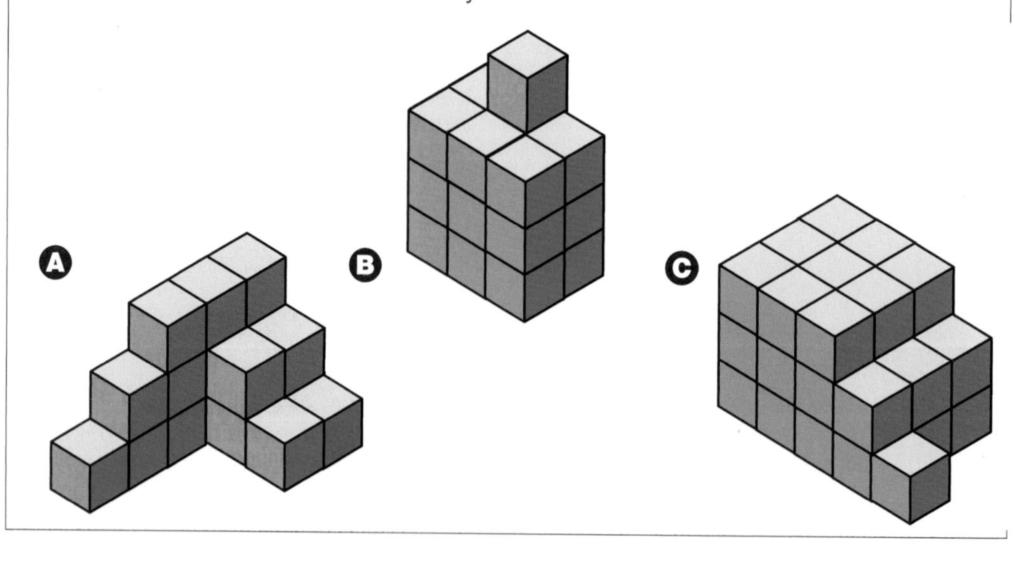

CHAPTER at a glance

Now that you have completed this chapter, try to do the following. If you cannot, go back to the sections indicated.

Make a strip of paper fail
- under tension
- by buckling under compression (15.1)

Press on a chunk of jelly, and find evidence that your compressive force is causing a shear failure in the material. (15.1)

Suppose that you need to remove part of a long nail that sticks out from a piece of lumber. Describe how you could use metal fatigue to break off the end of the nail. (15.1)

Use blocks of wood to model how a structure can fail from
- becoming unbalanced
- shear forces between parts
- the unsupported force of its own weight (15.1)

Find the centre of gravity of a small object, such as a tape cassette. Predict and then demonstrate what happens if you balance the object on top of a pop bottle with the thrust line passing
- through the top of the bottle
- outside the top edge of the bottle (15.2)

With a partner, demonstrate how a leaning structure (you!) can be stabilized by a balancing force provided when your partner acts as
- a buttress
- a guy wire (15.2)

Model a building's peaked roof with a piece of cardboard folded into a V. Show how extra weight on the model roof can create an outward force as well as a downward force. Use tape or string on your model roof to turn it into a truss structure that does not collapse so easily. (15.2)

Prepare Your Own Summary

Summarize this chapter by doing one of the following. Use a graphic organizer (such as a concept map), produce a poster, or write the summary to include the key chapter ideas. Here are a few ideas to use as a guide:

- Identify the key parts of a lever and the type of structure most likely to be damaged by lever action.
- Describe
  (a) one way that tension forces cause structures to fail
  (b) two ways that compression forces cause structural failure
  (c) two ways that torsion forces cause structural failure
- Identify one manufactured product that makes use of each behaviour: buckling, shearing, twisting.
- Define "thrust line," explaining
  (a) what it shows
  (b) how its position differs in stable and unstable structures
  (c) how it is connected to an object's centre of gravity
- Identify three ways that builders create firm foundations for their structures.
- If two balanced forces are acting on an object, explain what is special about
  (a) the size of the forces
  (b) the direction of the forces
  (c) the forces and the object's centre of gravity

CHAPTER

15 Review

Key Terms

fulcrum
input force
output force
shear
buckles
twist
metal fatigue
stable
centre of gravity
thrust line
foundation
pilings
footings
balanced (forces)
counterweight
guy wires
half-circle arches
buttresses
trusses
tie beam
truss girders
gyroscopes
spin stabilization

Reviewing Key Terms

If you need to review, the section numbers show you where these terms were introduced.

1. In your notebook, match each term in column A with the most closely related term in column B. Then write a sentence that uses both terms correctly.

A	B
• fulcrum (15.1)	• thrust line (15.2)
• input force (15.1)	• compression (14.3)
• buckle (15.1)	• lever (15.1)
• twist (15.1)	• foundation (15.2)
• centre of gravity (15.2)	• output force (15.1)
• bedrock (15.2)	• torsion (14.3)
	• stable (15.2)

2. Where could you find an example of each device listed below? (15.3)
   - **(a)** buttress
   - **(b)** tie beam
   - **(c)** guy wire
   - **(d)** counterweight
   - **(e)** gyroscope

Understanding Key Ideas

Section numbers are provided if you need to review.

3. Draw a diagram that shows how part of the base of a structure is pushed up when the structure itself tips over. (15.1)

4. Give an example of each type of failure and also a device or a situation in which it can be useful. (15.1)
   - **(a)** buckling
   - **(b)** shearing
   - **(c)** twisting
   - **(d)** snapping

5. Describe three differences between the two structures below that make structure A more stable than structure B. (15.2)

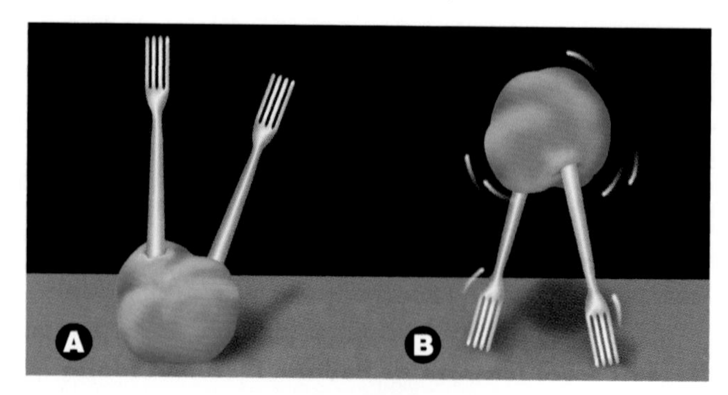

6. If young children balance a playground teeter-totter as shown, where is the centre of gravity in each situation? (15.2)

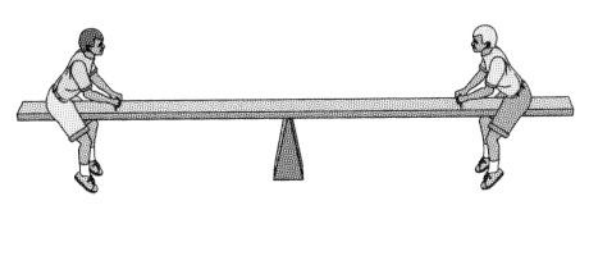

A B

Developing Skills

7. Create a concept map to summarize the ideas you studied in this chapter.

8. Use the following graph to answer the questions below.

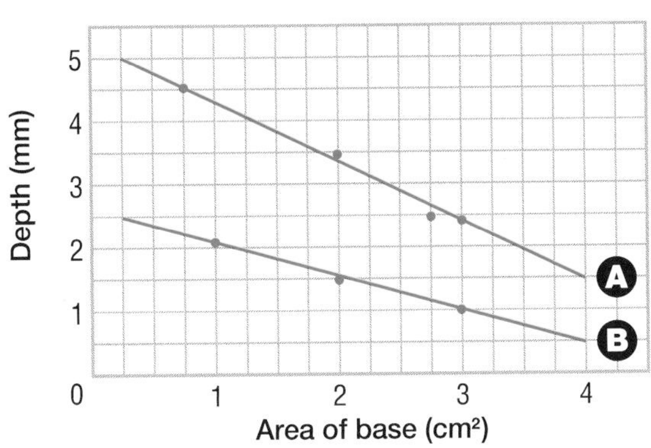

(a) Which soil sample was packed more tightly, sample A or sample B?

(b) How far would a wood block with an area of 2.0 cm^2 sink in loose soil?

(c) A nail was driven 1.5 cm deep into the packed soil. What was its base area?

Problem Solving/Applying

9. What type of foundation would you recommend for each structure listed below? Give a reason for each choice.

(a) a brick garden wall

(b) a boat dock

(c) a roof over the patio behind a house

(d) the concrete patio under the roof in part (c)

10. Cassette tapes and CDs are sold in brittle plastic cases that often crack and break.

(a) What type of material failure is happening when you step on a CD case and it cracks?

(b) Name three ways that you could strengthen the case while still making it out of the same plastic. Why do manufacturers not make these plastic structures stronger?

Critical Thinking

11. Metal fence posts are often set in wide holes that are filled with concrete. Identify two ways that this increases the stability of the fence posts.

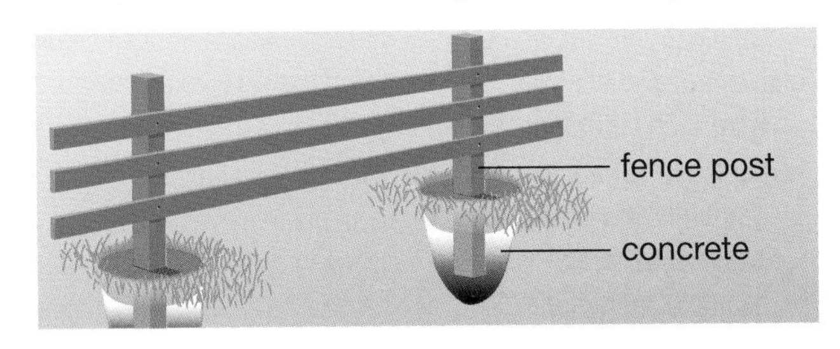

12. A tree branch can support your weight when you stand on it near the trunk of the tree. When you move farther toward its tip, however, it bends and breaks.

(a) Explain why your weight has such different effects on the branch.

(b) Sketch the bending branch. Mark the part that is in compression and the part that is in tension.

(c) Will the top of the branch snap or buckle when it breaks? Why? What happens to the bottom of the branch? Sketch the break in the branch to illustrate your answer.

Pause& Reflect

Go back to the beginning of this chapter on page 434, and check your original answers to the Getting Ready questions. How has your thinking changed? How would you answer those questions now that you have investigated the topics in this chapter?

5 Ask an Expert

It is your turn to take out the garbage. Halfway to the curb, the bag breaks, and its messy, smelly contents fall to the ground. Why did the bag break? What makes one brand of garbage bag stronger than another? Ask materials researcher Steven Haynes, supervisor of the product development division at a large materials testing company. He has helped plastics manufacturers test the strength of their products for nearly 20 years.

Q A garbage bag is plastic, but my telephone is, too. How can so many products be made from plastic, yet all look and feel different?

A Plastic products are formed in many different ways. Plastic can be melted and then pushed into a mould to make products such as margarine tubs, or it can be melted and pressed into large sheets to make products such as garbage bags.

Q Are these products very strong?

A They're usually strong in one direction. The little bits, called polymer molecules, that make up plastic are like long strings. When the melted plastic is pushed into sheets or into a mould, the strings line up so that most of them are running in the same direction, side by side. If you pull end to end on these strings, they are much stronger than if you pull sideways and separate them from one another. The plastic product is very strong in the direction that the strings run, but it is weaker and stretchier in the other direction.

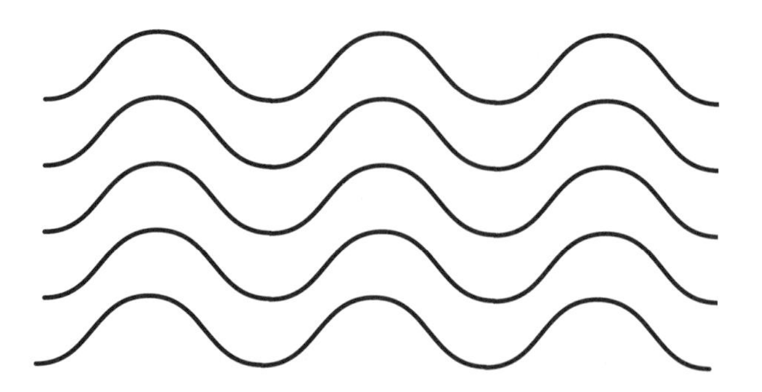

Q Are all plastic products stronger in one direction than in the other?

A No, it depends on how they are made. Some plastic products are placed in a mould that heats up the plastic and then squeezes it into the right shape. As a result, some of the strings, or molecules, "crosslink." This means that some run in one direction, while others run in other directions and join them together. Products made this way are strong in all directions.

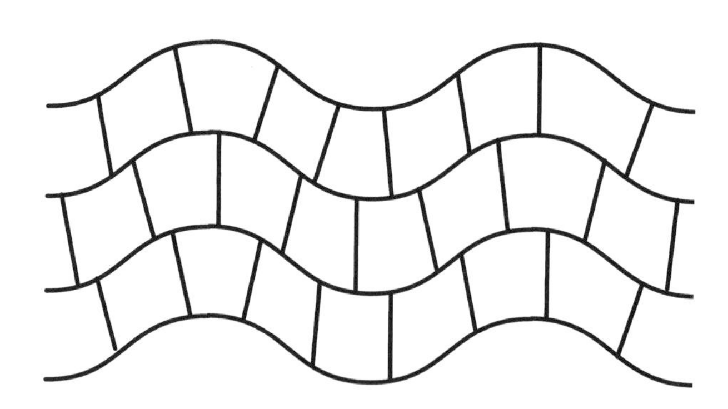

Q Shouldn't garbage bags be made this way, to be stronger?

A Well, this process is very slow. It's only used for products that need strength in both directions. Some products need flexibility instead. You wouldn't be able to get as much garbage in a garbage bag if it was rigid and not stretchy. A rigid bag would be more expensive, too, because it would take longer to make.

Q How do you test garbage bags for strength?

A Companies that bring a bag to us for testing want to know how much the bag will stretch and how easily it will break. To find out, we do a tensile strength test. We cut a strip from the bag, and place one end of the strip in a grip at the base of a universal testing machine. The other end of the strip goes in a grip above, which moves up and down. The machine pulls the upper grip up and measures how much force is needed to stretch the plastic and finally break it. Then we cut a strip from the bag in the other direction and test it the same way.

We also test how easily the edge of the bag will begin to tear and how easily it will keep tearing once it has started. Another test will tell us how easily a sharp object, such as a twig or a bone, can poke through. We give our results to the company so they can decide if their bag is strong enough to sell, or if they need to make changes.

Q Have you always liked science?

A I enjoyed studying science in high school, so I decided to take applied chemistry at Ryerson Polytechnical Institute. When I graduated, I got a job testing plastics. I didn't know much about them then, but I've learned a lot on the job and I've taken some other courses, too.

EXPLORING Further

Do Your Own Tensile Strength Test

With a partner, place one garbage bag on top of another, with the top bag turned sideways. Use a craft knife to cut a 20 cm x 2.5 cm strip through all four layers. Label the two horizontal strips "H" and the two vertical strips "V." Be sure that the edges of each strip are smooth, with no nicks.
CAUTION: Be extremely careful when cutting.

Face your partner. Each of you should hold one end of a vertical strip in one hand and one end of a horizontal strip in the other. Very gradually, one of you should begin to pull, equally, on the two strips. Which strip breaks first? Which stretches farthest? In which direction are the molecules aligned?

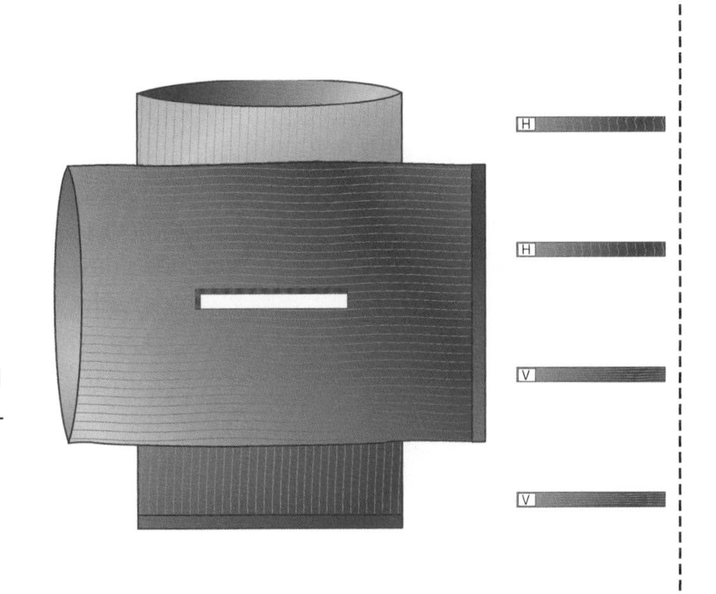

UNIT 5

An *Issue* to Analyze

A CASE STUDY

The Oldman River Dam

20 Years Ago

You are standing outside, overlooking a broad river valley like the one in the photograph. An old man and a ranch woman on horseback are with you.

You It sure is hot. The sun and breeze are just frying me. Looks as though they fried the river, too. It's just a trickle.

Ranch woman It's a lot bigger in the spring, when the snow is melting in the mountains. Water pours down these valleys, and when the three rivers join, it's a regular flood. Completely covers the lowlands. The mud and silt keep all those cottonwood and willow trees healthy. And they shelter the deer and birds — even some prairie falcon, and they're pretty rare. Fishing's good, too. This is one of the three best trout streams in Alberta. But everything will change if the government goes ahead with that dam. The water behind the dam will fill the whole valley.

Megaprojects, such as dams, affect both the natural and human environment. Often it is not clear whether the changes they cause bring more problems or more benefits.

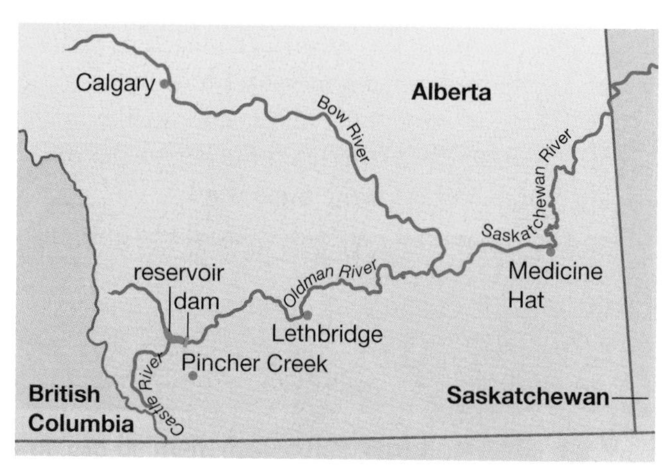

Southwestern Alberta near the Oldman River Dam

Old man My people, the Peigan, call the river "Napi," meaning "creator." Because he brings life to trees and wild creatures, Napi is the source of our life, too. This valley has been our summer hunting ground for thousands of years. There are sacred places and old burial grounds all along the river. One hundred years ago, we signed an agreement with the government, giving us rights to the land and water. There's been talk and study about building a dam since 1958. As long as they keep studying, the river is safe. And we would never agree to building anything that would destroy the valley.

Prepare and Present

With your group, discuss these questions. Prepare answers, with reasons, that you can share with the class.

1. Do you (or any people you know) have a special "wild place"? What is it like?
2. Should old laws, like the treaties with First Nations peoples, be able to stop modern developments?

The Oldman River Dam

Summer 1987

- 8 000 000 m^3 of earth and gravel
- 76 m high and 3000 m long
- $350 million to build
- reservoir holds enough water to cover Calgary to a depth of 30 cm
- built on the least costly site that would allow farmers to irrigate their land, prevent water shortages in the cities, and prevent flood danger

3 In your opinion, what kind of natural or historical feature would be important enough to prevent construction of a dam or another big project?

4 Is it always best to build a project for the cheapest price? Give reasons for your answer.

Thousands protest Oldman River Dam

June 1988

Last night, almost 10 000 protesters gathered at an outdoor concert above the Oldman River Valley to protest the building of the Oldman River Dam.

One organizer explained, "Maybe this will persuade the government not to finish the dam. Maybe the law courts can help. We're going to sue the Minister of Environment for destroying the fishing grounds along the river. That's illegal! If the Alberta courts don't agree, we'll go higher. The Federal Court of Appeal could order the government to stop construction."

A spokesperson for the Peigan nation said, "We are going to court, too. Some of the younger people are even talking about digging a new channel for the river. Then its water won't flow into the irrigation canals and the dam will be useless, even if it is finished. That would probably be against the law, but they're feeling pretty desperate."

5 Do you think rallies and marches are a good way to make people think about important issues, or are they a waste of time?

6 Do you think it is all right to break the law to change something that you believe is wrong? Think of an example to support your opinion.

Today

The dam was built, and is working as expected. The water supply to cities and to farmers is much more stable. A good warning system is in place for times when water must be released from the reservoir, putting low-lying areas at risk from flooding.

The reservoir has covered the valleys, and the trout are gone. There are other fish, but there is concern about the level of mercury in their flesh.

Today, such projects require much more extensive studies of their expected impact on wildlife, on people, and on communities before they are built.

Analyze

Pool your ideas on questions 2, 4, 5, and 6 with those of your classmates. For each question, hear two or three spokespeople with different ideas. Then vote on the questions. In your Science Log, record how you voted. If you changed your original opinion as a result of the discussion, record what made you change.

UNIT 5 PROJECT

Ready, Set ... Build!

These engineering students have designed and built a very unusual transportation device. Look closely at their racing toboggan. Can you tell that it is made of . . . concrete?

The racers are competing in the Great Northern Concrete Toboggan Race. They took a whole winter to create their entry. Like most design projects, it was divided into several stages.

INTERNET CONNECT

www.school.mcgrawhill.ca/resources/

The Great Northern Concrete Toboggan Race is hosted and often won by Canadian students. To find out more, go to the web site above. Go to the **Science Resources**, then to **SCIENCEPOWER 7** to know where to go next. If engineering students in your area enter the contest, you could contact them or their supervisor and find out details about their entry.

1. Identify criteria and specifications.

   According to the rules for the race, all entries must
   - be made out of concrete
   - have a safe, streamlined body shell
   - carry four riders
   - have mass less than 136 kg
   - be steerable
   - be able to make a controlled stop at the bottom of the race hill

   Prizes are given for speed, design, and team spirit.

2. Develop a work plan.

   The concrete toboggan race is held in February, but teams start working on their entries many months before. They plan when each task must be completed and who will be responsible for it, so the whole project will be finished on schedule.

3. Test a protoype and construct the toboggan.

   Models or prototypes of the body, steering, and braking systems are built, tested, and modified to improved the design. Then the racing toboggan itself is built and carefully finished.

4. Test and evaluate the toboggan.

   At the race site, the toboggans are inspected to make sure that they meet the specifications. They are judged for design. Finally, they are raced against each other.

5. Report results.

   Each team prepares a written summary of their design ideas and final results, along with ideas for improvement. The next year's competitors use this information to improve their entries.

Challenge

You can follow the same design steps used by the toboggan makers to build your own amazing transportation device — out of cardboard!

Design Criteria

A. As a class, decide on a type of transportation device to build. Decide whether it will be a model or large enough to carry a person.

B. Your teacher may have suggestions for cardboard boats and cars, or you may choose something original.

C. Make a list of specifications and criteria for judging, so the rules of the contest are clear.

Plan and Construct

1. In your group, develop a design that will meet the criteria. Use everything you learned in this unit.
2. Write a one-page product proposal which outlines the specifications your device will meet and contains a list of materials and a labelled sketch of your design.
3. Develop a work plan. Decide what each group member will do and when each task must be completed. Write your work plan in two forms: a time line and a list of tasks and deadlines for each group member. Leave space to record whether each task is completed on time by the proper person.
4. Construct and test a prototype device. Keep a project log as you work. Include information about problems that occur, how you decide to overcome them, and the changes you make to your original design as construction proceeds. On a separate page of your project log, keep a list of materials you use. Include the cost of any materials you buy.
5. Construct your transportation device.

Evaluate

1. Have your device inspected by your teacher to ensure that it meets all construction specifications.
2. Measure or calculate
   (a) dead load
   (b) live load
   (c) structural efficiency
3. Test your device against those built by classmates.
4. Record how your device placed in the competition. If any part had a structural failure, describe exactly how the part failed and how you could modify and improve it.
5. Use your product proposal, work plan, and your answers to 2 and 4 to prepare a final evaluation of your project. Use any format approved by your teacher.

Classifying Living Things

Over 2000 years ago, the Greek philosopher Aristotle developed a system of classification that grouped organisms according to whether they were plant or animal. Scientists used Aristotle's system for hundreds of years. As they discovered more and more living things, the system did not work well because it did not show probable relationships between similar organisms.

In 1735, Carolus Linnaeus produced a new system that also classified all organisms as plant or animal. This new system was, however, very different in other ways from Aristotle's system.

Linnaeus' system gives a two-word name to each type of organism. This system of naming

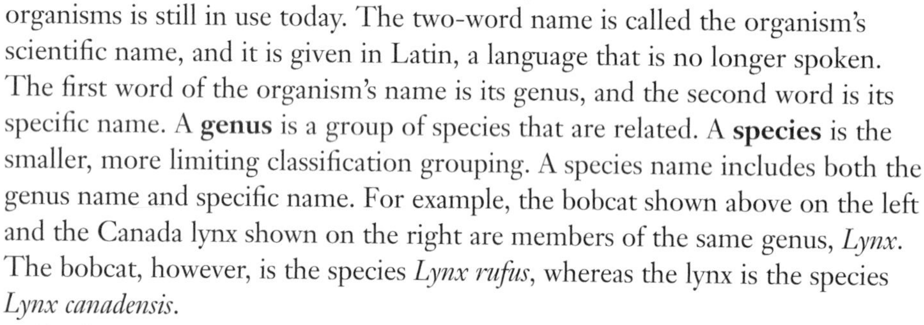

organisms is still in use today. The two-word name is called the organism's scientific name, and it is given in Latin, a language that is no longer spoken. The first word of the organism's name is its genus, and the second word is its specific name. A **genus** is a group of species that are related. A **species** is the smaller, more limiting classification grouping. A species name includes both the genus name and specific name. For example, the bobcat shown above on the left and the Canada lynx shown on the right are members of the same genus, *Lynx*. The bobcat, however, is the species *Lynx rufus*, whereas the lynx is the species *Lynx canadensis*.

By the 1900s, scientists had discovered a great diversity of organisms on Earth. Separating organisms into only two main groups or **kingdoms**, plant and animal, began to seem inadequate. For example, bacteria are just too different from either plants or animals to be grouped with either. Similarly, fungi such as bread mould, yeast, and the many kinds of mushrooms are very different from plants and animals. In 1969, Robert Whittaker proposed a system that classifies organisms into five different kingdoms. The illustrated table on the next page shows the major groups of organisms and their kingdoms.

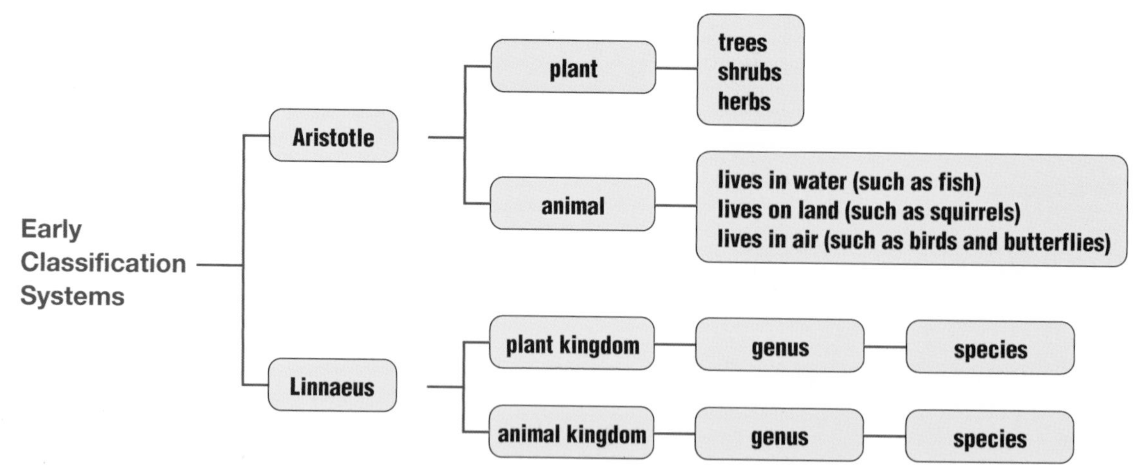

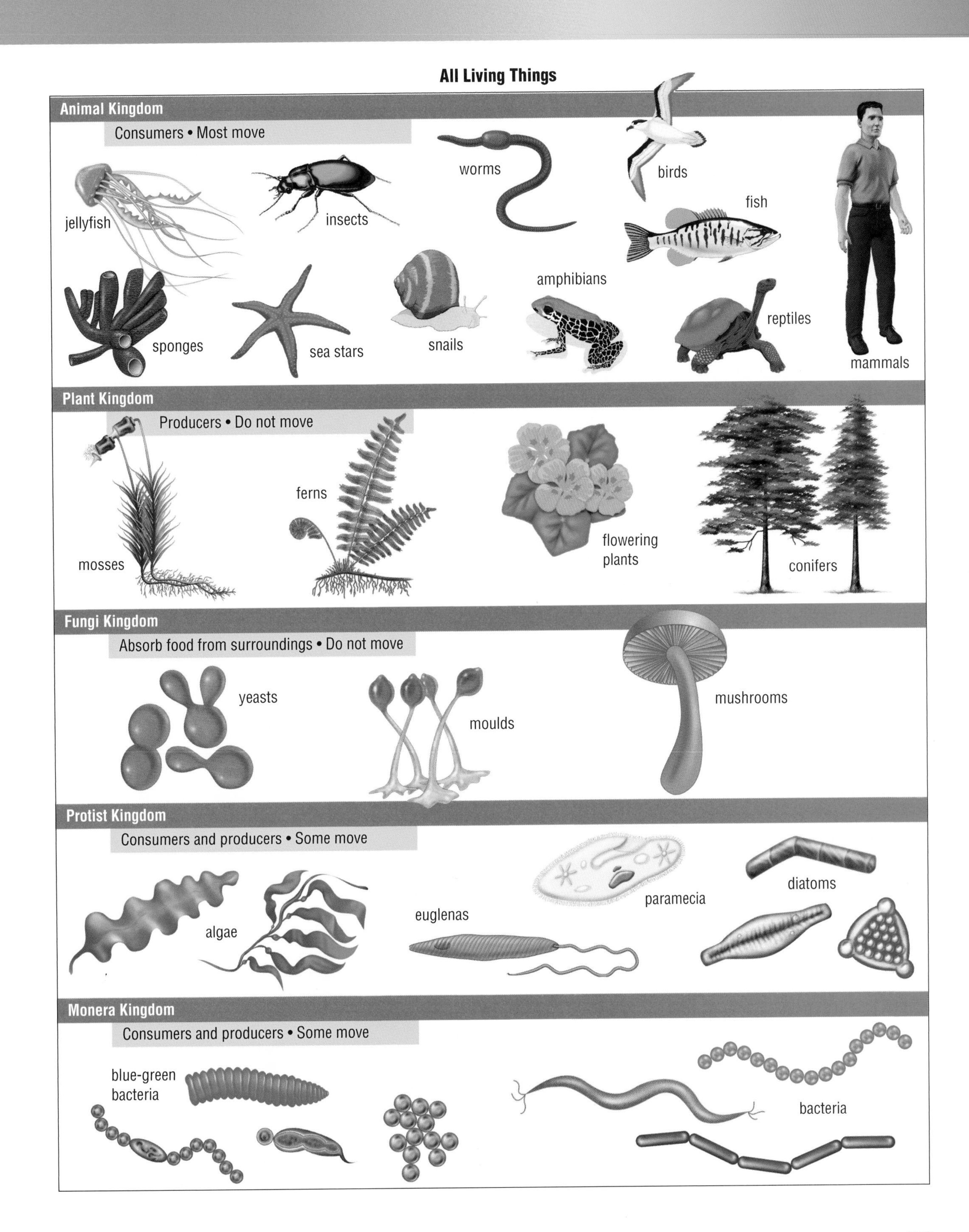
All Living Things
Animal Kingdom
Consumers • Most move
jellyfish
insects
worms
birds
fish
sponges
sea stars
snails
amphibians
reptiles
mammals
Plant Kingdom
Producers • Do not move
mosses
ferns
flowering plants
conifers
Fungi Kingdom
Absorb food from surroundings • Do not move
yeasts
moulds
mushrooms
Protist Kingdom
Consumers and producers • Some move
algae
euglenas
paramecia
diatoms
Monera Kingdom
Consumers and producers • Some move
blue-green bacteria
bacteria

APPENDIX

B

The Design Process

This book gives you an opportunity to design, plan, and construct your own devices, systems, and products. To help you develop and present your design concepts, you may find it useful to make two- or three-dimensional scale diagrams, a mock-up, or a prototype (model). In many projects, the construction of the end product is too costly or time-consuming. Diagrams and models, therefore, are simple, inexpensive ways of showing the final result of your ideas.

Start with Sketches

1. Outline objects when a two-dimensional view is sufficient, such as a pair of scissors.

2. As you are getting started, use printed material provided by your teacher to copy shapes you would like to use. Tracing is a great way to copy shapes from printed material.
3. Use plastic templates as guides for producing smooth curves.
4. If the end product is to be hand-held, cut out the full-size templates (patterns) to get the feel of it.

5. Apply drawing and communication skills throughout the project. Create graphics to accompany your design. You might want to produce an advertisement including graphics like the one shown. In fact, for some projects, the graphics may be your end product.

Design on Graph Paper

Use graph paper to produce your design. The grid will help you draw to scale. As well, use graph paper to produce clear flow diagrams.

6. Isometric grid paper is an excellent aid in making pictorial sketches. Use it to help plan your design.

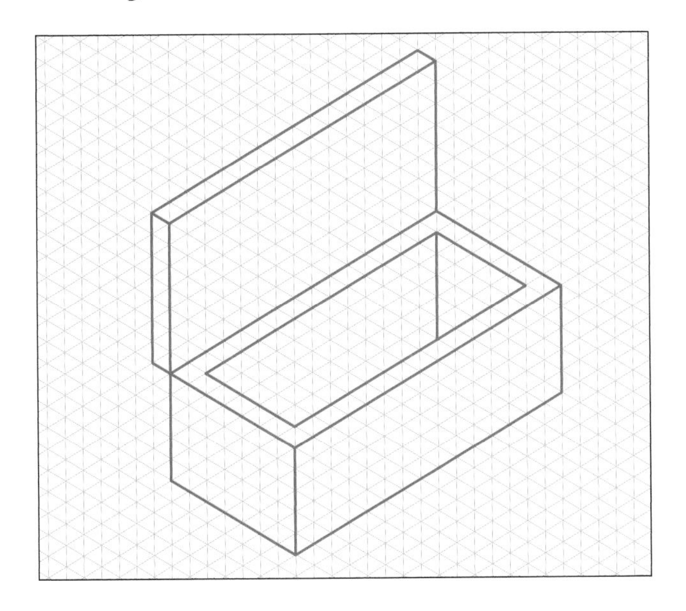

7. Sometimes showing exploded views of your planned project is useful. If you wish, you could draw these on graph paper for better accuracy.

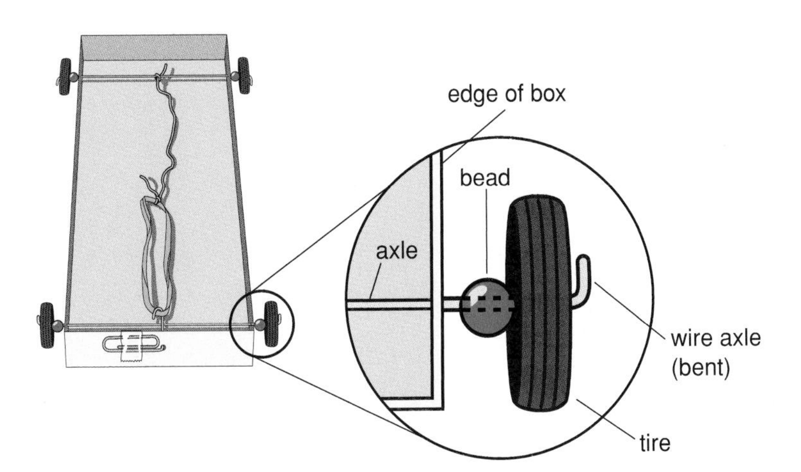

Plan, Construct, and Evaluate

8. Use a technological problem-solving approach like the one explained on pages IS-14 – IS-17 to carry out your design plan.

9. As you design, plan, and construct more projects, you will find that your sketching and drafting skills will improve. You can even use computer programs to develop parts of your projects if you have access to the necessary equipment and software.

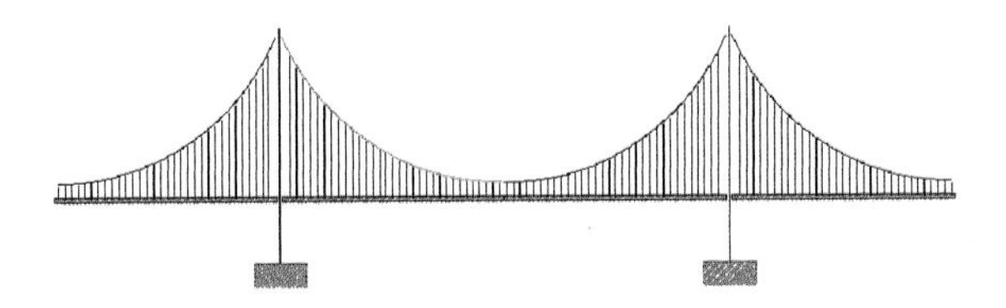

10. As you design, plan, and construct each project, be sure to continue thinking of how you can solve any problems you encounter in its development. The evaluation process at the end may take you in new directions, or you may decide to make only minor changes to your design.

Science and Technology Skills Guide

USING YOUR TEXTBOOK AS A STUDY TOOL

SCIENCEPOWER™ 7 contains a lot of useful information. It can help you to add to what you already know and to identify areas of inquiry that you might like to learn more about. How can you read your textbook effectively in order to help you accomplish these goals? This *SkillPower* will give you some ideas that will help you better remember what you read.

Organizing the Information in Your Textbook

Read all of the suggestions presented here. Use the learning methods that work for you, but try others as well. Doing something a different way can help you see ideas more clearly and better understand them.

1. When you are starting a new unit, read the *Chapter Title*, the *Key Ideas*, and the *Key Skills* in each chapter. They will help you to focus on what each chapter presents. Think about how the ideas fit into the "big picture" or main theme of the chapter. Try to predict some ideas you might learn about in each chapter. Write some of your own questions about the chapter topic.
2. Try rewriting the chapter section headings and subheadings as questions. Then look for the answer to the question in the section.
3. Think about what you are reading, and write brief notes to help you remember the information in each paragraph.
4. Look at the sample page shown on the right. It provides an example of how you can arrange your own notes as you work through the text. Look at all of the suggestions presented here. Use the learning strategies that work for you, but try others as well.

Unit: Earth's Crust
Unit Introduction: Main Idea
Forces and processes that change and shape Earth's crust, and form minerals, fuels, and soils that we use.
Chapter Titles:
10 Minerals, Rocks, and Soils
11 Earthquakes, Volcanoes, and Mountains
12 The Story of Earth's Crust
Chapter 10: What do I expect to learn?
How minerals and rocks form.
Chapter Introduction: Main Idea
What processes cause changes in Earth's crust, and how are the results of these processes used?

	Main idea	Other ideas
Section 10.1	Most minerals are rare	- Quartz and mica are common. - A mineral can be an element or a compound - Minerals have different properties that help identify them.
Section 10.2	A substance's "scratchability" Can be used to identify it.	- Friedrich Mohs developed the Mohs hardness scale. It can be used to compare objects according to hardness value. - Other properties, such as crystal structure, also help identify minerals.
Section 10.3	Lustre is a way of identifying minerals.	- Minerals can be shiny or dull.

Also look at any terms that are in bold (dark, heavy) type. You will find it helpful to pay special attention to these terms. They provide important definitions that you will need in order to understand and write about the information in the chapter. Make sure that you understand these terms and how they are used. Each boldfaced term appears in the *Glossary* at the back of this book.

Using Your Textbook Visuals

As you read, look at any photographs, illustrations, or graphs that appear on the page. Think about the information each visual provides, and note how it helps you to understand the ideas presented in the text. Visuals often help clarify or provide examples of these ideas. For example, look closely at the illustration on the right. What information does it give you?

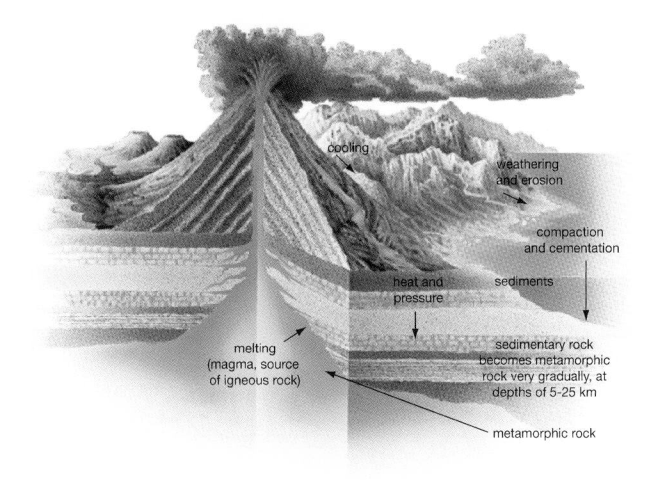

The rock cycle

Making Sure You Understand

At the end of every section and every chapter, you will find review questions. These questions are designed to help you reinforce your learning of the main ideas covered in each section and chapter. If you are unable to answer them, reread the text to find the answers.

If you feel that you do understand the chapter well, try using the questions provided at the end of the chapter under *Prepare Your Own Summary* to summarize the key ideas in the chapter.

Instant Practice

1. Go to the unit your teacher has told you that you will be studying, and try method number 1, on page 473.
2. In the first chapter of the unit, try method number 2.
3. Look for boldfaced terms in the first section of the first chapter of the unit. Record the terms and their meanings.

Graphic Organizers

A good way to organize the information you are learning is to use a **graphic organizer**. One kind of graphic organizer that you will find useful is a **concept map**. A concept map is a diagram that represents visually how ideas and terms are related. It can help you to clarify the meanings of the ideas and terms, and better understand what you are studying.

The following concept map is known as a **network tree**. Notice how some words are enclosed while others are written on connecting lines. The enclosed words are the main ideas or concepts. The lines in the map link related concepts, and the words that are written on the lines describe relationships between the concepts.

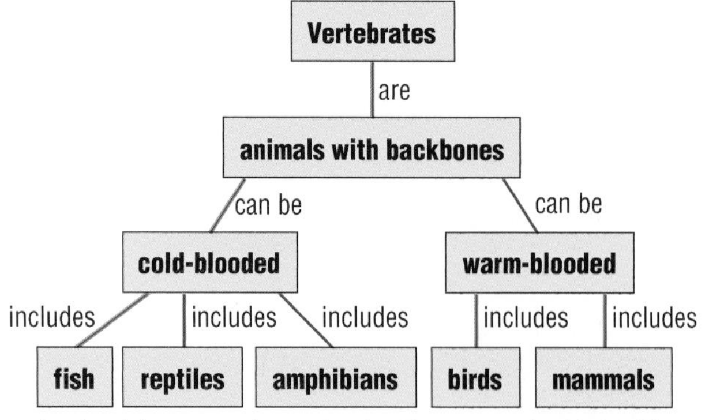

A network tree concept map showing the groups of animals with backbones.

As you learn more about a topic, your concept map will grow and change. Concept maps are just another tool for you to use. There is no single "correct" concept map. Your map is a record of your thinking and it shows the connections that make sense to you. Make your map as neat and clear as possible, and be sure that you have good reasons for suggesting the connections between concepts.

Although your map may contain many of the same concepts as other students' maps, the concepts on your map may be recorded and linked differently. You can use your map for study and review. You can refer to it when you want to recall concepts and relationships. At a later date, you can use your map to see what you have learned and how your ideas have changed.

An **events chain concept map** describes ideas in order. In science, an events chain can be used to describe a sequence of events, the steps in a procedure, or the stages of a process. When making an events chain, you first need to find the one event that starts the chain. This event is called the initiating event. Then you find the next event in the chain and continue until you reach an outcome or final result. Here is an events chain concept map that shows the events you might go through in a typical morning.

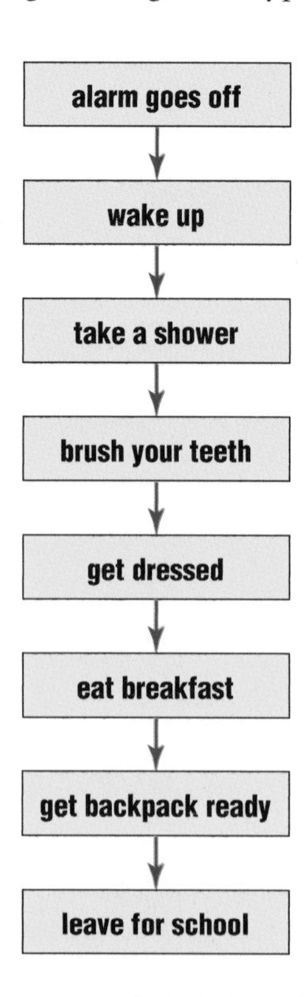

An events chain map of a typical morning routine.

A **cycle concept map** is a special type of events chain concept map. In a cycle concept map, the series of events does not produce a final outcome. A cycle concept map has no beginning and no end. To construct a cycle map, you first decide on a starting point, and then you list each important event in order. Since there is no outcome and the last event relates back to the first event, the cycle repeats itself. Look at the cycle concept map showing the stages of life that a frog goes through.

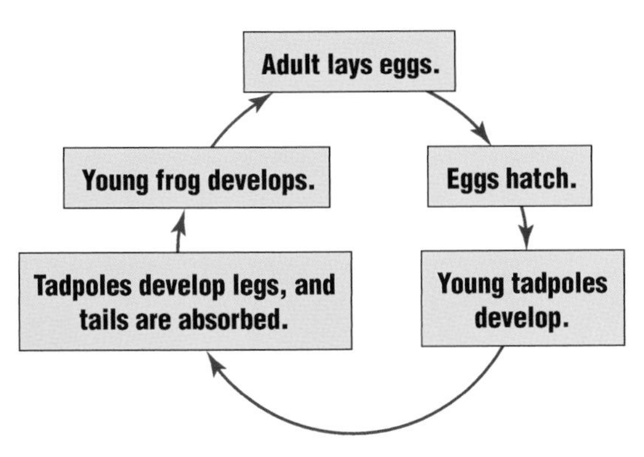

A cycle concept map showing the life history of a frog.

A **spider map** is a concept map that can be useful for brainstorming. You may have a central idea and several associated concepts, but they may not be directly related to each other. By placing these associated ideas outside the main concept, you can begin to group these ideas so that their relationships become easier to understand. Examine the following spider map showing what happens when light strikes an object.

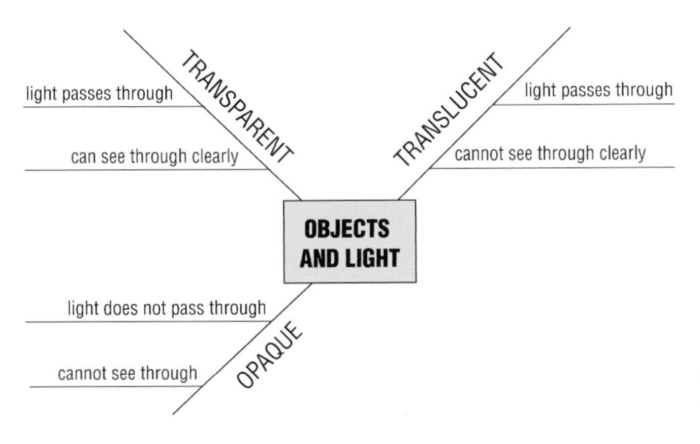

A spider concept map showing what happens when light strikes three different objects.

Comparing and contrasting is another way to help solidify your learning. When you compare, you look for similarities between two things. When you contrast, you look for differences. Thus you can compare and contrast by listing ways in which two things are similar and ways in which they are different. You can also use a graphic organizer called a **Venn diagram**, as shown below.

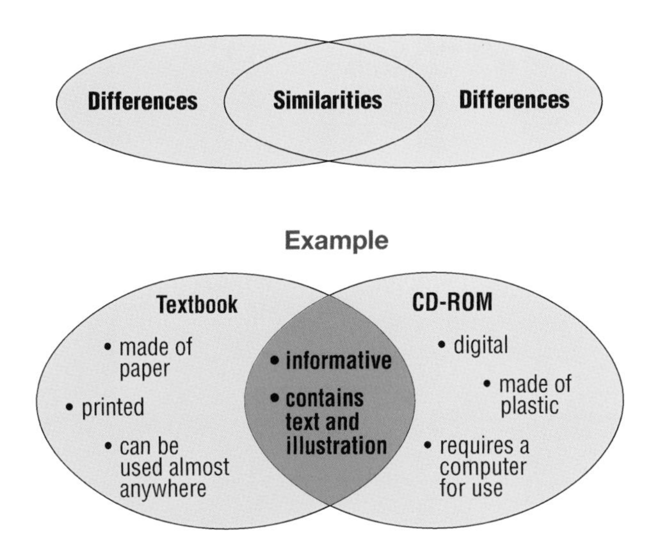

A Venn diagram is a useful tool to help you show similarities and differences.

Instant Practice

1. The following words are out of order: hockey, team sports, ice, diamond, field, bat, puck, hardball, cleats, ice skates, baseball, cleats, soccer ball, stick, soccer, feet. Use these words to produce a network tree concept map.
2. Produce an events chain concept map that starts with lunch and ends with your return home from school.
3. Produce a cycle concept map using the following words: summer, winter, fall, spring.
4. Make a Venn diagram to compare and contrast a CD and a cassette tape.
5. In a group, create a spider map based on one of the following topics:
   (a) food and nutrition
   (b) music
   (c) scientific discoveries
   (d) communication

HOW TO USE A SCIENCE LOG

Scientists keep logs — detailed records — of their observations, new data, and new ideas. You can keep a *Science Log* (or *Science Journal*) to help you organize your thinking.

In your *Science Log*, you can record what you already know about a topic and add to this information as you learn more. Your teacher might ask you to keep a *Science Log* as a special booklet or as a marked-off section of your science notebook. Whichever approach your teacher takes, you will find that writing about new ideas as you learn them will help you understand them better.

It is useful to consider what you already know about a topic. You may be surprised to find out that you know more about a topic than you realized. On the other hand, you may realize that a topic needs special attention because you do not know very much about it. The value of a *Science Log* is that you find out for yourself how clear (or unclear) your understanding is. You do not have to wait until your teacher assesses your understanding through a formal test or examination.

SCIENCEPOWER™ 7 makes sure that you can add to your *Science Log* effectively. For example, each chapter begins with a set of *Getting Ready* questions.

Getting Ready...

- What do a diver and a carbonated liquid have in common?
- Why do some materials dissolve quickly and some dissolve slowly?
- How is gold mined?

Can you answer any of these questions from your previous studies? You can write your answers, draw a sketch, or use whatever ways you find best to explain what you know. Also, feel free to record that you know very little about a particular topic. This will help you focus on some of the things you need to learn as you read through the chapter.

Pause& Reflect

Imagine that you are on a camping trip and you find a cave that looks as though it has recently opened up because it has only a crevice for an opening. Small enough to squeeze inside, you are immediately bedazzled by a wall of shimmering gemstones. In your Science Log, make up a myth to explain something about this special wall. Include a description of the gems in the cave, using the terms "lustre," "hardness," "colour," and "streak."

Throughout each chapter, *Pause & Reflect* features keep you thinking about what you have learned. These features are designed to help you make connections between ideas and organize your thoughts. Your teacher may ask you to use all these features regularly to record your new knowledge. On the other hand, your teacher may leave it up to you to decide how often you want to use them.

The final item in each Chapter Review is a *Pause & Reflect* feature that asks you to look at your original answers to the *Getting Ready* questions at the beginning of the chapter, and to write new answers to these questions. You may be amazed at how much your answers have changed, based on what you have learned by studying the chapter.

Here are some entries that you might want to include in your *Science Log*:

- questions that you would like to be able to answer
- sketches and notes about models and processes in science
- graphic organizers (see SkillPower 1 for a few examples)

- your thoughts on what is difficult for you, and your ideas on what you can do to overcome any barriers you have when learning a new topic
- notes about interesting items in the news that involve a chapter topic and that spark more questions or new answers to existing questions
- profiles of leading Canadian scientists or technologists that you learn about in the media
- profiles of careers related to science and technology that you find interesting
- connections among science and other subject areas that occur to you in the course of your learning

Your *Science Log* will help you become a better learner, so take the time to make entries on a regular basis.

Instant Practice

1. A balloon filled with helium will rise in air, but a balloon that you blow up yourself will sink. Scientific investigation can help answer why one balloon rises and the other sinks. Think about some of the questions you would need to ask to help you find out about the way the balloons behave. Record your questions and then exchange your set of questions with a classmate so that each of you can start your own *Science Log* using these questions.

2. Find and read a newspaper article that deals with technology. Write a short summary of the article in your *Science Log*. After you have done this, record why the subject of the article makes you think the article is about technology. Is science involved, as well? If so, record how it is connected to technology.

3. What area of science or technology are you most interested in? Record why you might like to pursue a career in this scientific or technological field. Write down three issues or problems facing society that you could help deal with by following such a career. For each issue, record how your chosen field can contribute to a solution or improvement.

Skill
POWER 3

WORKING IN GROUPS

Teamwork is usually the most effective and efficient way to accomplish a task. You have probably been part of a team yourself, perhaps in a sport, an orchestra, or a community project. In this course, you will often be asked to work in a group to complete a task. In a co-operative group, each member has one or more assigned tasks to carry out. This enables the group to make the best use of each member's special skills and to develop new skills in all the members. Groups often develop a special ability to work together, so each presentation or project that is completed together is an improvement over the last.

Working well in a group takes effort and practice. The best way to develop your teamwork skills is to consider carefully what makes a group succeed at a task.

Instant Practice

1. Think about your favourite team sport. Do all the teammates have the same role? If not, how are the roles different, and how does each teammate help the other? How can a team make the best use of a member's special skills?

2. Think about what working in a group means. Write down some of the skills you think are important in order to be a successful team member. Beside each skill, write why you need it.

3. After a few minutes, your teacher will ask you to move into a group. Discuss the skills you wrote down, and find out what others in the group wrote. Develop a list of six important skills that you all agree are necessary for teamwork, along with the reasons why they are necessary.

4. On your own, did you come up with the same teamwork skills as your group did? How did your group resolve any difficulties in agreeing on the six most important skills? Did all of you participate in the same way, or did some group members find it easier to participate than others?

5. Look at the list below, and think about the group in which you have just been working.
   (a) How would you rate your performance on a scale of 1 to 4, with 4 being excellent?
   (b) How would you rate the performance of the others in your group, using the same scale?
   (c) How could you improve your performance?
   (d) How might the others in your group improve their performance?
   (e) How could you help another group member improve his or her performance?
   (f) What kind of help could you use in learning to work better in a group?

Assessing Group Performance

These are some behaviours to aim for when doing group work in this course and other courses.

I share my ideas with others in the group.

I show others respect, even if I disagree with them.

I listen to another group member when she or he is speaking and I do not interrupt.

I encourage others to speak.

I stay on the group's task.

I help the group stay on task.

I do not allow myself to be distracted.

I keep my voice low enough that it will not distract other groups.

I allow others to present their ideas, even when I think I know the answers.

UNITS OF MEASUREMENT

Throughout history, groups of people have developed their own systems of numbering and units. When different groups of people began to communicate with each other, they discovered that their units of measurement were not the same, and this caused confusion. For example, imagine trying to report your height in ells and your weight in scruples, or going out to "buy a hogshead of strawberries!" Even the "foot" as a measurement took a long time to standardize.

For consistency, scientists throughout the world have agreed to use the same system of measurement, the metric system of numbers and units. The metric system is also the official system of measurement in Canada.

The Metric System

The word "metric" comes from the Greek word, *metron*, which means "measure." The **metric system** is based on multiples of ten. For example, the basic unit of length is the metre. All larger units of length are expressed in units based on metres multiplied by 10, 100, 1000, or more. Smaller units of length are expressed in units based on metres divided by 10, 100, 1000, or more. Each multiple of ten has its own prefix (a word joined to the beginning of another word). For example, *kilo-* means multiplied by 1000. Thus, one kilometre is one thousand metres.

$$1\text{ km} = 1000\text{ m}$$

The prefix *centi-* means divided by 100. Thus, one centimetre is one one-hundredth of a metre.

$$1\text{ cm} = \frac{1}{100}\text{ m}$$

The prefix *milli-* means divided by one thousand. Thus, one millimetre is one one-thousandth of a metre.

$$1\text{ mm} = \frac{1}{1000}\text{ m}$$

In the metric system, the same prefixes are used for nearly all types of measure, such as mass, weight, area, and energy. The following table lists the most commonly used metric prefixes.

Commonly Used Metric Prefixes

Prefixes	Symbol	Relationship to the base unit
giga-	G	1 000 000 000
mega-	M	1 000 000
kilo-	k	1 000
hecto-	h	100
deca-	da	10
–	–	1
deci-	d	0.1
centi-	c	0.01
milli-	m	0.001
micro-	μ	0.000 001
nano-	n	0.000 000 001

(**Note:** Time does not have a metric form of measure. Time is still measured in seconds, minutes, and hours. There are 60 s in 1 min, 60 min in 1 h, and 24 h in 1 d.)

Example 1

The length of Canada's longest river, the Mackenzie River, is 4241 km. How many metres is this distance?

Solution

$$4241\text{ km} = ?\text{ m}$$
$$1\text{ km} = 1000\text{ m}$$
$$4241 \times \frac{1000\text{ m}}{1} = 4\ 241\ 000\text{ m}$$

Example 2

There are 250 g of rice in a package. Express this mass in kilograms.

Solution

$$1\text{ kg} = 1000\text{ g}$$
$$250 \times \frac{1\text{ kg}}{1000} = 0.250\text{ kg}$$

You are probably most familiar with the units for length and mass. As you continue in your science courses, you will learn about other quantities of measurement. The following table lists most of the frequently used metric quantities that you will encounter in this course.

Frequently Used Metric Quantities, Units, and Symbols

Quantity	Unit	Symbol
length	nanometre micrometre millimetre centimetre metre kilometre	nm μm mm cm m km
mass	gram kilogram tonne	g kg t
area	square centimetre square metre hectare	cm^2 m^2 ha
volume	cubic centimetre cubic metre millilitre litre	cm^3 m^3 mL L
time	second	s
temperature	degree Celsius	°C
force	newton	N
energy	joule kilojoule	J kJ

Instant Practice

1. A can contains 0.355 L of pop. How many millilitres does the can contain?
2. The height of a table is 0.75 m. How high is the table in centimetres?
3. A package of chocolate-chip cookies has a mass of 396 g. What is the mass of the cookies in milligrams?
4. One cup of water contains 250 mL. What is the volume of one cup of water in litres?
5. The distance from your kitchen to the front door is 6000 mm. How far is the front door in metres?
6. A student added 0.0025 L of lemon juice to water. How much lemon juice did the student add in millilitres?

SI Units

In science classes, you will often be asked to report your measurements in **SI** units. The term SI is taken from the French name *Le Système international d'unités*. In SI, the base unit of mass is the kilogram, the base unit of length is the metre, and the base unit of time is the second.

Example 1

Convert 426 cm to the SI base unit.

Solution

The SI base unit of length is the metre.

$$1 \text{ m} = 100 \text{ cm}$$

$$426 \times \frac{1 \text{ m}}{100} = 0.426 \text{ m}$$

Example 2

Convert 1.7 h to the SI base unit.

Solution

The SI base unit of time is the second.

1 min = 60 s

1 h = 60 min

Therefore, 1.7 x 60 x 60 s = 6120 s

Instant Practice

Convert the following quantities to the SI base unit.

1. 7.02 g
2. 32 min
3. 8.13 km
4. 25 961 mm
5. 223 625 cm
6. 3.25 h

ESTIMATING AND MEASURING

Estimating

How long is your pen or pencil? How much time does it take you to return home from school? What is the width of your desk? You could probably answer all of these questions fairly quickly by estimating — making an informed judgment about a measurement. The estimate gives you an idea of the measure, but it is not totally accurate. Scientists also make estimates of measurements when exact numbers are not essential.

It is useful to be able to estimate as accurately as possible. For example, suppose that you want to know how many words are in your textbook. Counting every word on every page would be very time-consuming and unnecessary. Instead, you can count the number of words on one page of your textbook. Then you multiply this number by the total number of pages to find an estimate of the total number of words in the book.

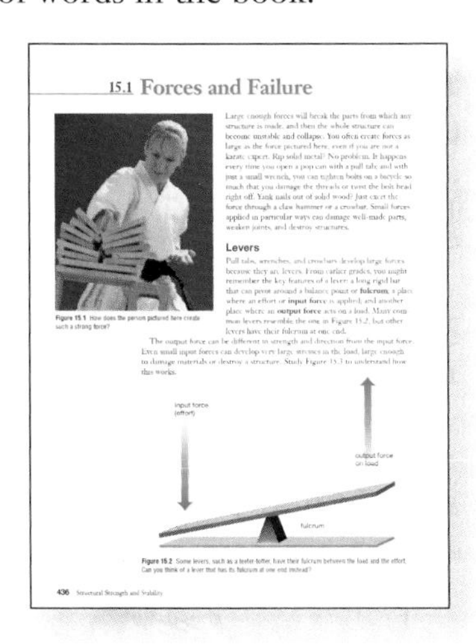

15.1 Forces and Failure

Large enough forces will break the parts from which any structure is made, and then the whole structure can become unstable and collapse. You often create forces as large as the force pictured here, even if you are not a karate expert. Rip solid metal? No problem. It happens every time you open a pop can with a pull tab and with just a small wrench, you can tighten bolts on a bicycle so much that you damage the threads or twist the bolt head right off. Yank nails out of solid wood? Just exert the force through a claw hammer or a crowbar. Small forces applied in particular ways can damage well-made parts, weaken joints, and destroy structures.

Figure 15.1 How does the person pictured here create such a strong force?

Levers

Pull tabs, wrenches, and crowbars develop large forces because they are levers. From earlier grades, you might remember the key features of a lever: a long rigid bar that can pivot around a balance point or **fulcrum**; a place where an effort or **input force** is applied; and another place where an **output force** acts on a load. Many common levers resemble the one in Figure 15.2, but other levers have their fulcrum at one end.

The output force can be different in strength and direction from the input force. Even small input forces can develop very large stresses in the load, large enough to damage materials or destroy a structure. Study Figure 15.3 to understand how this works.

Figure 15.2 Some levers, such as a teeter-totter, have their fulcrum between the load and the effort. Can you think of a lever that has its fulcrum at one end instead?

436 Structural Strength and Stability

Instant Practice

1. An African elephant has a mass of about 7500 kg. Students in your class have an average mass of about 45 kg. Estimate the number of students it will take to equal the elephant in mass. Calculate the exact number to see how close your estimate was.

2. The greatest distance across Canada is 5514 km. A jet airliner flies about 600 km per hour. Estimate how long it takes for the jet to fly across Canada. Calculate the time to see how close your estimate was.

3. A 1000 mL (1 L) jar is filled with candies. How can you make a good estimate of the number of candies in the jar?
   - **(a)** Decide how you can use a 100 mL container and a small amount of candies to estimate the number of candies in the larger jar.
   - **(b)** Carry out your plan.
   - **(c)** Compare your results with those of two or three classmates.
   - **(d)** About how many candies will a 0.5 L container hold? About how many will a 2 L container hold?

Measuring Length

You can use a metre stick or ruler to measure short distances. Metre sticks and rulers are usually marked off in millimetres and/or centimetres. You place the zero mark of the metre stick or ruler at one end of the distance to be measured and read the length at the other end.

Instant Practice

Use a ruler to measure the distance between the following points: A and D; C and E; B and F.

A •   • B

• C   • E   • D

• F

Measuring Area

Area refers to the amount of surface of an object. For example, you may want to know the area of a piece of paper, a rock, or even a shirt. Area is reported in square units, such as cm^2. When you want to calculate an area, you can use length measurements. For a square or a rectangle, you find the area by multiplying the length by the width.

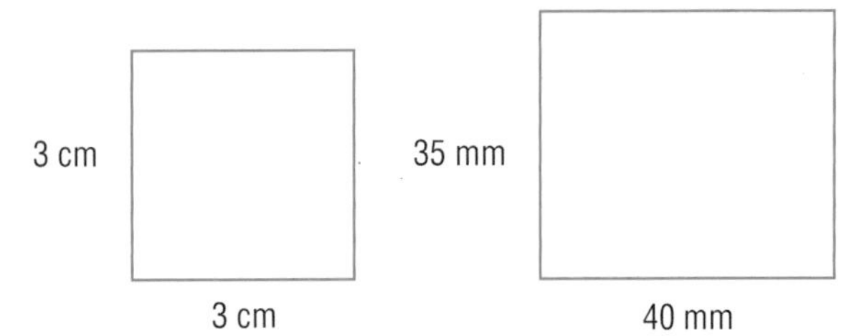

Area of square is 3 cm × 3 cm = 9 cm^2.
Area of rectangle is 35 mm × 40 mm = 1400 mm^2.

Instant Practice

Imagine that you are in charge of an art project that will transform one wall of your classroom into a large mural to show a diversity of materials. You may use as many different kinds of materials as you like, but each piece must be a 30 cm × 30 cm square. How many squares will you need to make your mural?

1. First decide what unit of measurement will be most practical for the area of the mural. Why would you not choose to measure the area in mm^2?
2. Measure the height and the width of the wall you will cover.
3. Calculate the area of the wall in the measurement you have selected.
4. Calculate how many 30 cm × 30 cm squares you will need to fill 1 m^2.
5. Multiply this number by the area of the wall in m^2.

Make sure that you always use the same units — if you mix centimetres and metres, your calculations will be wrong. Remember to ask yourself if your answer is reasonable. (You could make an estimate to check it).

Measuring Volume

The **volume** of an object is the amount of space that the object occupies. Why would you want to know volume? Think of an example, from your own experience, when knowing the volume of an object is useful. Did cooking or baking come to mind?

There are several ways of measuring volume, depending on the kind of object you want to measure. A cubic metre is the space occupied by a cube 1 m × 1 m × 1 m. This unit of volume is used to measure large quantities, such as the volume of concrete in a building. In this course, you are more likely to use cubic centimetres (cm^3) or cubic millimetres (mm^3) to record the volume of a solid object. You can calculate the volume of a cube by multiplying the length of its sides. For example:

$$\text{Volume} = 1\text{ cm} \times 1\text{ cm} \times 1\text{ cm} = 1\text{ cm}^3$$

You can calculate the volume of a rectangular solid if you know its length, width, and height.

$$\text{Volume} = \text{length} \times \text{width} \times \text{height}$$

The units that are used to measure the volume of a solid are called **cubic units**. If all the sides are measured in millimetres (mm), the volume will be in cubic millimetres (mm^3). If all the sides are measured in centimetres (cm), the volume will be in cubic centimetres (cm^3).

The units that are used to measure the volume of a liquid are called **capacity units**. The basic unit of volume for liquids is the litre (L). In this course, you also measure volume using millilitres (mL). Recall that 1 L = 1000 mL. You have probably seen the capacity of juice, milk, and soft drink containers given in litres and millilitres.

Cubic units and capacity units are interchangeable. For example:
$1\ cm^3 = 1\ mL$
$1\ dm^3 = 1\ L$
$1\ m^3 = 1\ kL$

As you can see in the diagrams below, the volume of a regularly shaped solid object can be measured directly.

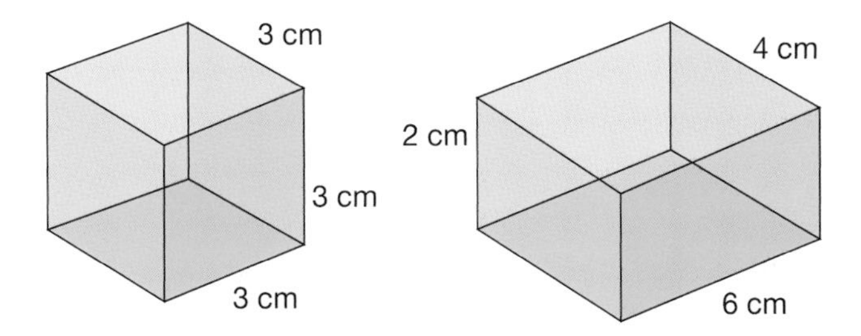

Measuring the volume of regularly shaped solids

The volume of a liquid can also be measured directly with a graduated cylinder, as shown below. Make sure that you measure to the bottom of the **meniscus**: the slight curve where the liquid touches the sides of the container. To measure accurately, make sure that your eye is at the same level as the bottom of the meniscus.

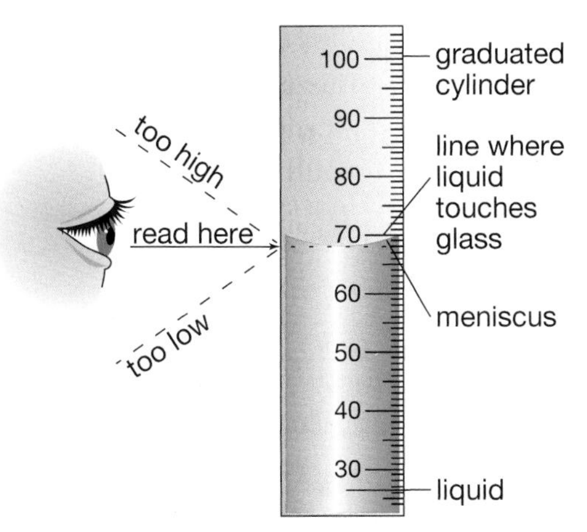

Take a reading from the bottom of the meniscus with your line of sight at the level of the meniscus.

The volume of an irregularly shaped solid o must be measured indirectly. You need to find out volume of liquid that the object will displace. As shown here, the volume of liquid that is displaced is equal to the volume of the object.

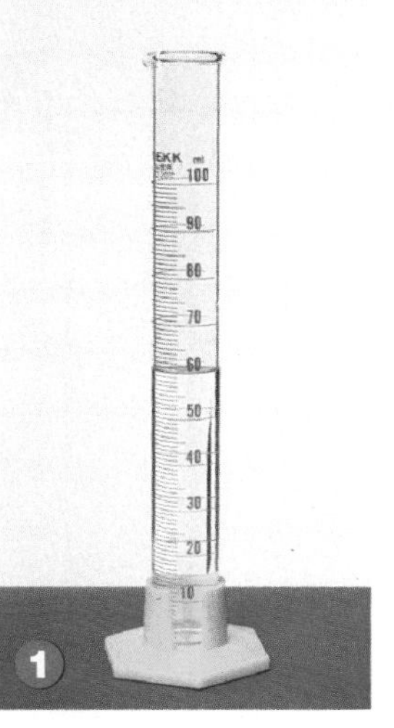

Record the volume of the liquid.

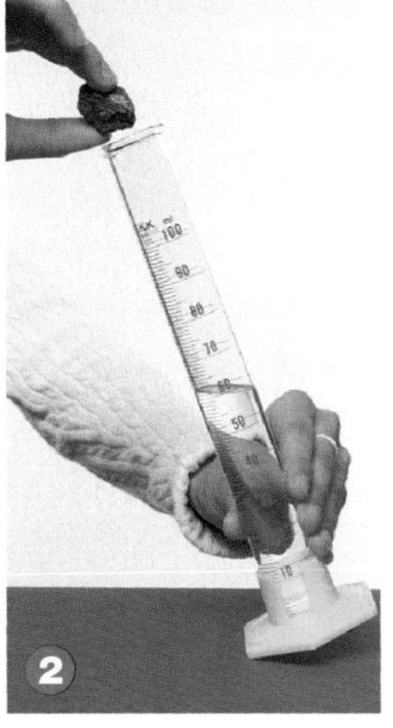

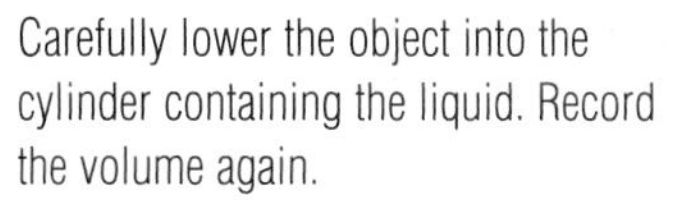

Carefully lower the object into the cylinder containing the liquid. Record the volume again.

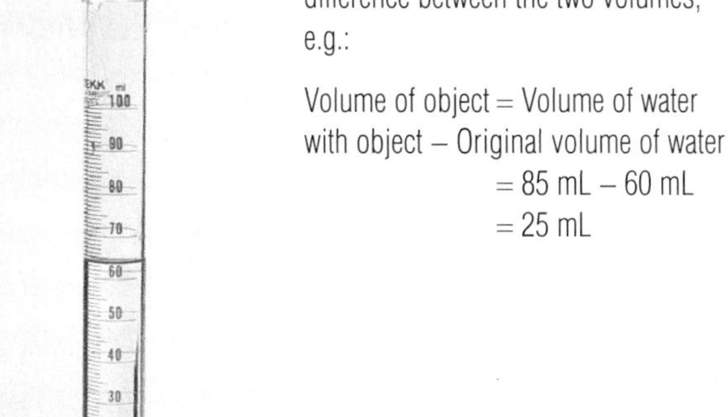

The volume of the object is equal to the difference between the two volumes, e.g.:

Volume of object = Volume of water with object – Original volume of water
= 85 mL – 60 mL
= 25 mL

Measuring the volume of an irregularly shaped solid

Instant Practice

1. Write an explanation of how to find the volume of an irregularly shaped solid.

2. Read the volume indicated by each graduated cylinder below.

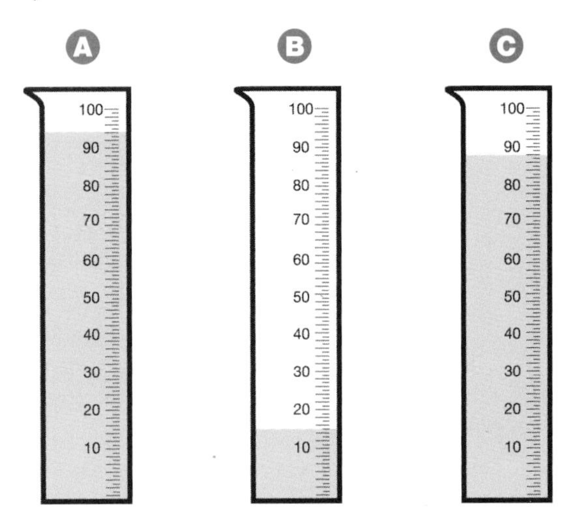

3. Determine the volume of a series of objects using the method of water displacement. Try to use objects that are different in shape and made of different materials, such as a pebble, a marble, and a coin.

Measuring Mass

Is your backpack heavier than your friend's backpack? This can be difficult to check by holding a backpack in each hand. You need a way to measure mass accurately. The **mass** of an object is the measure of the amount of material that makes up the object. Mass is measured in milligrams, grams, kilograms, and tonnes.

A balance is used for measuring mass. A triple beam balance, such as the one shown here, is commonly used. To measure the mass of a solid object, follow these steps:

1. Set the balance to zero. Do this by sliding all three riders back to their zero points. Using the adjusting screw, make sure that the pointer swings an equal amount above and below the zero point at the far end of the balance.

2. Place the object on the pan. Observe what happens to the pointer.

3. Slide the largest rider along until the pointer is just below zero. Then move the rider back one notch.

4. Repeat with the middle rider and then with the smallest rider. Adjust the smallest rider until the pointer swings equally above and below zero again.

5. Add the readings on the three scales to find the mass.

What if you add a quantity of a substance, such as sugar, to a beaker? How would you know the mass of the substance? First you find out the mass of the beaker, as shown. Then you pour the sugar into the beaker and measure the mass of the beaker and sugar together. Subtract the mass of the beaker from the mass of the beaker and sugar together. This gives you the mass of the sugar.

Mass of sugar = Mass of sugar and beaker together – Mass of beaker

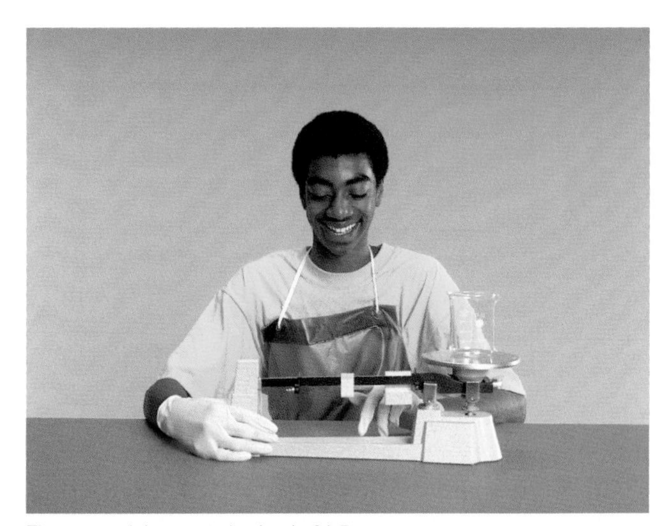

The mass of the empty beaker is 61.5 g.

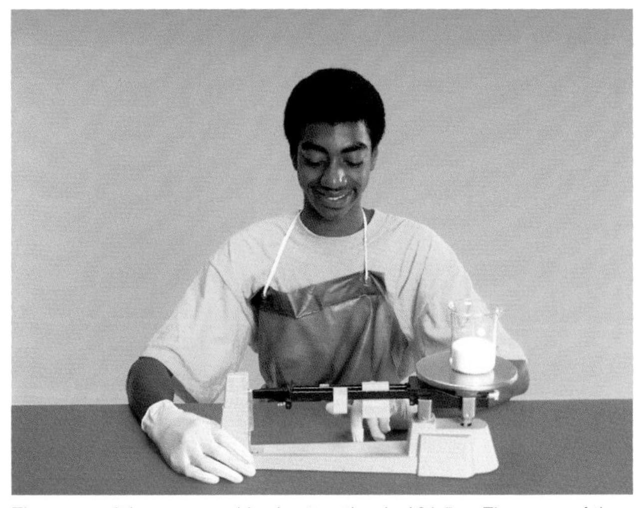

The mass of the sugar and beaker together is 161.5 g. The mass of the sugar equals the mass of the sugar and beaker together minus the mass of the beaker (161.5 g – 61.5 g = 100 g).

Instant Practice

1. Which takes more "muscle" to carry, your favourite paperback book or a calculator? Find out by using a balance to compare their masses.

2. Write the steps you would take to find the mass of the contents of a glass of pop.

Measuring Temperature

"Temperature" is a measure of the thermal energy of the particles of a substance. In the very simplest terms, you can think of temperature as a measure of how hot or how cold something is. The temperature of a material is measured with a thermometer.

For most scientific work, temperature is measured on the Celsius scale. On this scale, the freezing point of water is zero degrees (0°C), and the boiling point of water is one hundred degrees (100°C). Between these points, the scale is divided into 100 equal divisions. Each division represents 1 degree Celsius. On the Celsius scale, average human body temperature is 37°C, and a typical room temperature may be between 20°C and 25°C.

The SI unit of temperature is the Kelvin (K). Zero on the Kelvin scale (0 K) is the coldest possible temperature. This temperature is also known as absolute zero. It is equivalent to -273°C, which is 273 degrees below the freezing point of water. Notice that degree symbols are not used with the Kelvin scale.

Most laboratory thermometers are marked only with the Celsius scale. Because the divisions on the two scales are the same size, the Kelvin temperature can be found by adding 273 to the Celsius reading. Thus, on the Kelvin scale, water freezes at 273 K and boils at 373 K.

Tips for Using a Thermometer

When using a thermometer to measure the temperature of a substance, here are three important tips to remember:

- Handle the thermometer extremely carefully. It is made of glass and can break easily.
- Do not use the thermometer as a stirring rod.
- Do not let the bulb of the thermometer touch the walls of the container.

Celsius thermometer

Instant Practice

Your teacher will supply your class with three large containers of water, each at a different temperature.

1. Twelve students will each be provided with a thermometer. When your teacher says "now," the students will take temperature readings of the water in the different containers. Four students will be asked to take a temperature reading of the water in one container. Four others will take the temperature reading of the water in the second container, and four others will take a reading of the water in the third container. Each student should keep the temperature reading a secret until putting it on a class chart.

2. Make a class chart on the chalkboard to record each of the students' temperature readings. The three columns will be:

Container 1	Container 2	Container 3
Temperature Reading (°C)	Temperature Reading (°C)	Temperature Reading (°C)

3. Each student will record the temperature reading of the water in the container used.

4. Did each person get the same temperature reading of the water in the same container? If the temperature readings were not all the same, explain why you think this might be so.

ORGANIZING AND COMMUNICATING SCIENTIFIC RESULTS

If you skim through the pages of this textbook, you will find many tables and graphs. Textbooks present scientific information in tables and graphs because they are easy to read and understand. Scientists also present their data in tables and graphs to communicate with other scientists. To learn more about making and using tables and graphs, study the following examples and then practise what you have learned.

Making a Table

When you do investigations in your science courses, you will often be asked to record and present your own data in the form of a table. To learn how to make a data table, read through the following example. As you read the steps, examine the table to see how the steps were followed.

Example

Organize the data in the following paragraph in the form of a table:

White rats have an average mass of 0.15 kg and a resting heart rate of about 350 beats per minute. An average 12.0 kg dog has a resting heart rate of about 100 beats per minute. Adult humans have an average mass of about 70 kg and a heart rate of about 72 beats per minute. An elephant with a mass of 4000 kg has a heart rate of about 30 beats per minute.

1. To start making the table, determine the number of categories. In this example, there are three categories: animal, mass, and heart rate. Thus, your table needs three columns with the headings "Animal," "Mass," and "Heart rate."
2. Next, determine the number of items. There are four items: white rats, dogs, humans, and elephants. Leave one row for your headings, and make four rows for the different types of animals.
3. Write the names of the animals in the first column. Then fill in the second and third columns with the mass and heart rate of each type of animal.
4. Always give your table a title.

Table 1 Heart Rates and Masses of Four Animals

Animal	Mass (kilograms)	Heart rate (beats per min)
white rat	0.15	350
dog	12.0	100
human	70.0	72
elephant	4000	30

Instant Practice

Make a table from the data in the following paragraph:

Half a cup of cooked pasta has about 15 g of carbohydrate, 3 g of protein, and no fat. A small piece (32 g) of beef has no carbohydrate, 7 g of protein, and about 5 g of fat. Half a banana has 15 g of carbohydrate, no protein, and no fat. One cup of milk has 12 g of carbohydrate, 8 g of protein, and about 3 g of fat. One large tomato has about 5 g of carbohydrate, 2 g of protein, and no fat.

Making a Stem-and-Leaf Plot

Stem-and-leaf plots help you see trends in data at a glance. Values on the stem represent a range of data, and the leaves represent the exact value of each data point in this range. Therefore, the size of each leaf shows you the number of data points in each range. Study the example below to see how to read and make a stem-and-leaf plot. Be sure that you can see how the numbers in the plot fit the raw data.

Example

A student took a survey of 24 junior-high students. She asked them to keep track of the number of hours of television they watched in one week. Her results (in hours) were 35, 24, 14, 18, 42, 28, 12, 38, 26, 56, 9, 24, 35, 7, 28, 42, 23, 38, 11, 35, 14, and 21. Use these data to make a stem-and-leaf plot.

1. Start your stem-and-leaf plot by writing the headings, "Stem" and "Leaf."
2. Group the data in ranges of ten: 0 to 9, 10 to 19, 20 to 29, 30 to 39, 40 to 49, and 50 to 59. In the column labelled "Stem," write 0 in the first row to indicate that it will contain all of the numbers from 0 to 9.

3. Write the number 1 under "Stem" in the second row, because 1 is the first digit in all the numbers from 10 to 19.
4. Complete the stem by writing the numbers 2 through 5 to represent the first digit in each group of numbers.
5. In the first row under "Leaf," write, in order, all the numbers between 0 and 9 that are found in the raw data.
6. In the second row under "Leaf," write the second digit of all the numbers from the data that start with 1.
7. Complete the leaves by writing the second digit of each number in the data that correspond to the first digit in the stem.
8. Examine the plot, and estimate the average number of hours that the teens watch television in a week.

Weekly Television Watching (hours per week)

Stem	Leaf
0	7 9
1	1 2 4 4 8
2	1 3 4 6 6 8 8
3	5 5 5 8 8
4	2 2
5	6

Instant Practice

1. The student marks for a test on Unit 1 were as follows: 53, 65, 75, 84, 68, 72, 95, 52, 65, 77, 88, 85, 90, 60, 70, 74, 68, 78, 66, 76, 81, 74. Display the marks as a stem-and-leaf plot.
2. How many students wrote the test?
3. How many students scored 70 or higher on the test? How did your stem-and-leaf plot make it easier to answer this question.
4. After you have studied the next section on graphing, come back to your stem-and-leaf plot of the students' marks and produce a bar graph using the data.

Graphing

A graph is a very good way to display data. A graph can help you to see patterns and relationships among the numbers. Throughout your science courses, you will be using many types of graphs, such as line graphs, bar graphs, histograms, and circle graphs (pie charts). Here are some examples to help you use and understand graphs.

Drawing a Line Graph

Line graphs make it easy to see relationships between two sets of numbers. You can use a line graph to predict events that are not even on the graph. The following example demonstrates how to create a line graph from a data table.

Example

The following data, from Statistics Canada, give the population of Canada according to censuses taken during various years. To learn how to make a line graph from these data, examine the graph as you read the steps.

Table 2 Population of Canada

Year	Population (millions)
1901	5.37
1921	8.79
1941	11.5
1961	18.2
1981	24.9
1991	28.1
1996	30.0

1. With a ruler, draw an x-axis and a y-axis on a piece of graph paper. (The horizontal line is the x-axis, and the vertical line is the y-axis.)
2. Label the axes. Write "Year" along the x-axis and "Population (millions)" along the y-axis.
3. Now you have to decide on a scale to use. A graph usually looks best if the x and y-axes are similar in length. Each axis should be divided into 5 to 10 intervals. Since the x-axis must go from the year 1900 to 2000, it can be divided into 20-year intervals. The y-axis goes from zero to 30 (million). It divides nicely into intervals of 5 million. Be sure that your intervals are big enough to read easily.
4. To plot the first point, 5.37 million, carefully move a pencil up the y-axis until you reach a point just above 5. Now move right, slightly

beyond the y-axis, to represent the year 1901. Make a dot at this point. Repeat this procedure until you have plotted all the data points in the table. Finally draw a line from one point to the next.

5. Give your graph a title.

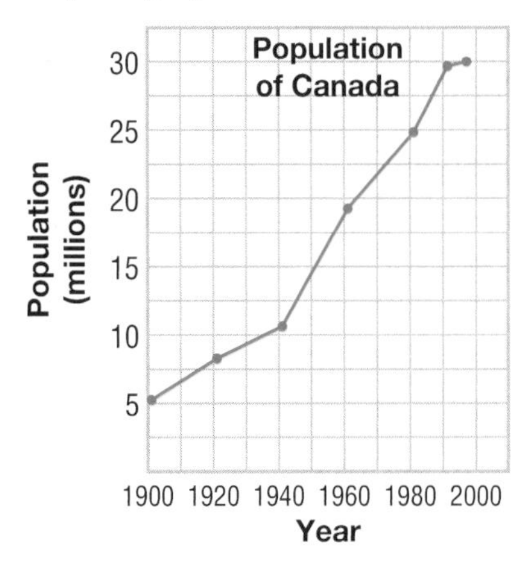

Instant Practice

Make a line graph using the following data on the development of baby teeth in infants and children. Put age on the x-axis and number of teeth on the y-axis. Be sure to include units in your labels.

Table 3 Number of Teeth and Baby's Age

Age (months)	Average number of teeth
8	5
10	9
13	12
16	18
20	25
28	28

Drawing a Bar Graph

A bar graph helps you to compare the number of items in one category with the number of items in other categories. The height of the bar represents the number of items in the category. Study the example, and then make your own bar graph.

Example

Make a bar graph from the data listed in the table on oil production. As you read the steps, examine the completed graph to see how the steps were followed.

Table 4 Oil Production in Five Leading Countries

Country	Crude oil produced (millions of tonnes)
Canada	93.9
Mexico	148.0
Norway	153.0
United Kingdom	122.0
United States	321.0

1. Draw an x-axis and a y-axis on a sheet of graph paper. Label the x-axis with the names of the countries. Label the y-axis with the title "Crude oil produced (millions of tonnes)."
2. Look at the data carefully in order to select an appropriate scale. Notice that the values go up to 321 million tonnes. Intervals of 50 million tonnes would be convenient. Write the numbers on your y-axis scale.
3. Decide on a width for the bars in the graph. They should be large and easy to read.
4. Mark the width of the bars on the x-axis. Leave the same amount of space between each bar.
5. To draw the bar for Canada, go to the centre of the bar on the x-axis, then go up until you nearly reach 100. Make a mark to represent 93.9 million tonnes. Use a pencil and a ruler to draw in the first bar.
6. Repeat the procedure for the other countries.
7. When you have drawn all the bars, colour them so that each one is different. Make a legend or key to explain the meanings of the colours.
8. Give your graph a title.

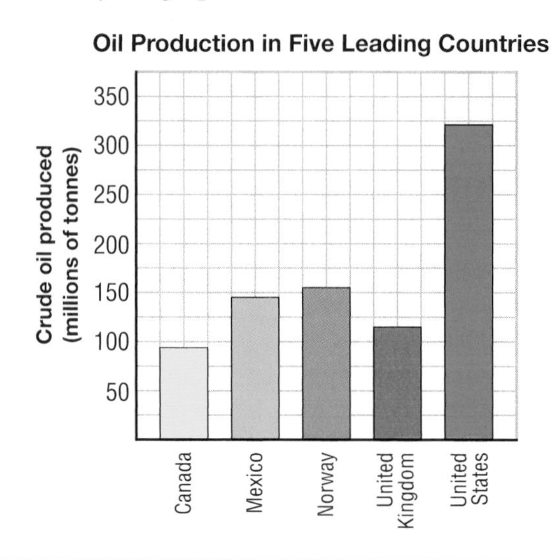

Instant Practice

In a survey of 84 students, aged 11 to 14 years, each student was asked to name her or his favourite sport. The responses are shown in the following table. Make a bar graph using the data.

Table 5 Favourite Sport of Students Aged 11 Through 14

Sport	Number of students choosing sport
baseball	16
basketball	11
hockey	13
cross-country skiing	3
downhill skiing	7
soccer	10
swimming	14
volleyball	10

Constructing a Histogram

The graph on this page is called a histogram. Compare the histogram to the bar graph on page 488. How is a histogram similar to a bar graph? How is a histogram different from a bar graph? You probably noticed that there is no space between the bars in a histogram but there is space between the bars in a bar graph. The reason for the space in a bar graph is that each bar represents a different item, such as a country or a sport. In a histogram, the x-axis represents one continuous item, divided into size categories. In the histogram, notice that the x-axis represents age, divided into ten-year categories. Thus the x-axis is quantitative, meaning that it contains numerical values. In a bar graph, the x-axis is qualitative and cannot be described with numbers. Follow the steps in the example that follows to see how to make a histogram.

Example

A student took a survey of 100 people who owned a personal stereo. The student included a question asking for the ages of the people. Table 6, called a frequency table, shows how many people in each age group owned a personal stereo. Make a histogram to display these data. As you read the steps, examine the completed histogram.

Table 6 Ages of People Who Own a Personal Stereo

Age range	Frequency (number of people)
1-9	19
10-19	26
20-29	17
30-39	13
40-49	7
50-59	6
60-69	5
70-79	4
80-89	3

1. On a piece of graph paper, draw an x-axis and a y-axis. Label the x-axis "Age ranges" and the y-axis, "Number of people (frequency)."
2. Separate the x-axis into nine equal segments. Label the segments using the age ranges listed in the frequency table.
3. Make a scale on the y-axis that goes just above 26.
4. Move your pencil up the y-axis to 19, and make a light mark. Then, using a ruler, make a bar that is the width of the 0–9 age range.
5. Repeat this procedure for each age range and corresponding number of people.

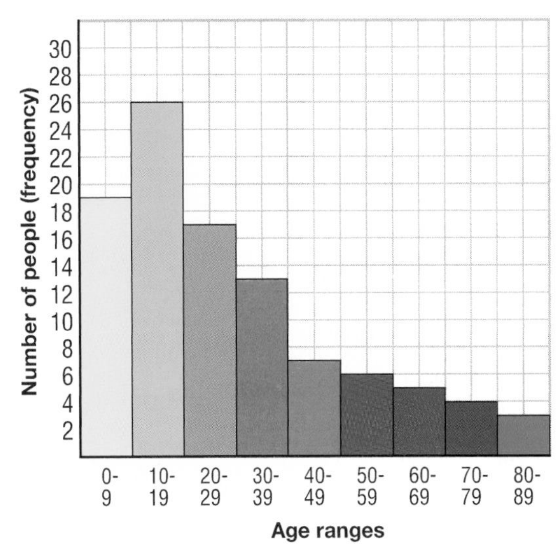

Instant Practice

The following frequency table gives the number of launches of the space shuttle in two-year intervals. Make a histogram using the data.

Table 7 Launches of the Space Shuttle	
Years	**Number of launches**
1981-1982	5
1983-1984	9
1985-1986	11
1987-1988	2
1989-1990	11
1991-1992	14
1993-1994	14
1995-1996	14
1997-1998	13
1999-2000 (planned)	10

Drawing a Circle Graph

A circle graph (also called a pie chart) is a very good way to present data as percentages of a total. To learn how to make a circle graph, study the following example.

Example

The table below lists the populations of the different regions of Canada in percent, according to Statistics Canada, prior to April 1, 1999. Make a circle graph from these data.

Table 8 Percent of Population in Regions of Canada		
Region	**Percent of total population**	**Degrees in "piece of pie"**
Atlantic Provinces	7.8%	**28.1°**
Québec	24.2%	**87.1°**
Ontario	37.7%	**136°**
Western Provinces	30.0%	**108°**
Territories	0.3%	**1.1°**

1. Copy the first two columns of Table 8 into your notebook. Add a third column to your table. Label it "Degrees in piece of pie."

2. Determine the number of degrees in the "piece of pie" that represents the population of each region, using the following formula:

   Degrees for "piece of pie" =

   $$\frac{\text{Percent for region}}{100\%} \times 360°$$

   For example, the calculation for the "piece of pie" representing the Atlantic Provinces is

   Degrees for Atlantic Provinces =

   $$\frac{7.8\%}{100\%} \times 360° = 28.1°$$

   Write the number of degrees in the third column of the table.

3. Use a compass to make a large circle on a sheet of paper. Put a dot in the centre of the circle.

4. Draw a straight line from the centre of the circle to the edge. Place a protractor on this line and make a mark at 28°. Draw a line from the centre of the circle, through the mark, to the circumference of the circle. This is the "slice of pie" that represents the fraction of the total population that lives in the Atlantic provinces.

5. To draw the next "piece of pie," representing the population of Québec, use the line that ended the first "piece of pie" as the starting place.

6. Repeat the procedure for the remaining regions of Canada.

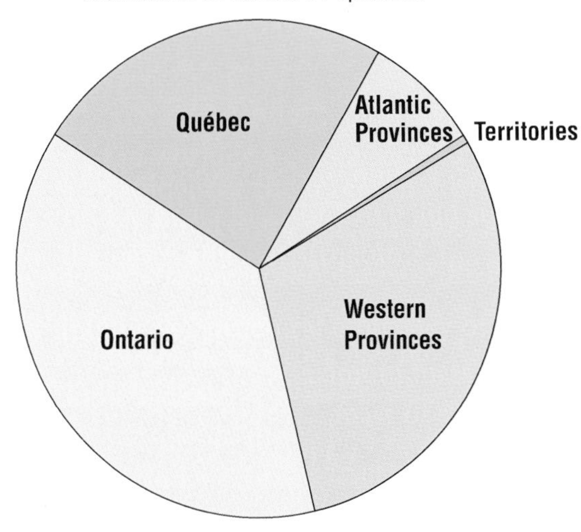

Instant Practice

1. Make a circle graph using the data in Table 9, "Percent of Area of Regions of Canada (prior to April 1, 1999)." When you have completed your circle graph, compare it with the circle graph for population. What significant differences do you see between the population size and the total area of the different regions of Canada?

Table 9 Percent of Area in Regions of Canada (prior to April 1, 1999)

Region	Percent of total area	Degrees in "piece of pie"
Atlantic Provinces	5.5%	
Québec	15.5%	
Ontario	10.7%	
Western Provinces	29.1%	
Territories	39.2%	

2. Identify the regions mentioned in Table 9 in the following map of Canada.

A map of Canada

SAFETY SYMBOLS

The following safety symbols are used in the *SCIENCEPOWER*™ 7 program to alert you to possible dangers. Be sure that you understand each symbol you see in an activity or investigation before you begin.

	Disposal Alert This symbol appears when care must be taken to dispose of materials properly.
	Fire Safety This symbol appears when care should be taken around open flames.
	Thermal Safety This symbol appears as a reminder to use caution when handling hot objects.
	Sharp Object Safety This symbol appears when a danger of cuts or punctures caused by the use of sharp objects exists.
	Electrical Safety This symbol appears when care should be taken when using electrical equipment.
	Skin Protection Safety This symbol appears when use of caustic chemicals might irritate the skin or when contact with micro-organisms might transmit infection.
	Clothing Protection Safety A lab apron should be worn when this symbol appears.
	Eye Safety This symbol appears when a danger to the eyes exists. Safety goggles should be worn when this symbol appears.
	Poison Safety This symbol appears when poisonous substances are used.
	Chemical Safety This symbol appears when chemicals that are used can cause burns or are poisonous if absorbed through the skin.
	Animal Safety This symbol appears whenever live animals are studied, and the safety of the animals and the students must be ensured.

WHMIS (Workplace Hazardous Materials Information System)

Look carefully at the WHMIS (Workplace Hazardous Materials Information System) safety symbols that are shown below. The WHMIS symbols are used throughout Canada to identify the dangerous materials that are found in all workplaces, including schools. Make sure that you understand what these symbols mean. When you see these symbols on containers in your classroom, at home, or in a workplace, use safety precautions.

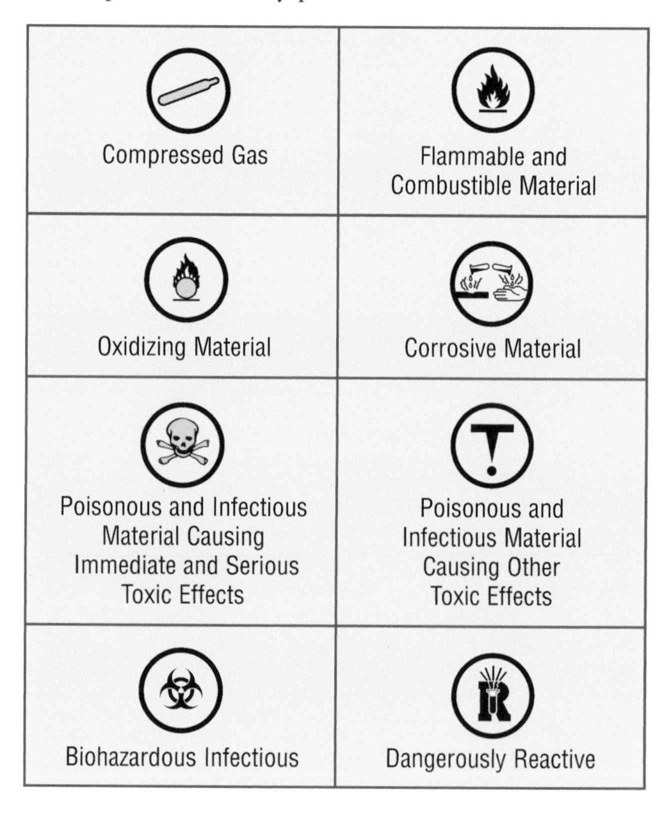

Instant Practice

1. Find four of the *SCIENCEPOWER*™ 7 safety symbols in activities or investigations in this textbook. Record the page number and title of the investigation or activity in which you found each symbol. What possible dangers that relate to the symbol are in the activity or investigation?

2. Find two of the WHMIS symbols on containers in your school, or ask your parent or guardian to look for two WHMIS symbols in a workplace. Record the name of the substance on each container and the place where you, or your parent or guardian, saw the container stored. What dangers are associated with the substance in each container?

USING MODELS IN SCIENCE

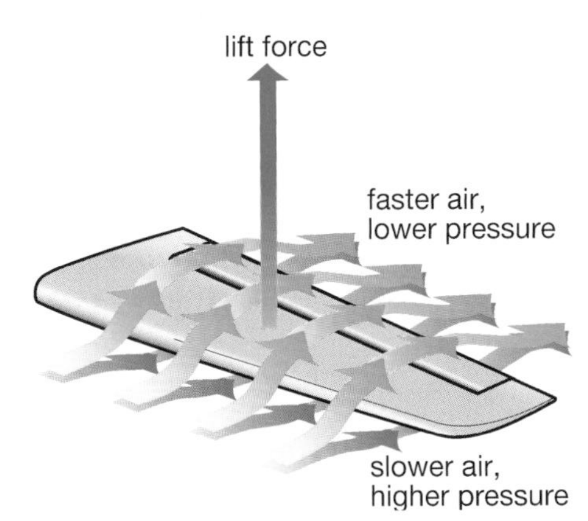

When the wing of an airplane cuts through the air, it pushes up the air above the wing. This causes the air above the wing to move faster than the air below the wing. As a result of the wing's shape, there is less air pressure above the wing than below it. The difference in air pressure creates the "lift" that allows an extremely heavy airplane to rise up into the sky.

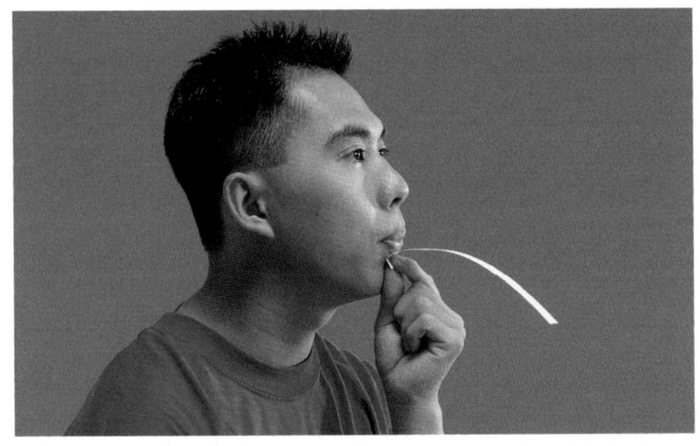

Cut a strip of paper 2 cm wide and 20 cm long. To make a tab, fold back one end of the strip 1 cm. Holding the paper by the tab, put it just below your lips. Blow smoothly. What happens to the paper?

When engineers design a new airplane, they build a scale model and test it in a wind tunnel. Huge fans blow air toward the model. Streamers of smoke make the flow of the air visible. These smoke patterns allow the engineers to test their predictions about the lift that the wings will provide.

Compare the photographs and the illustrations at the left. How are they similar? Each photograph or illustration models the scientific principles of flight. A scientific model is anything that helps you understand, communicate, or test an idea.

Textbooks use illustrations, diagrams, charts, and graphs to model scientific concepts. When you visualize or create a picture of an idea, the idea is often easier to understand. The first illustration shows how the air moves above and below an airplane wing. You may remember, from previous science studies, that air pressure under a wing creates lift.

Many people understand a concept better when they make a model that demonstrates the concept. Activities and investigations are learning tools. The student in the photograph is using a strip of paper to model an airplane wing. When the student blows over the top of the paper, the air above the "wing" moves faster than the air below it, just as the first illustration shows. Watching the paper lift up helps the student better understand the concept of "lift."

Scientists build models to test hypotheses or to find possible answers to questions. Engineers use models to test new designs. In the second photograph, aircraft engineers are testing the design of an airplane in a wind tunnel. Not only can they see how the air flows around the wings, they can also detect turbulence, or uneven flow of air, around all the parts of the model. Turbulence causes drag, which slows an airplane.

Examples

As you can see, models help scientists and students in many ways. Look at the following photographs to find more reasons for using models.

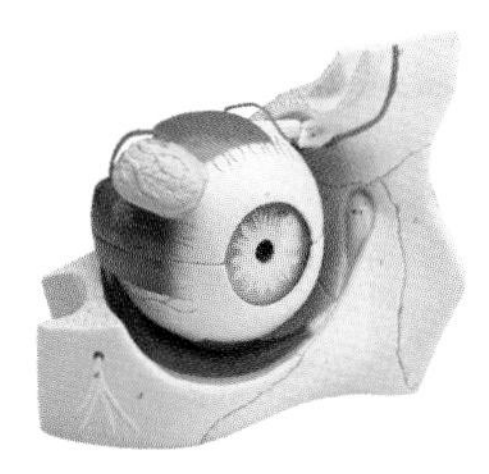

Sometimes important structures are hidden from view. Models allow you to see how these structures work. You can use this model of the human eye to learn how your own eyes focus on near and distant objects. As well, you can learn why some people are near sighted and others are far sighted. You can even see how glasses or contact lenses help to correct these vision problems.

From Earth, you cannot see all the planets in the solar system at the same time. Therefore, you cannot see how the planets are arranged. This model helps you see how the planets orbit the sun and how the solar system is arranged. Distances, however, are not to scale.

Instant Practice

Build a model of a lake and of dry land to learn how they affect climate.

You will need a pencil, a piece of paper, two baking dishes, potting soil (or dirt), water, two identical lamps with identical light bulbs (or direct sunlight), and two thermometers.

The pan of water models a large body of water, such as a lake or an ocean. The soil models dry land, a large distance from a body of water. The temperature of the air just above the water will tend to be similar to the temperature of the water. The temperature of the air just above the land will be similar to the temperature of the land.

Safety Precautions

Be careful when handling the thermometers. They are fragile and can break easily.

1. Fill one baking dish with water and the other with potting soil. Press down gently on the soil until it is fairly firm. The water and the soil should both be about 4 to 5 cm deep.
2. Place a strong lamp about 20 cm above each dish, or place the dishes in direct sunlight. Be sure that both dishes are getting the same amount of light.
3. Place a thermometer in each dish. The bulb of the thermometer should be just below the surface of the soil.
7. Read and record the temperature of the soil and the water as soon as you put the thermometers into the dishes. Then read and record the temperature about every 15 min for 1 h.
5. Examine the temperature readings. What would you infer about the climate near a very large body of water? What would you infer about the climate on land, a large distance from a lake or an ocean?
6. Wash your hands after completing this activity.

SCIENTIFIC AND TECHNOLOGICAL DRAWING

Have you ever heard the expression "a picture is worth a thousand words"? This expression means that a difficult idea can often be explained more easily by using a visual, such as a picture or a diagram. Have you ever used a drawing to explain something that was too difficult to explain in words? Perhaps you have sketched a map when giving directions to a friend. Think of an example, from your own experience, when a drawing helped you get an idea across to someone else or clarified a difficult idea for you.

A clear drawing can often assist or replace words in a scientific explanation. In science, drawings are especially important for explaining difficult concepts or describing something that contains a lot of detail. It is important that you make your drawings clear, neat, and accurate.

Example

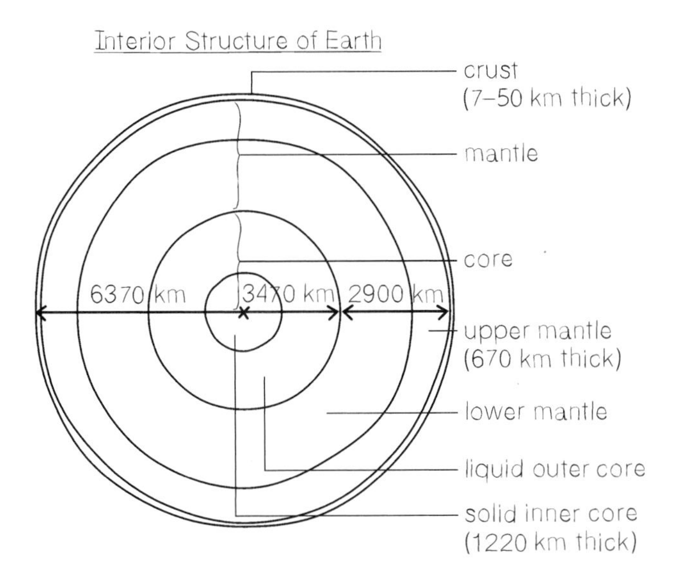

Examine the drawing shown here. It is taken from a Grade 7 student's notebook and summarizes the internal structure of Earth. The student's verbal description of Earth's interior included an explanation of how the various layers are arranged and gave the thickness of each layer. As you can see, a clear diagram can support or even take the place of many words of explanation. While the drawing itself is important, labelling it clearly is also important. If you are comparing and contrasting two objects, label each object, as well as the points of comparison between them.

Making a Scientific Drawing

Follow these steps to make a good scientific drawing.

1. Use unlined paper and a sharp pencil with an eraser.
2. Give yourself plenty of space on the paper. You need to make sure that your drawing will be large enough to show all the necessary details. You also need to allow space for labels. Labels identify parts of the object you are drawing. Place all of your labels to the right of your drawing, unless there are so many labels that your drawing looks cluttered.
3. Carefully study the object that you will be drawing. Make sure that you know what you need to include.
4. Draw only what you see, and keep your drawing simple. Do not try to indicate parts of the object that are not visible from the angle you are observing. If you think that you need to show other parts of the object, do a second drawing. On each drawing, indicate the view that is shown.
5. Shading or colouring is not generally used in scientific drawings. If you want to indicate a darker area, you can use stippling (a series of dots). Note the stippling in the illustrations of the hand-held drill. You can use double lines to indicate thick parts of the object.
6. If you do use colour, try to be as accurate as you can. Choose colours that are as close as possible to the colours in the object you are observing.
7. Label your drawing carefully and completely, using lower-case (small) letters. Pretend that you know nothing about the object you have just observed, and think about what you would need to know if you were looking at it for the first time. Place all your labels (if possible) to the right of the drawing. Use a ruler to draw a horizontal line from each label to the part you are identifying. Make sure that none of your label lines cross.
8. Give your drawing a title.

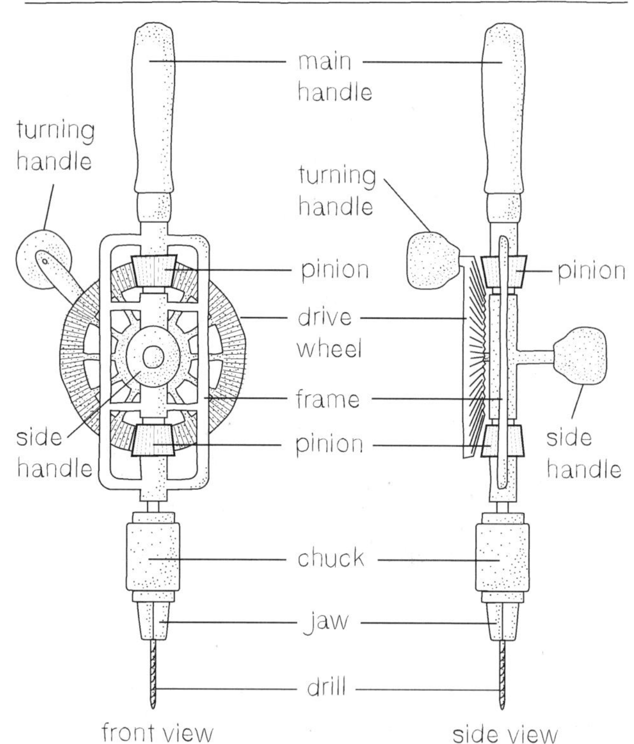

Tips on Technological Drawing

You will find that well laid-out drawings are a valuable learning tool. Please refer to "Appendix B: The Design Process", for ideas on creating illustrations and graphics for projects. You can also ask the advice of specialist teachers, engineers, or technology experts.

Instant Practice

1. Draw an object in your classroom. Use stippling as a technique to give it three dimensions.
2. Exchange drawings with a classmate to see if each of you can identify the other's "object." As well, give each other feedback on how you think the drawing could be improved for greater clarity.
3. Draw a spherical object. Use stippling to give the impression of curvature.
4. Select any mechanical system in your classroom or at home, such as a window blind or a door hinge. Show two different views of the system to help someone else understand how the system works.

USING RESOURCES AND THE INTERNET EFFECTIVELY

Scientists build on what others have learned in the past. In order to be part of the long chain of science inquiry, scientists keep careful records of sources of information they have used and of any new ideas they produce themselves.

In the same way, you need to do careful and thorough research and to keep careful records of what you observe, think, and discover. (*SkillPower 2: How to Use a Science Log*, on page 476, will help you do this). Probably your biggest task in doing research in this course will be finding your way through the vast amount of information that is available on every area of science. By practising the skills below, you will find it easier to obtain the information you need quickly and efficiently.

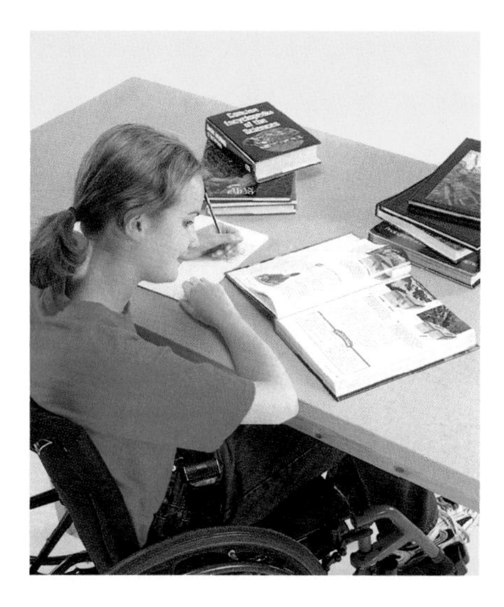

Using Resources Effectively

In your classroom, you probably have books and magazines that you can use to find out more about certain topics. For much of the information you need, however, you will want to use your school library or a local library. A library can be overwhelming, but if you approach it in an organized way, you can quickly and efficiently find what you need.

First, be aware of the huge amount of material you can access through a library: general and specialized encyclopedias, directories (telephone, city, postal code), almanacs, atlases, handbooks, periodicals (magazines), newspapers, government publications, pamphlets, tapes, videos, CD-ROMs, databases, and the ever-changing information on the Internet.

To make the best use of your time, and the resources that are available to you, ask yourself these questions before you start your research:

- *What* information do I need? How much detail do I need?
- *When* is the work due? (This might help you decide how much detail you need.)
- *Why* do I need the information? Am I preparing something for a particular audience (my teacher, another group, or another class)?
- *How* will I be presenting the information: as a written report, a poster, an oral presentation, a multimedia presentation?

In the library, you will find the materials organized under three headings — author, title, and subject. You can start by looking up any one of these headings in the library's card catalogue or computer system. When you find material that you think you can use, carefully record its identification number. The librarian will help you track down the materials.

What do you do when you have a stack of books with far more pages than you have time to read? Limit your search to what you really need to know. Imagine that this textbook is one of the library books you are consulting. You are writing a report about the formation of mountains. You already know enough about this textbook to realize that you do not need to read all of it to obtain the information you need. Look at the Contents at the beginning of the book to find out which chapter you should read to learn about mountain formation. The title of Chapter 11, "Earthquakes, Volcanoes, and Mountains," is a clue that you will find the information you need there. Sometimes, however, a seemingly unrelated topic can also include information on the topic you are researching. Therefore your next stop should be the book's index. An **index** is a list of a book's content that appears at the end of the book. It is arranged in alphabetical order and broken down into far more detail than the table of contents. The index provides a specific page reference for each entry. If you look in the index at the end of this book, you will see that there is also some information on mountains in Chapter 12, "The Story of Earth's Crust."

Now, suppose that you have a book that has several chapters relating to your topic. What is the best way to get the information you need from these chapters in a way that is useful to you? Use the same methods that you used to see the "big picture" in your textbook: look at the chapter titles, introductions, subheadings, photographs, illustrations, and graphs.

When you begin to read the sections that contain the information you want, you need to make notes. Do not just copy complete sentences. Making notes involves actively thinking about what you are reading and how all the information is connected. The more actively you make notes, the more you will get out of your reading. Write the main idea of each section you read. Quickly read each paragraph in the section, and write short notes below the main idea. Develop your own special "shorthand" to use when making notes. (Just make sure that you can understand your notes when you reread them!)

Using the Internet Effectively

The web site address for the publisher of this textbook is http://www.mcgrawhill.ca. Using this URL (Universal Resource Locator, or web site address) will take you to the headquarters of McGraw-Hill Ryerson Limited. You can use the McGraw-Hill Ryerson web site to obtain information about specific topics in this textbook. Beyond that, it is a good idea to ask an adult for help when you are looking for information on the Internet. Remember that anyone, anywhere, can develop a web site or post information on the Internet. Anyone can use the Internet to "publish" personal opinions. Sometimes it is difficult to distinguish accurate scientific information from these opinions. Always check the source of the information. Be wary of an individual who is publishing alone. Government sites and educational association sites tend to contain much more reliable information. Follow your own school's guidelines for "surfing" the Internet to do your research, and make good use of McGraw-Hill Ryerson's School Division web site at www.school.mcgrawhill.ca/resources/ to simplify your search.

What Is the Internet?

The Internet is an extensive network of interlinked yet independent computers. In less than two decades, the Internet went from being a highly specialized communications network used mostly by the military and universities to a massive electronic bazaar. Today, the network includes

- educational and government computers
- computers from research institutions
- computerized library catalogues
- businesses
- home computers
- community-based computers (called *freenets*)
- a diverse range of local computer bulletin boards.

Anyone who has an account on one of these computers can send electronic mail (e-mail) throughout the network and access resources from hundreds of other computers on the network. Here are some of the ways you will find the Internet most useful as a learning tool.

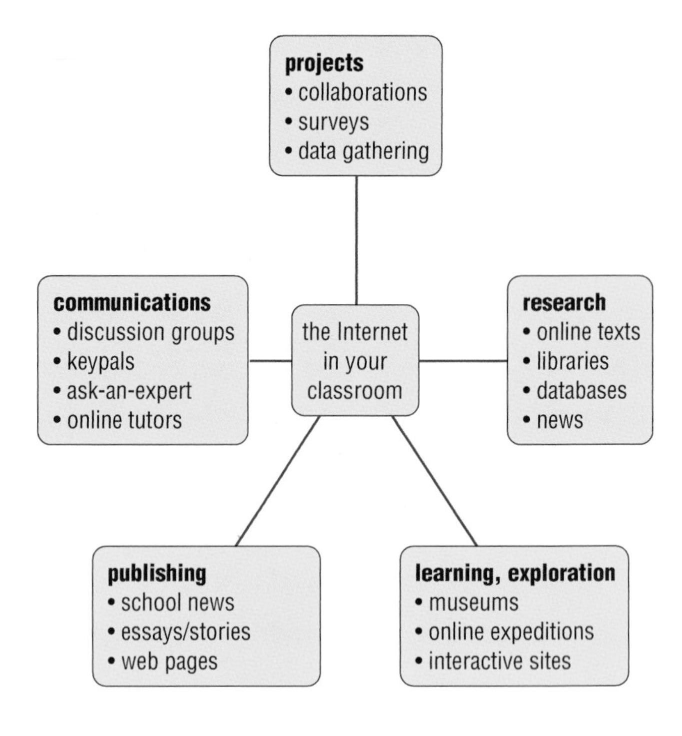

Answers to Reviewing Key Terms, page 340

- P wave
- epicentre
- S wave
- anticline
- seismologist
- composite volcano
- ash-and-cinder cone
- Richter scale
- surface wave
- syncline
- focus
- tsunami
- shield volcano

The McGraw-Hill Ryerson web site provides you with additional information for your science program.

Gue, David et al. SCIENCEPOWER™ 7: Science, Technology, Society, Environment. McGraw-Hill Ryerson Limited, Toronto, 1999.

At the end of every research project, record each source of information you used. Here is an example of the proper way to list a source in a bibliography. List your sources by author's last name, in alphabetical order.

Identify Your Resources

Before you put away each type of material, record the information that identifies it. You will need to let your audience know where you obtained your information, and you may need to go back to it. Make sure that you record

- the author's name (or the name of the group that provided the information)
- the name of the resource
- the name of the publisher or information source
- the city where the resource was published
- the publishing date
- the URL, if the information is from an Internet site

Most of this information can be found on the copyright page at the beginning of a book. The URL can be found in the address bar at the top of an Internet page. (Some URLs are very long and complex. Get a partner's help to make sure that you have copied the URL correctly.)

Above is an example of how to cite your source. (NOTE: et al. means "and others.")

INTERNET CONNECT

www.school.mcgrawhill.ca/resources/

You have looked at animal adaptations, and the photographs here show some plant adaptations. Find out about the special adaptations of cacti for their dry environments by going to the web site above. Go to **Science Resources**, then to **SCIENCEPOWER 7** to know where to go next. Write and sketch your findings.

This recurring feature in *SCIENCEPOWER™ 7* will take you to interesting web sites.

Glossary

How to Use This Glossary

This Glossary provides the definitions of the key terms that are shown in **boldface** type in the textbook. Other terms that are not critical to your understanding, but that you may wish to know, are also included in the Glossary. The pronunciations of terms that are difficult to say appear in square brackets after the terms. Use the following pronunciation key to read them:

a = mask, back
ae = same, day
air = stare, where
e = met, less
ee = leaf, clean
ih = ice, life
i = simple, this
o = stop, thought
oh = home, loan
oo = food, boot
uh = sun, caption
uhr = insert, turn
yoo = cute, human

abiotic [AE-bih-o-tik] a term applied to non-living things in the environment; for example, air, water, and soil are abiotic

absolute dating a method of determining the age of objects by measuring the radioactive decay of their smallest particles

active of a volcano, the stage when materials such as lava, smoke, and ash are released into the environment

adaptation an inherited characteristic that helps an organism survive in its environment

adhesive a sticky substance, such as glue or epoxy cement, that is used to hold objects or materials together

aesthetics [e-STHE-tiks] a branch of philosophy that studies the principles of beauty; the properties of an object that make it pleasing to the senses

aftershocks smaller ground movements caused by seismic waves moving outward from an earthquake's focus

anticline an upfold of rock layers in sedimentary rock

arch a curved structure, or part of a larger structure, that is often used to distribute weight downward

area the amount of surface; measured in square units such as cm^2

ash-and-cinder cone the smallest of the three main types of volcanoes, having steep sides formed by layers of ash and rock

atmosphere in the biosphere, the air surrounding the hydrosphere and the lithosphere

balance a device to measure mass; many balances work by using the force of gravity

balanced forces forces that are equal in size, opposite in direction, and pointing along the same straight line

bar graph a diagram consisting of horizontal or vertical bars that represent (often numerical) data

beam a horizontal support used to hold the weight of the material above

bed a layer of sedimentary rock

bedrock the part of Earth's crust that is made up of solid rock, lying beneath the soil and looser rocks

bimetallic strip a device made up of two thin layers of two different metals joined together; used in scientific experiments to observe the effects of heating or cooling; also used in thermostats

biological community *see* community

biological evidence the type of evidence obtained from living or non-living organisms

biological population *see* population

biological weathering the break-up or disintegration of rocks through the physical or chemical effects of living organisms

biomass the total mass of living matter; often expressed in terms of dry weight per unit area

biome a region of land that contains certain kinds of organisms, particularly plants; determined by climate; examples are desert, grassland, and forest

biosphere the thin area around Earth that can sustain life; made up of the atmosphere, the hydrosphere, and the lithosphere

biotic [bih-O-tik] a term applied to living things in the environment, such as humans, plants, birds, animals, and insects

boiling rapid vaporization occurring at a specific temperature called the boiling point

boiling point the temperature at which a liquid begins to boil and change into a gas or a vapour

brace a device used to add strength to a structure, usually by forming a rigid triangle at the point where pieces come together at a right angle

brazing a process in which a melted material is applied to a different type of material; the melted material hardens when it cools, forming a rigid joint that holds the other material in place

bromthymol blue [BROM-thih-mol] a chemical indicator that changes colour (from blue to green to yellow) when carbon dioxide is present

buckle of a material, to fold under a compressive force

buttress a slanted brace that supports part of a structure, transferring its thrust line to the ground along an angle

calibrated of thermometers, marked with a number scale to indicate temperature

capacity the largest amount that can be held by a container (usually measured in litres or millilitres)

capacity unit unit used to measure the volume of liquids; an example is the litre (L)

carbon dioxide and oxygen cycle the process by which carbon dioxide and oxygen are cycled and recycled through the atmosphere

carbonaceous film [car-bon-AE-shuhss] a type of fossil found in sedimentary rock when organic material is compressed, leaving a thin carbon film

carnivore an animal that eats other animals; examples are lynx, wolf, hawk

catastrophism [cat-AS-tro-phis-uhm] a theory developed in the seventeenth and eighteenth centuries; according to this theory, sudden catastrophes, such as volcanic eruptions and earthquakes, created mountains and other features of Earth

Celsius (C) scale the most common scale for measuring temperature; on the Celsius scale, water at sea level boils at 100° and freezes below 0°

Cenozoic Era [sen-oh-ZOH-ik E-ruh] the fourth and current era on the geologic time scale; the era in which humans evolved

centre of gravity the point at which all of the gravitational force of an object may be considered to act

chemical change a change in matter that causes at least one new substance, with new properties, to be formed

chemical energy a type of potential energy that is stored in chemicals and released when the chemicals react

chemical weathering the break-up or disintegration of rocks through the effects of chemical reactions upon them

circle graph a circle divided into sections (like pieces of a pie) to represent data; also called a pie chart

classification (or biological key) a list of alternatives (e.g., backbone or no backbone) used by scientists as an aid in identifying an unknown plant or animal

cleavage of a mineral, the characteristic of splitting along smooth, flat planes

climate the overall weather that a large area has over many years

cogeneration [coh-jen-uhr-AE-shuhn] a method of energy conservation by which waste heat or energy from one industry is used by another industry

cold-blooded of an organism, having a body temperature that varies with the temperature of its environment

column an upright, cylindrical structure that is used to support another structure; also used as ornamentation

commensalism a symbiotic relationship between two different types of organisms in which one partner benefits and the other neither benefits nor loses

community an association of different populations of organisms in a particular environment or geographic area

compaction the process by which sedimentary rock is formed from sediment, through the weight and pressure of water and other sediment

complex mountains mountains that are formed by the combined processes of folding and faulting

composite of materials, made up of several different materials, with different properties, to fulfil a specific purpose

composite volcano a volcano formed in a subduction zone when rock, which is forced under other rock, melts into magma and pushes its way through vents to the surface

compost the part of soil composed of dead plant matter

compression force a force that compacts or squeezes a material

compressive strength a measure of the largest compression force that a material can withstand before changing shape or breaking apart

concept map a diagram comprising words or phrases in circles or boxes and connecting lines; used to show various relationships among concepts; can also contain references to events, objects, laws, themes, classroom activities, or other items or patterns related to the concepts

condensation the process of changing from a gas or vapour to a liquid; clouds, fog, and dew are examples of condensation

condensation point the temperature at which a gas or vapour begins to change into a liquid; the condensation point of a gas is the same temperature as the boiling point of the material in its liquid state

conservation of energy the law stating that the amount of energy within a system always remains the same if the system is left undisturbed

consumers organisms that eat the food made by producers; can be either herbivores, carnivores, or omnivores

continental drift a theory about Earth's structure; according to this theory, the continents have slowly changed their positions over time; the slow movement of continents

continental shelf the gradual, sloping edge of a continent that extends out beneath the ocean

contract of substances, to shrink or decrease in volume

control in a scientific experiment, a standard to which the results are compared; often necessary in order to draw a valid conclusion; ensures a fair test

controlled variable in an experiment, a condition that is not allowed to change

convection current continuous circulation of a fluid (either a liquid or a gas), in which thermal energy is transferred from hotter, less dense fluid to colder, more dense fluid

convergent boundary an area on Earth's crust where two plates are pushing against each other

co-ordinate graph a grid that has data points named as ordered pairs of numbers; for example (4,3)

core the innermost part of Earth; made of iron and nickel in solid and liquid form

corrugated [KOHR-ruh-gae-ted] of materials, to be made of layers, with a middle layer that is folded into a series of triangles to provide strength

counterweight a device used to balance potentially dangerous forces on an object or a structure

criteria a set of standards or expectations; specifications for a design

crust the thin, outermost layer of Earth

crystal the building block of minerals; crystals occur naturally and have straight edges, flat sides, and regular angles

cubic units the units used to report the volume of a substance; for example, cm^3

cycle concept map an events chain map in which a series of events does not produce a final outcome; this type of concept map has no beginning and no end

data facts or information

database an organized or sorted list of facts or information, usually generated by computer

dead load the weight of a structure upon itself

decomposers organisms that break down the cells of dead or waste materials and absorb their nutrients; many bacteria and fungi are decomposers

dependent (or responding) variable in an experiment, a condition that is changed as a result of changes to independent variables

deposition the process in which eroded material is deposited in another area

desalination [dee-sal-i-NAE-shuhn] a process for removing the salt from salt water

desertification [de-zuhrt-i-fi-KAE-shuhn] the process in which deserts are formed through the erosion of nutrient-rich topsoil; after desertification the soil is no longer able to support plant life

dew point the temperature at which air becomes saturated with water vapour, causing precipitation

dilute to weaken the strength of a solution by increasing the amount of solvent

dilute solution a solution that contains relatively little solute

direction of energy transfer the tendency of energy to move from a concentrated source; for example, thermal energy always moves from hotter objects to cooler ones

dissolving mixing a solute completely with a solvent to form a solution; the distinct properties of each of the materials combine into one set of properties

distillation a process for separating the parts of a liquid solution; the solvent is heated to change it into a gas, then converted back to a liquid state through condensation

divergent boundary an area of Earth's crust where two plates are pulling apart from each other

diversity a measure of how many different species live in an ecosystem; an ecosystem with many species has greater diversity than an ecosystem with only a few species

dormant of a volcano, a stage when no eruption is occurring

double cantilever a design used to support a great deal of weight; consists of beams that are braced on both ends and held up by a strong column at the centre

earthquake a disturbance and movement of Earth's crust due to a build-up of stress

ecologist [ee-KOL-oh-jist] a scientist who studies interactions between the abiotic and biotic parts of the environment

ecology the study of how organisms interact with each other and their environment

ecosystem all the interacting parts of a biological community and its environment

elastic energy a type of potential energy that is stored in an object, the shape of which is changed by bending, twisting, or compressing

electromagnetic radiation (EMR) energy that is transferred in the form of electromagnetic waves; examples of EMR include radio waves, X rays, and microwaves

element a type of pure substance (made of one type of particle or atom) that cannot be broken down into simpler parts by chemical means and that has a unique set of properties

energy the ability to do work and to cause change (chemical or physical)

energy flow the movement of energy, which originally comes from the Sun, from one organism to another

energy source an object or material that can transfer energy to other objects

epicentre [E-pi-sen-tuhr] the area on the surface of Earth that is directly above the focus, or source, of an earthquake

era one of the four longest subdivisions in the history of Earth

erosion the process of moving soil and rock from one place to another; mechanical weathering caused by the effects of wind and/or water

erupt of a volcano, to become active and release materials such as lava, smoke, and ash

evaporation the process by which a liquid, such as water, changes into a gas or a vapour

evaporative cooling a process in which the faster-moving particles on the surface of a liquid evaporate and escape into the air; the slower-moving particles, which are left behind, have lower kinetic energy, decreasing the temperature of the remaining liquid and the surface on which it is resting

events chain map a concept map used to describe a sequence of events, the steps in a procedure, or the stages of a process

expand of substances, to increase in volume

external force a force exerted on an object from outside the object

extinct of a species, no longer existing

extrusive rock the type of igneous rock formed when magma (lava) cools and solidifies above Earth's crust

fair test an investigation (experiment) carried out under strictly controlled conditions to ensure accuracy and reliability of results. In a fair test, all variables are controlled except the one variable under investigation.

fastener a type of rigid joint that attaches materials together and holds them in place; examples of a fastener include a nail, a staple, and a rivet

fault an area where two very large rock surfaces move against each other

feedback information that is gained from outside a particular system and returned to it for the purposes of modifying a behaviour or a process

fertile of soil, containing the nutrients needed for plant growth

fibres thread-like materials that make up plant and animal tissue, and some manufactured materials

fixed-continent model a theory about Earth's structure; according to this theory, the continents and the oceans have always occupied the same positions

fluids materials (liquids or gases) that lack a definite shape and can flow from one place to another

focus the place, deep in Earth's crust, where an earthquake begins

food chain a sequence of feeding relationships among living organisms, as they pass on food energy

food web the network of feeding relationships among organisms

footing a base for a wall in the foundation of a structure; a footing is wider than the wall to spread the weight over a larger area

force a push or pull, or anything that causes a change in the motion of an object

force diagram a drawing that uses arrows to represent the direction and strength of one or more forces

fossil any trace or remains of once-living organisms

fossil fuels fuels made of decomposed plants and other organisms that have been hardened or fossilized; fossil fuels take millions of years to develop; examples of fossil fuels include coal, oil, and natural gas

foundation the solid base of a structure

fractional distillation a process in which a solution is vaporized and condensed into several different products; for example, petroleum is vaporized and condensed to produce gasoline, diesel fuel, and kerosene

fracture the breaking of a mineral split into jagged, rough pieces where breakage is not controlled by cleavage

frame structure a type of structure in which a skeleton of materials supports the weight of the other parts

freezing point *see* melting point

frost wedging a process of mechanical weathering that occurs when water goes through a cycle of freezing and thawing; the water expands and contracts in the cracks of a rock, eventually breaking the rock apart

fulcrum the part of a lever that does not move

function of a structure or object, its main purpose

gas one of the phases or states of matter; a gas has no particular shape or size and can be compressed; a gas is sometimes known as a vapour

genus a group of related species

geothermal energy energy generated in the interior of Earth

global warming the gradual increase in the temperature of Earth's atmosphere; some scientists think that global warming results from a surplus of greenhouse gases in the environment and that it may have harmful effects on life on Earth

Gondwanaland the southern part of the supercontinent Pangaea, which split off approximately 200 million years ago

graphic organizer a visual learning tool that helps clarify the relationship between a central concept and related ideas or terms

gravitational energy a type of potential energy that an object has when it is above Earth's surface

gravitational force the force exerted by gravity on an object; measured in newtons (N); the preferred scientific term for the everyday term "weight"

greenhouse gases gases, such as carbon dioxide, that result from the burning of fossil fuels or wood;

greenhouse gases prevent heat from leaving the atmosphere, increasing the temperature of the atmosphere

groundwater the water contained in the lithosphere or Earth's crust

gusset a device used to add strength to a structure, usually by forming a rigid triangle at the point where two pieces come together at a right angle

guy wire a wire used to give stability to a structure by counterbalancing potentially dangerous forces

gyroscope [JIH-roh-skohp] a circular device with a heavy outer rim that spins at a very fast rate, stabilizing the axis so that the axis always points in the same direction

habitat the location where an organism lives

half-circle arch another term for a dome, such as are often found on planetariums or orthodox churches; their weight is transferred directly to the structure's foundation

half-life the amount of time that a given amount of radioactive substance takes to be reduced by one-half

hard water water that contains a high proportion of dissolved materials

heat thermal energy transferred from one object or substance to another because of a temperature difference

heat capacity the thermal energy needed to raise the temperature of 1 kg of a substance, such as water, by 1°C

heat insulators materials that slow the transfer or conduction of thermal energy from one object to another; examples of heat insulators include fibreglass and Styrofoam™ cups

herbivore an animal that eats only plant material; examples are grasshopper, beaver, and moose

heterogeneous [het-uhr-oh-JEEN-ee-uhs] of a mixture, made up of parts that retain their own properties, even if these properties are not visible to the unaided eye

histogram a type of bar graph in which each bar represents a range of values and in which the data are continuous

homogeneous [hoh-moh-JEEN-ee-uhs] of materials, having only one set of properties

host the organism that a parasite lives and feeds on

hot spot an area under Earth's crust where the temperature is much hotter than normal, forcing magma toward the surface

humus [HYOO-muhs] the dark-coloured part of soil that is rich in nutrients, such as nitrogen, phosphorus, potassium, and sulfur

hydrosphere the part of the biosphere that contains all of Earth's water

igneous rock [IG-nee-uhs] the type of rock that is formed by the solidification of hot magma; it is defined as either intrusive or extrusive

independent (or manipulated) variable in an experiment, a condition that is selected or adjusted to see what effect the change will have on the dependent variable

index fossil a type of fossil that can be used to determine the age of the material in which it is found

individual a single organism

infrared radiation (or heat radiation) a type of electromagnetic radiation that has a wavelength just greater than the red end of the visible light spectrum

input the materials or forms of energy that are used by a system to do work or to produce new materials (output)

input force of a lever, the external force or effort that is applied to it

insoluble of a substance, meaning not able to be dissolved in a particular solvent

internal force a force that acts on an object from the inside

intrusive rock the type of igneous rock formed when magma cools and solidifies below Earth's crust

joule (J) the standard SI unit for measuring energy

Kelvin scale a scale used for measuring temperatures in scientific experiments; on the Kelvin scale, pure water freezes at 273 K and boils at 373 K; the coldest possible temperature (also known as absolute zero) is 0 K

kilogram the primary measurement of mass in SI, equal to 1000 g; 1 kg is the primary standard for mass

kinetic energy [kin-E-tic] energy that is released or transferred by the motion of an object or its particles

kingdom one of five main groupings for classifying living things on Earth; the five kingdoms are: animal, plant, fungus, protist, and monera

lamination a process in which a layer of material is pressed or glued onto other layers

landfill site an area where garbage is deposited and eventually buried

Laurasia the northern part of the supercontinent Pangaea, which split off approximately 200 million years ago

lava the term used for magma when it breaks through Earth's crust, as in a volcanic eruption

law in science, a statement of a pattern, action, or condition that has been observed so consistently that scientists are convinced it will always happen

leaching the process by which materials from soil are dissolved and carried away by water

lift upward force on a forward-moving object that results when the air flow over the top of the object is faster than the air flow beneath it

limiting factor in an ecosystem, an abiotic or biotic factor that maintains balance by limiting the number of individuals in a population

line graph a diagram that shows how one value depends on or changes according to another value; produced by drawing a line that connects data points plotted in relation to a *y*-axis (vertical axis) and an *x*-axis (horizontal axis)

liquefaction [lik-we-FAK-shuhn] the process of changing solid material into a liquid-like substance, such as quicksand

liquid one of the states or phases of matter; in the liquid state, a material has a specific size or volume but not a specific shape

lithosphere in the biosphere, the solid mineral material that covers Earth; Earth's crust

live load the force or forces that act in or on a structure but are not part of the structure; examples of a live load include the wind, the weight of people, and a collision

loam a type of soil that is good for plant growth; made up of sand, silt, and clay

lustre the light-reflecting properties, or "shininess," of minerals

magma melted rock, formed under Earth's crust by high temperature and pressure; magma occasionally escapes to Earth's surface as lava

magnetometer [mag-net-O-met-uhr] a device that detects the direction and strength of a magnetic field

mantle the middle layer of Earth, located between the crust and the core, and made of rock

manufactured structure an object or a structure that is made by humans

mass the amount of matter in a substance; often measured with a balance

mass structure a structure, natural or manufactured, that is made by the piling up of materials; examples of a mass structure include a pyramid and a snow fort

matter anything that takes up space, has mass, and is made up of particles

mechanical energy the energy in a moving object or in moving parts of an object

mechanical mixture a substance made of more than one kind of material, in which the different materials can be easily identified

mechanical weathering of rocks, the break-up or disintegration by the actions of physical forces such as wind, water, and gravity

melt to change from a solid to a liquid

melting point (or **freezing point**) the temperature at which solid matter begins to change to liquid

Mesozoic Era [mes-oh-ZOH-ik E-ruh] the third era on the geologic time scale; the era in which dinosaurs were the dominant life form on Earth

metal fatigue a weakening of metal due to stress, resulting in an accumulation of small cracks

metallic ores rocks that contain a high proportion of metals and metal oxides

metamorphic rock a type of rock made when high pressure and heat act on another type of rock and change it into a new form

meteorological evidence the type of evidence that is obtained by studying climate change

metric system a system of measurement based on multiples of ten and in which the basic unit of length is the metre

micro-organisms organisms that are too small to be seen by the human eye without the aid of a microscope

mineral an inorganic, naturally occurring solid material; minerals can be either elements (pure substances) or compounds (two or more substances combined)

mixture a material made up of several different types of materials; in a mixture, each material retains its own properties

mobile joint a joint that is designed to allow movement; examples of a mobile joint include a door hinge and an elbow

model a verbal, mathematical, or visual representation of a scientific structure or process, which allows scientists to construct and test inferences and theories (e.g., the particle theory of matter)

Mohs hardness scale in geology, a scale that compares the hardness of ten minerals; talc has a hardness value of 1 (the softest) and diamond has a hardness value of 10 (the hardest)

mountain a large, naturally occurring formation of Earth's surface that rises sharply above the surrounding area

mutualism [MYOO-choo-al-is-uhm] a symbiotic relationship between two different types of organisms that is beneficial to both organisms

natural structure an object or structure not made by people

network tree a concept map in which some terms are circles while other terms are written on connecting lines

neutral axis the direction in which a structure, such as a beam, carries the least load

newton (N) the standard unit of force in the *Système international d' unitès* (SI)

niche [NEESH] the role or characteristic activity that is undertaken by an organism in an ecosystem; one organism may fill several different niches

normal fault a vertical fault in which rock moves downward

nuclear energy the energy released when the smallest particles (called atoms) of a substance break apart or fuse together; also known as atomic energy

omnivore an animal that eats other animals and plant material; examples are bear, raccoon, people

organic sedimentary rock that is largely made up of once-living matter; limestone is an example

organism any type of living creature

output the final materials and energy forms that a system produces by applying energy to raw materials (input)

output force of a lever, the force exerted by it upon another object

Paleozoic Era [pae-lee-oh-ZOH-ik E-ruh] the second era on the geologic time scale; the era in which the first plants and animals appeared

Pangaea [pan-JEE-uh] the name of the second supercontinent thought to have existed approximately 350 million years ago; Pangaea included all the present continents

parasite an organism that lives on or in another organism (the host) and feeds on it

parasitism a symbiotic relationship between two different types of organisms in which one of the partners is harmed and the other benefits

parent rock the original rock that was acted on by high pressure and heat to form a metamorphic rock

particle size of soil, the average size of the particles of various materials of which the soil is made

particle theory of matter a scientific model of the structure of matter; according to the particle theory, all matter is made up of extremely tiny particles, and each pure substance has its own kind of particle, different from the particles of other pure substances

period on the geologic time scale, a subdivision of an era

permeate of water, to drain through soil

pesticide a substance used to control insects or other organisms that are harmful to plants or animals

petrochemical a product that is produced from petroleum; there are over 500 000 different petrochemicals

petroleum a type of oil found in rock formations in Earth's crust; petroleum is refined into products such as gasoline or jet fuel

phases of matter the different forms (solid, liquid, or gas) that matter can take; also known as states of matter

photosynthesis [foh-toh-SIN-the-sis] the process by which plants make their own food using sunlight

phytoplankton [fih-toh-PLANK-ton] plankton that use photosynthesis to make their own food

piling a large, cylindrical structure used to carry the weight of a structure to a solid foundation material

pillar an upright support

plankton the general name for microscopic plants, algae, and other organisms that float in oceans and other bodies of water

plasma a phase or state of matter rare on Earth, in which individual particles of a material break down into smaller pieces called electrons or ions; for matter to transform into plasma, extremely high temperatures are required

plate one of the large sections into which Earth's crust is divided

plate tectonics a theory about Earth's structure; according to this theory, Earth's crust is made up of very large pieces, called plates, that are always moving very slowly on Earth's mantle

plateau on a graph, a flat, horizontal region where data remain constant

pollution a collective term for the different types of harmful materials that are released into the environment through human activities

population a group of organisms of the same species found in a particular geographic area

potential energy stored energy

Precambrian Era the first of the four eras on the geologic time scale

precipitation the water (in its liquid or solid state) that falls to Earth; rain, snow, sleet, hail, etc.

predator an organism that catches and eats other organisms of a different species

prey an organism that is caught and eaten by another organism of a different species

primary standard the name given to a small cylinder of metal on which the kilogram (kg) is based; equivalent to 1 kg

primary (P) waves the fastest moving of the three types of seismic waves that are produced by an earthquake, originating from its focus; can pass through solids, liquids, and gases

principle of superposition a geological theory; according to this theory, in undisturbed layers of rock, the oldest layers will be on the bottom and the youngest layers will be on the top

producers plants that use energy from the Sun to make nutrients they need to survive; includes some bacteria that transfer energy from particles

properties the characteristics of materials; every material has its own unique set of properties; examples of properties include colour, odour, and density

protozoa [proh-toh-ZOH-uh] one-celled, animal-like organisms that live in or on other organisms

pure substance a material that is composed of only one type of particle; examples of a pure substance include gold, oxygen, and water

pyramid of numbers the number of individual organisms at each level of a food chain; the number of organisms decreases with each level higher in the food chain (there is a greater number of organisms at the bottom of the food chain than at the top)

quadrat a square tool used to measure and observe a small and representative section of a larger area

qualitative data information gathered by observations in which no measurements take place

qualitative property a characteristic of a substance that can be described but not measured

quantitative data data that consist of numbers and/or units of measurement; obtained through measurement and through mathematical calculations

quantitative property a characteristic of a substance that can be measured

radiant energy energy that is transmitted via electromagnetic waves; radiant energy can be absorbed and reflected by objects, and it moves through empty space at 300 000 km/s

radiation the transfer of energy in the form of electromagnetic waves

radiocarbon dating a method used to determine the age of organic remains by measuring the relative amount of radioactive carbon found in the remains

radiometric dating the process of determining the age of a geological specimen by measuring the relative amounts of radioactive particles that are present in the specimen

rate of dissolving the speed at which a solute dissolves in a solvent

recycling the process of using the same item over again; recycling can either use the item as it was originally used or find new uses for it, perhaps by changing its composition

refining the processing of petroleum to separate it into its parts, such as asphalt or kerosene

relative dating determining the order in which geological events occurred and the relative age of rocks by their positions in rock layers

respiration in the cells of living things, the process in which oxygen is used to get energy from food and is converted into carbon dioxide

reverse fault a vertical fault in which rock moves upward

Richter scale [RIK-tuhr] a scale on which the magnitude, or strength, of an earthquake is measured

rigid joint a device designed to fix an object into place; a joint that allows no movement; examples of a rigid joint include a nail and a screw

rock a natural material composed of one or more minerals

rock cycle the naturally occurring process in which rocks continue to change form over long periods of time

Rodinia the name of the earliest supercontinent thought to have broken apart approximately 750 million years ago; Rodinia included all the large land masses

sampling in population studies, a method used to estimate population size in ecosystems by finding out the number of individuals in a portion (that is, the sample) of the population and then calculating the total number for the population as a whole

saturated solution a solution in which no more of a solute is able to be dissolved at a particular temperature

scale a series of equally divided sections that are marked and numbered for use in measurement (e.g., centimetres, litres, or grams)

scavenger an organism that eats dead or decaying plant or animal matter; a carrion beetle is an example of a scavenger

science a body of facts or knowledge about the natural world, but also a way of thinking and asking questions about nature and the universe

science inquiry the orderly process of asking concise and well-focussed questions and designing experiments that will give clear answers to those questions

scientific investigation an investigation that involves the systematic application of concepts and procedures (e.g., experimentation and research, observation and measurement, analysis and sharing of data)

sea floor spreading the process in which an ocean floor slowly increases in size over time because of the formation of new igneous rock along a fault

secondary succession the gradual growth of organisms in an area that was formerly home to many different species; the regeneration of a burned forest is an example

secondary (S) waves the second fastest moving of the three types of seismic waves that are produced by an earthquake, originating from its focus; can pass through solids but not liquids or gases

sediment loose material such as bits of rock, minerals, and plant and animal remains

sedimentary rock the most common type of rock on Earth's surface; formed by the compacting of sediment (loose materials, such as minerals and organic remains)

seismic waves [SIHZ-mik] the energy waves (either primary, secondary, or surface) that are released by an earthquake and travel outward from its focus

seismograph [SIHZ-moh-graf] a sensitive machine that is attached to bedrock in order to measure the strength of earthquakes

seismologists [sihz-MOL-oh-jists] scientists who study earthquakes

shadow zone an area on Earth's surface that is not reached by primary waves after an earthquake, due to the bending of P waves as they pass through Earth

shear of a section of compressed material, to slide over another section along a weak point

shear force a force that bends or tears a material by pushing parts of it in opposite directions

shear strength a measure of the largest shear force that a material can withstand before tearing apart

shield volcano the largest of the three main types of volcanoes; formed above an area, called a hot spot, where the temperature under the crust is much hotter than elsewhere, causing lava to be forced upward through vents

shell structure a type of structure that obtains its strength from a thin, carefully shaped outer layer of material and that requires no internal frame; examples of a shell structure include an igloo and an egg

shrinking apple theory a nineteenth-century theory about Earth's structure; according to this theory,

Earth was once a hot mass, which cooled and shrank over time; the theory compared Earth to an apple that dried up, causing wrinkles (mountains) and valleys between the wrinkles (oceans and lakes)

SI (from the French *Le Système international d'unités*) the international system of measurement units, including such terms as kilogram, metre, and second

society a group of people united by common goals and interests

soft water water that contains few dissolved minerals

soil a mixture of weathered rock, organic matter, mineral fragments, water, and air

soil profile a description of the characteristics of the different layers that make up a particular soil

soldering [SO-duhr-ing] a process in which a melted material is applied to a different type of material; the melted material hardens when it cools, forming a rigid joint that holds the other material in place

solid one of the states or phases of matter; in the solid phase, materials keep a specific shape and size

solubility the limit to how concentrated a solution can become, before it becomes a saturated solution at a particular temperature; for example, no more than 35 to 37 g of salt will dissolve in 100 g of cold (0°C) water

soluble of a substance, able to be dissolved in a particular solvent; something that is soluble is called a solute

solute a substance that can be dissolved in a solvent; for example, salt is a solute that dissolves in water

solution a homogeneous mixture of two or more substances; the distinct properties of the different substances that make up the solution are combined into one set of properties

solvent a substance into which a solute may be dissolved; for example, water is a solvent that dissolves sugar

sonar (sound **n**avigation **a**nd **r**anging) a technology that bounces sound waves off an object to determine its distance from the source of the waves

species a narrow classification grouping for organisms; e.g., a wolf is the species *Canis lupus*, while a dog is the species *Canis familiaris*

specific heat capacity of a material, the energy change that is required to warm or cool a standard amount of the material (1 g or 1 kg) by 1°C

specifications a set of standards or expectations; criteria

spider map a concept map used to organize a central idea and a jumble of associated ideas that are not necessarily related to each other

spin stabilization the tendency of an object that is spinning on its axis to move in a predictable manner; an example of spin stabilization is the motion of a bicycle wheel

spring scale a device used to measure the force of gravity exerted on an object by measuring the object's weight in newtons

stable of a structure, tending to maintain its shape and position

states of matter the different forms (solid, liquid, or gas) that matter can take; also known as phases of matter

stem-and-leaf plot an organization of data into categories based on place values; values on the stem represent a range of data, while the leaves indicate the data points that lie in that range; when turned on its side, resembles a bar graph

streak the colour of a mineral in powdered form; a property useful in the identification of minerals

stress an internal or external force that acts on an object, perhaps causing it to move or change shape

structural efficiency a number that expresses the relative ability of a structure to support a load; calculated by dividing the maximum mass of the supported material by the mass of the structure; the higher the number, the more weight the structure can bear compared to its own weight

STSE an abbreviation for the interrelationships among science, technology, society, and the environment

subduction zone a place on Earth's crust where high pressure pushes one very large piece of rock below another; earthquakes are often formed in subduction zones

sublimation the process in which a solid changes directly into a vapour without passing through the liquid stage

succession the process by which new species gradually replace old species in an ecosystem

supersaturated solution a solution that contains more solute than would normally dissolve at a particular temperature

surface area the amount of surface of an object; measured in square units such as cm^2

surface waves the slowest moving of the three types of seismic waves that are produced by an earthquake, originating from its epicentre; surface waves do the most damage of the three types of waves

symbiosis [sim-bih-OH-sis] an interaction between organisms of different species living in close proximity to each other in a relationship that lasts over time

syncline a downfold of a rock layer in sedimentary rock

system a set of things that are organized and interact with each other to such an extent that they may be described as a single unit

table an orderly arrangement of facts set out for easy reference; for example, an arrangement of numerical values in rows or columns

technology the design and construction of devices, processes, and materials to solve practical problems and to satisfy human needs and wants

temperature a relative measure of how hot or cold something is, measured on a scale; the average kinetic energy of the particles in a substance

tensile strength a measure of the largest tension force that a material can withstand before changing shape or breaking apart

tension force a force that pulls on a material and stretches it apart

texture of soil, how it feels to the touch; texture is affected by the size of the particles in the soil

theory an explanation of an event that has been supported by consistent, repeated experimental results and has therefore been accepted by many scientists

thermal conduction the direct transfer of thermal energy from one particle or object to another through contact or collision

thermal energy the energy generated by the movement or vibration of particles; the total kinetic energy of all the particles in a substance

thermal pollution a warming of the environment that results from human activities, such as the burning of fossil fuels

thermogenic [THUR-moh-jen-ik] of plants or animals, able to raise their own temperature

thermograph a thermometer that records temperature

thermometer a device used to measure temperature

thrust line the line that runs downward from an object's centre of gravity, through which force is transferred

tie a device used to add strength to a structure, usually by forming a rigid triangle at the point where the pieces come together in a right angle; a type of rigid joint, such as a piece of rope, that is used to pull objects or materials together and hold them in place

tie beam the beam that ties the ends of an arch together to form a truss

topsoil the topmost layer of soil, which is dark-coloured and rich in humus

torsion force a force that acts on a material by twisting its ends in opposite directions

torsion strength a measure of the largest torsion force that a material can withstand and still be able to return to its original shape

transform boundary an area of Earth's crust where plates are sliding past each other

transformation the changing of a substance or material with a particular set of properties into a new substance (or substances); a change in the characteristics of something

transpiration the process in which water that is taken in by a plant or an animal evaporates from the organism

truss an arch with ends that are tied together by a tie beam

truss girder a structural support made from a truss

tsunami [soo-NA-mee] a very large wave caused by an earthquake under a large body of water

twist of a material, to change shape through the application of torsion forces

uniformitarianism a theory developed in the eighteenth century; according to this theory, geological processes that occur in the present are the same processes that occurred in the past

unifying theory a single theory that explains many different natural phenomena, events, objects, or processes

unsaturated solution a solution in which more of a solute can be dissolved at a particular temperature

variable a condition or factor that can influence the outcome of an experiment

Venn diagram a graphic organizer consisting of overlapping circles; used to compare and contrast two concepts or objects

vent an opening in Earth's crust through which magma can escape, forming lava

vertical fault a fault in which rock moves up or down

volcano an opening in Earth's crust that can release materials such as lava, smoke, and ash; volcanoes can be either active (releasing materials) or dormant (not releasing materials)

volume the measurement of the amount of space occupied by a substance; measured in litres or cubic units such as cubic centimetres (cm^3)

warm-blooded of an organism, maintaining a relatively consistent body temperature regardless of the environment; all mammals are warm-blooded

waste heat energy that is transferred outside the system in which it is generated, without doing any useful work

waste management consultant a professional who advises industries on how to reduce waste products or how to dispose of them in a way that is less harmful to the environment

water cycle the continuous movement of water through the biosphere; the water cycle consists of evaporation, transpiration, condensation, and precipitation

water-holding capacity the ability of a soil to retain water; soils with low water-holding capacity allow a great deal of water to permeate through them

weather the conditions of the air layer that surrounds Earth in a local area, over a short period of time

weathering the process in which rocks are broken down and sediment is formed by mechanical, chemical, or biological means

weed a plant that grows where it is not wanted

weight the force of gravity exerted on a mass

welding a process in which pieces of metal or plastic are fused together by the application of heat

WHMIS an acronym that stands for Workplace Hazardous Materials Information System

Index

The page numbers in **boldface** type indicate the pages where the terms are defined. Terms that occur in investigations (*inv.*) and activities (*act.*) are also indicated.

Photo Credits

iii inset J.A. Wilkinson/Valan Photos; **vii left** © Corel Corporation; **viii top, middle** © Corel Corporation; **ix top left** Digital Vision Ltd. All Rights Reserved., **bottom** Maximilian Stock Ltd./Science Photo Library; **x top** NCC/CNN, **bottom** NASA; **xi top left** Comstock Photofile Ltd., **middle left** Doran Weisel/Masterfile, **bottom** Courtesy Dr. J.W. Sheridan; **xii top left** Canadian Press CP/Photographer Kevin Frayer; **middle** © Corel Corporation, **bottom** Comstock Photofile Ltd./Colin Quirk; **xiii** NASA, Langley Space Center; **xxviii top left** G. Van Loocke/ Agence D.P.P.I./Publiphoto, **bottom left** Courtesy Westcoast Energy Inc./Photographer Dolores Baswick; **top right** John Solhden/Visuals Unlimited, **bottom right** J.M Petit/Publiphoto; **IS-1** Gamma Liaison/ Ponopresse 620.534; **IS-2 top left** Canada in Stock/Ivy Images, **top right** Bob Semple, **middle right** William C. Jorgensen/Visuals Unlimited, **bottom left** Kjell B. Sanved/Visuals Unlimited, **bottom right** Bill Ivy/Ivy Images; **IS-5 top left** Betty Sederquist/Visuals Unlimited, **top right** Courtesy of Mark Ludbrook/Photographer David Balfour; **bottom right** Mike De Mocker/Visuals Unlimited; **IS-6 top** Lawrence Berkeley National Laboratory/SNO, **middle left** Golf Data Collection, Human Performance Laboratory, University of Calgary, **middle right** Scott Camazine/Photo Researchers; **IS-7 top right** SUI/ Visuals Unlimited; **IS-8 top** Thomas Kitchin/First Light, **bottom left** © Corel Corporation, **bottom right** Comstock Photofile Ltd./Gordon Graham; **IS-9 top left** Comstock Photofile Ltd./Photot E. Otto, **top right** © Corel Corporation; **IS-11** Ivy Images; **IS-14 top, both** Courtesy Bell Canada Historical Collection; **bottom** Sforzesco, Museo di Strumenti Musicali/Art Resource, NY.; **IS-16 top** S.W. Ross/Visuals Unlimited; **IS-19 left** John D. Gordon/ Visuals Unlimited, **right** Thomas Kitchin/First Light; **IS-20 right** K. Straiton/First Light, **left** Darwin Wiggett/First Light; **2-3** J. David Andrews; **3 inset** The Hamilton Spectator/John Rennison Photographer; **4-5** Larsson Kjell-Arne/Publiphoto; **5 inset** © Corel Corporation; **6 top** LSHTM/Tony Stone Images, bottom BY PERMISSION OF RACHEL CARSON HISTORY PROJECT; **7 top left** Dr. Jeremy Burgess/Science Photo Library, **middle** Joe McDonald/ Visuals Unlimited, **top left** © Corel Corporation; **8** Comstock Photofile Ltd./Denver Bryan; **9 lion teeth** H. Pooley/Animals Animals, **horse teeth** C.Rouso/ Publiphoto, **pelican's beak** T. Urlich/H. Armstrong Roberts/Comstock Photofile Ltd., **eagle's beak** Thomas Kitchin/First Light, **snake** Daniel Heuhlin/Publiphoto, **alligator** Steven David Miller/Animals Animals, **duck foot** David Young-Wolff/PhotoEdit, **falcon talons** © Corel Corporation; **11** Pam Hickman/Valan Photos; **12** J. M. Labat/Jacana/ Photo Researchers; **14 left** Wayne Lankinen/Valan Photo, **right** Gerald & Buff Corsi/Visuals Unlimited; **19** Bowles & Sober/Career Connections: Great Careers for People Who Like Being Outdoors, published by Trifolium Books Inc./Weigl Educational Publisher Limited.; **20 top left** Philip Norton/Valan Photos, **middle left** Aubrey Lang/Valan Photos; **22 top left** ©Parks Canada/W. Lynch, **top right** Francis Lepine/Valan Photos, **middle left** P.G. Adam/Publiphoto, **middle right** Ivy Images, **bottom** Simon Fraser/Science Photo Library; **23 top right** Fletcher & Baylis/Photo Researchers, **bottom** Orangutan Foundation International; **28** Norbert Wu/www.norbertwu.com; **29 top right** Bill Ivy/Ivy Images, **middle right** Thomas Kitchin/First Light; **30** © Robert Bateman. Reprinted with permission of Mill Pond Press. Reproduction courtesy of Boshkung Inc.; **36-37** Tom & Pat Leeson/Photo Researchers; **37 inset** Canadian Press SKTN; **40** John de Visser/Masterfile; **42** Steve Maslowski/Photo Researchers; **43 left** Val Wilkinson/Valan Photos, **middle** Canadian International Grain Institute, **right** Norseman Plastics; **47** Bert Luit-Photographer/University of Calgary; **48 top** Robert & Linda Mitchell Photography, **middle left** Tim Flatch/Tony Stone Images, **middle right** Robert Calentine/Visuals Unlimited; **49 top** J. R. Page/Valan Photos, **left** J.A. Kraulis/Masterfile, **right** Tony Joyce/Masterfile; **50 left** Luiz Marigo/Ivy Images, **right** Dave B. Fleetham/Visuals Unlimited; **55 top** Mark Tomalty/Masterfile, **bottom** 1990, Scott Camazine/Photo Researchers; **57** V. McMillan/Visuals Unlimited, **58 top** John Anthony Scullion/Valan Photos, **bottom** Dan Ginding/Tourism Newfoundland and Labrador; **62-63** CORBIS/AFP; **63 inset** John Boykin/PhotoEdit; **65** John Fowler/Valan Photos; **69** © Corel Corporation; **73** Y. Beaulieu/Publiphoto; **77** B. Milne/First Light; **78 middle** Astrid & Hanns-Frieder Michler/Science Photo Library, **bottom** Erberhard E. Otto, Toronto/Comstock Photofile Ltd.; **79 middle left** ©Parks Canada/R. Grey, **middle right** Tom J. Ulrich/ Visuals Unlimited, **bottom right** ©Parks Canada/T. Grant; **81 top right** Robert Simpson/Valan Photos, **bottom right** Christine Case/Visuals Unlimited; **83** C.A. Girouard/Publiphoto; **85** Mark Lewis/Tony Stone Images; **90** Courtesy Hilda Ching; **92** Michael Newman/PhotoEdit; **93** Janet Dwyer/Ivy Image; **94** David Young Wolf/PhotoEdit; **96-97 all** © Corel Corporation; **98-99** Comstock Photofile Ltd./MALIK, **99 inset** Ivar Mjell/Masterfile; **101 left** D. Robert Franz/Masterfile, **right** Barrett & MacKay/Tourism Newfoundland and Labrador; **102 top** Jeannie Kemp/Valan Photos; **bottom left** G. Biss/Masterfile, **bottom right** Dick Hemingway; **103** © Corel Corporation; **110 left** Henry Birks & Sons, Ltd.; **111** Shelagh Audette; **116** Richard Nowitz/Valan Photos; **117 top** D. Robert Franz/ Masterfile, **bottom** Barrett & MacKay/ Tourism Newfoundland and Labrador; **120-121** Ivy Images, **120 inset** T. Bonderud/First Light; **126** © Tom Van Sant/Geosphere Project, Santa Monica/Science Photo Library; **127** Christine Case/Visuals Unlimited; **131 bottom** Francoise Sauze/Science Photo Library; **132** D. Auclair/Publiphoto; **133** The Sun Syndicate; **134** Michel Julien/Valan Photos; **138** "Indian Sugar Camp", by Seth Eastman, Courtesy Musee Canadien Des Civilisations/Canadian Museum of Civilization No: 86757; **139 top right** Peter Miller/Photo Researchers, **bottom** V. Last/Geographical Visual Aids; **141** Marc Pokempner/Tony Stone Images; **142** J. P. Danvoye/Publiphoto; **146-147** Comstock Photofile Ltd./Jim Kozmiuk; **150 both** Richard Megna/Fundamental Photographs; **160 left** Photo nimatallah/Art Resource, NY, **right** Archivi Alinari, Firenze/Art Resource, NY; **163** Canadian Press CP; **164 top left** Will Crocker/The Image Bank, **top right** Royal Ontario Museum NS 5560, **bottom left** The New Brunswick Museum, Saint John N.B.; **bottom right** Royal Ontario Museum 981.43.836; **165** First Light; **167** Inga Spence/Visuals Unlimited; **168** Don Haddy/Career Connections: Great Careers for People Interested in How Things Work, published by Trifolium Books Inc./Weigl Educational Publisher Limited.; **169** Ken Lucas/Visuals Unlimited; **170 top** 58-1026-1569 Bunnell Album; University of Alaska Fairbanks, Rasmuson Library Archives; bottom CORBIS-BETTMAN/ Publiphoto; **172** Courtesy of Dusanka Filipovic; **173** Ken Lucas/Visuals Unlimited; **176-177** Courtesy of Art Colebourne/Canadian Grain Commission; **178** Comstock Photofile Ltd.; **179** Ron Watts/First Light; **180** Digital Vision Ltd. All Rights Reserved., **182-183** Maximilian Stock Ltd./Science Photo Library, **183 inset** Jeff J. Daly/Visuals Unlimited; **184-185** Comstock Photofile Ltd./Gordon Graham, **185 inset** Canadian Press CGYH; **187 left** CORBIS/David Lees; **right** The Granger Collection; **188 top** Science Photo Library, **middle** Lynda Powell, **right** Science Photo Library; **189** Science Photo Library;**194** Paul Chelsey/Tony Stone Images; **196** John D. Cunningham/Visuals Unlimited; **197 top** Adam-Hart Davis/Science Photo Library/Photo Researchers, **bottom** Stephen Frisch/Stock, Boston, PNI; **198** © Corel Corporation; **202** B. Templeman/Valan Photos; **204** University of British Columbia; **207 top** Thomas Kitchin/First Light, **bottom** Science VU/Visuals Unlimited; **211** © Corel Corporation; **212-213** CORBIS/Richard Hamilton Smith; **213 inset** Dr. Jeremy Burgess/Science Photo Library; **217** Paul Shambroom/Photo Researchers; **221 left** Charles D. Winters/ Photo Researchers; **right** Larry Miller/Photo Researchers; **222** Rafael Macia/Photo Researchers; **227 bottom right** NASA; **231** NCC/CNN; **232** NASA/Science Photo Library; **233** Jeff Greenberg/Visuals Unlimited; **234 bottom right** James R. Page/Valan Photos; **240-241** NASA; **241 inset** Jon Becker; **242 left** Lynette Cook/Science Photo Library, **right** Pete Turner/The Image Bank; **244** Amwell/Tony Stone Images; **246 top** Jean Claude Hurni/Publiphoto, **bottom** Science Kit & Boreal Laboratory Gas Convection Apparatus 61229-01, 800-828-7777; **247** Reg Johnson/Career Connections: Great Careers for People Interested in How Things Work, published by Trifolium Books Inc./Weigl Educational Publisher Limited.; **257 top** Artbase Inc., **bottom** V. Last/Geographical Visual Aids; **258** The Granger Collection; **259** Arnulf Husmo/Tony Stone Images; **260 top** Jeff Lepore/Photo Researchers, **bottom** J. Sylvestor/First Light; **261**P.G. Adam/Publiphoto; **262** The Hamilton Spectator; **266** Thomas Kitchin/First Light; **269** Artbase Inc.; **267** © Corel Corporation; **270** Courtesy of Mario Patry; **271** Courtesy of Mountain Equipment Co-op; **272** Leland Bobbe/Tony Stone Images; **273** Myrleen Ferguson/ PhotoEdit; **274 right** Science Kit and Boreal Laboratory, Solar Oven, 800-828-7777, **left** Paolo Koch/Photo Researchers; **276-277** Alberto Garcia/SABA, **276 inset** NASA; **278-279** Comstock Photofile Ltd./Eric Hayes, **279 inset** © Corel Corporation; **280 top left** © Corel Corporation, **top middle** Deborah Vansley/Valan Photos, **top right**

Biophoto Associates/ Photo Researchers, **middle left** A. J. Copley/Visuals Unlimited, **middle right** Courtesy of the Geological Survey of Canada (photo number GSC KGS2359L); **281** 1993 Paul Silverman/Fundamental Photographs; **282** Comstock Photofile Ltd.; **283 top, bottom left** Doug Martin; **bottom right** Science Kit and Boreal Laboratory, Obsidian – 60272, 800-828-7777; **286 left** © Corel Corporation, **middle** Charles D. Winter/Photo Researchers; **right** Bombardier Aerospace; **287 top left** W. Gregoire/Valan Photos, **top right** Terry Domico/Earth Images, **bottom** Henry Birks & Sons, Inc.; **289** Daryl Benson/Masterfile; **292** Comstock Photofile Ltd./Bob Rose, **293 top** © Corel Corporation, **bottom left** Val Wilkinson/Valan Photos, **middle, bottom left** Joyce Photo/Valan Photos; **294 top left** Dr. Morely Read/Science Photo Library/Photo Researchers; **top left centre** Chuck Keeler/Tony Stone Images; **top right centre** Jeannie R. Kemp/Valan Photos, **top right** Westlight/First Light, **bottom right** Joyce Photo/Valan Photos, **bottom left** Mark A. Schneider/Visuals Unlimited, **bottom left** John Eastcott/Yva Momatiuk/Valan Photos, **bottom right** Reprinted with Permission from the United States Department of Agriculture, Natural Resources Conservation Service from "Soil Erosion by Water," April, 1990.; **297** Francois Gohier/Photo Researchers; **298** © Corel Corporation; **299 top** © Corel Corporation, **bottom** Pam E. Hickman/Valan Photos; **301 top** John S. Conway/Soils Research Group, **bottom** Courtesy Jean-Serge Vincent/Geological Survey of Canada/with permission of Sylvia Edlund; **304** Artbase Inc., **305** Victoria Vincent/Career Connections: Great Careers for People Interested in the Past, Published by Trifolium Books Inc./Weigl Educational Publisher Limited.; **306** Grant Heilman/Comstock Photofile Ltd.; **308** Phillip Norton/Valan Photos; **309** © Corel Corporation; **312-313** Associated Press AP/CP Archive, **312 inset** Alberto Garcia/SABA, **314 middle** David R. Frazier/Photo Researchers, **bottom** The Natural History Museum, London; **316** Michael Townsend/Tony Stone Images; **322 top** Reprinted By Permission of Prentice Hall, Inc., Upper Saddle River, NJ., **bottom left** Tom and Susan Bean, Inc./DRK PHOTO; **324 top left** Courtesy Westcoast Energy Inc./Photographer Dolores Baswick, **top right** Ron Watts/First Light, **bottom** Francois Gohier/Science Source/Photo Researchers; **325 left** Francois Gohier/Photo Researchers, **right** Giraudon/Art Resource, NY; **326** R.E. Wilcox/U. S. Geological Survey; **327** Doran Weisel/Masterfile; **328** Bob Rose/Comstock Photofile Ltd., **329** Werner Forman/Art Resource, NY; **334** J. R. Page/Valan Photos; **335** Photo E. Otto/Comstock Photofile Ltd.; **339** The Natural History Museum, London; **342-343** *Tyrannosaurus* and *Edmontosaurus* image by Eleanor M. Kish with permission of the Canadian Museum of Nature, Ottawa, Canada; **343 inset** Morin, Tanimoto, Zhang, Yuen; **344 left** The Bettman Archive/CORBIS-BETTMAN, **right** Tony Stone Images; **346** Dick Hemingway; **347** The Bettman Archive/CORBIS-BETTMAN; **349** Courtesy Dr. J.W. Sheridan; **352 bottom left** World Ocean Floor, Bruce C. Heezan and Marie Tharp, 1977, Marie Tharp 1977. Reproduced by permission of Marie Tharp, 1 Washington Ave., South Nyack, NY 10960, **right** Scripps Institution of Oceanography, UCSD; **354 left** Rod Catonach/Woods Hole Oceanographic Institution, **right** Alexander Malahoff/University of Hawaii/U.S. Geological Survey; **355 top** The Ontario Science Centre, **middle** Dudley Foster/Woods Hole Oceanographic Institution, **bottom** Woods Hole Oceanographic Institution; **357** Woods Hole Oceanographic Institution; **358** © Corel Corporation; **359** W.A. Padgham, photo; **360 top left** Vaughan Fleming/Science Photo Library/Photo Researchers, **top right** Val Wilkinson/Valan Photos, **bottom left** Sinclair Stammers/Science Photo Library/Photo Researchers, **bottom right** Vaughan Fleming/Science Photo Library/Photo Researchers; **363** Courtesy of the Geological Survey of Canada, (Photo -GSC 204478); **364** Reuters/David Mercado/Archive Photo; **368** Courtesy of Charlotte Keen; **370** J. Eastcott/Y. Momatiuk/Valan Photos; **373** Artbase Inc.; **373 stamp** © Canada Post Corporation, 1992. Reproduced with Permission; **374-375** © Corel Corporation; **375** Canadian Press CP/Photographer Kevin Frayer; **376-377** J. Heguy/First Light; **376 left inset** Lester Lefkowitz/Masterfile, **right inset** © Corel Corporation; **377 top inset** © Corel Corporation, **bottom inset** D. Pedku/H. Armstrong Roberts/Comstock Photofile Ltd.; **379** Courtesy of the National Research Council Canada; **380** David R. Frazier, Photolibrary/ Photo Researchers; **382** CORBIS-BETTMAN; **383** G. Petersen/First Light; **384 top** © Corel Corporation, **bottom** Peter Aprahamian/ Science Photo Library/Photo Researchers; **385** Courtesy of Marie-Anne Erki; **387 left** Space Frontiers/Masterfile, **top inset** Gabor Geissler/Tony Stone Images, **bottom inset** Anne-Marie Weber/FPG/Masterfile; **388** Carl Bigras/Valan Photos; **389** Thomas Kitchin/First Light; **390 top** RUBE GOLDBERG reprinted by permission of United Feature Syndicate, Inc.; **391 both** © Corel Corporation; **393** 1995 Di-Lun Kwan/Courtesy of the Vancouver Public Library; **395** Mike Dobel/ Masterfile; **397 top** David Sutherland/Tony Stone Images, **bottom right** Valan Photos; **400** Mark Lewis/Tony Stone Images; **402** RNHRD/NHS Trust/Tony Stone Images; **403-404** © Corel Corporation; **406-407** Canadian Press/MACLEANS; **406 inset** Harold & Esther Edgerton Foundation, 1999, Courtesy of Palm Press, Inc.; **409 car** Artbase Inc., **motorcycle** Benn Mitchell/The Image Bank, **boxer, turkey** Artbase Inc., **lime** © Corel Corporation, **nickel** Coin image courtesy of the Royal Canadian Mint/Image de la piece courtoisie de la Monnaie royale canadienne., **raisin** Ken Wagner/Phototake Inc./PNI, **blueberry, vitamin pill** Artbase Inc., **stamp** Courtesy Canada Post, **bottom four** © OHAUS Corporation; **410 top** Comstock Photofile Ltd./Henry Georgi, **bottom** NASA; **411** NASA; **418 both** Gillian Bartlett/Career Connections: Great Careers for People Interested in Art & Design, published by Trifolium Books Inc./Weigl Educational Publisher Limited. **419** Canadian Press CP; **420 left** © Corel Corporation, **bottom right** Hideo Kurihara/Tony Stone Images; **425** Steve Satushek/The Image Bank; **426 left** Robert McCaw, **top right** Ray Nelson/Phototake/PNI; **431** NASA; **434-435** Paramount Pictures/Fotos International/Archive Photos; **435 inset** NASA; **436** Comstock Photofile Ltd.; **438** Y.R. Tymstra/Valan Photos; **439** Lawrence Livermore National Laboratory/ University of California/Science Photo Library/Photo Researchers; **440 top** Martyn Goddard/Masterfile; **442** Dr. Jeremy Burgess/Science Photo Library/Photo Researchers; **443** National Archives of Canada PA-26427; **445 top** © Corel Corporation, **bottom** UPI/CORBIS-BETTMAN; **446 sumo** Canadian Press/CP, **gymnast** Bill Ross/First Light, **baseball** Jed Jacobsohn/ALLSPORT, **ballet** The National Ballet of Canada/Photo: D. Street, **football** R. Krubner/H. Armstrong Roberts/Comstock Photofile Ltd.; **450** The Hamilton Spectator/Sheryl Nadler; **451** Deborah Davis/Tony Stone Images; **455 top left** " IMTEK IMAGINEERING/Masterfile, **top right** S. Lissau/H. Armstrong Roberts, Inc./Comstock Photofile Ltd., **bottom** Comstock Photofile Ltd. Barbara DeWitt; **456 top left**, Comstock Photofile Ltd., **middle left** Comstock Photofile Ltd./Robert Hall, **middle right** P. Wysocky/Explorer/Photo Researchers, **bottom** Bert Studios, Inc./Comstock Photofile Ltd.; **457 left** Comstock Photofile Ltd./Colin Quirk, **right** Stuart Hunter/Masterfile; **459** Hideo Kurihara/Tony Stone Images; **462** Courtesy of Steven Haynes; **463** Courtesy of Steve Nazar/ Cambridge Materials Testing; **464** Al Harvey/The Slide Farm; **465** Department of Infrastructure - Government of Alberta; **466** Carleton Engineering GNCTR 1999 (Great Northern Concrete Toboggan Team); **470 both** Daniel J. Cox/Tony Stone Images; **477** Shirley Collier and Gary Grant; **478 top** David Young-Wolff/ PhotoEdit; **493 bottom left** NASA, Langley Research Center; **bottom right** Science Kit and Boreal Laboratory, Eye Cross Section 65058, 800-828-7777; **494 top** Science Kit and Boreal Laboratory, Copernician Solar System Model 63878-0, 800-828-7777. **Back Cover, Unit 1** J. David Andrews; **Unit 2** © Corel Corporation, **Unit 3** Maximilian Stock Ltd/Science Photo Library; **Unit 4** Alberto Garcia/SABA; **Unit 5** © Corel Corporation.

Text Credits

IS-2 Galbraith et al., Analyzing Issues: Science, Technology, & Society, Trifolium Books Inc., 1997, portions adapted with permission. **470-471** "Appendix B: The Design Process." Adapted from By Design: Technology Exploration & Integrations, Metropolitan Toronto School Board, published by Trifolium Books Inc. **497-499** "Skillpower on the Internet." Adapted from Heide, Ann, and Stilborne, Linda. The Teacher's Complete & Easy Guide to the Internet [added], 2nd Edition, published by Trifolium Books Inc., 1999.

Illustration Credits

21 Morgan-Cain and Associates. From *Life Science* by Lucy Daniel et al, © 1997 Glencoe/McGraw-Hill; **40** Barbara Hoopes-Amber, p.496. From *Life Science* by Lucy Daniel et al, © 1997 Glencoe/McGraw-Hill; **46, 51, 60** Pond and Gilles. p.498 top, bottom. From *Life Science* by Lucy Daniel et al, © 1997 Glencoe/McGraw-Hill; **54** Chris Forsey, p. 510. From *Life Science* by Lucy Daniel et al, © 1997 Glencoe/McGraw-Hill; **59** Morgan-Cain and Associates. p. 492. Adapted from *Life Science* by Lucy Daniel et al, © 1997 Glencoe/McGraw-Hill; **65** Chris Forsey, p. 518. From *Life Science* by Lucy Daniel et al, © 1997 Glencoe/McGraw-Hill; **75** Morgan-Cain and Associates, p.560. From *Life Science* by Lucy Daniel et al, © 1997 Glencoe/McGraw-Hill; **83** Chris Forsey, p. 563. From *Life Science* by Lucy Daniel et al, © 1997 Glencoe/McGraw-Hill; **286** From Earth Science by Ralph Feather Jr., 1997 Glencoe/McGraw-Hill.; **295** Preface, Inc, from Earth Science by Ralph Feather Jr., 1997 Glencoe/McGraw-Hill.; **322** Edwin Huff from Earth Science by Ralph Feather Jr., 1997 Glencoe/McGraw-Hill.; **326 top** John Edwards from Earth Science by Ralph Feather Jr., 1997 Glencoe/McGraw-Hill.; **328 top** Gary Hinks/Publisher's Graphics from Earth Science by Ralph Feather Jr., 1997 Glencoe/McGraw-Hill.; **351** Precision Graphics from Earth Science by Ralph Feather Jr., 1997 Glencoe/McGraw-Hill.; **353** Leon Bishop from Earth Science by Ralph Feather Jr., 1997 Glencoe/McGraw-Hill.; **358** Tom Kennedy from Earth Science by Ralph Feather Jr., 1997 Glencoe/McGraw-Hill.; **359** Dartmouth Publishing, Inc., from Earth Science by Ralph Feather Jr., 1997 Glencoe/McGraw-Hill.; **361** Jim Shough from Earth Science by Ralph Feather Jr., 1997 Glencoe/McGraw-Hill.; **362** Dartmouth Publishing, Inc. from Earth Science by Ralph Feather Jr., 1997 Glencoe/McGraw-Hill.; **365** Jim Shough from Earth Science by Ralph Feather Jr., 1997 Glencoe/McGraw-Hill. "Scientific supplies for experiments provided courtesy of Exclusive Educational Products, 243 Saunders Road, Barrie, Ont. L4N 9A3 (705-725-1166)."